CRUSTAL EARTH MATERIALS

CRUSTAL EARTH MATERIALS

Loren A. Raymond
Appalachian State University

Neil E. Johnson
Virginia Polytechnic Institute and State University

WAVELAND PRESS, INC.
Long Grove, Illinois

For information about this book, contact:
Waveland Press, Inc.
4180 IL Route 83, Suite 101
Long Grove, IL 60047-9580
(847) 634-0081
info@waveland.com
www.waveland.com

Copyright © 2018 by Waveland Press, Inc.

10-digit ISBN 1-4786-3263-1
13-digit ISBN 978-1-4786-3263-4

All rights reserved. No part of this book may be reproduced, stored in a retrieval system, or transmitted in any form or by any means without permission in writing from the publisher.

Printed in the United States of America

7 6 5 4 3 2 1

We dedicate this book
To our professional colleagues and mentors,
who inspired, challenged, and educated us:
James R. Craig, Chuck Guidotti, Marshall E. Maddock,
F. K. McKinney, Eldridge Moores, Stanley Skapinsky,
Fred Webb, Jr., and Henry E. Wenden;
To our spouses who supported us throughout the project:
Margaret and Kay;
and
To the rising generations of learners represented by
Maya Mochizuki Raymond

Contents

Preface xi

1 The Context of Crustal Earth Materials 1
Introduction 1
Minerals, Rocks, Fluids, and Soils 2
The Structure and Chemistry of Earth 4
Plate Tectonics 8
Earth Systems at and Near the Surface 11
■ SUMMARY 12
■ SELECTED REFERENCES 12 ■ NOTES 12

2 Fundamentals of Physics and Chemistry Relevant to Earth Materials 13
Introduction 13
Some Concepts of Physics 13
    Measurement 13
    Derivative Measures 14
Some Concepts of Chemistry 16
    Atoms and Their Structure 16
    Ions and Ionization 18
    Chemical Bonding 18
    Chemical Reactions, Stability, and the Direction of Chemical Reactions 22
    Phase Diagrams 23
Nuclear Reactions 30
    The Reaction Process 30
    Geologic Uses of Isotopic Systems 31
■ SUMMARY 31
■ SELECTED REFERENCES 32 ■ NOTES 32

3 Fluids in and on the Earth 33
Introduction—The Nature and Distribution of Fluids 33
Flow of Fluids 34
    Streams, Stream Channels, and Flow of Water over the Surface 34
    Flow of Fluids Through Porous Rocks 38
    Flow of Fluids in Fractures and Conduits 39

Currents in the Oceans, Seas, and Lakes 41
Kinds and Sources of Fluids and Fluid-Mediated Processes in the Earth 42
- ■ SUMMARY 46
- ■ SELECTED REFERENCES 46 ■ NOTES 46

4 Crystallization and Crystallography 49

Introduction 49
Order, Disorder, Crystallinity, Nucleation, Crystal Growth, and Morphology 49
Solid Solution 54
The Properties of Symmetry 54
- Symmetry in One Dimension 54
- Symmetry in Two Dimensions 56
- Symmetry in Three Dimensions 60

Crystal Features 65
- Lines and Planes within Crystals: Miller Indices 65
- Faces, Morphologies, and Forms of Crystals 67

- ■ SUMMARY 71
- ■ SELECTED REFERENCES 71 ■ NOTES 71

5 Minerals: Their Classification and Identification in Hand Specimens 73

Introduction 73
Mineral Classification 73
Identifying Minerals in Hand Specimens 76
- Properties Related to Mineral Growth 76
- Properties Related to Mineral Interactions with Light 83
- Properties Related to Mineral Strength 87
- A Property Related to Mineral Mass: Specific Gravity 90
- Other Mineral Properties 91

- ■ SUMMARY 92
- ■ SELECTED REFERENCES 93 ■ NOTES 93

6 Advanced Methods of Studying Minerals and Rocks 95

Introduction 95
Historical Methods 95
Optical Mineralogy 96
- Light Interactions with Matter 96
- Design of the Petrographic Microscope 98
- Use of the Petrographic Microscope 99

Powder X-Ray Diffraction (XRD) 101
- X-Rays and Diffraction 101
- XRD Analysis Procedures and Applications 102

Other Common Analytical Methods 103
- Electron-Beam Microprobe (EM) 104
- X-Ray Fluorescence (XRF) 106

- ■ SUMMARY 106
- ■ SELECTED REFERENCES 107 ■ NOTES 107

7 Igneous Rocks 109

Introduction 109
The Origin of Magmas 109
Movement of Magmas 113
Modification of Magmas 114
Emplacement and Eruption of Magmas and the Formation of Igneous Structures 117
Crystallization and Solidification of Magmas and the Resulting Textures 121
Chemical Composition and Classification of Igneous Rocks 124
Basalts and Their Significance 135
Igneous Rock Associations 137
   Mid-Ocean Ridge Rocks 138
   Rock Associations of Subduction Zones and Orogenic Belts 139
   Anorogenic Igneous Rocks and LIPs 140
Some Final Comments 141
- SUMMARY 141
- SELECTED REFERENCES 142 - NOTES 142

8 Weathering and Soils 145

Introduction 145
Weathering Processes 145
The Structure, Chemistry, Mineralogy, and Physics of Soils 150
   Chemistry of Soils 151
   Physical Properties of Soils 153
Soil Classification 155
   The USDA Classification 156
   Unified Soil Classification 156
Origins of Soils 156
Paleosols 158
- SUMMARY 159
- SELECTED REFERENCES 160 - NOTES 160

9 Sedimentary Rocks 161

Introduction 161
Erosion, Transportation, and Deposition of Sediments 161
Sedimentary Environments, Textures, Structures, and Compositions 164
   Sedimentary Environments 167
   Sedimentary Textures 170
   Sedimentary Structures 174
   Sedimentary Rock Compositions 181
Classifications of Sedimentary Rocks 183
The Stratigraphies of Sedimentary Environments 193
   Continental Environments and Rock Associations 193
   Transitional Environments and Rock Associations 196
   Marine Environments and Rock Associations 197
Diagenesis and Lithification of Sediments 201
- SUMMARY 203
- SELECTED REFERENCES 203 - NOTES 203

10 Metamorphic Rocks   205
Introduction   205
Metamorphism and Metamorphic Processes   205
Metamorphic Rocks and Their Chemical Compositions   207
Metamorphic Structures and Textures   208
Classification of Metamorphic Rocks   218
Metamorphic Facies, Facies Series, and Depictions of Metamorphic Mineral Assemblages   220
Metamorphic Sites, Associations, and Special Types   225
   Contact and Hydrothermal Metamorphism   225
   Static (Burial) Metamorphism and Metamorphism in Subduction Zones   226
   Metamorphism in Arcs and Moderate-to-High-Temperature Metamorphic Belts   228
   Metamorphism of Ultrabasic Rocks   228
   Metamorphism Along Faults   230
■ SUMMARY 232
■ SELECTED REFERENCES 233   ■ NOTES 233

11 Resource Geology   235
Introduction   235
Resources and Deposits   235
   Resource Value   236
   Resource Knowledge and Accessibility   236
Classification, Types, and Origins of Resource Deposits   238
   Magmatic Process Deposits   239
   Hydrous Fluid Deposits   243
   Sedimentary Deposits   256
Other Minerals, Rocks, and Energy Resources   262
■ SUMMARY 263
■ SELECTED REFERENCES 265   ■ NOTES 265

12 An Overview and Summary   267
Introduction   267
Petrotectonic Assemblages and Plate Settings   268
■ SUMMARY 271
■ SELECTED REFERENCES 272

Appendix A: Determinative Tables for Mineral Identification   273
Appendix B: Identifying Common Rocks in Hand Specimens   305
Appendix C: Optical Mineralogy and Petrography in Thin Sections   317

Glossary   321
References   335
Figure Credits   355
Index   357

Preface

Earth materials include the minerals, rocks, melts, and other fluids of the Earth, plus surficial soils that blanket much of the continental areas. In addition, fossils, particularly those composed of original body parts or their replacements, are included among the Earth materials. This book is designed for sophomore-junior level university students (but not introductory students) as an exploration of the nature and origin of Earth materials, excluding the fossils, found at or near the Earth's surface. A background in both introductory physical geology, including plate tectonics and common minerals and rocks, and basic chemistry is assumed, but a review of some fundamentals of chemistry and physics is provided in chapter 2 to assure that students have the proper background as they proceed through the text. This volume may also be useful to laypersons with some background in science and a serious interest in understanding the materials that make up the landscapes they see around them.

We used an early version of this text in our team-taught, sophomore-level course in Earth Materials at Appalachian State University. Interactions with students in that course guided the further development of the text materials and gave us a sense of the needs and capabilities of students at the intermediate level of university studies who had future goals in geology, sustainable development, public school teaching, and ancillary sciences such as biology.

Chapters of *Crustal Earth Materials* include discussions of physical characteristics, processes of formation, and behaviors of each of the major types of Earth materials found at or near the surface of Earth, including the fluids and soils. Appendices provide useful mineral and rock identification keys as well as information on optical examination of minerals and rocks—the latter a subject area that may not be included in some courses on Earth materials. The appendices are considered useful for simple exercises and preliminary identification work, but the book is not intended to be a laboratory manual. As far as we know, none is available. For laboratory work on rocks and minerals, we recommend laboratory manuals such as Raymond (2009) and guides and compendia of systematic mineralogy such as Mottana et al. (1978), Phillips and Griffen (1981), Yardley et al. (1990), Deer et al. (1992), and Perkins and Henke (2004).

From a pedagogical perspective we have included problems within the text, notes, references to supplementary information, the aforementioned appendices, and online resources (waveland.com/Raymond-Johnson). Abundant illustrations help clarify and reinforce the concepts presented. While the book is dominated by black-and-white images, we provide a 32-page insert of color images. Moreover, the online resources feature color images and supporting, descriptive materials. *Crustal Earth Materials* also contains a glossary, since terms introduced in earlier chapters are used in later ones. These key terms are boldfaced in the chapters. *Italicized* text represents words or phrases highlighted in order to emphasize them to the reader, but these do not appear in the glossary. With regard to references in the text, we feel that by the sophomore level of university study, students need to understand that each of us in science builds upon the framework and background provided by earlier research. Few ideas result in paradigm shifts and are entirely new. Rather, each scientist extends the limits of the known by first understanding what is known and then building upon it. For this reason, we have included a modest number of references in the text. Students are encouraged to examine these and the references in the notes. Faculty are encouraged to create course activities that go beyond this text by using those reference materials.

Our knowledge of Earth materials derives both from the education provided to us by many former professors, colleagues, and authors, and from experience in geological research. We owe each of those who contributed to our knowledge our gratitude for their contributions to our understanding. Reviews of early drafts of specific chapters were provided by Callum Hetherington, Mary Keskinen, Brady Rhodes, Bill Size, and James Wilson, and the penultimate draft incorporates the suggestions of three anonymous reviewers. We thank all reviewers for their helpful feedback. Any errors in this text, however, are our responsibility. We also wish to acknowledge the assistance of Debbie Bauer, Marg McKinney, Anthony Love, and Tom Terranova in technical matters; the support of Appalachian State University, the Sonoma State University Geology Department, and Virginia Tech; the guidance of Editor Don Rosso at Waveland Press; and the invaluable assistance of Debi Underwood (graphics) and Laurie Prossnitz (editorial) of Waveland Press.

The Context of Crustal Earth Materials

INTRODUCTION

The Earth sciences—such as geology, hydrogeology, and soil science—like all sciences, challenge those who do science to illuminate the reality of nature and to solve problems involving our understanding of natural processes. That is the essence of science. The natural world is available for observation. Careful observations raise questions such as "What is this (or that) thing?" and "How do natural processes "operate?" Geology, for example, begins outdoors, where we can observe, describe, and measure a variety of natural objects (things) that constitute the Earth (figure 1.1a). Those observations and measurements usually lead us to the laboratory, where we do optical, chemical, X-ray, computer simulation, and other studies to expand our descriptions and to answer the questions (solve the puzzles) that nature provides for us (figure 1.1b).

In this book, we illuminate some fundamental characteristics of Earth and its parts and address puzzles specifically provided by *Earth materials*. **Earth materials** include all of the solid, liquid, and gaseous substances that constitute our planet. Those materials can be divided into crustal Earth materials, inner Earth materials, and enveloping Earth materials. The crust is the outer layer of the solid Earth above a seismic (earthquake wave) boundary called the "Moho" (short for Mohorovicic Discontinuity). It is the uppermost part of the rigid lithosphere, the lower part of which consists of inner Earth materials and overlies the plastic inner Earth asthenosphere below. Crustal Earth materials are the rocks, minerals, sediments, some fluids, and soils that constitute the crust. Inner

(a) The first author in the field collecting data.

(b) The second author in the lab, using the X-ray diffractometer for geological analyses.

Figure 1.1
The authors doing geologic research.

Earth materials include all of those materials that occur inside the Earth, that is, below the crust. Enveloping Earth materials are (1) the gases of the *atmosphere*, with its intermixed liquids (condensed water droplets that are also part of the hydrosphere), and (2) the *hydrosphere*, which includes all the waters of the atmosphere; the waters of lakes, rivers, and oceans on the surface of Earth; plus the groundwater within the crust. Intermixed with the enveloping and upper crustal Earth materials are the organisms that make up the *biosphere*. If we consider the overall shape and structure of Earth, then, it is a crude sphere that is composed of the somewhat concentric subspheres named above (figure 1.2).

This text is about inorganic (nonliving) crustal materials and the associated fluids that are important to crustal processes. Only where they play an important role in the formation of crustal materials are biospheric materials discussed. In addition, where inner or enveloping Earth materials interact with crustal materials and where an understanding of them facilitates understanding of the crustal materials, these inner or enveloping Earth materials are also discussed. Together, the crust and various closely associated spheres, plus the internal parts of the Earth, constitute Earth systems.

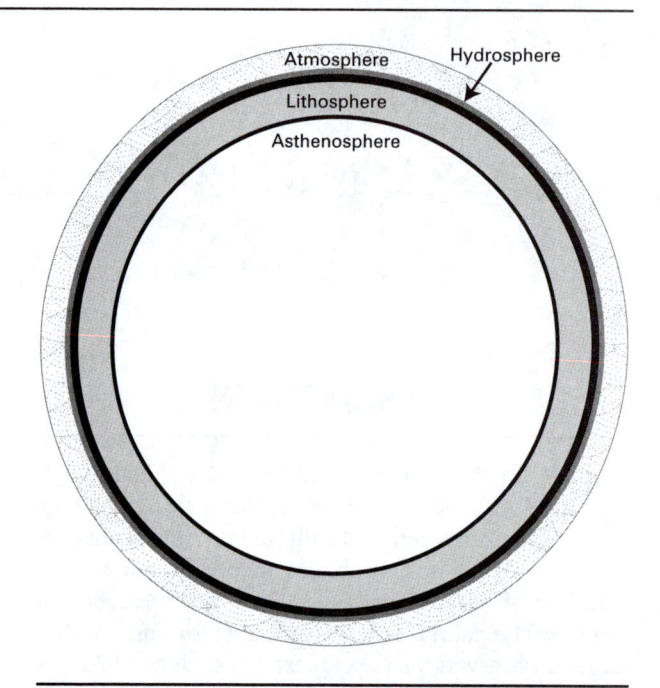

Figure 1.2
Cross-section (not to scale) of the various spheres of the outer levels of Earth. Levels below the asthenosphere are omitted. The black layer is the crust, which forms the upper layer of the lithosphere.

MINERALS, ROCKS, FLUIDS, AND SOILS

The fundamental crustal Earth materials covered in this book are minerals, rocks, fluids, and soils. Minerals have traditionally been defined as naturally occurring, solid, inorganic, crystalline compounds or elements, with a fixed or limited range of chemistry. That traditional definition was modified by a commission of the International Mineralogical Association to include some amorphous and organic materials (Nickel, 1995). Thus, following this modification, a **mineral** is considered to be an element or chemical compound that, *in general*, but with some exceptions, is a solid, inorganic, naturally occurring, and crystalline material produced by geological processes.

The various criteria for a material being classified as a mineral preclude some materials encountered in daily life from being minerals. The restriction that minerals be naturally occurring materials or materials produced by geological processes eliminates all materials made by humans from the category of materials we call minerals—even synthetic gems such as ruby and diamond that have compositions more or less identical to those of natural rubies and diamonds. The original restriction that minerals be solids eliminated all liquids and gases from the mineral category, including liquid mercury found in mines, even though such materials are naturally occurring. Today, however, liquid mercury is accepted as a mineral species, as one of the exceptions to the general criteria. Most minerals are crystalline, which means that the atoms that constitute them must possess what we call a *long-range internal order* (i.e., the atoms are arranged in repeated patterns). Yet, under the new definition, not all minerals have that atomic order, and some amorphous (noncrystalline) materials have been accepted as minerals. In addition, some naturally occurring organic materials are now included as minerals, such as hydrocarbon compounds that crystallize without the aid of organisms, and biogenic components of limestones derived from organisms (but not those directly precipitated as shells). For practical purposes and generalized analyses, students may use the traditional definition as a "working definition" of a mineral, as long as it is realized that a few exceptions have now been made for some naturally formed materials that are amorphous, organic, or liquid.

Minerals may be composed of a single element, but are commonly a chemical compound. For example, gold (element symbol = Au) is a mineral, as are the compounds silicon dioxide (SiO_2), which is the mineral quartz, and $Ca_2Mg_5Si_8O_{22}(OH)_2$, the mineral tremolite. Generally, minerals have a fixed or limited range of chemistry. Hence, minerals such as quartz consist of a single compound. While it is true that quartz may have impurities, such as the iron that is important in making some quartz purple (purple quartz is called *amethyst*), it is also true that no other elements are *required* components in quartz, nor are significantly different ratios of silicon to oxygen allowed in the mineral quartz. In contrast, minerals such as olivine have a range of chemistry, albeit a limited range. The chemical formula for olivine can be written $(Mg,Fe)_2SiO_4$. Any combination of the percentages of Mg and Fe totaling 100% is allowed in olivine. Hence, Mg_2SiO_4 is olivine, Fe_2SiO_4 is olivine, and minerals with Mg and Fe divided evenly, 75% to 25%, or in some other combination that totals 100%, are still olivine. If we substitute Ca for Fe, however, the mineral is no longer olivine, *sensu stricto*—it is a different mineral, monticellite. To reemphasize the point, the range of chemical variation within each mineral is generally limited.

Rocks can now be defined. A **rock** is a naturally occurring solid, composed of crystalline mineral grains, glass, fragmented or altered organic matter, or combinations of these materials (Raymond, 2007, p. 2). Thus, rocks are naturally occurring and many are composed predominantly of crystalline solids. Yet, several types of materials we typically refer to as rocks—including most limestones (some composed entirely of materials precipitated by organisms), obsidian (volcanic glass), and solid organic materials (e.g., coal)—are either organic in origin or are not dominantly crystalline.

Three classes of rocks exist. These are igneous, sedimentary, and metamorphic rocks. Igneous rocks are formed by the solidification of melted Earth material, in some cases at or near the surface of Earth (the volcanic rocks) and in others, below the surface (the plutonic rocks). Sedimentary rocks form from accumulated masses of rock and mineral fragments, for example in the form of sand or gravel, and from materials that precipitate or crystallize from solutions, with or without the assistance of organisms (e.g., so-called "lime mud" on the shallow ocean floor). Metamorphic rocks are igneous, sedimentary, or previously formed metamorphic rocks that have been changed by fluids, pressures, temperatures, or stresses different than those under which the rock formed.

Rocks and minerals (in most cases) are solids. In solids, atoms and molecules are limited in their movements to very small volumes of space, making the materials rigid. Some Earth materials, however, can flow and are called liquids and gases, or collectively, **fluids**. In liquids, molecules and atoms are somewhat restricted in their movements, because they are crowded by other atoms or molecules. The atoms and molecules do not have a fixed position within the body of material, and the molecules can slide past one another, allowing the liquid, as a whole, to flow. In gases, the molecules are not much constrained by neighboring atoms, because those neighboring atoms are relatively few in number. As a consequence, gases are quite compressible: they can change volume (i.e., dilate). Adding pressure to a gas will change its volume considerably, a fact quantified in Boyle's Law.[1] Liquids do not dilate or compress significantly. At high pressure (P) and temperature (T), however, the distinction between gases and liquids disappears, because the behavior of high P,T fluids is intermediate between that of gases and liquids.

The most important fluids in considerations of crustal Earth materials are fluids that are dominantly water-based, called hydrous fluids. In some settings, however, the fluids may be CO_2-rich, such as at some spots deep in the crust or in areas in which minerals containing chemically bound CO_3 are quite abundant. In other areas—especially those associated with oceanic sediments and sedimentary rocks containing natural organic hydrocarbons (petroleum)—methane (CH_4) may be a dominant component of the fluid. Melted rock is also a fluid. Fluids important in the crust are discussed in chapter 3.

Soils, one of the important crustal Earth materials discussed in this text, are derived from rocks through weathering (so called "decay") of mineral materials. Strictly speaking, **soil** is defined as weathered and disaggregated rock and associated organic materials, *unmoved* from and overlying the bedrock. Soil is important for a variety of reasons, not the least of which is that soil is the precursor or "parent" material of many rocks. When transported from its site of formation, soil becomes the sediment moved by rivers and streams, gravity, ice, and wind. Technically, after movement, it is no longer soil, because it has been moved from its site of formation. Ultimately, the sediment may be deposited as the precursor to sedimentary rocks.

The Structure and Chemistry of Earth

Recall that we can consider the overall, general shape and structure of Earth to be a crude sphere, composed of somewhat concentric subspheres (figure 1.2). As discussed in greater detail below, the solid Earth consists of both (1) a rigid outer layer, the lithosphere, composed of mantle materials underlying shallower crustal materials, (2) deeper mantle materials of the asthenosphere and mesosphere—layers that occur beneath the lithosphere, and (3) a two-layered, solid-liquid core. The asthenosphere, between the lithosphere and mesosphere, is plastic and the outer core is liquid nickel-iron containing other minor components. The atmosphere forms an enveloping sphere of gases that intermix with the waters (the hydrosphere) of Earth. The hydrosphere is a diffuse sphere, inasmuch as waters are mixed into lower levels of the atmosphere and the upper levels of the lithosphere, as well as forming the concentrated masses of the oceans, seas, lakes, and streams on Earth's surface.

Internally, *Earth is divided into two different kinds of groups of layers*, each group based on different seismic wave properties. Traditionally, the solid Earth was divided into a group of compositional layers that are primarily defined on the basis of the *velocities* of seismic waves that pass through them. The outermost layer, the **crust**, is a relatively rigid and brittle layer. The crust beneath the oceans is different than that of the continents. The oceanic crust is a single major layer that, downward, has gradually increasing wave velocity. Within this single layer, the oceanic crust contains a series of sublayers composed of different kinds of igneous rocks, capped by a generally small amount of sediment, facts predicted earlier and confirmed by decades of seismic studies and drilling in the ocean crust (Ewing, 1969; Cann et al., 1983; Robinson et al., 2000). The oceanic layers are dominated by rocks rich in **si**licon, **i**ron, and **ma**gnesium and in the past were therefore called *sima* (and the crust *simatic*). The continental crust has two main layers, separated by a relatively abrupt increase in wave velocity called the **Conrad Discontinuity** (figure 1.3). The continental crust is not as well layered internally as the oceanic crust and consists of a mix of denser rocks at depth, here referred to as the *complex lower crust*, and an upper part dominated by rocks such as granite, various sedimentary rocks, and metamorphic rocks. These continental crustal rocks, considered together, tend to be rich in **si**licon and **al**uminum and thus this crust has been referred to generally as *sial* and *sialic* crust. The thickness of the crust varies in space and over time.[2]

The crust is underlain by the mantle and is separated from it by another zone of rapid and relatively abrupt velocity increase called the **Mohorovicic Discontinuity** or "Moho" (figure 1.3). The **mantle**, as well as the **core** at the center of Earth, can be subdivided into layers. The mantle can be divided into the upper, middle (transition zone), and the lower mantle, as well as additional sublayers, all based on properties revealed by seismic waves. Beneath the mantle, a liquid **outer core** is succeeded inward by a solid **inner core**. The materials of these lower mantle and core layers are never exposed at the surface of Earth, but exert strong influences on rock formation.

The second subdivision of the outer parts of Earth is a threefold subdivision based on the *rigidity* of the layers. The outermost layer, the **lithosphere**, is a rigid layer that varies in thickness from about 70 to about 150 km or more (Lillie, 1999; Rychert and Shearer, 2009). It includes two parts—lighter, relatively brittle and rigid crustal rocks and heavier, underlying layers of dense rocks assigned to the upper mantle (figure 1.3). Beneath the lithosphere is the **asthenosphere**, a plastic, less rigid zone in the mantle. Although the

---

Problem 1.1

The temperature of Earth generally increases with depth. The increase in degrees per kilometer is called the *geothermal gradient*, and geothermal gradients vary from place to place. A crudely average value for the continents is 20°C/km.

(a) What is the average temperature at the base of normal continental crust of 20 km thickness?

(b) At the same geothermal gradient, what would be the temperature of the basal crust thickened to 50 km in a mountain belt?

(c) If the geothermal gradient at a mid-ocean ridge flank were 75°/km, what would be the temperature at a depth of 10 kilometers, the base of the normal oceanic crust?

(d) If melting occurs at temperatures between 600°C and 1300°C, depending on rock composition, pressure, and abundance of fluids, would any of the sites in (a), (b), or (c) be a possible site of melting?

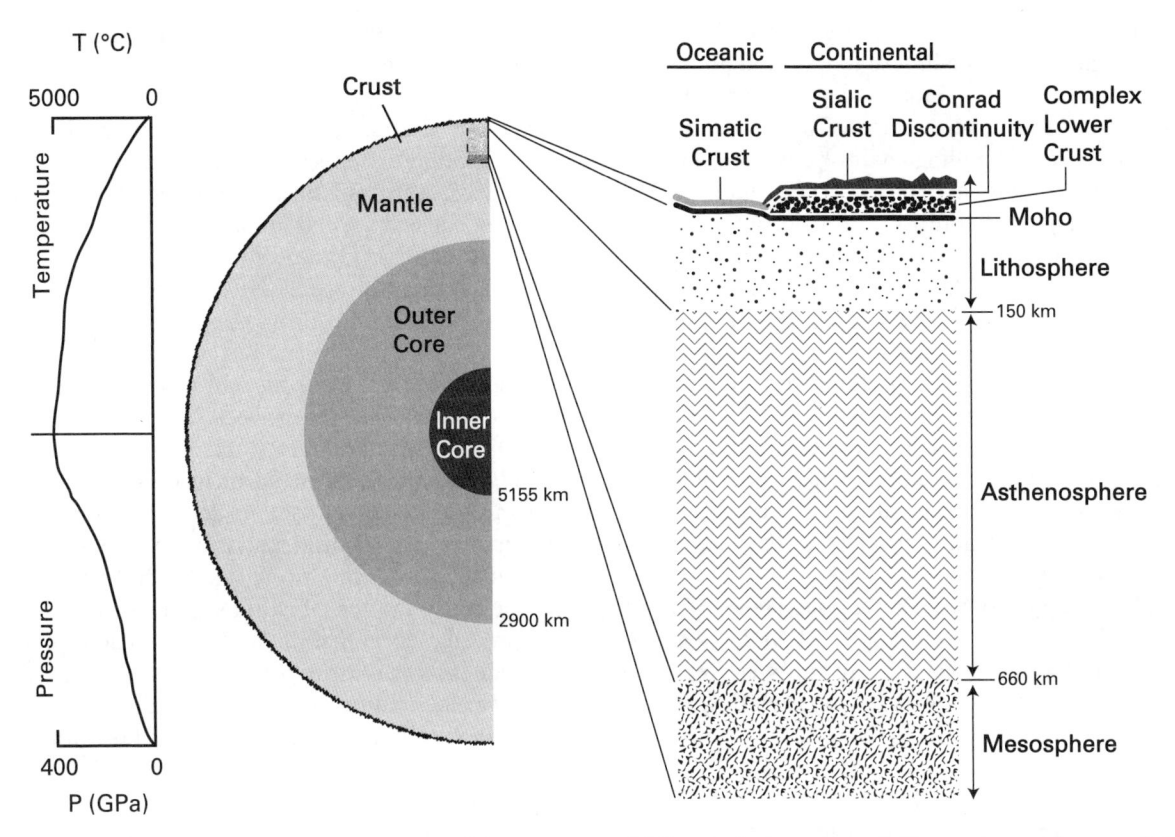

(a) Cross-section showing the major layers of the Earth with the generalized pressure and temperature profiles through the center of Earth.

(b) Expanded section of the uppermost layers of the Earth, showing the positions of the Conrad and Mohorovicic (Moho) discontinuities.

Figure 1.3
Generalized sketch of the interior of Earth.
(*Sources:* Modified from Raymond, 2007, figure 1.1; numerical data from Bolt, 1982; Lillie, 1999; Bass and Parise, 2008.)

crustal and mantle parts of the lithosphere, considered together, are relatively rigid, the lithosphere is mobile, because it is pushed, dragged, and pulled by the moving, plastic asthenosphere beneath it. In response to those specific driving mechanisms (push, pull, and drag), the curved, relatively thin slabs of the lithosphere move over the underlying mantle materials of the asthenosphere, which itself overlies a more rigid part of the mantle called the **mesosphere**.

The rigid lithosphere and the two zones beneath it are recognized primarily on the basis of particular seismic wave properties. Those properties combined with observations, theoretical analyses, and laboratory studies indicate that the layers have both different physical properties and different compositional characteristics. We can observe the compositions of the crustal rocks at the surface of Earth. We cannot observe, directly, the character of the asthenospheric or mesospheric rocks, but we can infer the particularly important fact that the asthenosphere contains small, widely distributed volumes of melted rock that contribute to its plasticity. The underlying mesosphere is a more rigid and dense layer predominantly composed of heavy iron- and magnesium-rich minerals, plus more localized amounts of melted rock. Investigations called seismic tomography studies reveal that the mesosphere also contains many pieces of lithosphere that have been pushed or carried down into it (subducted) and have sunk (Hutko et al., 2006; Karason and van der Hilst, 2000). In summary, the lithosphere, asthenosphere, and mesosphere are distinguished primarily by their relative degrees of rigidity revealed by the behavior of seismic (earthquake) waves that pass through them, but in addition, each zone has other properties.[3]

Plate tectonic theory (discussed below) is based on the concept that the lithosphere is rigid and is broken into several thick, broadly curviplanar slabs that move across the upper surface of the asthenosphere and interact by (1) pulling apart or diverging, (2) sliding past one another along fault zones, and (3) colliding or converging (figure 1.4; see also color plate 1a). Currently accepted theory suggests that a combination of forces—including currents called convection currents within the moving hot, plastic asthenospheric rocks flowing beneath the lithosphere, and gravitational drag as plates go down into the mantle—cause the plates to move (Elsasser, 1971).[4] The currents are, in turn, driven by heat escaping from the interior of Earth. In addition, where one plate descends beneath another, it can exert a dragging force that pulls the plate away from adjoining plates and into the mantle.

The compositions of the various layers of Earth are clearly different (Krauskopf and Bird, 1995, ch. 21).[5] Considering the solid Earth in its entirety, it is dominated by eight elements — iron (Fe), oxygen (O), silicon (Si), magnesium (Mg), nickel (Ni), sulfur (S), calcium (Ca), and aluminum (Al) (figure 1.5). All other elements comprise less than 1% of the Earth. Considering only the outermost layer, the crust, we observe that the elements O, Si, Al, Fe, Mg, Ca, potassium (K), and sodium (Na), in order of decreasing abundance, are the most important elements. In contrast, in the innermost part of Earth, the core, Fe and Ni are dominant, whereas in between, in the mantle layers, O, Si, Mg, Fe, Al, and Ca are most abundant.

Not only is the solid Earth separated into layers that have various compositions, so too the atmosphere, consisting predominantly of gaseous nitrogen (N) and oxygen (O), and the hydrosphere, consisting almost entirely of water (H_2O), differ in composition from the lithosphere. From the perspective of the solid Earth, a more important fact is that the Earth as a whole is not homogeneous, but has layers of different compositions that have been separated from one another. In scientific terms, this means that the Earth is *differentiated*. Early in Earth history, the planet was chemically heterogeneous in detail, but more or less homogeneous, if considered as a whole, meaning that it is likely there were no layers. Several processes led to the concentration of various elements in layers that became the core, mantle, and crust. These processes include sinking of iron-rich materials to form the core, and partial melting of rising mantle rocks (partial melting is melting of a part, but not all of a rock). Partial melting of mantle rocks to form magmas, and processes of magma modification, such as fractional crystallization (discussed in chapter 7), are particularly important in ongoing differentiation because they facilitate concentration of lighter elements such

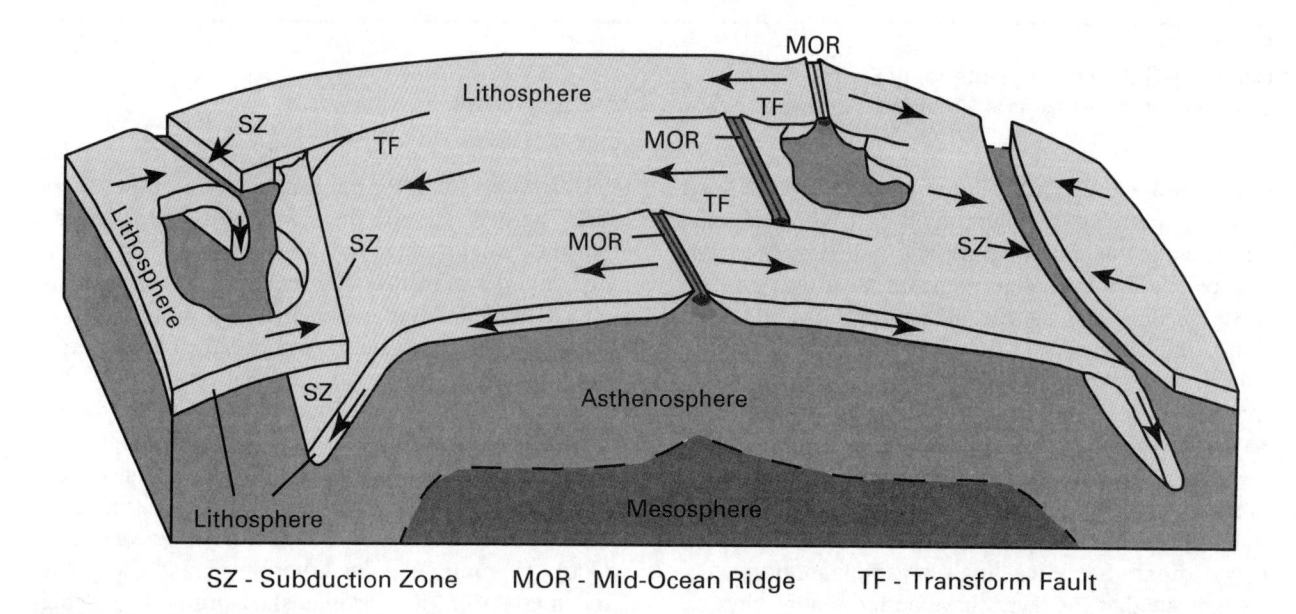

SZ - Subduction Zone   MOR - Mid-Ocean Ridge   TF - Transform Fault

Figure 1.4
Schematic block diagram showing the three types of plate boundary. Mid-ocean ridges represent spreading centers.
(*Source:* Raymond, 2007, p. 5; modified from Isacks et al., 1968.)

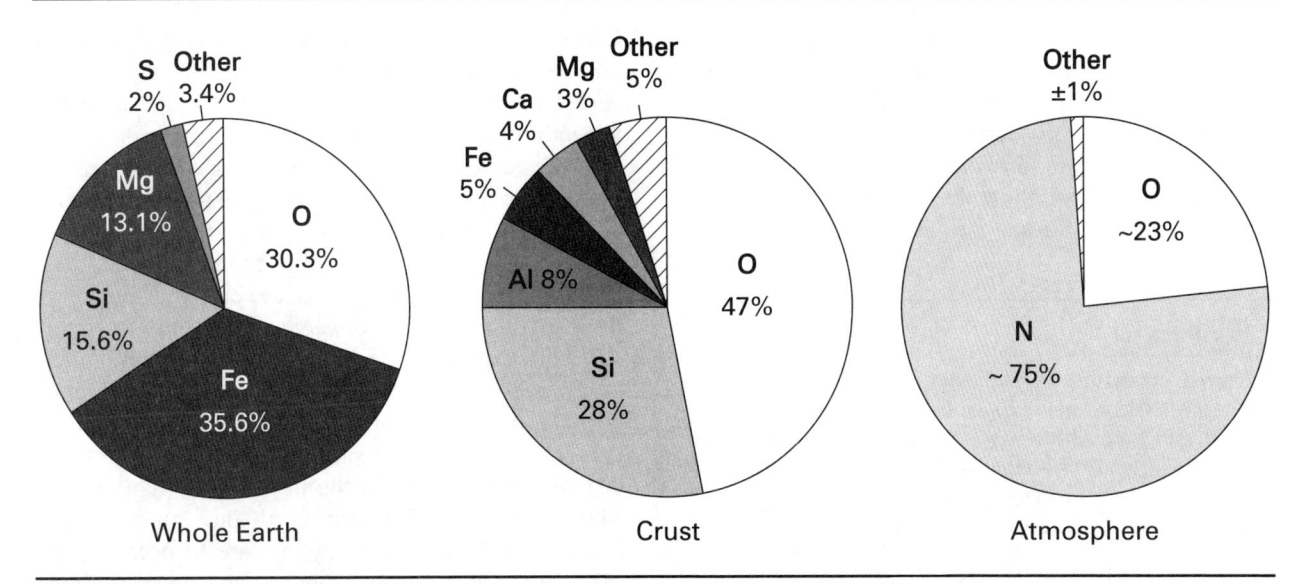

Figure 1.5
A comparison of the elemental abundances in weight percent for the whole Earth, the continental crust, and the atmosphere.
(*Sources:* Data from Press and Siever, 1986; Kump et al., 1999; Rudnick and Gao, 2005; Dyar et al., 2008.)

as Si, Al, Na, and K in the melts that rise and crystallize to form crustal minerals and rocks.

The chemical differences among Earth's layers and the physical conditions under which those layers exist, such as the pressures and temperatures, control the formation of different minerals within the rocks of each layer. The outer core, under high pressure and temperature, is liquid Fe-Ni with other very minor constituents; whereas the inner core, which is under even greater pressures, is primarily solid, Fe-Ni-alloy minerals containing minor additional elements. The mantle, dominated by O, Si, Mg, and Fe in a high-pressure environment, is composed predominantly of distinctive, generally dense, magnesium-iron silicate minerals in which the atoms are densely packed. The crust, however, consists mostly of lighter minerals composed of the abundant crustal elements O, Si, Al, Fe, Mg, Ca, K, and Na, minerals characteristic of rocks exposed at the surface. Manganese, titanium, phosphorous, hydrogen, and carbon are also important in some relatively common crustal minerals. These crustal minerals include, in order of decreasing abundance, the feldspars, quartz, pyroxenes, amphiboles, micas, olivines, clays and chlorites, and carbonate minerals (figure 1.6: for mineral images see color plates 7–20). These are the minerals described and discussed in the chapters that follow. Rare and less abundant elements contribute to the compositions of the more than

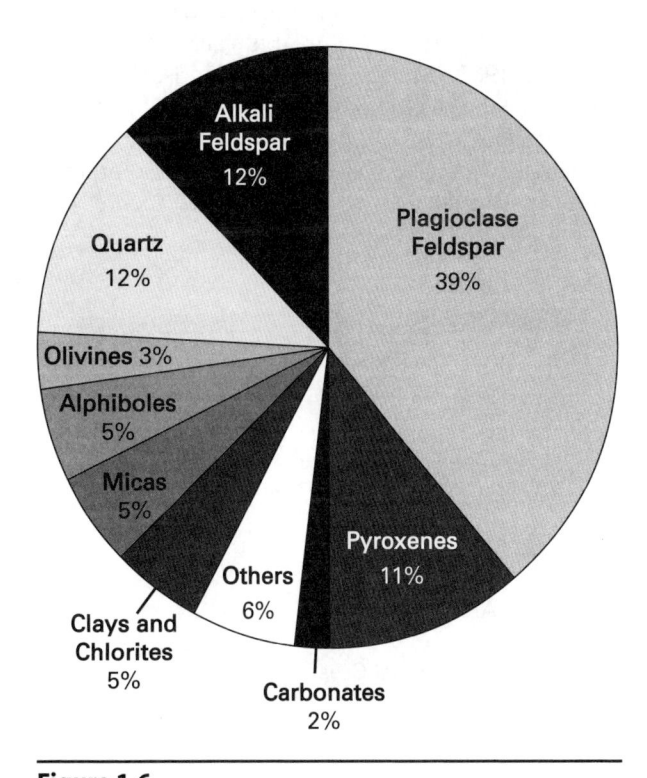

Figure 1.6
The crustal abundances in weight percent of the rock-forming minerals.
(*Source:* Data from Ronov and Yaroshevsky, 1969.)

4700 other minor to rare minerals that occur in the crust.[6] The rare minerals are particularly important from a practical point of view if they contain elements valuable to modern society—such as copper, zinc, manganese, titanium, and palladium—and if they are abundant enough to allow the mining and extraction of significant quantities of the valuable elements.

> **Problem 1.2**
>
> Draw a rectangular graph showing the *abundances* of oxygen, iron, silicon, and nitrogen for the core, continental crust, and atmosphere on the ordinate and the three *layers* on the abscissa. Write a brief explanation of why this distribution makes sense in terms of the differentiation of Earth.

PLATE TECTONICS

Plate tectonics is a theory that explains the origin of the major structural features of Earth—the ocean basins, continents, mountain ranges, and great faults—in terms of the interaction of large slabs of lithosphere (Wilson, 1965; Morgan, 1968, 1972; Isacks et al., 1968; Le Pichon et al., 1973). Recall that the curved slabs of lithosphere, called **plates**, interact in three fundamental ways. The plates slide past one another in zones of shear called **transform faults** (figure 1.4).

Zones where plates separate and diverge from one another are called **spreading centers**, whereas convergent zones where plates collide and one plate descends beneath the other are called **subduction zones**.

Where two plates collide—e.g., two oceanic plates, two plates bearing continents, or an oceanic plate and a continent-bearing plate—the resulting tectonic features, especially the mountain belts and included rock types that form, are partly a function of the particular style of convergence or collision (figure 1.7).[7] For example, plate edges containing only simatic oceanic crust may converge, with one subducting beneath another, yielding an arcuate row of volcanoes composed of generally silica-poor volcanic rocks—an oceanic *volcanic arc*—on the upper plate (figure 1.7a). The Caribbean arc between North and South America is an example. Alternatively, a plate of oceanic rock may collide with and subduct beneath one containing a continent at its edge, producing an active continental margin with a volcanic arc on the continent (figure 1.7b). Such arcs typically contain a range of rock compositions from silica-poor to silica-rich rocks. The Cascade Range of volcanoes in western North America, from northern California to southern British Columbia, formed in this way. Another possibility is one in which two plates bearing continents collide (figure 1.7c). For example, India collided with Asia to form the Himalaya Ranges in such a process. These types of convergence possibilities are further complicated where plates collide at angles in oblique convergence.

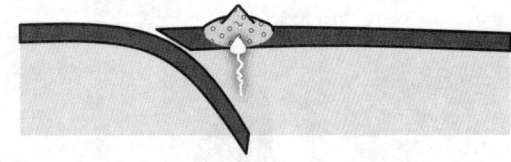

(a) Oceanic plate - oceanic plate convergent zone with volcanic arc on the overriding plate.

■ Oceanic Crust

▫ Arc Crust

▨ Continental Crust

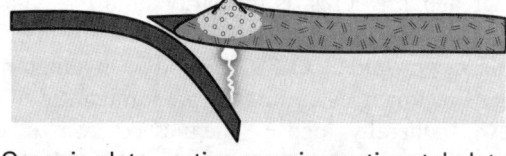

(b) Oceanic plate - active margin continental plate convergent zone with volcanic arc.

(c) Active margin continental plate - passive margin continental plate convergent zone with mountain belt on overriding plate.

Figure 1.7
Three types of plate convergent zones.

The variations yield differences in the rock-forming processes. Subduction/collision zones ultimately become orogenic belts (i.e., mountain belts) composed of diverse rocks formed in the region of the collision.

Two other features important to plate tectonic processes and the plate tectonic framework of the Earth are the *mantle plumes* and the *triple junctions*. **Triple junctions** are sites at plate corners, where three plates join. In general, triple junctions migrate and complicate the mountain-building and rock-forming processes at plate margins. The **mantle plumes** are huge, crudely cylindrical, rising masses of plastic mantle material. The tops of plumes melt, yielding magmas that rise through the asthenosphere and lithosphere to produce volcanic eruptions at the surface of Earth. Mantle plumes underlie the Yellowstone region of North America and the southeast end of the Hawaiian Islands. Complex plumes composed of a mix of melted and unmelted materials also form above subduction zones.[8]

The Earth's lithosphere is divided into seven large plates and several smaller ones (figure 1.8). Each plate is bounded by some combination of spreading centers, transform faults, and subduction zones. The margins of the African plate, for example, are dominantly spreading centers and transform faults, whereas the Philippine Sea plate is surrounded largely by subduction zones. As noted, the types of margins bounding a plate determine the type of tectonic activity (faulting, folding, and mountain building) and the mineral- and rock-forming processes that occur at the plate margins and, in part, within plate interiors. Plate interior and climatic conditions, governed by factors such as the size of continents, the relative distribution of oceans and continents, the distribution and size of mountain ranges, and the latitude, control the formation of soils and partly control the formation of sedimentary rocks on the continents. Fluids of various types, on and below the surface, move in response to many of these controls and also move in response to thermal and atmospheric controls. The fluids interact with the minerals, rocks, and soils that previously formed in and on the Earth.

In the chapters that follow, where we focus on rock-forming processes, we will focus on rocks that form at various sites within the plates and at plate margins. The group of rock types formed at a particular location is called a **petrotectonic assemblage** (Dickinson, 1971). Figure 1.9 depicts, in generalized terms, various rock types and petrotectonic assemblages that form at various sites, including one type of convergence zone (in this case at an ocean plate/active margin, continent-bearing plate convergence

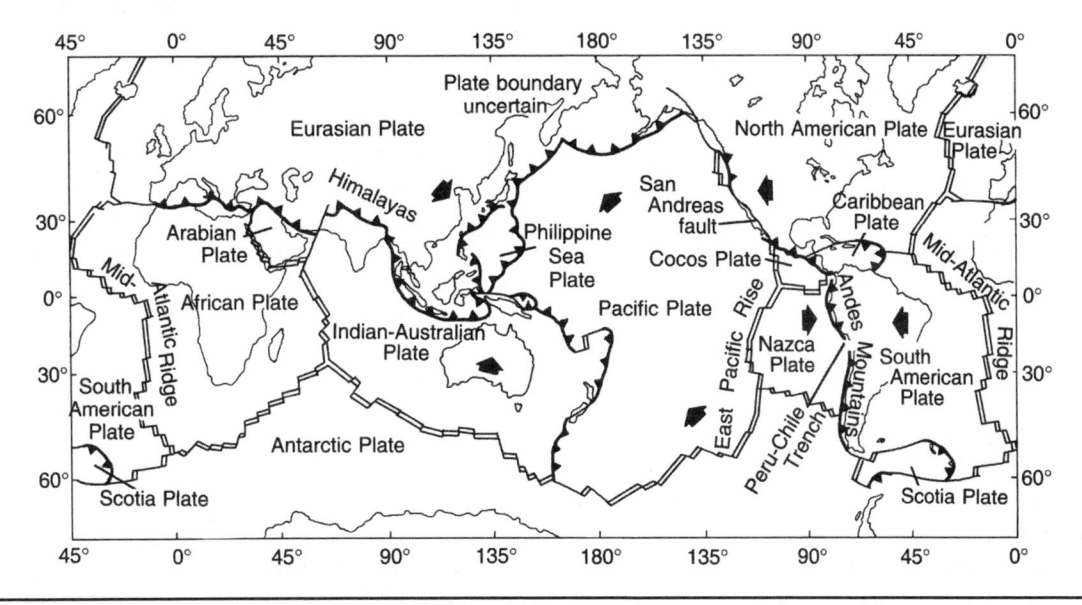

Figure 1.8
Map of the world showing the major lithospheric plates. Double lines indicate spreading centers (divergent boundaries), heavy barbed lines are subduction zones or orogenic belts (convergent boundaries), and single lines along plate boundaries are transform faults (shear boundaries).
(*Source*: W. Hamilton, U.S. Geological Survey.)

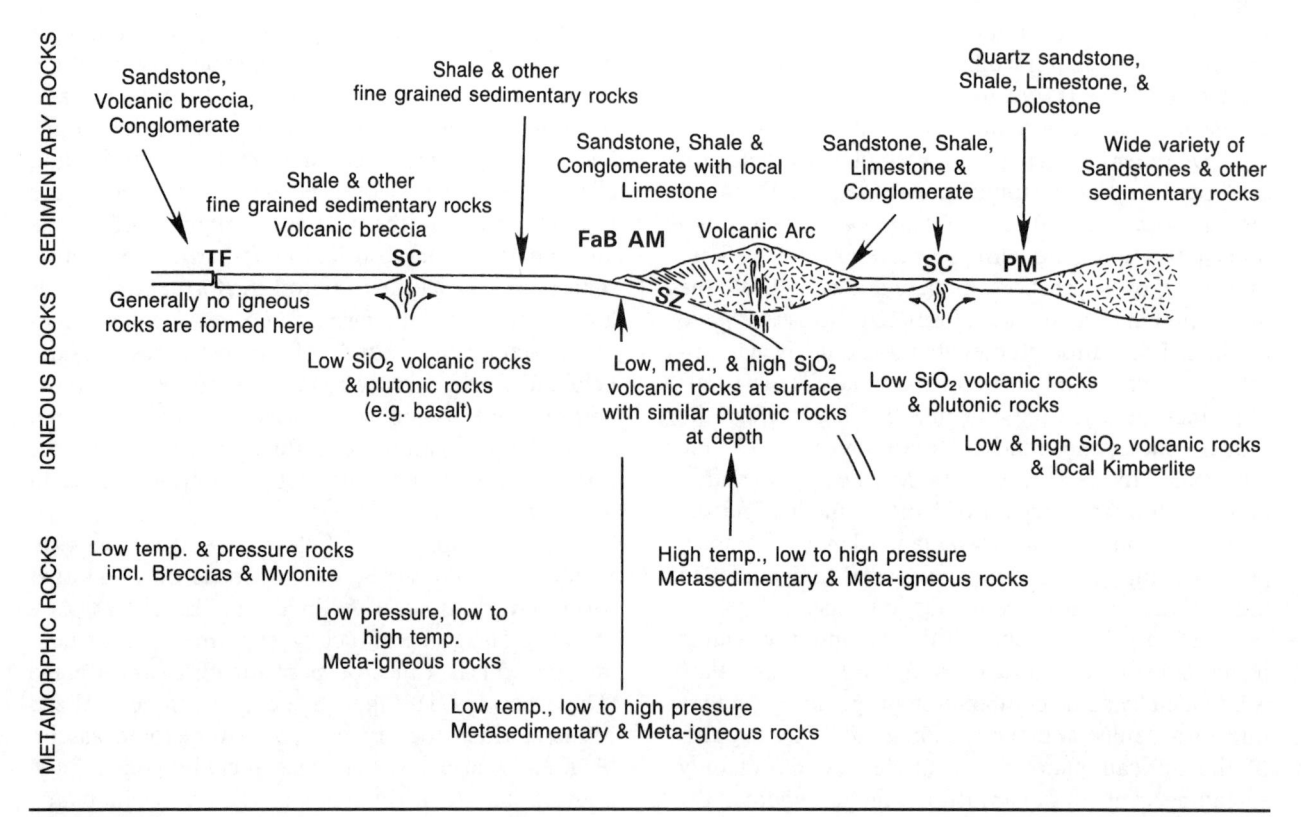

Figure 1.9
Diagram of the generalized petrotectonic assemblages characteristic of plate boundaries and plate interiors. TF = Transform Fault, SC = Spreading Center, FaB = Forearc Basin, AM = Active Margin, SZ = Subduction Zone, PM = Passive Margin.
(*Source:* Raymond, 2007, p. 5.)

zone). Aspects of these petrotectonic assemblages, discussed in the following chapters, are summarized briefly here.

Silica-poor igneous rocks are dominant at mid-ocean ridges. At the spreading centers, such as at mid-ocean ridges or in back-arc basins behind an arc, tension stretches the plates, causing thinning and the development of normal faults and down-dropped blocks (i.e., graben). Thinning of the crust induces mantle rocks to rise toward the surface and, in rising, they melt, generating Si-poor (basaltic) lavas not radically different from those of the Hawaiian volcanoes formed above a mantle (hot spot) plume. The lava flows spread along the rift valley graben formed by the faulting. Beneath the basalt flows, coarse-grained igneous plutonic rocks crystallize from melts that do not make it to the surface. Rock rubble and sand that will become the rocks called breccia and sandstone are eroded from the fault scarps, and finer-grained sediments (mud and chemical precipitates) falling from the water column accumulate atop the lava flows.

Along transform faults, shearing changes the mid-ocean ridge rocks to fault breccias near the surface, and to highly deformed metamorphic rocks called *mylonites* at depth, where the pressures and temperatures are elevated. If a transform fault pulls apart a bit, as the sliding occurs, a "transtensional" fault zone is formed, which produces a rift basin. In this basin, not only are the deformed rift rocks present, but sedimentation will produce gravels (later to become conglomerates and breccias), sands (future sandstones), and muds (future mudrocks) deposited atop these deformed rocks. If tension is significant, some volcanic eruptions will occur in the zone, and low-silica (basaltic) volcanic rocks will form.

At convergent margins, the complexities that may arise are many. Thrust faults and folds are common, especially where plates carrying continents collide to form mountain belts. Subduction of one plate beneath another causes melting in the mantle rocks above the subduction zone. The melts rise toward the surface and are modified to yield many different rock types that define the core of a curvilinear volcanic "arc," while heating the surrounding rocks to form a variety of metamorphic rocks. Erosion of the mountains produced in these zones of rising melt and converging plates yields sediments that blanket the mountain slopes and wash into the basins surrounding the mountains. On the seaward side of volcanic arcs of continental margins, such as seaward of the Pacific coast west of Washington and Oregon, subduction continues to form an "active margin" containing *forearc basins* that receive sediment from the arc and continent. On continental margins where major tectonic activity no longer occurs, such as along the eastern coast of the United States and the western coasts of Norway and Namibia, sediment accumulates in "passive margin" basins. Within the continents, a wide variety of sediments form, produced by the erosive and depositional actions of water, wind, and glacial ice. Rare volcanic activity, triggered by rising materials within the mantle, yields volcanic rocks and underlying plutonic rock types.[9] In any of these settings, chemically active fluids moving from one rock type to another or from one set of pressure-temperature (P,T) conditions to another may alter the rocks or deposit veins of minerals, some of which have economic value. On the surface, weathering and erosion yield soils and landscapes. We address each of these processes in the chapters that follow.

EARTH SYSTEMS AT AND NEAR THE SURFACE

At the surface of Earth, the various spheres intersect. Energy arriving from the sun drives atmospheric circulation and heating, with the circulation acting to redistribute accumulated thermal energy.[10] Circulation also controls precipitation and the distribution of water across the atmosphere and surface of Earth. The precipitation feeds streams, glaciers, groundwater reservoirs, and ultimately the seas and oceans of the hydrosphere. At the surface of the lithosphere, atmospheric gases interact with rain and other precipitation to yield acids that are important in weathering of rocks, which produces soils. Plants and animals of the biosphere—the growth of which is largely supported by solar energy, nutrients in the rocks and minerals of the lithosphere, water of the hydrosphere, and carbon dioxide in the atmosphere—all contribute to weathering. Biological agents such as algae and bacteria produce organic acids, which, together with acids produced by the interaction of the atmosphere and the hydrosphere, also contribute to soil formation. As they grow, plants create biomass that becomes a part of the soil. Once soils are produced, movement of hydrospheric constituents—specifically running water (streams) and moving ice (glaciers)—erodes and moves the soil, converting it to sediment. Moving air (wind) and gravity contribute to these processes of erosion and sediment formation. Erosion by streams, wind, and glaciers, as well as gravity, also sculpt and construct the various topographic features of the landscape—such as linear mountains, semicircular valleys, sand dune fields, and plateaus—that give each region its distinctive character. Finally, eruption of volcanoes at the surface of the lithosphere, and sedimentation of materials moved from higher to lower places on the surface by forces of erosion and transportation, ultimately produce new rock masses.

The hydrologic cycle—the movement of water between atmospheric, hydrospheric, and lithospheric realms (Fetter, 1994, ch. 1)—contributes in a variety of ways to Earth processes. Precipitation is necessary for erosion by streams and glaciers. Groundwater facilitates weathering and erosion. Waters that move deeper into the lithosphere become heated and enriched with chemicals that, as the waters move, induce changes in the minerals and chemistry of the rocks. In special circumstances, these thermal (hot) fluids create deposits of unusual and valuable minerals. Thus, moving waters are important elements of Earth systems.

While the rocks and minerals of the lithosphere are the main focus of this book, it is clear that interacting Earth systems are important to the origins of those rocks and minerals. The many processes operating

---

Problem 1.3

Using the Internet:

(a) Find the definition of *carbon sequestration*.

(b) Assemble a list of minerals of both organic and inorganic character that might be used for carbon ion sequestration.

(c) Discuss the pros and cons of using at least two of these minerals for carbon sequestration.

(d) What minerals likely formed as the result of recent experiments on carbon sequestration in basalts of Iceland?

within the various spheres of the Earth produce constant change in the nature, distribution, and configuration of the crustal Earth materials. For example, in the case of volcanism fed by materials from the mantle below the lithosphere, new rock and mineral material are added to the lithosphere. As another example of the interaction of the spheres, weathering involving the atmosphere, hydrosphere, and biosphere changes the nature of materials already existing in the lithosphere. Together, all of the *processes* that produce crustal Earth materials and the nature of the *materials* produced are the subjects of this book.

SUMMARY

All of the solid, liquid, and gaseous substances that comprise the Earth constitute the Earth materials. The fundamental *crustal* Earth materials are minerals, rocks, fluids, and soils. Minerals are elements or chemical compounds that are generally solid, inorganic, and crystalline, and additionally are naturally occurring and produced by geological processes. Most have limited ranges of chemical composition. Rocks are similarly solid and naturally occurring, but are composed of crystalline mineral grains, fragmented or altered organic matter, glass, or combinations of these materials. Rocks and minerals constitute much of the solid Earth, a sphere divided into inner and outer parts. The solid Earth is enveloped by the atmosphere and hydrosphere that contain a dispersed biosphere. The Ni-Fe, solid inner core and liquid outer core are enveloped successively by the mesosphere (lower and middle mantle), asthenosphere (plastic middle mantle), and lithosphere (upper mantle and crust). Curved slabs of lithosphere called plates move over the plastic asthenosphere, driven by convection currents in the underlying mantle and by additional forces. Plates interact to form new crustal minerals and rocks and larger crustal components, such as mountain ranges, all of which are observed at the surface. Variations in the materials formed at specific plate boundaries (the plate edges) yield diverse and distinctive petrotectonic assemblages that are keys to past Earth history.

SELECTED REFERENCES

Dickinson, W. R. 1971. Plate tectonics in geologic history. *Science*, v. 174, pp. 107–113.

Dyar, M. D., Gunter, M. E., and Tasa, D. 2008. *Mineralogy and Optical Mineralogy*. Mineralogical Society of America, Chantilly, VA. 708 p.

Krauskopf, K. B., and Bird, D. K. 1995. *Introduction to Geochemistry* (3rd ed.). McGraw Hill, New York. 646 p.

Kump, L. R., Kasting, J. F., and Crane, R. G. 1999. *The Earth System*. Prentice Hall, Upper Saddle River, NJ. 351 p.

Le Pichon, X., Francheteau, J., and Bonnin, J. 1973. *Plate Tectonics*. Elsevier Scientific, Amsterdam. 300 p.

Lillie, R. J. 1999. *Whole Earth Geophysics*. Prentice Hall, Upper Saddle River, NJ. 361 p.

Raymond, L. A. 2007. *Petrology: The Study of Igneous, Sedimentary, and Metamorphic Rocks* (2nd ed.). Waveland Press, Long Grove, IL. 720 p.

NOTES

1. Sienko and Plane (1961, p. 132).
2. For example, see Lillie (1999, p. 92ff) and Mantle and Collins (2008).
3. For more details of the internal structure of Earth see standard geophysics texts such as Lillie (1999) and Bolt (1982), and see Condie (2001) and Bass and Parise (2008).
4. See Moores and Twiss (1995, p. 80ff) for a review of other forces.
5. Also see Wyllie (1971, p. 104), Ronov and Yaroshevsky (1969), and Rudnick and Gao (2005).
6. For example, see *Dana's Manual of Mineralogy* (Gaines et al., 1997) or Dyar et al. (2008).
7. See Moores and Twiss (1995, ch. 9) for more information on the varieties of plate collisions.
8. Gorczyk et al. (2007).
9. See Raymond (2007) and the references therein for information on rock formation at petrotectonic sites.
10. Kump et al. (1999, ch. 4).

Fundamentals of Physics and Chemistry Relevant to Earth Materials

INTRODUCTION

Beginning geology students often are surprised by the emphasis placed on knowledge of chemistry, physics, and mathematics in geology because it is the prospect of doing science outdoors that drew them to geology. Yet our planet is made up of rocks, rocks are made up of minerals, and minerals are naturally occurring chemical compounds that interact in accordance with some basic physical principles. In a broader perspective, the world is a chemical one that follows physical laws that can be described mathematically. Thus, a full understanding of the materials of Earth requires a basic understanding of the physics and chemistry of those materials.

In order to discuss and quantify particular parts of the natural world, it is often necessary to limit our discussion to a part of that world. To do so, we define systems. A **system** is any part of the natural world that we wish to isolate for the purposes of analysis or discussion, either physically in real time or conceptually in our minds. Thus, a can of soda can be considered a system, whether we have it in hand or are just thinking about it as a volume within the natural world. Systems may be rather large—the solar system—or rather small, like a microscopic crystal in a rock. As we discuss physical and chemical principles, it is often convenient to discuss systems, so that we can limit the complications in the discussion.

SOME CONCEPTS OF PHYSICS

Physics is the study of matter, energy, time, and the relations between them. Perhaps more than any of the other natural sciences, physics is a quantitative science that involves measurement and calculation. Certain kinds of mathematics, such as calculus, were created to solve the problems of physics.

In this section, we examine principles of physics important to understanding Earth. We cover the concepts of measurement, force, and velocity, as well as some additional derivative measurements such as stress. Because thermodynamics is essential to understanding chemical reactions in minerals and rocks, we also introduce its basic principles.

Measurement

Physical analyses require measurements. Important fundamental units of measure of interest to us here are units of time, length, and mass. With regard to time, the most common unit of measure used is the second—a unit defined in terms of the vibration of an atom. Geologists, of course, also use much longer periods of time, with a million years being a fundamental unit. About 31.5 million seconds (31.5×10^6 seconds) make a year, and one million times this value is a million years, one basic unit of geologic time.

Units of length are defined in terms of wavelengths of light. One traditional measure is the **meter**, defined as a distance equal to 1,650,763.73 wavelengths of orange-red light of krypton-86.[1] The meter is commonly divided into 100 parts, that is, 100 centimeters; 1000 parts or 1000 millimeters, and smaller units, such as the Angstrom (Å)—one 10 billionth of a meter ($1Å = 10^{-10}$ m). The multiple of the meter most commonly used is the kilometer, equal to 1000 meters. For comparison to English units, one meter is 3.28 feet and one kilometer is equal to 0.6214 miles.

The standard unit of mass is the **kilogram**, defined as the weight of a standard platinum-iridium cylinder.[2] A kilogram equals 1000 grams; and one gram equals 1000 milligrams. One kilogram is 2.2 English pounds.

Another less definable characteristic of natural materials and systems that is often measured is **temperature**. Qualitatively, we feel things as being hot or cold. In a more quantitative sense, we would say that two systems with the same temperature have no measurable tendency to change with regard to temperature and are in "thermal equilibrium." In a circular sort of reasoning, we can then define temperature as that property of a system that renders it in thermal equilibrium with another system.[3] Again, at equilibrium there is no tendency for the system to change.

The units of temperature are degrees Kelvin (K) and degrees Celsius (C), or the occasionally used alternative, degrees Centigrade.[4] Water, ice, and water vapor can coexist at one point called the triple point (usually plotted in a graph of temperature versus pressure). This point is used as the standard for the Kelvin temperature scale and is given the value 273.16°K. Degrees Celsius are related to degrees Kelvin by the equation

$$K - 273.15° = °C$$

so that the triple point of water at 273.16°K is 0.01°C [i.e., 273.16°K − 273.15 = 0.01°C]. The Fahrenheit scale, commonly used in the United States, is related to the Celsius scale by the relation

$$°F = 9/5°C + 32°F$$

Problem 2.1

Convert 500° Celsius, a typical metamorphic temperature in the Earth's crust, to both (a) a Kelvin and (b) a Fahrenheit temperature.

Derivative Measures

Using basic units, we can obtain several other units derived from the basic units, units we call derivative measures. These include area, distance (displacement), velocity, acceleration, force, and stress. **Area** is, of course, a two-dimensional measure of surface, e.g., consisting of the product of two lengths ($l_1 \times l_2$) measured perpendicular to one another, or the product of the radius of a circle (r) squared times pi (π), i.e., πr^2 (where pi = π = 3.1416). A standard metric unit is square meters (m^2). Distance, like length, is measured in meters or kilometers, but **distance** (D) is the difference in *position* of a body at two different times (figure 2.1). The position (p) (the length from a

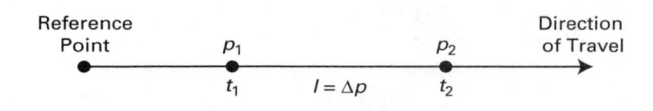

Figure 2.1
Diagram illustrating the difference in position 1 (p_1) and position 2 (p_2) from a reference point (difference = ΔP), yielding a length (l) or distance (D) ($l = D = \Delta p$).

reference point) at a later time (time 2) minus the position (again, the length from the same reference point) at an earlier time (time 1), gives a distance or displacement, i.e.,

$$p_2 - p_1 = \Delta p = l = D$$

Velocity is likewise a derivative unit, but is derived from the distance and a difference in time. In short, the velocity (v) is the distance between p_2 and p_1 (Δp) divided by the time required ($\Delta t = t_2 - t_1$) for a body to move between p_1 and p_2, i.e.,

$$v = \Delta p / \Delta t$$

Deformation of rocks involves derivative units such as force. **Force** is a measure of the mass (m) times acceleration (a). Acceleration derives from velocity. In a general situation, velocity will change in magnitude over time, so that the velocities at two different times, $v_2 - v_1$ divided by the difference in time (Δt) defines what is called the average acceleration (= $\Delta v / \Delta t$). At any instant in time, there is an *instantaneous equivalent* of the average acceleration, and that instantaneous acceleration (a) is defined by the calculus expression

$$a = dv/dt$$

Using this instantaneous acceleration, we may define force (F), using the mass (m), as

$$F = ma \text{ [or } F = m(dv/dt)]$$

In geological cases, the acceleration is commonly the acceleration of gravity on Earth, which has a value of 9.8 meters per second squared. We define this special acceleration as g and

$$g = 9.8 \text{ m/sec}^2 = 980 \text{ cm/sec}^2$$

The standard unit of force is called the newton (N). A newton is the force required to accelerate a one kilogram mass at an acceleration of one meter per second squared (N = kg · m/sec^2).

Problem 2.2

Calculate the force on one cm² at the base of a continent produced by a load of rock that is 20 kilometers thick and has a density of 2.70 gram/cm³. You will find it useful to remember the acceleration of gravity and to know that a measure of force is the dyne, which equals one gram · cm/sec².

Problem 2.3

Calculate the stress at the base of the continent described in problem 2.2 if the continent has a basal area of 1000 km × 2000 km. Give your answer in gigapascals (GPa). Note: 1 GPa = 10^9 Pa. 1 dyne/cm² = 10^{-1} Pa.

Stress is another important derivative physical measure of what is called the "state of a system." By definition, stress (σ) is force applied over an area, i.e.,

$$\text{stress} = \sigma = F/A = ma/A$$

The standard unit of stress is the pascal (Pa), which is one newton applied over an area of one meter squared, i.e., 1 N/m². In other words, 1 Pa = 1 kg/m/sec² or 1 Pa = 1 kg · m⁻¹ · sec⁻².

The acceleration of gravity will exert a downward force on any mass (for example, a mass of rock or a mass of water) in the Earth's crust (figure 2.2). For a standard unit of area (e.g., one cm² or one m²), then gravity creates a downward stress. If the surface of the Earth is considered to be horizontal (locally), the stress is perpendicular to the surface and is a *normal stress*. In a more general situation, where the surface is not horizontal, the stress created by the downward force of gravity can be divided (or as we say, "resolved") into a normal stress (σ_n) perpendicular to the surface and a *shear stress* (τ) parallel to the surface (figure 2.3). Shear stresses are important in causing water to flow downhill, in generating landslides that contribute to erosion, and in influencing the flow of fluids in channels.

There are various types of stresses. As we discuss below, materials and systems are considered to be stable if there is no tendency to change. Now, if we consider an extremely small—an infinitesimally small—cube of quartz as a system, we can define the conditions of stress under which that system is stable (figure 2.4). There will be three normal stresses that are directed perpendicular to each of the sides of the cube, stresses designated as sigma (σ). Considered in a coordinate framework of X, Y, and Z axes, subscripts on σ indicate the direction of the stress and the face of the cube perpendicular to the axis under consideration. Thus, in figure 2.4, σ_{xx} refers to a stress acting on the face perpendicular to the X-axis and directed parallel

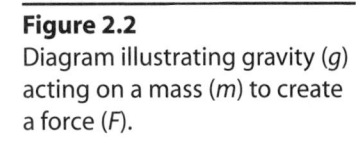

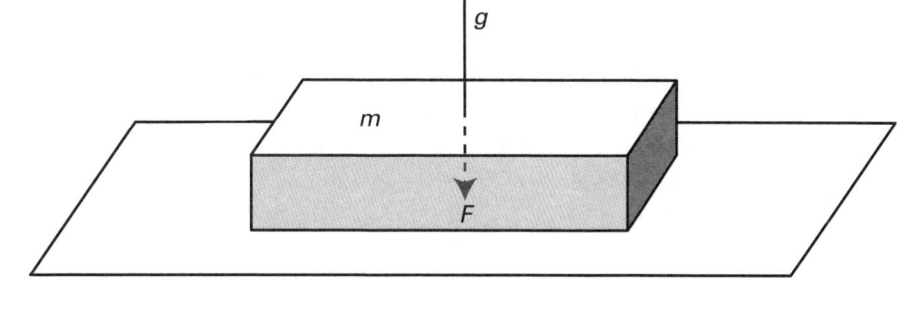

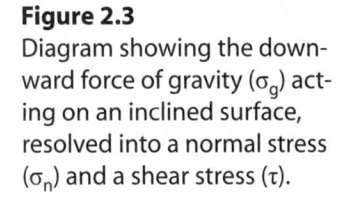

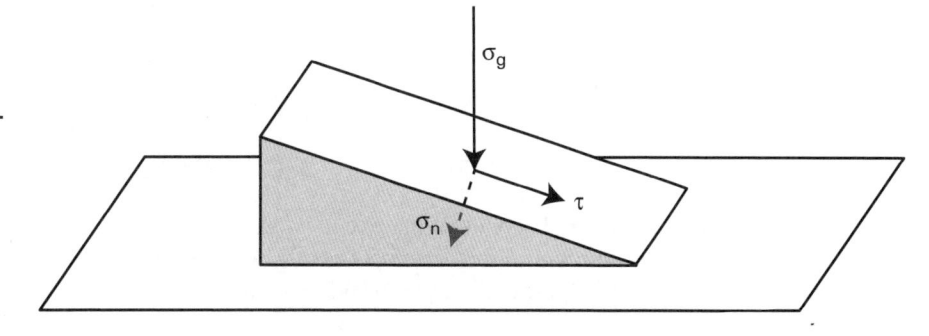

Figure 2.2
Diagram illustrating gravity (*g*) acting on a mass (*m*) to create a force (*F*).

Figure 2.3
Diagram showing the downward force of gravity (σ_g) acting on an inclined surface, resolved into a normal stress (σ_n) and a shear stress (τ).

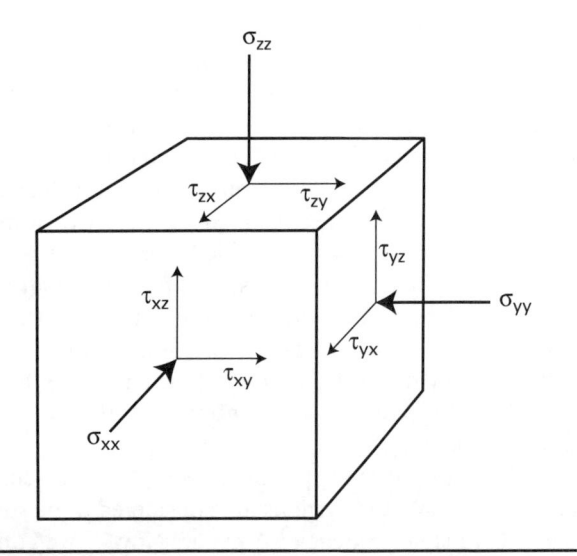

Figure 2.4
Block diagram showing three normal stresses and the associated shear stresses acting on surfaces perpendicular to the X, Y, and Z axes of a cube. See text for discussion.

to the X-axis. Similarly, σ_{yy} refers to a stress acting on the face perpendicular to the Y-axis and directed parallel to the Y-axis, and σ_{zz} refers to a stress acting on the face perpendicular to the Z-axis and directed parallel to the Z-axis. In addition, each face has two nonredundant shear stresses acting on it. These are τ_{xy}, τ_{xz}, τ_{yx}, τ_{yz}, τ_{zx}, and τ_{zy}. The τ_{xy} stress refers to a stress acting on the face perpendicular to the X-axis, but acting in the direction parallel to the Y-axis. The other shear stresses can be deciphered accordingly.

Now, if the infinitesimally small cube is at equilibrium—that is, it has no tendency to change position or rotate—then one of two conditions must apply. Either all of the shear stresses must equal zero,

$$\tau_{xy} = \tau_{xz} = \tau_{yx} = \tau_{yz} = \tau_{zx} = \tau_{zy} = 0$$

or

$$\tau_{xy} = \tau_{yx}$$

and

$$\tau_{yz} = \tau_{zy}$$

and

$$\tau_{xz} = \tau_{zx}$$

When such a condition of stability exists, the shear stresses are nullified and the condition of stress is described by three mutually perpendicular normal stresses (or stress directions). Lines depicting these directions are called the principal stress axes: σ_1, σ_2, and σ_3. When these three are of equal size ($\sigma_1 = \sigma_2 = \sigma_3$), the stress is called hydrostatic stress and is referred to as **pressure**. If one or two of these stresses is not equal to the third, there is a stress that deviates from the mean stress (the average of the three), and this stress is called the **deviatoric stress**. Both pressures, such as the pressure provided at depth by an overlying pile of rock (P_{LOAD}) and deviatoric stresses, such as that produced by the deforming stress of a continental collision, are important in fields such as structural geology and metamorphic petrology.

Some Concepts of Chemistry

Atoms and Their Structure

Everything we see is made of something. The nature of that "something" is truly an ancient question. Perhaps the best-known, early answer was supplied by pre-fourth-century BC Greek intellectuals. They believed there are four fundamental materials called elements and that all other materials are made up of varying proportions of these elements. Over time, early chemists applied the word "element" to various materials they discovered that could not be chemically divided into more fundamental parts having the same properties. Today, that basic definition applies; that is, an **element** is defined as one of the more than 100 basic chemical constituents of matter that cannot be separated into simpler substances having the same properties as the parent constituent.

Determining the nature of these elements was a fundamental challenge that evolved into arguments focused on whether matter is continuous, even at an infinitely small scale, or discontinuous, and made up of tiny, discrete units. The Greek philosopher Democritus, who favored the discontinuous view, referred to the hypothetical discrete units of elements as **atoms** and we use this term today to define the individual particles of the chemical elements.[5] Atoms are the basic components of matter and consist of a nucleus of one or more protons, usually with neutrons, and a surrounding array or "cloud" of electrons. Each element consists of atoms, all of which, by definition, have the same number of protons in the nucleus. If an atom has a different number of protons in the nucleus, it is an atom of a different element.

Early on, whether atoms themselves were divisible or indivisible remained a speculative question, but late in the 19th century observations of negatively charged (cathode) and positively charged (anode) rays led to the discoveries of (subatomic) particles that made up those rays—**electrons** and **protons**, respectively. An early hypothesis, that atoms are small but homogeneous masses of positive and negative charges, was changed as a result of a series of experiments by Rutherford, who demonstrated that the protons were concentrated in the centers of atoms, in what he called the **nucleus** (Garrett et al., 1972). Rutherford postulated that electrons orbited the nucleus in a fashion reminiscent of planets in a system around a star, and later Bohr applied the idea of quantized atomic energy levels to the orbits of electrons in hydrogen, postulating that the electrons only absorb or emit specific amounts (quanta) of energy when they move between orbits (figure 2.5).

Our current understanding of the nature of atoms is that they are made of three different types of particles: electrons, protons, and **neutrons** (table 2.1). In the nucleus of each atom, there is a specific number of protons, and that number is called the *atomic number* of the element. A matching number of negatively-charged electrons that balances the positive charge of the protons surrounds the nucleus, making the atom electrically neutral. The nuclei of all atoms, except hydrogen, also contain neutrons that form a nuclear "glue" keeping the protons from flying apart in response to the mutual electrostatic repulsion of their charges. The nucleus has nearly all of the mass of the atom, as each electron has just a very small fraction of the mass of a proton or neutron.[6]

Instead of simply circling the nucleus, like a satellite circles a planet, electrons "circle" at a series of energy levels called *orbitals*, outward from and surrounding the nucleus, as suggested by Bohr. These orbitals, in general, are not spherical, but have shapes determined by complex mathematical functions.[7] Electrons fill these energy levels beginning at the lowest energies and continuing to higher energies. There are different levels and sublevels within the orbitals and these are described by a series of quantum numbers (table 2.2). Using this terminology, the electronic configuration of a hydrogen atom would be $1s^1$ (one electron at the lowest energy level, 1s). The next largest element, helium, has atoms with a $1s^2$ configuration (with two electrons in the first orbital), whereas the larger lithium atoms are configured with $1s^2\ 2s^1$, and so on.

The various orbitals are limited in the number of electrons they can hold and, as noted, inner orbitals are generally filled before outer orbitals. Listed in order of increasing atomic number, each successive atom on the list has one more proton (i.e., an atomic number, one number larger) than the one before it, balanced by an equal number of electrons. Hence, with only one electron added at a time, the outer orbitals generally are not filled. As an example, the element lithium has three electrons, two filling orbital one and only one in orbital two, where eight are needed to fill that orbital. Electrons in the outermost orbitals of any particular atom are involved in chemical bonding and are referred to as

Table 2.1 Subatomic Particles of Importance to Earth Materials

Name	Electric Charge	Mass (amu*)
Electron	−1	1/1837
Proton	+1	1
Neutron	0	1

*atomic mass unit = 1/12 of ^{12}C mass

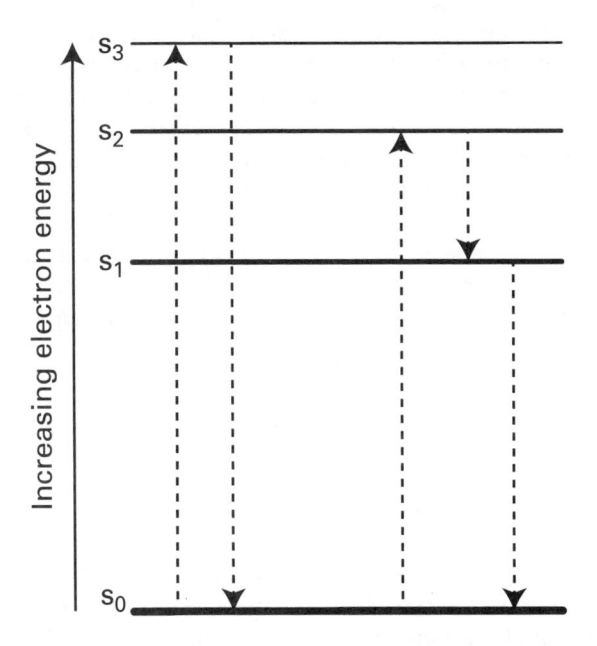

Figure 2.5
Schematic diagram of orbital energy levels (s_0, s_1, etc.). Electrons can move to higher or lower energy levels, but only in discrete steps, much as one can ascend or descend the rungs of a ladder.
(*Source:* Garrett et al., 1972.)

Table 2.2 Labels, Values, and Important Characteristics of the Quantum Numbers

Quantum number	Level labels	Values	Characterization
Principal (n)	K, L, M...	1, 2, 3...	Electron energy levels, distance of orbital from nucleus
Azimuthal (*l*)	s, p, d, f	0, 1... (n – 1)	Electron angular momentum, shape of orbital
Magnetic (m)	—	–1, 0, +1	Orbital spatial orientation
Spin (s)	—	–1/2, +1/2	Electron spin

valence electrons. Atoms with outer orbitals that are completely filled (such as helium, neon, and argon) do not bond with other atoms, and such atoms constitute elements referred to as noble gases.

Given the large number (totaling 92) of naturally occurring elements, chemical elements would seem to be a somewhat chaotic collection if there were not some means of classifying them. The most common and successful method of organization and classification is to list the atoms in order of increasing atomic number on a chart, in which atoms are grouped according to how the orbitals fill with electrons. This results in the familiar Periodic Table of the Elements (figure 2.6). The rows correspond to the principal quantum number of the orbital being filled and the columns represent the subshell levels. One particular benefit of this table is that the elements in a given column tend to be chemically similar, owing to the fact that they have similar valence electron configurations.

Ions and Ionization

Atoms are not electrically neutral all of the time. Electrons can be gained or lost from the outer orbitals. Gain or loss of electrons creates an imbalance in the ratio of electrons to protons and yields a net charge in the atom. An atom with a charge is called an **ion**. If the net charge is positive, it is a **cation** (because it is attracted to a negatively charged cathode), and if the net charge is negative, the atom is an **anion**. In general, the orbitals with valence electrons are not filled, so it is easiest for electrons to be added to or subtracted from these orbitals. Elements whose valence orbitals are mostly *unfilled* are metals and their valence electrons are easily given up. Nonmetals have valence orbitals that are mostly filled and these elements tend to attract electrons. The noble gases, whose valence orbitals are completely filled, ionize only with great difficulty. Again, using the terminology of table 2.1, sodium metal (Na), for example, has an electronic configuration of $1s^2\ 2s^2\ 2p^6\ 3s^1$ with only one valence electron ($3s^1$)—seven short of a filled shell. Chlorine (Cl), in contrast, has a configuration of $1s^2\ 2s^2\ 2p^6\ 3s^2\ 3p^5$, with seven valence electrons (i.e., $3s^2\ 3p^5$ or 2+5), requiring only one to fill the outer shell of 8. As a result, sodium readily gives up its one electron to become Na^{+1} (a sodium ion having a positive charge of +1) and chlorine just as readily gains an electron to become Cl^{-1} (a chlorine ion with a negative charge of –1) (figure 2.7 on p. 20). In both cases, ionization of the atoms results in filled outer shells. The sodium electron configuration now matches that of neon; and the chlorine electron configuration matches that of argon (figure 2.6); but of course, the number of protons in each nucleus remains the same after the electron exchange, so that the atoms remain as sodium and chlorine atoms, now in ionic form.

The particular process of ionization in which an electron is lost is called **oxidation,** and the process in which an electron is gained is called **reduction**. The combined process of reduction and oxidation is often referred to as **redox**. The charge on each ion is its **oxidation state**. Many atoms can have more than one oxidation state. For example iron, in addition to its neutral state, can be Fe^{+2} (ferrous) or Fe^{+3} (ferric). Because each atom has a different electronic configuration, each will have a different tendency to gain or lose electrons, which will clearly affect the type of bonding that occurs during chemical reactions.

Chemical Bonding

It is common knowledge that the Earth is not composed only of pure elements. Atoms of the chemical elements combine in myriad ways to form molecules and compounds, so the heart of chemistry is the study of how atoms form chemical bonds. Why does bonding occur at all? One answer relates to the stability of the electron shells surrounding the atoms. An atom with a filled shell is more stable than one with an unfilled shell. A second answer is related to the fact that as two atoms approach each other, their orbitals will begin to overlap, and the electrons of one atom will repulse the negative charges of the electrons of the other atom.

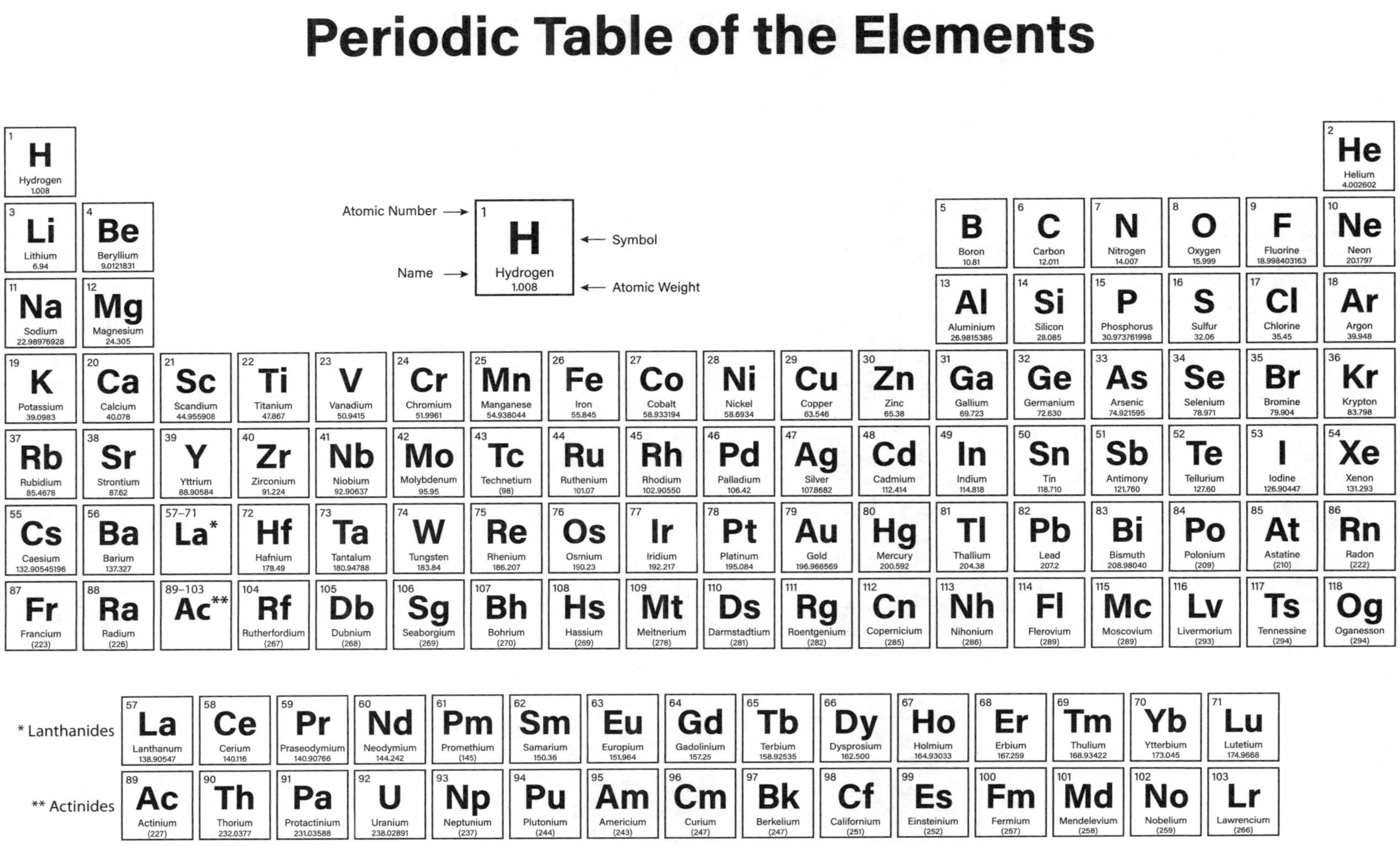

Fundamentals of Physics and Chemistry Relevant to Earth Materials

Periodic Table of the Elements

Figure 2.6
Periodic Table of the Elements.
(*Source*: Modified from Shutterstock.)

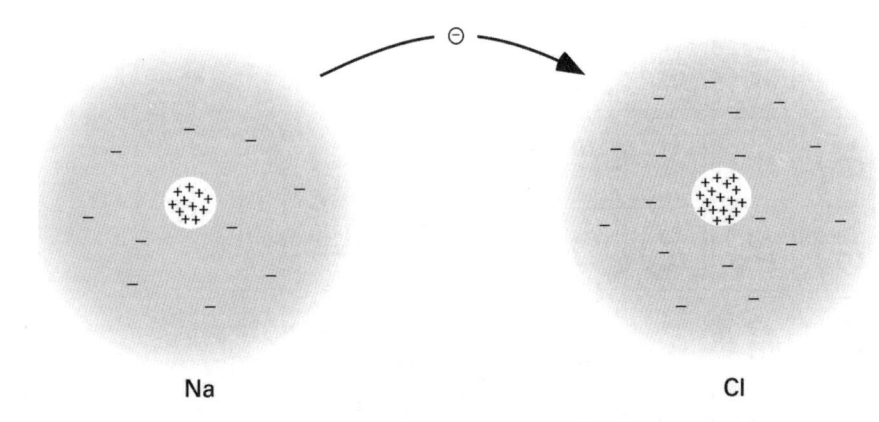

Figure 2.7
The neutral sodium atom (11 protons and 11 electrons) gives up one electron to the neutral chlorine atom (17 protons and 17 electrons), resulting in filled orbitals for both atoms. The resulting positively charged sodium (11p, 10e) is electrostatically attached to the resulting negatively charged chlorine (17p, 18e).

Atoms minimize these repulsions by changing their individual atomic orbitals into combined bonding orbitals, thereby lowering the overall energy.[8]

Ionic Bonding Versus Covalent Bonding

Perhaps the most conceptually simple type of bond comes about from simple electrostatics: a positively charged object is attracted to a negatively charged object. Using the example of Na and Cl atoms described above, recall that a sodium atom gives up one electron to become a cation with a filled outer shell, and a chlorine atom absorbs one electron to become an anion with a filled outer shell. The cation and anion are attracted to each other (and to other Na$^+$ and Cl$^-$ ions), because of these opposing charges. The attraction causes them to form a solid composed of connected NaCl that we know as the mineral halite (common salt). A simplistic view of an **ionic bond**, then, is one of "theft": the greedy chlorine steals an electron from the generous sodium. This, however, is a rather poor representation of reality.

A more appropriate description is that the electrons are highly localized about each of the nuclei and the excess positive charge of the sodium nucleus is attracted to the excess negative charge in the filled outer electron shell of the chlorine (figure 2.8a). This is consistent with the observations that: (1) there is little electron density between the two ions that does not belong to one or the other, and (2) the bonds do not show any appreciable directionality; that is, there is no particular angle that is formed between bonded atoms. The lack of directionality accounts for some particular characteristics of ionically bonded substances (table 2.3).

Not all compounds are made from elements at the opposite ends of the periodic table. Elements closer to the center of the table have greater difficulty in attracting or giving up the electrons necessary to complete an energy shell. These elements are inclined to form cova-

Table 2.3 Comparison of the Mineral Characteristics that Derive from Metallic, Covalent, Dipole, and Ionic Bonding

	Bond Type			
	Metallic	**Covalent**	**Dipole***	**Ionic**
Electron Density Between Nuclei	High	Medium	Low	Very Low
Characteristic				
Transparency	Opaque	Opaque to Partial	Partial to Full	Partial to Full
Thermal Conductivity	High	Low	Low	Low
Electrical Conductivity	High	Low	Low	Low
Crystal Symmetry	Very High	Low	Low	High
Melting Point	Low to High	High	Very Low	Medium to High
Hardness	Low	High	Low	Medium
Fracture Resistance	Medium	Very High	Very Low	High

*Van der Waals, Hydrogen

Fundamentals of Physics and Chemistry Relevant to Earth Materials 21

lent bonds. As the name implies, **covalent bonds** represent "cooperation" between these elements to solve the problem of incomplete energy shells. Some valence electrons are shared between the atoms, and this covalency allows them to fill their outermost shells, at least part of the time. These types of bonds are considerably stronger, much less polar, and more directional than ionic bonds, accounting for some particular properties of covalently bonded materials (table 2.3). From a quantum perspective, the shared electrons are less

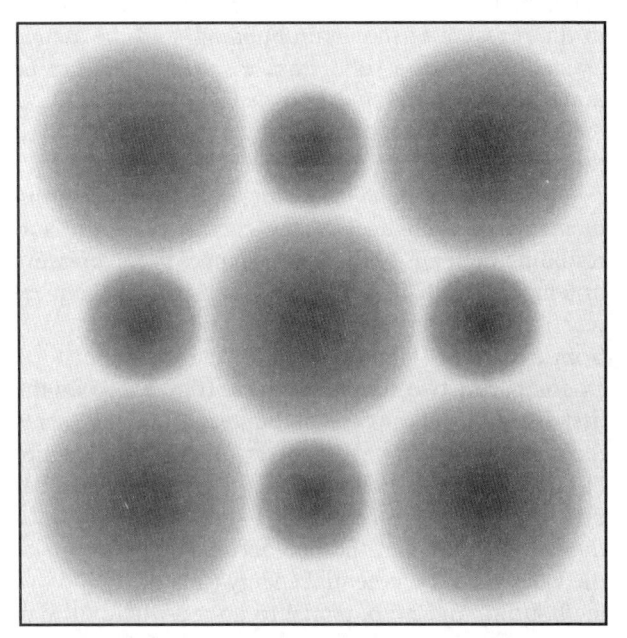

(a) Section through NaCl crystal.

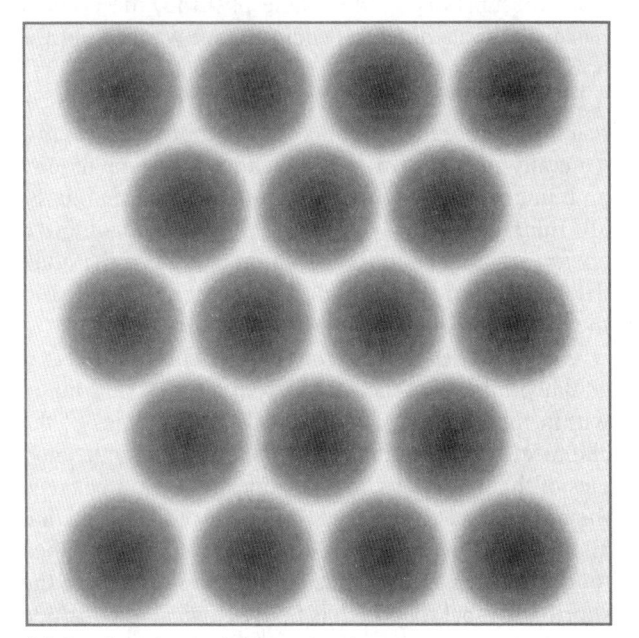

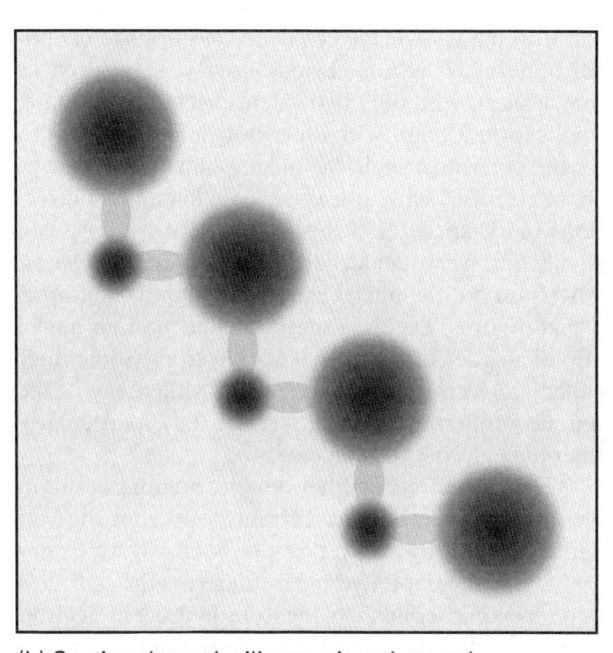

(b) Section through silicate mineral crystal.

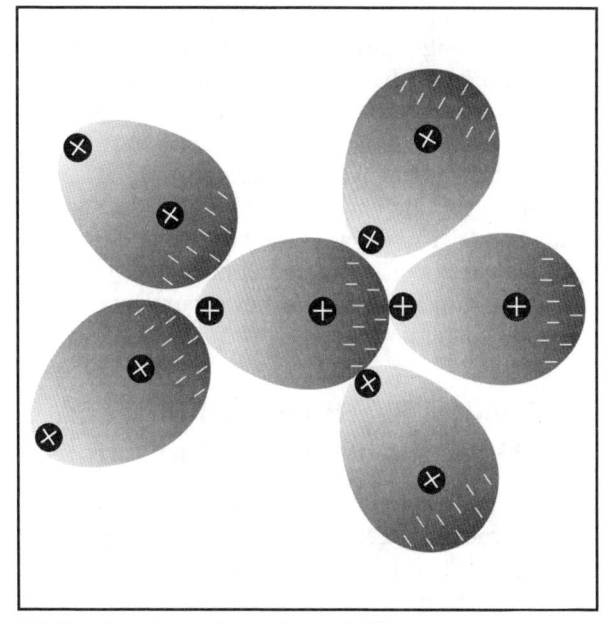

(c) Section through Mg crystal.

(d) Section through condensed HF.

Figure 2.8
Electron density maps for materials with different bonding types. Darker shading corresponds to more electrons.

localized on the bonded nuclei, spending an appreciable amount of time between them. This explains the greater density of electrons found within a covalent bond (figure 2.8b), as well as their greater directionality.

Metallic Bonding and Dipolar Interactions (Van der Waals and Hydrogen Bonds)

Given a knowledge of valence electrons in chemical bonding, it is not obvious how a metal such as magnesium, with only two valence electrons to spare, might form a bond with other magnesium atoms. To fill the outermost shell, the magnesium atoms have to share electrons with at least seven others, in all directions in 3D-space. If such sharing occurs, it creates a significant, very delocalized electron density everywhere around the nuclei, making the bonds nondirectional (figure 2.8c). In a sense, the magnesium nuclei "float" in a sea of shared valence electrons (sometimes called an electron gas) that helps provide many of the unique properties of metals (table 2.3), particularly their high electrical conductivity.

The side effects of other types of bonding result in weak interactions between certain atoms, creating Van der Waals and hydrogen types of bonds. If we examine a molecule of hydrogen fluoride, the HF has strongly ionic bonds, so the sole hydrogen electron spends most of its time attached to the fluorine atom. The HF molecule then has a nearly bare proton on one end and an excess of negative charge on the opposite end, making it *dipolar* in nature. The positive pole of this molecule then attracts the negative poles of other molecules (and vice versa), binding them together in what are called **hydrogen bonds** (figure 2.8d). When a similar effect occurs for atoms other than hydrogen, the bonds are referred to as **Van der Waals bonds**. These types of bonds are much weaker than other types of bonds and although they are not prominent in minerals, where they occur they result in zones of structural weakness.

Chemical Reactions, Stability, and the Direction of Chemical Reactions

Bonding between atoms represents the simplest case of a chemical reaction. A chemical reaction is an interaction between atoms, molecules, and/or compounds in which there is no change in the number of any of the kinds of atoms involved. Minerals represent chemical compounds formed by natural chemical reactions (as compared to compounds synthesized by humans). Yet, the same chemical processes or reactions that yield minerals can be investigated in the laboratory and (with some caveats) used to understand their natural counterparts. In nature, the chemical evolution of a magma prior to a volcanic eruption, the deposition of valuable metal-bearing minerals, the weathering of mine spoils to contaminate an adjacent watershed, the change in mineral content in response to the pressure of mountain building, and the simple crystallization of a solid from a solution, all can be understood given some knowledge of the chemical reactions that occur within and between minerals and the fluids with which they interact.

Minerals respond to changes in their environment, be they changes in the chemistry of the surrounding minerals or fluids, or changes in the pressure or temperature. These changes take place via chemical reactions that represent movement toward or away from a state of stability.[9] Stability can be understood by comparing two simple diagrams (figure 2.9). In the first, a boulder at the edge of the steep slope represents the unstable case. By virtue of its position, the boulder has a great deal of potential energy and any perturbation will send it crashing to the bottom of the slope. In the second diagram, the boulder is at the bottom of the valley with no potential energy (relative to the valley bottom) and no potential to go anywhere: it may be considered to be in a stable condition. By moving from the unstable to the stable position, the boulder has given up its potential energy. If we want to put the boulder back on top of the hill, we must add energy to it by pushing it upward. By analogy, some chemical reactions proceed spontaneously and produce energy, generally in the form of heat. Other reactions require an input of energy to force them to proceed. Similarly, all minerals then have some amount of internal energy, analogous to the potential energy of the boulder above, and the amount of this energy changes as the mineral is included in a chemical reaction.

Where does this internal energy come from? It is actually the sum of the energy from all possible sources within the mineral. These include the energies of the chemical bonds, vibrations of the atoms, kinetic energies of their electrons, gravitational attraction between the atoms, and so on. It even includes the energy that allows the mineral to resist any pressure that is applied to it. The term used for this grand sum is **enthalpy**, the shorthand label for which is **H**. It is logical to conclude that this measure of energy would allow us to determine the stability of a mineral in a reaction, but it is not quite that simple, because some of this energy is unavailable to participate in chemical reactions.

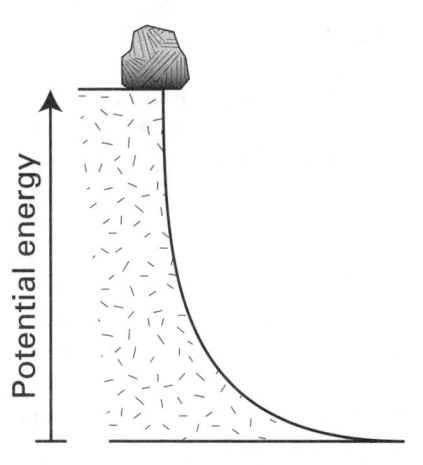

(a) Sketch of a rock, unstable at the edge of the steep slope, with a great deal of potential energy. A small activation will cause this potential energy to be released.

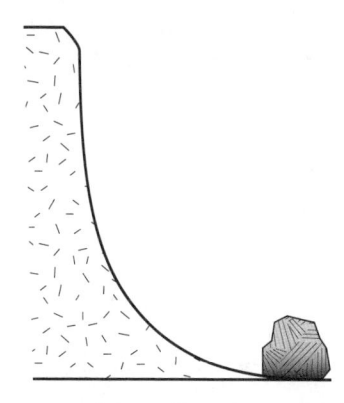

(b) At the bottom of the valley, the rock is in a stable position with no potential energy. In order to add potential energy to the rock, the rock would have to be pushed uphill.

Figure 2.9
Simple illustrations showing the states of instability and stability.
(*Source:* Adapted from Ehlers, 1972.)

Consider again the boulder on the edge of the slope. If a device measures the impact of the boulder when it reaches the valley floor, it would seem obvious that we could use that measured impact to calculate the original potential energy. But as it turns out, some of that potential energy has been lost on the way down. The boulder encountered friction along its path, which absorbed some of the potential energy. Impacts along the way produced heat and other vibrations in the boulder and the ground, drawing off more energy. Similarly, not all of the enthalpy is available for chemical reactions, and this unavailable enthalpy is called the **entropy**, which is labeled **S**.[10] The energy available for chemical reactions is the enthalpy (H) minus the entropy (S) multiplied by the temperature (T) (since entropy increases as a function of temperature) or:

$$H - TS = G$$

The symbol **G** is used for this available energy and is called the **Gibbs free energy**. It is this quantity that allows us to calculate the stabilities of minerals in chemical reactions (see box 2.1). In general, this is done by examining differences in H, S, and G. The differences between the enthalpy (ΔH), entropy (ΔS), and Gibbs free energy (ΔG) for a given mineral and the materials from which the mineral formed (usually the pure elements[11]) are what are tabulated for use. In most geochemical work, the value of interest is the ΔG of a reaction. By convention, if the ΔG is negative, it means that the reaction will proceed spontaneously; i.e., the boulder will roll downhill. If the ΔG is positive, the reaction will not proceed without external assistance (the boulder must be pushed uphill), and if the ΔG is zero, then the materials are at equilibrium.

Problem 2.4

There are two forms of calcium carbonate ($CaCO_3$) that are commonly found as minerals: aragonite and calcite. Using the free energy values below, determine which of these two is stable under normal surface conditions [temperature = 25°C, pressure = 1 atmosphere or 101325 Pa, aragonite ΔGf = –269.55 kcal/mole, calcite ΔGf = –269.80 kcal/mole] (Faure, 1998).

Phase Diagrams

Although it is possible to perform calculations like the calculation of ΔG of a reaction under many different conditions, keeping track of tables of these results would be particularly arduous. Instead, if the results of such calculations or equivalent measurements are plotted on a graph, it becomes a simple mat-

Box 2.1 Sample Calculation of Mineral Stability Using Gibbs Free Energy Data

Consider the reaction:

(1) $\quad FeOOH \rightarrow Fe + O_2 + H_2O$

which, when balanced, is:

(2) $\quad 4FeOOH \rightarrow 4Fe + 3O_2 + 2H_2O$

To calculate if this reaction will proceed of its own accord or whether it will require an input of energy to drive the reaction, we must assemble the Gibbs free energy values (in kJ/mol) for each of the phases (Eby, 2004):

(3) $\quad \Delta G^0_f(FeOOH) = -488.55$

(4) $\quad \Delta G^0_f(Fe) = 0.0$

(5) $\quad \Delta G^0_f(O_2) = 0.0$

(6) $\quad \Delta G^0_f(H_2O) = -237.14$

The convention for calculations of this type is the sum of all the values for the products minus the sum of all the values for the reactants, so the equation becomes:

(7) $\quad \Delta G_r = (4 \times 0.0) + (3 \times 0.0) + (2 \times -237.14) - (4 \times -488.55)$

(8) $\quad \Delta G_r = 0 + 0 + (-474.28) - (-1954.2)$

(9) $\quad \Delta G_r = +1479.92$

Because the result is positive, the reaction as shown in (1) requires external energy to be added for it to proceed. This is analogous to the situation of the boulder at the bottom of a valley, which must be pushed uphill.

If we take the opposite reaction:

(10) $\quad Fe + O_2 + H_2O \rightarrow FeOOH$

and redo the calculations, the result is:

(11) $\quad \Delta G_r = -1479.92$

This negative value indicates that the reaction will proceed spontaneously, or that the boulder will roll downhill.

This result is intuitive to most anyone. The mineral goethite (named after the author Goethe and pronounced "guhtite") is a component of rust, and energy must be added to convert rust to iron, oxygen, and water. However, iron, oxygen, and water obviously react spontaneously to produce rust.

---

ter to see how and where changes occur. Such a diagram, one that illustrates what is stable under what conditions, is called a **phase diagram** (a **phase** is nothing more than a unique substance that can, in principal, be physically separated from one or more other substances). Figure 2.10 is a simple example of such a diagram—a plot of the phases of water as a function of temperature. The areas (or fields) of stability and the boundaries between these fields reflect differences in free energy. Below 0°C, the stable phase (the one with the lowest G) is ice. At 0° (actually very slightly above 0°), both water and ice are stable, meaning that the ΔG of the reaction from one to the other is zero.

A one-dimensional diagram such as this is of limited use. A more usable phase diagram of water includes a second variable (figure 2.11) and is called a P-T (pressure-temperature) diagram. As before, the single phase fields represent the areas where the particular phase has the lowest G, and the two-phase boundaries mark the lines where the ΔG of reaction is zero. In figure 2.11, there are two points of particular interest highlighted: A, the **triple point** of water, is the point at which all three phases are stable at the same temperature and pressure, whereas B is the **critical point**, a point marking the pressure and temperature beyond which there is no difference between the liquid and gaseous phases.

Another diagram of considerable value is called a **binary eutectic diagram**, one type of T-X (temperature-composition) diagram (figure 2.12a on p. 26), so called because it has two compositional components, A and B. The components occur as the phases a and b.[12] The leftmost edge of the diagram represents the composition of 100% A, the rightmost edge the composition of 100% B, and all points between the two sides are some mixture of A and B that totals 100%. Every point in the diagram represents a unique combination of a temperature value and a ratio of components A and B. For example, the point labeled 1 is at a temperature of 950°C and a ratio A:B of 83:17; the point labeled 2 is at a temperature of 725°C and a ratio A:B of 58:42.

As before, the diagram is divided into separate fields that indicate what phases are stable under what

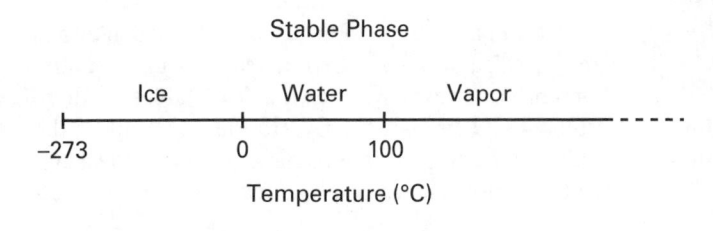

Figure 2.10
The phases of water as a function of temperature. At 0° ice and water coexist stably, meaning that the ΔG of the reaction from ice to water (or water to ice) is zero. The same situation occurs between water and steam at 100°.

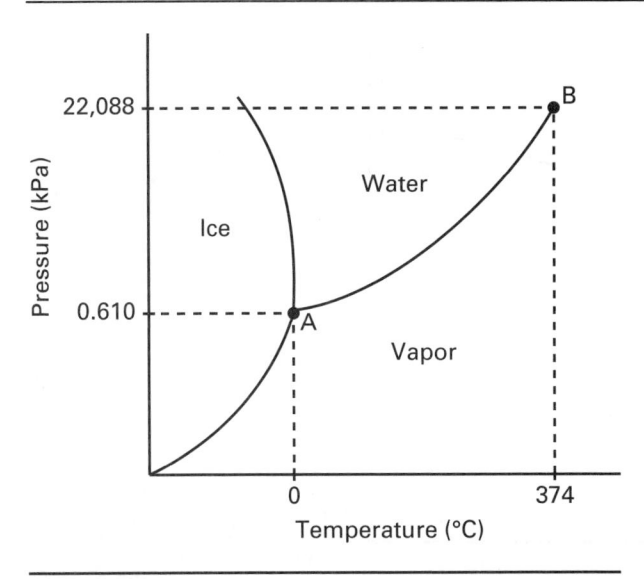

Figure 2.11
Phase diagram showing the phases of water as a function of temperature and pressure. The solid curved lines mark the boundaries where two phases stably coexist. The triple point (A) is the single temperature and pressure at which three phases coexist.
(*Source:* Adapted from Ehlers, 1972. Used with permission.)

conditions, and the fields are separated by lines that mark where phase changes occur. At the highest temperatures, the stable phase is a single liquid called "melt." The exact composition of this liquid may vary from 100% A to 100% B.[13] The lower boundaries of this field are a pair of curving lines that separate the liquid-only field from the fields in which liquid and crystals coexist. Each of these lines is labeled as a **liquidus**. For any given composition, the liquidus may be considered as the lowest temperature possible for a liquid-only state, or alternatively, the highest temperature possible for a mixture of liquid and crystals.

At the lowest temperatures, crystals of solid *a* (composition 100%A) and solid *b* (composition 100%B) exist in a physical mixture. The relative amounts of *a* and *b* (the "bulk composition") of the mixture are (is) depicted as a point (at a particular temperature) in the diagram. The upper boundary of this solids-only field is called the **solidus**, and by analogy with the liquidus description above, it may be considered as the lowest possible temperature for a mixture of liquid plus crystals, or the highest possible temperature for a solid-only state. The liquidus and solidus intersect at a single point, which is called the **eutectic**. Between the liquid-only and solid-only fields, there are two fields in which crystals of either *a* or *b* coexist with a molten liquid.

Figure 2.12b is identical to figure 2.12a, but with the labels abbreviated or removed, and this figure can be used to understand what happens to a liquid melt as it cools. Point 1 represents a liquid at a temperature of 1000°C with an A:B ratio of 38:62. As this liquid cools below 1000°C, the path of the liquid as it cools can be depicted as a straight line extending downward from the original point (of specific composition and temperature) to the point that it reaches the liquidus at point 2. At this point, crystals of *b* begin to precipitate from the liquid. Crystals of *b* are composed entirely of component B. The loss of B from the liquid means that the liquid becomes enriched in component A. A line showing this enrichment would trend toward the A side of the diagram, but since the temperature simultaneously is falling, the path followed (represented by the line from 2 to 3) is a curve. As the system continues to cool, and crystals of *b* continue to precipitate out, the liquid composition follows the downward curve of the liquidus line. By the time the liquid has reached point 3, enough *b* has been precipitated from the liquid that its composition has an A:B ratio of 52:48. This process continues until the liquid composition reaches the eutectic (point 4, with an A:B ratio of 58:42), whereupon crystals of *a* begin to precipitate from the liquid along with the crystals of *b*. Crystals of *a* and *b* precipitate out at the eutectic in proportions appropriate to keep the composition of the liquid remaining at the eutectic composition (not moving toward the A or B sides). Under equilibrium conditions of crystallization, the heat given up as crystals form is lost to the surroundings at a rate that keeps the temperature of the system at the eutectic temperature until all of the liquid disappears. All of the liquid disappears when the ratio of *a* and *b* crystals is 38:62, exactly the same as the ratio of A:B in the original melt at point 1.

Additional insight into the cooling behavior of binary eutectic diagrams can be gained by plotting a graph of time versus temperature for the cooling path just discussed (figure 2.13a on p. 27). The liquid begins cooling at a fixed rate starting at point 1 and continues until it reaches point 2, which is on the liquidus. At point 2, the rate of cooling slows, as shown by the decreased slope of the cooling curve. The reason for this slowdown can be understood by considering what begins to occur at point 2. A chemical reaction (liquid = liquid + crystals of *b*) begins to proceed spontaneously, which means the ΔG of reaction

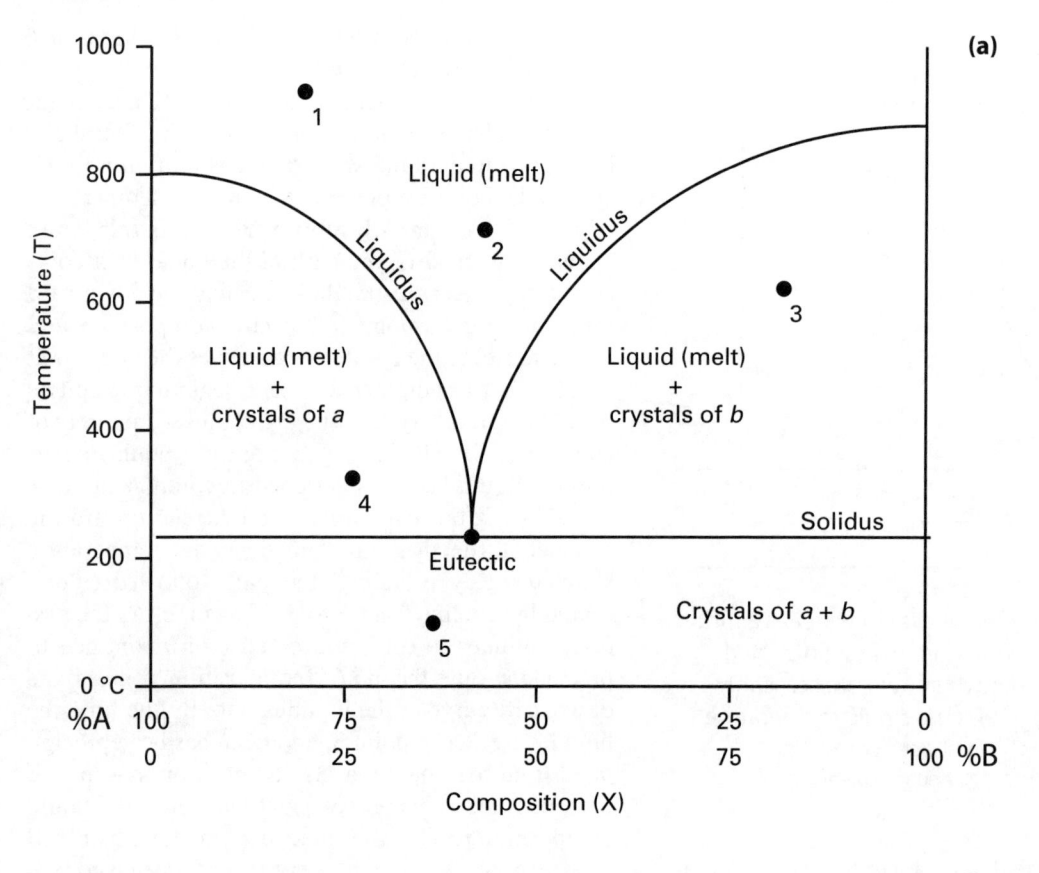

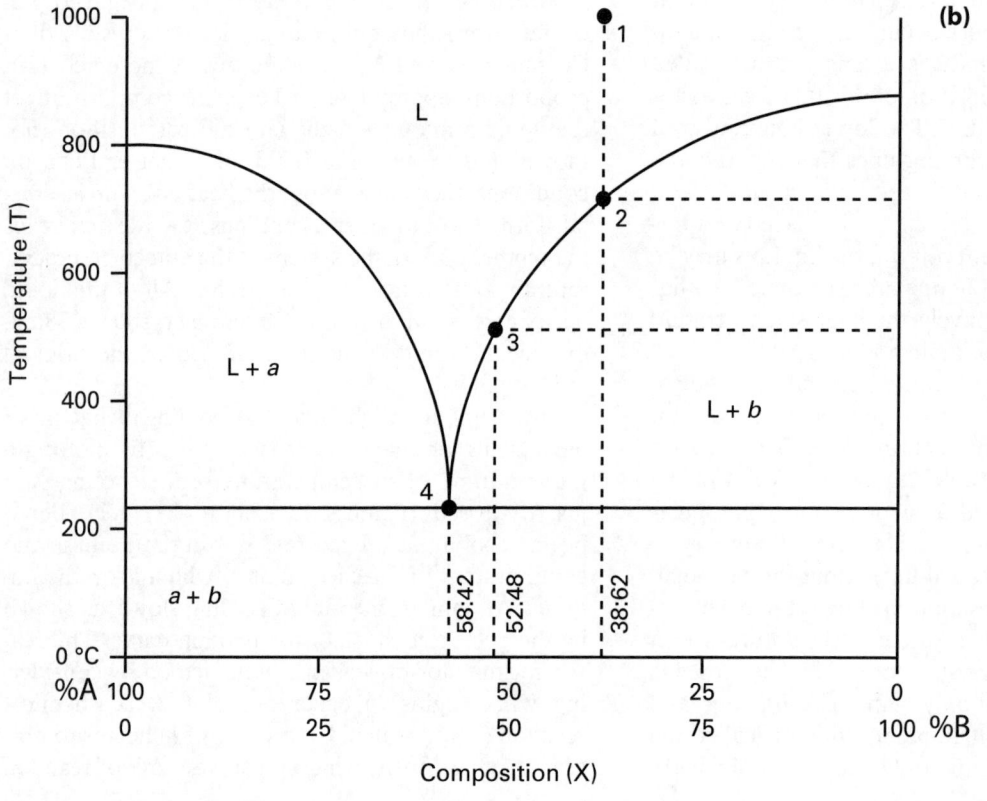

Figure 2.12 Binary eutectic diagram. (a) A generalized temperature-composition (T-X) phase diagram with a binary eutectic point. See text for discussion of numbered points. (*Source*: Adapted from Raymond, 2007.) (b) The same diagram as in (a), with some labeling abbreviated or omitted for clarity, showing equilibrium crystallization of a melt of composition 1. See text for discussion.

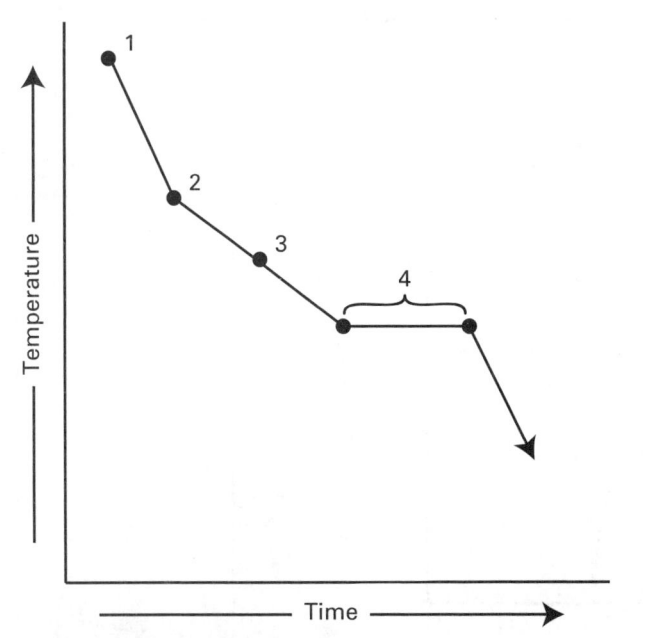

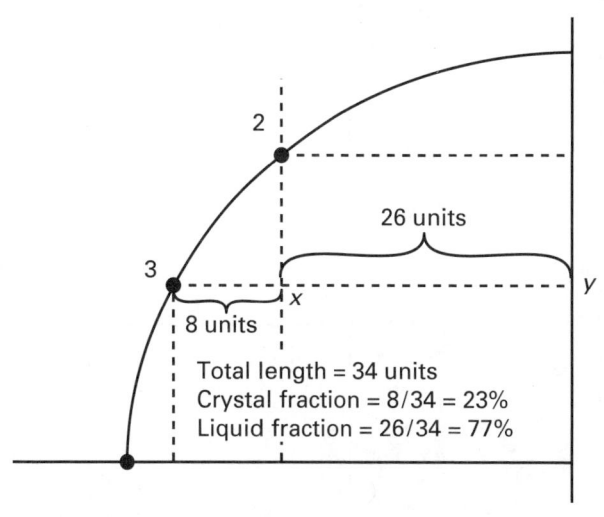

Fundamentals of Physics and Chemistry Relevant to Earth Materials 27

(a) Time-temperature diagram corresponding to the cooling sequence illustrated in Figure 2.12b and described in the text.

(b) Enlarged portion of Figure 2.12b demonstrating how the lever rule can be used to determine the fraction of crystals or liquid at a given temperature.

Figure 2.13
Analysis of time-temperature paths and component ratios in binary diagrams.
(*Source:* Adapted from Raymond, 2007.)

is less than zero. The crystallization of crystals of b produces energy, referred to as the **latent heat of crystallization**, and the cooling of the liquid slows as this heat is added to the system.

The cooling and crystallization of crystals of b continue along this new curve through point 3 until the eutectic is reached at point 4, whereupon the curve changes slope once more. At the eutectic, a second chemical reaction begins to proceed (liquid = crystals of a + crystals of b). The crystallization of a and the crystallization of b both contribute latent heats, and together they provide sufficient heat to balance the heat lost to cooling. Under conditions of equilibrium, the temperature of the system does not change until the liquid has completely crystallized, and once this occurs, cooling proceeds at the original fixed rate.

The horizontal dashed lines in figure 2.13b that connect points on the liquidus to points on the rightmost edge of the diagram are called **tie lines** and have a special significance. Because the lines are horizontal, the points they connect are at the same temperature, so tie lines within a single field indicate what compositions coexist at a given temperature. The two horizontal dashed lines in figure 2.13b demonstrate that crystals of b can coexist with the two different liquid compositions indicated by the leftmost ends of the tie lines. The solidus can also be considered as a tie line, as it links crystals of b with the eutectic liquid compositions and, beyond it, crystals of a, indicating that the two crystals can stably coexist with the eutectic melt.

The relative proportions of crystals and liquid can also be determined by using the tie lines. Consider when the cooling liquid first intersects the liquidus at point 2. At this moment there are no crystals of b, hence the system consists of 100% liquid. As cooling occurs, the liquid composition migrates down the liquidus, losing heat (moving to lower T) and component B. As more and more crystals of b are precipitated, the ratio of crystals (of b) to liquid continually increases. The proportions of any two phases along any tie line (i.e., at any temperature) can be determined using the **lever rule**. This rule makes use of the fact that the point where the bulk compositional line intersects the tie line (labeled as x in figure 2.13b) can be treated like the fulcrum of a lever. To maintain balance on a lever, the largest mass must be closest to the fulcrum, just as the

Problem 2.5

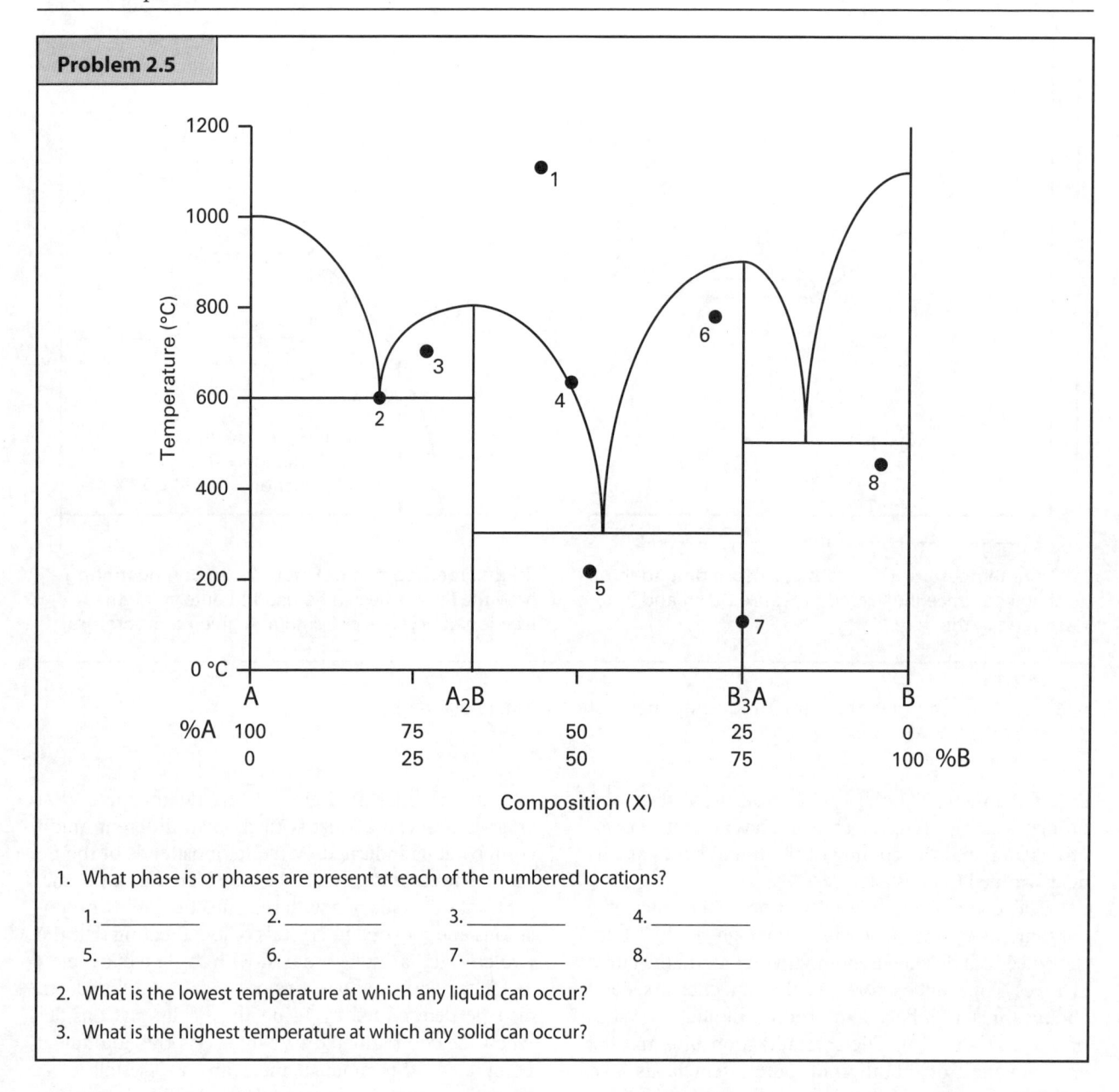

1. What phase is or phases are present at each of the numbered locations?

   1. _____ 2. _____ 3. _____ 4. _____
   5. _____ 6. _____ 7. _____ 8. _____

2. What is the lowest temperature at which any liquid can occur?

3. What is the highest temperature at which any solid can occur?

heaviest rider on a teeter-totter must sit closest to the center. Similarly, the phase in the greatest proportion along the tie line must be the one closest to point x. It is easy to see that the distance from point 3 on the liquidus to point x is shorter than the distance from point y on the rightmost side. Hence, the more abundant phase on this tie line must be the liquid rather than the crystals of b. The exact proportion of liquid and b crystals is determined from the ratios of the lengths of the two line segments. As shown by the calculation in the lower right, the crystals of b make up 23% of the system along this line and the remaining 77% is the molten liquid (of composition 52% A and 48% B).

Binary Solid Solution Diagrams

A second type of binary diagram involves the phenomenon of **solid solution,** found naturally in minerals such as olivine $(Mg,Fe)SiO_3$. In olivine, the iron and magnesium can freely substitute for each other in the mineral structure. A binary solid solution diagram is illustrated in figure 2.14. Just as in the binary eutectic diagram, the upper curve is the liquidus and the

the crystal composition (12% c, 88% d; a ratio of 12:88) that stably coexists with the liquid (point 3a on the liquidus) of composition 57% C, 43% D. In cooling from point 2, both the precipitated solid and the surrounding liquid changed in composition in the general direction of pure C (on the left). For points 3a/3b, note also that the total amount of liquid has decreased and the total amount of solid has increased. Eventually, under equilibrium conditions, the entire system cools sufficiently so that the stable crystal composition on the solidus is directly beneath point 1, and the last of the liquid (of final composition 76% C, 24% D) has reacted with the precipitated crystals. The now solid mass of (c,d) has a composition of 25% c, 75% d (25:75), the initial composition of the melt at point 1, and this solid subsequently follows a vertical path as it cools.

Eutectic Systems, Earth History, and Rock Formation

The eutectic behavior of systems containing two or more minerals is important to Earth history and igneous rock processes. Both the cooling process depicted in the examples above and the melting process, not described, are important. A full understanding of the issues requires that the systems be expanded to eutectic diagrams involving three or more phases, resulting in complexities that are beyond the scope of this text. Yet, two important points will emphasize the significance of these diagrams in understanding Earth history and rock formation. First, recall that in chapter 1 we noted that from a beginning in which the Earth had no layers, it became differentiated, in part by partial melting. In a mix of several common minerals, combinations of quartz, potassium feldspar, and muscovite mica are among the first mineral combinations to melt and the melts that form do so at relatively lower temperatures than do melts derived from other combinations of minerals. As described below, the melt is a partial melt and will not have the same composition (in general) as the whole rock.

To begin to understand this, reexamine the simple binary eutectic diagram in Figure 12b. If we reverse the cooling process and instead heat a rock, an initial understanding of the partial melting process can be realized. If we begin with a rock composed of crystals of a plus crystals of b in a ratio of 62% b and 38% a at 200°C, a rock in the solid field at the bottom of the diagram, and begin to heat that rock, a partial melt will form. This is a consequence of the geometry of the diagram. As the temperature rises, the rock reaches the eutectic temperature. Using tie lines, we can see that the first melt to form is the eutectic melt

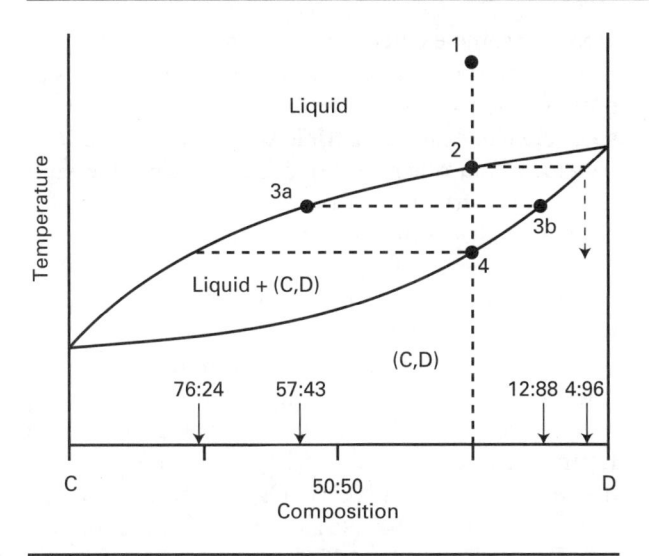

Figure 2.14
A generalized, binary solid solution temperature-composition (T-X) phase diagram with complete solution between components C and D. Upper curve is liquidus curve. Lower curve is solidus curve. See text for discussion of lines and points.
(*Source:* Adapted from Ehlers, 1972. Used with permission.)

lower curve is the solidus (replacing the straight line of the eutectic diagram). At point 1, a homogenous melt (liquid) of composition 25% C and 75% D begins to cool. The liquid follows a vertical cooling path until it reaches the liquidus at point 2. At the liquidus, crystalline solids begin to precipitate, but the composition of the crystals is neither pure C nor pure D. Instead, the composition of the first-formed crystals is found by drawing a tie line from point 2, across the liquid + crystal field, to the curved solidus, then downward to the composition axis. For this example, the crystals (of $c,d1$) that form at this point are of composition 4% c and 96% d (ratio 4:96), and these crystals coexist with the liquid of composition 25% C and 75% D.

As cooling progresses, the liquid composition moves down the slope of the liquidus, just as was the case for a simple binary eutectic diagram. The solid that has crystallized, however, does not remain fixed in composition. The solidus also is a curve, as is the liquidus, meaning that the crystals of c,d react with the liquid, simultaneously exchanging some component D for component C in the liquid and crystallizing new c,d crystals, as the temperature falls. The compositions and temperatures for one point in the cooling history are illustrated by the points labeled 3a and 3b at the two ends of a tie line. On the solidus, point 3b represents of composition 58% A and 42% B, a melt much different in composition than the initial rock composition of 38:62. In short, the whole rock does not melt. Only a small partial melt forms initially. In the case of the binary eutectic in figure 12b, the partial melt forms at the eutectic and if that melt is separated and crystallized as a rock, that rock would be composed of 58% *a* and 42% *b*. The rock would be enriched, in this case, in component A relative to the parent rock.

In rocks, the first partial melt is enriched—relative to the parent rock that is being melted—in light elements such as silicon, aluminum, and potassium. Because of early melt enrichment in certain elements, partial melting in the early Earth tended to produce lighter element-enriched melts that rose toward the surface, where they crystallized as new rocks. The new rocks consisted of minerals composed of the elements in the cooling melts. These new rocks together formed new (differentiated) crustal layers. While this is a simplified explanation involving only two components, the general ideas involved apply to more complex Earth systems.

A second example of the role of eutectic diagrams in geologic systems can similarly be explained using the same binary eutectic A-B (figure 12b). Analysis of the ocean crust across the globe reveals that, in general composition, the ocean crust everywhere is surprisingly similar. The reason for this is partial melting. The mantle beneath the oceanic crust, like the oceanic crust, has a broadly similar composition around the globe. The mantle is relatively poor in silica and it consists of minerals rich in magnesium and iron, such as the minerals olivine and pyroxene. Partial melting of mantle rock at a eutectic, like melting of the rock *a* + *b* at a temperature a little bit higher than 200°C in figure 12b, yields a melt enriched in the more easily melted components. Since the oceanic crustal mantle is broadly similar in composition everywhere, the partial (eutectic) melt formed from it is everywhere generally similar in composition and is higher in silica and alumina than the mantle parent rock. The rock formed from this melt is called **M**id-**O**cean **R**idge **B**asalt or MORB. Again, this is a simplified explanation of a more complex process, but the general principle of partial melting to produce a rock of new composition from a parent rock of different composition is fundamental to understanding the origins of many igneous rocks.

Solid Solutions and Elemental Compatibility

The discussion of the phase diagram for a binary solid solution raises a point of considerable importance in understanding the genesis of rocks. If some atoms of some elements can substitute for each other in a mineral structure, it stands to reason that atoms of other elements cannot do so. To put it another way, some elements are **compatible** with particular mineral structures and other elements are **incompatible** with those same structures. Consider again the case of a mineral crystal forming from a melt. Atoms of compatible elements will be absorbed into the growing structure, whereas atoms of incompatible elements will remain behind in the melt.

This concept of compatibility and incompatibility leads to a classification of elements according to their tendencies to accumulate in different parts of the Earth.[14] The four classes are: Siderophile ("iron-loving" elements such as Ni, Co, Os, and Ir); Chalcophile ("copper-loving" elements such as Se, As, Zn, and Cd that concentrate in sulfide minerals); Lithophile ("rock-loving" elements such as Rb, Ba, Nb, Th, and U that concentrate in silicate minerals); and Atmophile ("air-loving" elements such as N, Ar, and Xe).

The factors that lead to incompatibilities are most commonly associated with the relative sizes of the atoms or their ions, and the net charge on these ions. Elements that are incompatible due to their large size and low charge are referred to as low field-strength elements (LFSE) or more commonly, large-ion lithophile elements (LILE). These include such elements as K, Rb, Sr, and Ba. Elements whose incompatibility is a function of their high charge relative to their size are called high field-strength elements (HFSE) and include Th, U, Zr, Hf, and most of the rare-earth elements (REE).

NUCLEAR REACTIONS

The Reaction Process

When modern atomic concepts were beginning to develop, one argument against the idea of atoms was the fact that the measured elemental masses were nonintegral numbers. It was argued that if elements were in fact made of small particles, the masses should prove to be whole numbers. This problem was resolved by the discovery of neutrons and the determination that there are varieties of specific elements—that is, there are atoms (by definition, with identical numbers of protons) that have different numbers of neutrons in their nuclei. Such atoms, with the same numbers of protons but different numbers of neutrons, are called **isotopes**. For example, naturally occurring carbon has an atomic weight of 12.011 atomic mass units (amu), which results from a mixture of different amounts of several different isotopes (figure 2.15). The weighted average of the masses, based on the abundance of each isotope in nature, gives carbon an average mass of 12.011. Elemental isotopes may occur naturally or be produced artificially in nuclear reactors.

Most naturally occurring isotopes are stable; that is, have no inherent tendency to break down. In some isotopes, however, the neutron/proton ratio is such that the forces holding the nucleus together are not sufficient. These unstable nuclei randomly and spontaneously change, capturing or emitting radiation or particles, and in some cases, splitting entirely. This process of nuclear breakdown is called radioactive decay, and the energy released via **radioactivity** provides internal heat to the Earth. Nuclear breakdown processes are nuclear reactions, reactions in which the kinds or nature of the atoms in the reaction changes during the reaction. A typical nuclear reaction is the decay of one isotope of uranium, U^{238}. The nucleus of a U^{238} atom will emit an alpha particle and the atom becomes transmuted into an atom of thorium, Th^{234}. This thorium isotope[15] is also unstable and decays into protactinium, Pa^{234}. The chain of *transmutation* via nuclear reactions continues until a stable isotope is produced. For the U^{238} decay chain, the stable end product is an isotope of lead, Pb^{206}.

Although it is unknown exactly when any given unstable nucleus will decay, the *rate* at which a large number of these nuclei will decay can be determined very precisely. Because these reactions occur within nuclei, they are not affected by other physical or chemical factors or by chemical reactions, and the rate of decay remains constant.[16]

Geologic Uses of Isotopic Systems

The fact that radioactive isotopes decay at a fixed rate provides an obvious means for measuring the absolute ages of the minerals that contain these isotopes, and by extension, the rocks that contain the minerals. Although in subsequent chapters some use will be made of radiometrically obtained dates, the details and techniques are beyond the scope of this text.

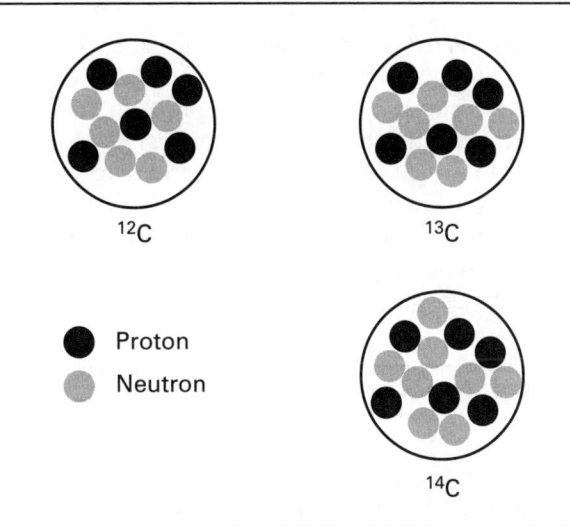

Figure 2.15
Nuclei of three isotopes of carbon. All three nuclei have the same number of protons (6), but ^{12}C has six neutrons, ^{13}C has seven, and ^{14}C has eight, which creates the differences among their masses. The ratio of the three isotopes in nature is 99:1:0.0001, so the weighted average is the atomic mass for carbon, 12.011.

Some stable isotopes are particularly useful in investigations of geologic systems, notably the isotopes of oxygen and sulfur. Because of the relatively low masses of these elements, their isotopes respond slightly differently to environmental conditions.[17] The O^{18}/O^{16} ratio of seawater is a sensitive indicator of ocean temperatures, hence this ratio in marine fossils provides data on paleothermometry. The S^{34}/S^{32} ratio is similarly used to determine the origins of sulfide minerals in ore deposits.[18]

Finally, there are analytical uses of isotopes in mineral studies. Certain isotopes are sensitive to radiation or to magnetic fields, and when these isotopes are incorporated into the crystalline arrangements of minerals, spectroscopic techniques such as Mössbauer and Nuclear Magnetic Resonance (NMR) analyses provide information about mineral structures that is not readily obtainable by other means.[19]

Summary

The minerals, rocks, fluids, and soils of Earth are chemicals, chemical compounds, or collections of chemical compounds. They behave in response to the physical and chemical laws that govern nature. Earth materials respond to various pressures and stresses imposed within the Earth, where pressure is defined as the focus of three equal and mutually perpendicular normal stresses. Like all chemical compounds, those in Earth materials are composed of atoms, building blocks of natural materials called elements, each of which is defined as one of the more than 100 basic chemical constituents of matter that cannot be separated into simpler substances having the same properties. Various chemical reactions that form compounds via bonding between atoms take place in response to physical conditions relating to matter, energy, and time. Bonds include ionic, covalent, and van der Waals types, each of which is a function in some way of forces resulting from the sharing of electrons between atoms. The stability of crystalline materials can be depicted on phase diagrams in which phases are substances that can, in principal, be physically separated from one or more other substances. The reactions and the physical changes in Earth materials may all be explained, ultimately, through mathematical expressions of conditions of state and change.

Selected References

Anderson, G. M. 1996. *Thermodynamics of Natural Systems.* John Wiley, New York. 382 p.

Barrett, J. 2002. *Structure and Bonding.* Wiley Interscience, New York. 181 p.

Ehlers, E. G. 1972. *The Interpretation of Geological Phase Diagrams.* W. H. Freeman, San Francisco. 280 p.

Faure, G. 1986. *Principles of Isotope Geology* (2nd ed.). John Wiley, New York. 608 p.

Faure, G. 1998. *Principles and Applications of Geochemistry* (2nd ed.). Prentice-Hall, Upper Saddle River, NJ. 600 p.

Garrett, A. B., Lippincott, W. T., and Verhoek, F. H. 1972. *Chemistry: A Study of Matter* (2nd ed.). Xerox Publishing, Lexington, MA. 674 p.

Gill, R. 1996. *Chemical Fundamentals of Geology.* Chapman and Hall, London. 290 p.

Goldschmidt, V. M. 1923. Geochemical Laws of the Distribution of the Elements. *Norske videnskaps-akademi i Oslo. Matematisk-naturvidensapelig klasse.* v. 2, 117 p.

Hawthorne, F. C. (ed.). 1988. *Spectroscopic Methods in Mineralogy and Geology,* in *Reviews in Mineralogy* 18, Mineralogical Society of America, Washington, DC. 698 p.

Nordstrom, D. K., and Munoz, J. L. 1986. *Geochemical Thermodynamics.* Blackwell Scientific, Oxford. 477 p.

Raymond, L. A. 2007. *Petrology: The Study of Igneous, Sedimentary, and Metamorphic Rocks* (2nd ed.). Waveland Press, Long Grove, IL. 720 p.

Sears, F. W., and Zemanksi, M. W. 1964. *University Physics.* Addison-Wesley, Reading, MA. 1028 p.

Winter, M. J. 1994. *Chemical Bonding.* Oxford University Press, New York. 92 p.

Notes

1. Sears and Zemanksi (1964).
2. Sears and Zemanksi (1964).
3. Sears and Zemanksi (1964).
4. Sears and Zemanksi (1964).
5. Historical references in this and the following paragraph are from Garrett et al. (1972).
6. The mass of an electron is 9.11×10^{-28} grams, whereas protons have a mass of 1.673×10^{-24} grams and neutrons have a mass of 1.676×10^{-24} grams (Garrett et al., 1972; Heisse et al., 2017).
7. Winter (1994).
8. Winter (1994).
9. This discussion derives from those of Gill (1996), Anderson (1996), Nordstrom and Munoz (1986), and Faure (1998).
10. See the references listed in note 9 for more rigorous definitions of entropy, including its relationship to statistical mechanics.
11. By definition, pure elements in their standard state are assigned a ΔG of formation of 0.0. See Faure (1998).
12. Strictly speaking, there is a difference between the chemical component A and a crystal a that is made up of A (see Raymond, 2007).
13. It is possible for a system to contain multiple liquids, and these liquids are not necessarily miscible. See Ehlers (1972).
14. Goldschmidt (1923).
15. Faure (1986).
16. Faure (1986).
17. The difference in mass between O^{18} and O^{16} is about 11.1%, whereas the difference in mass between U^{238} and U^{235} is about 1.3%.
18. Faure (1986).
19. See Hawthorne (1988).

Fluids in and on the Earth

3

INTRODUCTION—THE NATURE AND DISTRIBUTION OF FLUIDS

An urban myth has it that writer Mark Twain once remarked that "Whiskey is for drinkin' and water is for fightin' over." Indeed, access to water may be the cause of regional conflicts in this century. Yet, water is but one fluid important to society and is one of many important to geological processes.

Water and other fluids are critical to the operation of many Earth processes. We know this at some level of understanding because we observe erosion by running water, we know that rainfall is an important element of weather and climate, and we read about the serious problems related to contamination of groundwater. Yet, fluids are much more important than most people realize.

Fluids exist in the atmosphere and lithosphere as extensions of the hydrosphere. The biosphere would not exist without them. Beyond the obvious fluids at the surface and in the atmosphere, fluids in the lithosphere—including groundwater, igneous fluids (especially magmas), hydrothermal fluids, and metamorphic fluids—are vital to many geologic processes. In this chapter we examine the nature of fluid flow and introduce some important roles fluids play as an Earth material in Earth processes. Fluids in the atmosphere are beyond the scope of this text.

The two most abundant kinds of fluids in and on the Earth are hydrous (water-based) fluids and magmas (melted rock materials). The fluid of the hydrosphere—water—is predominantly stored at the surface in the oceans. This saline ocean water comprises 97% of the hydrosphere's water.[1] Fresh surface waters constitute only 0.009% of the hydrosphere. Groundwaters comprise 0.61% of the total. The other abundant fluid, magma, is transient at the surface because it begins to solidify upon reaching the surface. Some silica-poor magmas do flow many kilometers over the surface before solidifying, but they too are transient as fluids. Most commonly, magmas flow in fractures and in porous zones within the lithosphere. Because magmas have high density, high viscosity, and high temperature, they differ substantially from other fluids and are discussed separately.

Fluids in the Earth possess several properties and exist under a variety of conditions that control their behavior. The properties include density, viscosity, and chemical composition, whereas the conditions include low to high temperatures, low to high pressures, and dispersed versus concentrated distribution. The **density** (ρ) of a fluid (or any material) is defined as its mass divided by the unit volume. By definition, the density of water at surface conditions of 25°C and an atmospheric pressure of 1 atmosphere (10^5 Pa) is 1.00 gram/cm^3. Addition of chemicals to water will increase the density, as will increases in pressure, whereas increases in temperature will generally decrease the density. The same is also true for other common Earth fluids.

In contrast to water, with its density of 1.00 g/cm^3, the density of a basaltic magma is about 2.63 g/cm^3 at 10^5 Pa and it rises to 2.9 g/cm^3 at 2.0 GPa (Fujii and Kushiro, 1977; Kushiro, 1980). **Viscosity**, another property of fluids, can be thought of as an internal *resistance to flow*, a kind of internal friction. Highly viscous materials, like magma, flow with difficulty, whereas materials with low viscosity flow readily. The units of viscosity are the poise (1 dyne·sec/cm^2), one hundredth of which is a centipoise.[2] Water has a viscosity of 1.005 centipoise (0.01 poise) at 20°C and 10^5 Pa, whereas basaltic magma has a viscosity of 56 poise at 10^5 Pa and 1350°C, a viscosity that decreases to 25 poise at 2.0 GPa and the same temperature.[3] Clearly, increased pressure reduces the viscosity and so does increased temperature. Just as heating syrup makes it flow better, increases in the temperature of a magma will make it more fluid. Experiments show that water, too, decreases viscosity. Dry granitic magmas have viscosities on the order of 109 to 1012 poises, dramatically higher than basaltic magmas.[4] Adding water to a granitic magma may reduce the viscosity by up to six orders of magnitude, but hydrous granitic magmas still are generally 10,000 times more viscous than water, even at the high temperatures (600–800°C) at which they exist.

In terms of composition, hydrous and other fluids in the Earth carry a variety of ions in solution. Although the dominant chemical component of a hydrous fluid is H_2O, the overall composition of such a fluid is characterized both by the ions and isotopes within it and by the concentrations of those chemical constituents. Igneous fluids also contain various ions and isotopes. CO_2 and the ions Cl^{-1}, S^{-2} and SO_4^{-2}, in addition to H_2O, are important volatile constituents of igneous fluids (De Vivo et al., 2005). A variety of other ions are important in soil systems and in other geological settings. The ions in solution change the density and viscosity of the fluid (in most cases, just slightly), and they make the fluids chemically active—that is, prone to react with other materials that they encounter. The chemical activity of a fluid is also controlled by the pH (the acid to basic character) and the oxidation potential, the Eh, of the solution. Positive values of Eh reflect oxidizing conditions, whereas negative values of Eh reflect reducing conditions.[5] The pH is the negative logarithm of the hydrogen ion concentration. Neutral fluids have a pH of 7. The pH of near-surface waters in soils and rocks normally is in the range of 4 to 9, ranging from acidic to basic (Krauskopf and Bird, 1995, p. 331; Brady and Weil, 1999). Both the Eh and pH can affect the surrounding minerals, causing reactions that change the minerals and simultaneously alter the Eh and pH of the fluid.

Depending on the position of a fluid in the lithosphere—for example in the soil versus in the deep crust—it will exist under a particular set of conditions. Temperatures range from those below freezing of water, to temperatures in the mantle of more than 1000°C. Pressures also vary. Surface pressures of 105 Pa are replaced in the deep crust and upper mantle by pressures of 0.7 to 2.0 GPa. Fluids also exist as collected masses, such as an ocean; as dispersed fluid in the spaces between grains (the pores) of rocks below the groundwater table; or as highly dispersed fluid in the soil above the level of saturation (i.e., above the groundwater table).

FLOW OF FLUIDS

We learn from an early age that fluids flow. The most common fluids that exist in and flow through the many environments in and on the Earth are hydrous—they are dominantly water. Hydrous fluids flow across the surface in streams and reside on the surface in lakes, seas, and the oceans. In the lakes, seas, and oceans, fluid flow takes place in currents. Fluids in the subsurface occupy spaces (pores and fractures) within a volume of rock, a volume referred to as a **reservoir**. Subsurface crustal fluids are usually dominated by water, but locally the main components are carbon dioxide (CO_2), methane (CH_4), or, rarely, other gases. Each of these fluids may exist at depth in a supercritical state.[6] Fluids in subsurface reservoirs generally *flow* through either porous rocks or through fractures.

Streams, Stream Channels, and Flow of Water over the Surface

Water that flows over the surface of Earth *in streams* is called **runoff**. Stream water is derived directly from rain, from the melting of snow and ice, and from groundwater that flows back into the surface streams via springs and seepage. All of the land area drained by a stream and its tributaries through a particular discharge point is called the **drainage basin**. In the drainage basin, precipitation that (1) is not used immediately by organisms, (2) does not evaporate, or (3) does not infiltrate into the ground to become a temporary part of the groundwater, becomes part of the runoff.

Most of us have experience with streams, but we may not be fully aware of their properties. Here we summarize a number of observations made about water flowing in streams and about the nature of the stream channels through which it flows (Morisawa, 1968; Leopold, 1997).

1. Water flows in stream channels of a variety of sizes. Anyone who has waded or fished in a brook and also has crossed the Mississippi River knows this fact. During a rain, water begins its flow across the surface as a sheet in *overland flow*, but soon flows into small channels called *rills*, from rills into creeks (brooks, branches), and eventually into rivers of small to large size.

2. Sometimes water seems to flow smoothly in parallel lines in what is called **laminar flow**, whereas at

Fluids in and on the Earth   35

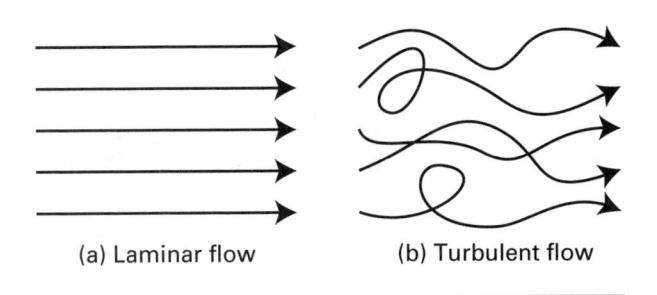

(a) Laminar flow   (b) Turbulent flow

Figure 3.1
Sketches showing the difference between laminar flow (a) and turbulent flow (b).
(*Source:* Modified from Raymond, 2007, p. 297.)

other times the water swirls about, producing **turbulent flow** (figure 3.1).

3. As water flows in a stream channel, we note that the flow along the sides of the stream is slower than the flow in the center of the channel. In fact, in straight sections of channel, maximum *flow velocity*, the distance traveled per unit of time, occurs at the surface or just below the surface in the middle of the stream channel.[7] In asymmetrical channels, the zone of maximum velocity of flow shifts toward the deeper part of the channel (figure 3.2).

Maximum flow velocity

(a) Symmetrical   (b) Asymmetrical

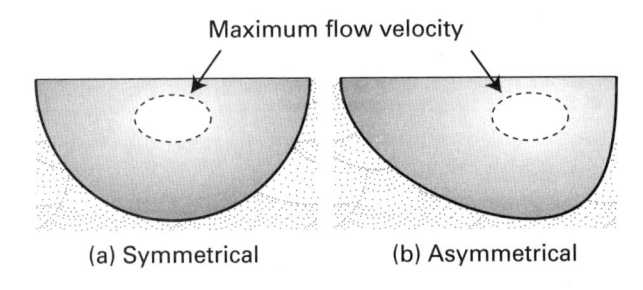

Figure 3.2
Cross-sectional views of streams showing the zone of maximum velocity of flow in symmetrical straight channels (a) and asymmetrical, curved channels (b).
(*Source:* Concept based in part on Leighley, 1934, reported in Morisawa, 1968.)

4. In map view, stream channels may be straight, sinuous, meandering, or braided (figure 3.3).
5. Streams erode materials from their channel margins to form sediment that is transported by the streams downstream towards sea level.

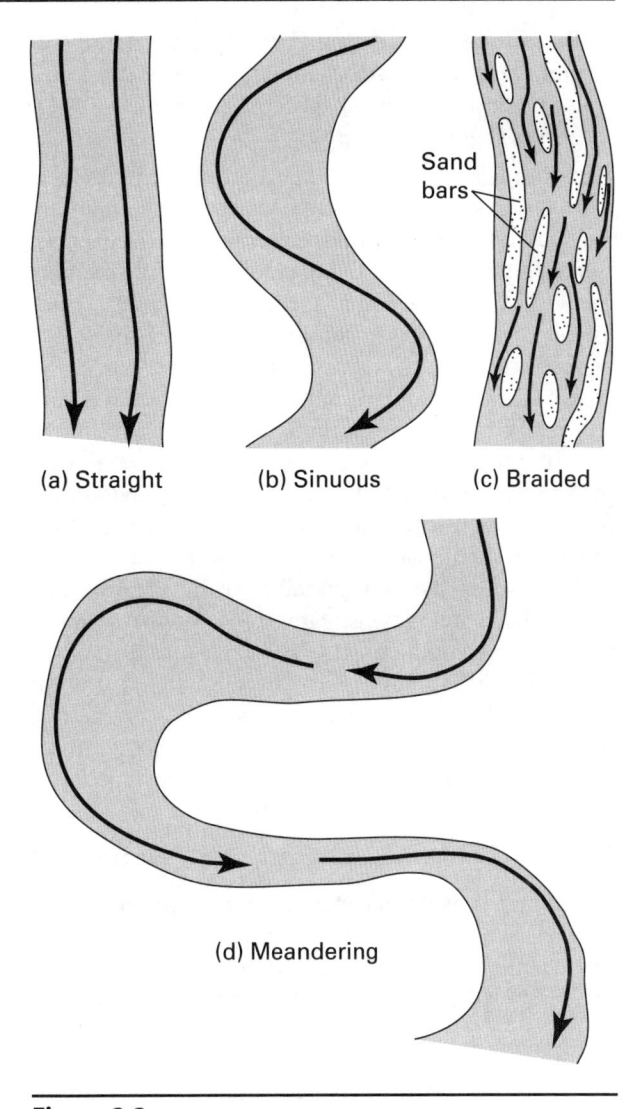

(a) Straight   (b) Sinuous   (c) Braided

Sand bars

(d) Meandering

Figure 3.3
Stream channels of various configurations. The line with the arrowhead drawn within the stream channel is the surface trace of the *thalweg*, a line connecting the deepest parts of the channel and which also defines the general path of flow.

6. Stream channels develop a set of pools, which are deeper areas of more laminar flow, and riffles, the areas of turbulent flow underlain by bars of sand or gravel sediment (Leopold, 1997, p. 65; Keller, 2000, ch. 5).
7. Stream channels exhibit a variety of cross-sectional shapes. The braided channels are broad and shallow, with much sediment distributed in bars, whereas other channels are deep and narrow with fewer bars of sediment.

8. Bars of sediment, or point bars, develop on the convex banks of stream channels, whereas erosion is particularly active on the concave banks of such curves (figure 3.4).

Each of these observations has importance to the geologic processes related in some way to Earth materials, particularly to the processes of erosion, the transportation of sediment, and the deposition of the sediment that ultimately becomes sedimentary rock. Each aspect of stream flow and stream channel shape can be quantified. Now we focus on some aspects of stream flow particularly relevant to the general processes of erosion, transportation, and deposition.[8]

Streams flow in response to the gravitational force pulling water down (figure 3.5). The gravitational force can be resolved into two parts, a normal stress acting perpendicular to the bottom of the channel and a shear stress that acts parallel to the channel and drives the water downstream. Flow of the water is resisted by friction along the bottom (and sides) of the channel; that is, it is resisted by the boundary shear stress. Recall from chapter 2 that velocity (v), the rate of flow in this case, is defined as *the distance traveled* (the difference between initial and final positions, $p_f - p_i = \Delta p$) *divided by the elapsed time* ($t_2 - t_1 = \Delta t$): Thus,

$$v = \Delta p/\Delta t \qquad \textbf{(Eq. 3.1)}$$

In laminar flow, the change in velocity of the stream (Δv) relative to the change in height above the base of the channel (Δh) is related to the shear stress (τ_f) divided by a value called the dynamic viscosity (μ):

$$(\Delta v)/(\Delta h) = (\tau_f)/\mu \qquad \textbf{(Eq. 3.2)}$$

The dynamic viscosity depends on how much sediment is mixed in the stream water. The more sediment, the higher the dynamic viscosity, and therefore the greater the resistance of the stream to flow. Obviously, clean water flows faster; or stated another way, higher dynamic viscosities result in lower velocities. It is also true that the greater the friction (i.e., the greater the boundary shear stress), the lower the velocity. In general, stream flow is not strictly laminar, except near the bottom and sides of the stream channel, and equation 3.2 cannot be applied generally to natural stream flow.[9] Rather, flow in natural streams is predominantly turbulent.

Stream velocity is not only a function of dynamic viscosity, it is also a function of the channel slope, the channel roughness, and a feature called the hydraulic radius. This relationship is expressed by the Manning formula,

$$v = (1/n)\, R^{2/3}\, S^{1/2} \qquad \textbf{(Eq. 3.3)}$$

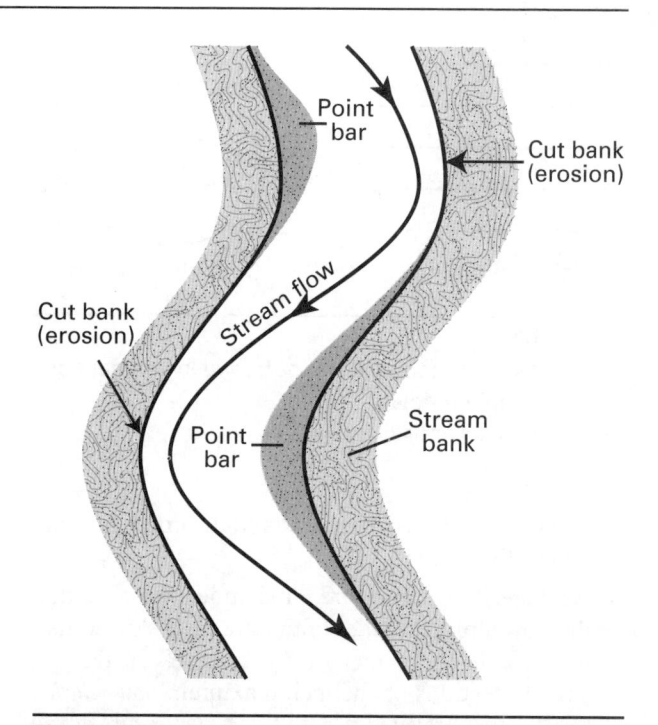

Figure 3.4
A sinuous stream channel showing areas of side erosion (cut bank) and deposition of sediment (point bar). The line with the arrowhead drawn within the stream channel is the surface trace of the thalweg, connecting the deepest parts of the channel and marking the general path of flow.

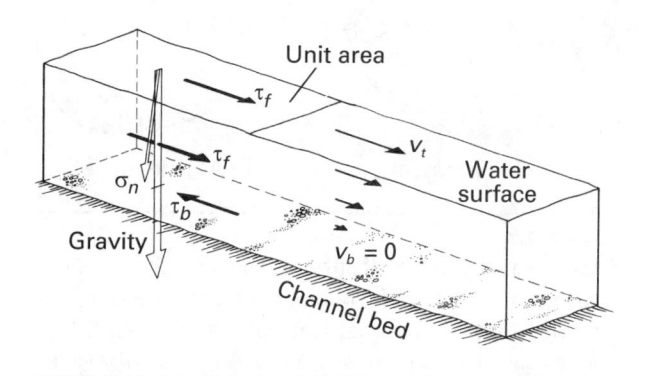

Figure 3.5
Diagram showing the forces acting on the water in a stream on a slope. σ_n = normal stress component of gravity. τ_f = shear stress component of gravity in the fluid. τ_b = boundary shear stress representing frictional resistance to flow along the channel base and sides. v_t = flow velocity at the top of the stream. v_b is the flow velocity (= 0) at the bottom surface of the stream.
(*Source:* Raymond, 2007, p. 295.)

where:
- v = velocity (in m/sec)
- n = roughness factor
- S = slope (in m/km)
- R = hydraulic radius (using m²/m)

The hydraulic radius is equal to the cross-sectional area (A) divided by the linear distance, in cross section, of the wetted perimeter (p); that is, the trace of that part of the channel sides and bottom that is under water (figure 3.6) (Morisawa, 1968, p. 37). Thus,

$$R = A/p \text{ (m}^2\text{/m)} \quad \textbf{(Eq. 3.4)}$$

Roughness is determined empirically—it must be measured for each stream—but values range from 0.03 for clean straight streams to 0.15 for streams with trees and tree trunks.[10] Clearly, roughness is a function of obstacles in the stream bed, and roughness increases with increasing abundance and size of those obstacles. Obstacles range from sand grains to cobbles, boulders, tree trunks, riffles, and bridge abutments—and temporarily from people playing in the water.

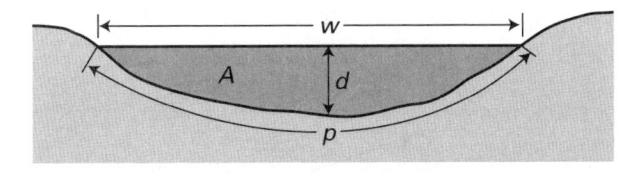

Figure 3.6
Cross-section of a stream showing the cross-sectional area (A), the width (w), the depth (d), and the wetted perimeter (p).

The obstacles not only control roughness, and therefore velocity, they also increase turbulence. Water flowing over and around obstacles is changed from a pattern of flow that is more laminar to one that is more turbulent. Turbulence is characterized by *eddies*, the curling patterns in the flow lines of the water. The size of the eddies is dependent, of course, upon the size of the obstacle. No natural stream channel is without some obstacles and that is why natural streams are characterized by turbulent flow. Turbulent flow results in erosion of the channel walls and floor.

The amount of water that flows past any point in a stream channel of a drainage basin is a function of both the velocity and the cross-sectional area (A). This amount of water is called the discharge (Q). Hence,

$$Q = Av \quad \textbf{(Eq. 3.5)}$$

Problem 3.1

The Manning formula may be stated in terms of metric units or English units. The metric formula is given in the chapter as Eq. 3.3. In routine engineering work in the United States, English units are typically used and the formula is written as

$$v = (1.49/n) \, R^{2/3} \, S^{1/2}$$

where
- v = velocity (in ft/second)
- n = roughness factor
- S = slope (in ft/ft)
- R = hydraulic radius (in ft, derived from ft²/ft calculations)

Using the English unit form of the equation, calculate (a) the velocity (v), and (b) the discharge (Q), of a stream with a roughness n = 0.10, a slope of 15.0 feet per mile, a cross-sectional area of 28.3 feet, and a semicircular perimeter (p) of 14.1 feet.

In general, in humid climates, velocity and discharge increase downstream (Leopold, 1997, p. 57). In arid climates, evaporation rates may be high enough to exceed runoff contributions from tributaries and groundwater seepage, with the result that streams decrease in discharge and velocity downstream, sometimes disappearing altogether. In climates somewhat between arid and humid ones, for example in Mediterranean climates like those of California, discharge may vary between summer and winter by a factor of 1000 or more. Discharge in the Russian River north of San Francisco, for example, may drop below 100 cfs (cubic feet per second) in the summer and rise to over 100,000 cfs in the winter (U.S. Geological Survey Water Data Report, 2013). Increases in velocity (and discharge) represent increases in energy. Increases in energy create an increase in the load carrying capacity of the stream. In short, faster flowing streams are able to erode and carry more sediment.

Streams erode their channels through turbulent flow and by using sediment as an abrasive agent. In general, channels have a crudely rectangular form, although in curved stretches of the stream, the cross-sectional shape is more like a sideways teardrop. In the curved stretches, erosion is concentrated on the concave bank or cut bank (as viewed from above), where the channel is deeper and the velocities are higher close to the bank (figure 3.4). Deposition occurs on the convex bank (viewed from above) forming a point bar, where velocities are lower and the channel is shallower.

Broadly speaking, the sediment eroded and transported by streams is eroded from the rocks and soils of the lithosphere. This erosion begins with overland (sheet) flow, where water flows across the surface, and is increased where velocities increase as water collects in rills and streams. At least part of that sediment is deposited in basins and ultimately becomes sedimentary rock.

Flow of Fluids Through Porous Rocks

Recall from chapter 1 that rocks consist of grains of mineral, rock, volcanic glass, and organic material. These grains are either interlocking in crystalline textures or they are stuck together in clastic textures.[11] Between the grains of the clastic-textured rocks, there are spaces or **pores**. We define the percentage of the volume of a rock that is made up of pores as the **porosity**. Crystalline-textured rocks may also have porosity, but the porosities are generally quite low. If the pores of a rock are connected enough to allow flow of fluid through the pores, the rock is said to be *permeable*. The **permeability** is a measure of the ability of a fluid to flow through a rock.

The flow of water through permeable rock is described by **Darcy's Law**, named for a French engineer who first studied (laminar) flow through porous media (Fetter, 1994, p. 94). Basically, Darcy, using a sand-filled pipe, found that the flow of water through a porous layer of material—specifically the discharge (Q) through the layer—is proportional to the difference in height between the two ends of the layer multiplied by the cross-sectional area and is inversely proportional to the length of the layer through which flow occurs (figure 3.7). Mathematically,

$$Q \propto \Delta h(A)/L$$

where
- Q = discharge
- Δh = difference in height between the two ends of the layer
- A = cross-sectional area of the layer
- L = length of layer through which flow occurs

Converting to an equality, the equation becomes

$$Q = KA(\Delta h)/L \quad \text{or} \quad \text{(Eq. 3.6)}$$

$$Q = KA(h_x - h_y)/L \quad \text{(Eq. 3.7)}$$

where
- h_x = the end of the bed where the initial flow measurement is taken
- h_y = the end where flow measurement ends[12]

The coefficient K is a constant called the *hydraulic conductivity*, a number that has the dimensions of length/time (= velocity). K is a function of properties of both the porous rock and the fluid flowing through it. In particular, the discharge is partly a function of grain size and shape.[13] Just as was the case for laminar surface flow, the dynamic viscosity (μ) also controls flow (see Eq. 3.2), so that discharge is inversely proportional to μ. The higher the dynamic viscosity of the fluid, the lower the discharge.

One of the processes that results from fluid flow through rock pores is chemical reaction between the fluid and materials along the flow path. Fluids become chemically active by virtue of the fact that they have a composition different from that of the rocks through which they are moving. If they are chemically active, they tend to react with the minerals around them. For example, a fluid with a high Si content will tend to react with minerals along the path of flow, converting some of them to more Si-rich minerals, or it will precipitate SiO_2, lowering the silica content of the fluid. Similarly, a fluid of low pH (an acidic fluid) will tend to readily dissolve minerals such as calcite, creating new pores or voids—a *secondary porosity* in the rock. In each case, the chemically active fluid

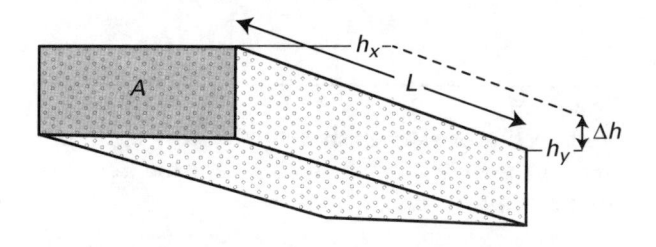

Figure 3.7
Diagram showing the parameters important in evaluating fluid flow through a porous rock layer using Darcy's Law. A = cross-sectional area; h_x = height of upper end; h_y = height of lower end; L = length of flow; and Δh is the difference in height between the two ends of the layer.

Problem 3.2

Using Darcy's Law, calculate the discharge (Q) for a 100-meter-long section of bed experiencing porous flow, if the height (h_x) at one end is 1000 meters and the height at the other end (h_y) is 998 meters, the cross-sectional area is 150 m², and the hydraulic conductivity is 16.7 m/sec.

reacts with the rock in some way to create void-filling cement, to change mineral chemistry, or to dissolve minerals—all processes that tend to change the character of the fluid and the rock or soil through which the fluid is passing.

Like low-density fluids, (high-density) magmas also flow through porous rocks. This is particularly evident in mantle rocks below mid-ocean ridges (Kelemen et al., 1995),[14] but such flow likely occurs elsewhere in the mantle as well. The mathematics and physics of melt flows in porous mantle rocks are more complex than are those used to describe the flow of low-density hydrous fluids between rigid sand grains, because the fluids have both high density and high viscosity and they are flowing through higher density but plastic (very high viscosity) rocks that may themselves flow and change in porosity and permeability over time (Spera, 1980; Turcotte and Phipps Morgan, 1992; Spiegelman, 1993).[15] Nevertheless, in the case of magmatic flow, flow is still governed, in part, by dynamic viscosity and permeability. In addition, the rate of flow is proportional to pressure differences between the bottom and top of the zone of flow and to the density differences between the plastic rocks and the melt: the larger the pressure difference and the greater the density difference between melt and surrounding rocks, the greater the velocity of flow.[16] Grain size distributions may also control flow (Wark and Watson, 2000).

Flow of Fluids in Fractures and Conduits

Subsurface fluids not only flow through the pores of rocks, they flow through fractures. In geologic parlance, fractures are usually called *joints*, but in studies of groundwater and flow of magma in the mantle and lower crust, the term *fracture* is used. Joints or **fractures** are curviplanar openings in rock parallel to which no significant movement has occurred, but across which there is a separation and loss of cohesion (in simple terms, fractures "open"). Nearly all rock masses contain numerous fractures oriented in three-dimensional space (striking and dipping) in various directions (figure 3.8; color plate 1b). The sizes of the fractures and their spacing, orientation, and nature vary and these characteristics depend on rock type, tectonic history, and uplift history. Both uplift and the tectonic forces of mountain building create stresses that can lead to the formation of fractures. In addition, fluids can create microcracks that coalesce into larger fractures through which they may flow (Green and Jung, 2005).

Fluids flowing in open fractures will only follow Darcy's Law if the flow is laminar, a condition that exists only at very low flow rates, in narrow fractures, and at the very edges of an open fracture. In general, flow in open fractures is fast enough to generate turbulent flow. Here, discharge is akin to surface flow in that equation 3.5 applies, i.e. $Q = Av$. Whether or not flow is laminar or turbulent can be revealed by the Reynolds number (R), a dimensionless number that relates various properties of a fluid to one another. The Reynolds number is given by the expression

$$R = \rho v d/\mu \qquad \textbf{(Eq. 3.8)}$$

where
- v = velocity
- ρ = fluid density
- d = diameter of the fracture
- μ = fluid viscosity

Low Reynolds numbers (<1000) reflect laminar flow, whereas high Reynolds numbers (>3000) reflect turbulent flow.[17] Flow between these values is variable. As an example of a Reynolds number and flow, if we calculate that the Reynolds number for a basaltic magma of density 2.9 g/cm^3 and viscosity of 25 poise flowing through a 100-cm, tube-like fracture with a velocity of 50 cm/sec, the Reynolds number would be 580 and the flow would be laminar.

$$R = (2.9 \text{ g/cm}^3)(50 \text{ cm/sec})(100 \text{ cm})/25 \text{ poise}$$

Remembering that a poise is a dyne·sec/cm^2 and that a dyne is a g·cm/sec^2,

$$R = 0.580 \times 10^3 \text{ g/cm}^3 \cdot \text{cm/sec} \cdot \text{cm} \cdot \text{cm}^2\text{sec}^2/\text{g} \cdot \text{cm} \cdot \text{sec}$$

$$R = 580$$

In general, fractures are thin and planar (i.e., slot-like): they are large in two dimensions and small in the third. As noted, if the fracture is thin, flow is generally laminar. Some fractures may have significant "obstacles" along their edges, increasing the likelihood of turbulence (Emerman et al., 1986). Furthermore, in larger fractures—those with greater area— the likelihood of turbulence is increased because

---

Problem 3.3

What is the Reynolds number for a magma flowing through a circular lava tube with a diameter of 3 meters, if the velocity of the lava is 100 cm/sec, the density is 2.59 g/cm^3, and the dynamic viscosity is 168 poise?

Figure 3.8
Photo of nearly vertical tensional joints (fractures) and subhorizontal exfoliation joints in granite. Acadia National Park, ME.
(*Source:* Photo by Loren A. Raymond.)

velocity (one of the factors in R) is partly a function of the size of the fracture, increasing with increasing size of the fracture. Equations for both laminar and turbulent flow in fractures take into account such additional factors as magma pressures, strike and dip of the fracture, and gravity.[18]

The fluid flow through rocks is important in a number of different contexts. First, there are both economic and practical significances for the study of porosity and permeability in rocks. Porous rocks are some of the most productive sources of groundwater. In crystalline (igneous and metamorphic) rocks, fractures provide the only reservoirs in some areas underlain only by these rocks. Petroleum and natural gas are extremely important and economically significant natural resources that are also fluids stored in rock pores (and reservoirs). Like other fluids, these fluids move via flow through porous rocks and fractures. The controversial method of hydrocarbon extraction involving hydraulic fracturing ("fracking") of rock greatly increases the number of fractures available for fluid flow in subsurface rocks. Hydrothermal (hot water) fluid flow, driven by the heat of bodies of igneous rock or melt, provides hydrothermal energy resources (e.g., steam to drive electricity-producing turbines). Hydrothermal fluids are also responsible for formation of some ore mineral deposits (Kesler, 2005).

From a more theoretical perspective, fluids entrapped in the pores of sediment at the time of sedimentation may initiate and facilitate **diagenesis**, the transformation of sediment into rock and the alteration of rocks at low temperatures and pressures (ch. 9). Fluids *moving through* rock are major agents of diagenetic change. For example, most dolostones are derived from limestones by conversion processes involving moving fluids containing Mg ions.[19] Finally, hydrothermal fluids and metamorphic fluids containing a variety of elements are important in processes of metamorphism (Ferry et al., 1998; Spear, 1993, ch. 19). In metamorphic rocks, the volumes of fluid may control the rates and completeness of metamorphic reactions. Metamorphic fluids containing a variety of ions promote and facilitate metamorphic reactions. Thus, understanding fluid flow in porous and fractured subsurface rocks is important in Earth systems, in economic geology, and in understanding the origins of some Earth materials.

Problem 3.4

Search the Web for information on fractures and "fracking" and write one or two paragraphs summarizing how fractures play a role in concerns about environmental damage resulting from fracking for natural gas.

CURRENTS IN THE OCEANS, SEAS, AND LAKES

The oceans and seas consist of huge masses of interconnected fluid. Within these masses of fluid, there are currents, both at the surface and at depth. The fact that there are currents means that some parts of the water mass move with elevated velocity through other parts. The currents are of two types, *wind driven* (surface) currents and **thermohaline** (temperature-salinity) **currents**, the latter of which has both surface and deep-water components.[20]

Wind currents are created, as the name implies, by wind blowing across the surface of the ocean, sea, or lake. For that reason, the dominant flow is horizontal. The friction between the air and water allows the wind to drag water along the surface, creating the current. Because of wind directions and the rotation of Earth, major ocean currents in the northern hemisphere form a clockwise rotating *gyre* (crudely circular current pattern). Currents like the Gulf Stream flow north along the western edges of the ocean and south along the eastern side of the ocean (figure 3.9). In the southern hemisphere, the gyres are counterclockwise.

Flow of ocean currents is turbulent and is influenced by a number of factors.[21] The viscosity, or internal friction in this turbulent flow, is called *eddy viscosity*. The coefficients of eddy viscosity can be divided into coefficients A_h and A_z for horizontal mixing and vertical mixing, respectively. Movements of matter with respect to the rotating Earth, including ocean currents, are subject to the Coriolis effect, the rotation to the right by moving masses north of the equator and the movement to the left by masses south of the equator. The Coriolis force is given by the equation

$$C_F = mvf \qquad \textbf{(Eq. 3.9)}$$

where

m = mass
v = velocity of the mass
f = the Coriolis parameter

The Coriolis parameter is

$$f = 2\omega \sin\phi \qquad \textbf{(Eq. 3.10)}$$

where ω is the angular velocity of the Earth about its axis and ϕ is the latitude. In a homogeneous, infinite ocean (an ideal case), the velocity of a surface current (v_s) will be

$$v_s = \tau / (A_z \rho f)^{1/2} \qquad \textbf{(Eq. 3.11)}$$

where ρ is the density of seawater and τ is the surface wind stress. Inasmuch as the ω and ϕ are known, and v, ρ, τ, m, and A_z can be measured or calculated, the (ideal) velocity of surface currents can be calculated.

Thermohaline currents produce vertical flow. Thermohaline currents are driven primarily by differences in density in parts of the ocean water, differences that result from differences in salinity (the salt content) and temperature. Cold water is heavier than warm water and it sinks. At high latitudes, where freezing creates high salinity and high density water, the dense waters sink, initiating flow. If water goes down, other water must rise to take its place, thus completing the cycle of flow in the current.

One additional, but smaller scale, type of current that occurs in the ocean is important from an Earth materials perspective. This is the turbidity current. Turbidity currents are density currents that owe their higher densities to transported sediment. Because they are heavy, once initiated, they flow down slopes, notably on the continental margins, and flow across the seafloor. "Triggering" of a turbidity current usually results from either overloading of sediment on the shelf edge and subsequent sudden sliding and flow of the sediment

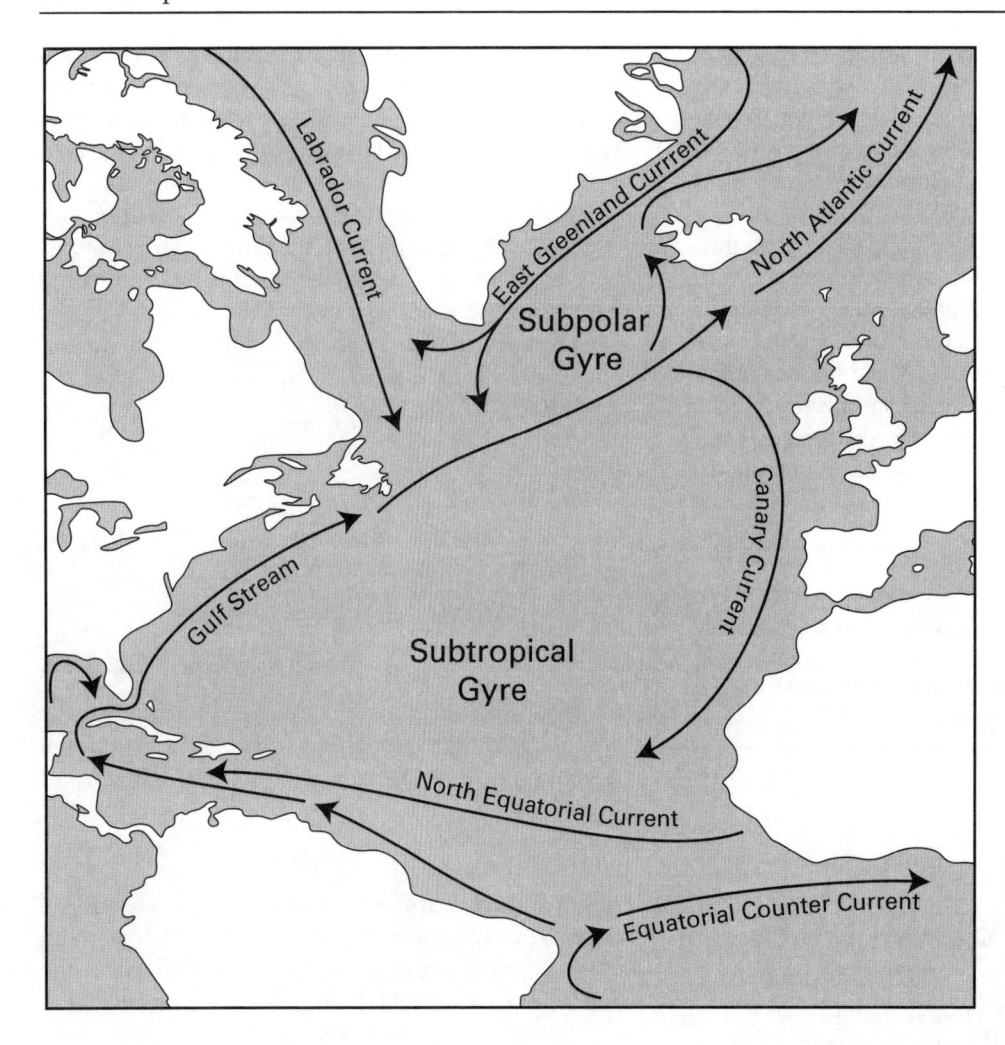

Figure 3.9
Sketch showing circular flow pattern of major ocean currents in the Atlantic Ocean.

down the slope, or sudden failure of the slope as a result of an earthquake. In either case, the sliding mass of sediment mixes with water to become a turbidity current.

Major ocean currents redistribute heat within the oceans by advection—that is, they carry the heat from one place to another. Equally important from a solid Earth materials perspective is the fact that marine currents both erode and transport sediment, eventually depositing the sediment as a parent material for a future sedimentary rock. Turbidity currents have also transported and deposited large amounts of sediment, especially in trenches, slope basins, and abyssal plains in the oceans. Wind driven and thermohaline currents transport and redistribute sediment on the slope, rise, and abyssal plains.

Currents exist in lakes as well as in oceans. As is the case with oceans, winds can generate wind-driven currents in lakes. Turbidity currents also occur here, especially where lakes are fed by sediment-laden waters.

KINDS AND SOURCES OF FLUIDS AND FLUID-MEDIATED PROCESSES IN THE EARTH

Various comments in the discussions above have alluded to the various sources and kinds of fluids in and on the Earth. Fluids consist of:

- magmas—melted rock
- magmatic fluids—fluids that come from the interior of the Earth as a component of a magma
- meteoric fluids—water derived from precipitation
- groundwater—meteoric water that has entered the soil or rock of the lithosphere as part of the groundwater system

- hydrothermal fluids—hot fluids, in part derived from magmas, heated by magmas or hot rocks, and usually derived from groundwater or a mix of magmatic fluids and groundwater
- metamorphic fluids—usually hydrothermal, chemically active fluids (so a special type of hydrothermal fluid), but including such other fluids as carbon dioxide (CO_2) and methane (CH_4)-based fluids—that migrate through rocks under the elevated P,T conditions of metamorphism

Each of these fluid types has particular properties such as distinctive viscosities, densities, major element and isotopic chemistries (including dissolved ions and elements), temperatures, and pressures. These properties and the origins and sources of the fluids distinguish them. Furthermore, each fluid contributes to processes in the lithosphere that give distinctive characteristics to other Earth materials. In particular, fluids facilitate crystallization of rocks and minerals, transport heat and ions that transform preexisting minerals and rocks into new minerals and rocks, contribute to the weathering of rocks to make soils, erode soils and rocks to yield sediments, and facilitate and contribute to the transformation of sediment into rock.

Magmas, the melted rock materials that ultimately crystallize to form the igneous rocks, dominantly form by the partial melting of rocks within the upper levels of the mantle or at the base of the crust (chapter 7). The compositions of the melts that develop depend on the compositions of the rock being melted (the parent rock) and the conditions of melting (including the degree of partial melting), but most magmas have between 45% and 75% SiO_2. The overall compositions of magmas are further discussed in chapter 7. The compositions of rocks and the high temperatures required to melt rocks (usually 700°C to 1300°C or more), result in silicate magmas of high temperature, high viscosity, and high density. The high viscosities and densities are further elevated as magmas begin to cool and crystallize, resulting in the formation of crystals that are dispersed through the melt.

As noted earlier, magmas move via both porous rock flow and flow through fractures. As magmas move, especially the lower silica magmas, they are particularly good at transporting (advecting) heat. Moving magmas also transport elements such as Si, Al, Na, and K toward the upper levels of the lithosphere, as was noted in chapters 1 and 2, facilitating differentiation of the Earth. The magmas that are more siliceous, aluminous, and alkali rich than the mantle and lower crustal rocks from which they were derived, crystallize to form the many igneous rock types that occur in the crust.

In contrast to magmas, **magmatic fluids** are fluids that arrive in the crust as part of a magma, but become separated from the parent magma, and have low density and low viscosity. They do, however, advect heat and transport some elements in ionic and molecular states, from the parent magma into the lithosphere and the atmosphere. Magmas have H_2O fluid contents that typically range from < 0.5% to 6% or more.[22] During crystallization of the magma, these fluids are concentrated and separated from the magma. If concentrated in zones or fractures within or closely associated with a granitic magma body, they may crystallize to form a very coarse-grained rock called *pegmatitic granite*.[23] Concentrated magmatic fluids commonly flow out into the rocks surrounding the magma body through fractures, but may also move to some degree through porous flow. In many cases, the magmatic fluids contain small amounts (in the parts per million range) of CO_2, SO_2, HCl, HF, and a wide variety of ions of unusual size and chemical affinity. As the fluids move away from the magma, they become hydrothermal fluids. From these fluids, minerals crystallize in veins to make ore deposits. The fluids diffuse out into surrounding rocks causing metamorphism and alteration, and as such, they may become metamorphic fluids.

The origins of magmatic fluids are not fully understood. It is likely that many of the magmatic fluids are recirculated, in the sense that they are carried into the mantle in the minerals of subducting plates and become a part of new magmas formed by flux melting.[24] Water and hydrogen may possibly be carried into the lower mantle.[25] What, if any, fluid can exist in or escape from the core is an unanswered question.

Meteoric and **groundwater fluids** are essentially impure waters. Rainwater forms after evaporation of surface waters lifts water vapor into the atmosphere, from which it is then precipitated back to the surface. The precipitation that infiltrates into the subsurface below the level of saturation becomes part of the groundwater. Chapter 9 addresses the importance of these waters in weathering processes. In short, these waters react with minerals, partially dissolving some to release Si, Na, Ca, and other elements to the fluid, and in the process, contribute to the breakdown of the rock. Some of the elements in solution re-precipitate to create new hydrous minerals. Thus, weathering breaks down rock and in doing so, produces soil, an important Earth material. The soil usually becomes sediment, the predecessor of sedimentary rocks. Clearly, meteoric water and groundwater contribute to the formation of other Earth materials.

As noted above, **hydrothermal fluids** are hot fluids, heated by magmas or hot rocks. They are usually derived from magmas, traditional groundwater, groundwater formed as seawater trapped within sediments, or groundwaters with fluids derived from metamorphic reactions (i.e., fluids that become metamorphic fluids). If the fluid is a hot, water-based fluid that is migrating, it is considered to be a hydrothermal fluid. As we show in chapter 11, hydrothermal fluids are important agents in the formation of many ore deposits. They fill this role by advecting heat, dissolving minerals in rocks, transporting ions as a variety of soluble complex ions and molecules, and redepositing those ions and molecules in the concentrated forms we call ore minerals and ore deposits. Hydrothermal fluids also transport the components of common minerals such as quartz and calcite and deposit them within rocks as somewhat tabular accumulations called **veins**.

Notable among the hydrothermal fluids are the fluids associated with mid-ocean ridges (Rona et al., 1983). These fluids are generated primarily by circulation of seawater down through fractures in the oceanic crust, where the seawater comes in contact with hot, recently crystallized igneous rocks. The fluids then rise through the crust, dissolving a variety of elements and causing metamorphism along the way. At the surface, the hydrothermal fluids come in contact with cold seawater as they spew forth from vents on the seafloor called "black smokers." Precipitated minerals rich in a variety of elements (such as copper) cloud the fluids as they escape and build small mineral mounds on the seafloor. These sites serve as incubators for localized microbial mats and unusual animal communities.

We can get some idea of the compositions of hydrothermal (and metamorphic) fluids by examining tiny bits of fluid—called fluid inclusions—trapped in the crystals of rocks formed from the fluids (table 3.1).[26] Water is the most abundant component in most cases. As noted in the discussion of metamorphic fluids below, water actually gives rise to several additional chemical species including OH^- and H^+ ions. In addition, in many ore-forming fluids, NaCl (salt) is a major component. Other components include molecules with S, K, Pb, Zn, Cu, Sn, Hg, and other elements.

Metamorphic fluids are derived from magmatic fluids, hydrothermal fluids, groundwater, and metamorphic reactions that yield fluid from the breakdown

> **Problem 3.5**
>
> Hydrothermal fluids are important sources of renewable energy. (a) Why are commercial sites of U.S. geothermal energy generation in the western United States? (b) Considering the chemicals carried in hydrothermal fluids, what problems might develop in transporting these fluids through pipes to power generators? Use the Internet to check and expand on your answers to this question.

of minerals.[27] As such, H_2O is the dominant component of most metamorphic fluids. The H_2O of the fluid gives rise to additional chemical species, including OH^-, H^+, H_2, O_2, and H_3O^+. In addition to oxygen- and hydrogen-bearing species, CO_2 is a major component of some fluids deep in the Earth and those associated with metamorphism of carbonate rocks, such as limestone. Carbon in the system also allows the formation of additional species such as CH_4 and CO. The sources of magmatic, hydrothermal, and groundwater fluids, which may become a part of metamorphic fluids, were discussed above.

Mineral reactions that yield fluids are typically dehydration reactions of the type:

Serpentine + 2 Brucite →
    4 Forsterite (olivine) + 6 Water

$Mg_6Si_4O_{10}(OH)_8 + 2\,Mg(OH)_2 \rightarrow$
    $4\,Mg_2SiO_4 + 6\,H_2O$

and decarbonation reactions like the reaction:

Calcite + Quartz →
    Wollastonite + Carbon Dioxide

$CaCO_3 + SiO_2 \rightarrow$
    $CaSiO_3 + CO_2$

In reactions such as these, which occur in specific rock types as metamorphic temperatures increase, water or carbon dioxide are added to the fluid during metamorphism. This fluid, which generally flows from place to place, may then interact with the minerals in other rocks to yield new minerals.

Metamorphic fluids are very important in many metamorphic reactions. Most reactions involve the formation of new minerals (as is the case in the two reactions shown above). In these reactions, metamorphic fluids serve to both bring in and carry away ions that participate in the reaction. In the cases above, only a fluid may leave the reaction site, but in many reactions, that fluid carries various ions into or out of the zone of the reaction.

There are several types of evidence that indicate that fluid flow is important in metamorphic reactions. These include the reactions themselves (like the ones above), fluid inclusions in metamorphic minerals, veins in metamorphic rocks, changes in the isotopic composition of rocks, and metasomatic zones—zones where there has been a clear and significant change in the overall chemical composition of rocks that have undergone metamorphism (Spear, 1993, ch. 19).

Table 3.1 The Compositions of Selected Fluids in the Earth

	Basaltic Magma (Glass) [in weight %]	Rhyolite Glass (Obsidian) [in weight %]	Hydrothermal Ore Fluid [in weight %]	Groundwater [Water in weight %; others in ppm]	Soil Water [Water in weight %; others in ppm]
H_2O	n.r.	n.r.	63.7	99+	99+
SiO_2	49.34	76.21	—	—	—
TiO_2	0.87	0.07	—	—	—
Al_2O_3	16.65	12.58	—	—	—
Fe_2O_3	n.d.	0.3	0.4	—	—
FeO	7.99†	0.73		—	—
MnO	0.16	0.04		—	—
MgO	9.23	0.03		—	—
Mg^{2+}		—		1.8	22
CaO	13.41	0.61		—	—
Ca^{2+}		—		13.5	77
Na_2O	2.15	4.05		—	—
Na^+		—		4.4	61
K_2O	0.15	4.72		—	—
K^+		—		1.2	3.2
P_2O_5	0.01	0.01		—	—
Cr_2O_3	—	—		—	—
CO_2	—	—	0.2	—	—
CO_3^{-2}		—		tr	tr
HCO_3^-		—		36	290
SO_4^{-2}		—		9.2	46.5
CH_4	—	—		—	—
NaCl	—	—	35.4	—	—
Cl^-		—		0.5	3.5
$CaSO_4$	—	—	0.3	—	—
Other		—			
TOTAL	100	99.87	100	100	100

tr = trace

† = Total iron as FeO

n.d. = not determined; n.r. = not reported

Sources: Basaltic magma (basalt glass from melt inclusion in plagioclase), Juan De Fuca Ridge, Sours-Page et al. (1999), sample O-2-7; Obsidian, Mono Craters, California, from Carmichael (1967), table 5, no. 18; Hydrothermal fluid from fluid inclusion trapped in copper deposit mineral, Butte, Montana, Roedder (1971); Groundwater from Wang et al. (2014), North Carolina Blue Ridge sample ASU-1D; Soil water from Wang et al. (2014), North Carolina Blue Ridge sample Th-11.

Summary

Surface waters, groundwater, igneous fluids (including magmas), hydrothermal fluids, and metamorphic fluids are vital to many geologic processes in the lithosphere. Earth fluids come from precipitation at the surface, evolve from formation and recirculation of surface waters—including those yielded by subsurface processes—and develop from primary and secondary sources in the mantle. The two most abundant kinds of fluids in and on the Earth are magmas and hydrous fluids. These fluids exist under varied conditions and exhibit a variety of characteristics that control their behavior, including density, viscosity, chemical composition, and chemical character—including Eh and pH. Magmas, for example, are two to three times as dense and 2500 to 10^5 times as viscous as water.

Surface waters flow in response to gravity. Flow in surface streams can be described by a group of mathematical equations. Streams erode their channels through turbulent flow and by using sediment to abrade the stream channel. Sediment eroded and transported by streams is eroded from the rocks and soils of the lithosphere and these sediments may become sedimentary rocks after they are deposited in various basins. The oceans and seas, in contrast to surface streams flowing over and within solid materials of the landscapes, contain currents that flow within the huge masses of water that constitute these water bodies. The currents are of two types: (1) wind-driven surface currents and (2) surface to deep water, temperature-controlled and salinity-controlled *thermohaline* currents.

In the subsurface, fluids flow through permeable rocks and fractures or "conduits." The flow of water through permeable rock is described by Darcy's Law, $Q = KA(\Delta h)/L$, where Q is the discharge, Δh is the difference in height between the two ends of the layer, A is the cross-sectional area of the layer, and L is the length of layer through which flow occurs. In general, flow in fractures is turbulent, a function of the Reynolds number, $R = \rho v d/\mu$, where v is the velocity, ρ is the fluid density, d is the diameter of the fracture, and μ is the fluid viscosity.

Selected References

Fetter, C. W. 1994. *Applied Hydrogeology* (3rd ed.). Prentice Hall, Englewood Cliffs, NJ. 691 p.

Hargraves, R. B. (ed.). 1980. *Physics of Magmatic Processes*. Princeton University Press, Princeton, NJ. 585 p.

Krauskopf, K. B., and Bird, D. K. 1995. *Introduction to Geochemistry* (3rd ed.). McGraw Hill, New York. 646 p.

Leopold, L. B. 1997. *Water, Rivers and Creeks*. University Science Books, Sausalito, CA. 185 p.

Morisawa, M. 1968. *Streams: Their Dynamics and Morphology*. McGraw-Hill, New York. 175 p.

Sears, F. W., and Zemansky, M. W. 1964. *University Physics* (3rd ed.). Addison-Wesley, Reading, MA. 1028 p.

Notes

1. The data in this paragraph are from Fetter (1994, p. 4).
2. Sears and Zemansky (1964, p. 320).
3. The data for water are from Sears and Zemansky (1964, p. 320) and for basaltic magma are from Kushiro (1980). Additional data on viscosity is available in Scarfe et al. (1987).
4. Baker (1996, 1998) and Scaillet et al. (1996) discuss magma viscosities and Holtz et al. (2001) discuss water in granitic magmas. Also see Sobolev and Chaussidon (1996) for data on water in basaltic magma and Zhang (1999) for data on water in rhyolitic magma.
5. See chapter 5 for additional discussion of Eh and pH. For more information on fluid pH and Eh values, see the diagram from Bass Becking et al. (1960) in Krauskopf and Bird (1995), and the associated discussion.
6. Krauskopf and Bird (1995, p. 484ff.)
7. Lane (1937; in Morisawa, 1968); Fetter (1994, ch. 3).
8. See hydrology and hydrologic summary texts such as Morisawa (1968) and Leopold (1997) for more complete discussion of the quantitative aspects of surface water flow.
9. Morisawa (1968, p. 31).
10. Both English unit and metric unit forms of the Manning equation can be used for velocity calculations (see problem 3.1) (Fetter, 1994). The form shown in the text is metric, but see problem 3.1 for the alternative. Roughness values are given by Morisawa, 1968, p. 31.
11. See chapters 5, 6, and 8 for discussions of textures.
12. Heath and Trainer (1981) and Fetter (1994, ch. 4). This equation is sometimes written with a negative sign to indicate that the flow is in the direction of decreasing hydraulic head, the total mechanical energy per unit weight measured as length (e.g., Fetter, 1994, p. 95, p. 134).
13. Fetter (1994, p. 95-96).

14. Also see Sparks and Parmentier (1994).
15. Also see Daines and Kohlstedt (1993).
16. Spera (1980).
17. Sears and Zemansky (1964, p. 328) note that flow for R up to 3000 may be unstable and switch back and forth between turbulent and laminar type flow. See Sears and Zemansky (1964, p. 327-329) or other university level physics texts; Nelson (1986); and Fetter (1994, p. 143) for discussions of Reynolds numbers.
18. See Rubin (1995) and Lister and Kerr (1991) for additional discussions and relevant equations.
19. See the summary in Raymond (2007, p. 304-305).
20. Thurman (1997, p. 202).
21. This discussion is based on The Open University Course Team (1989) text on ocean circulation.
22. Sobolev and Chaussidon (1996); Wallace and Anderson (2000). Holtz et al. (2001) suggest that water contents of granitic magmas locally may be as high as 20%.
23. See Raymond (2007, ch. 10) for a review of this process.
24. See Krauskopf and Bird (1995, ch. 18) for a more thorough discussion of the relationship between volatiles and magmas.
25. Ohtani (2005) reviews water in the mantle.
26. For example, see Touret and Dietvorst (1983) and the review by De Vivo et al. (2005).
27. See Spear (1993, chs. 18 & 19) and also Touret and Dietvorst (1983).

Crystallization and Crystallography

Introduction

An old story has a man showing up at the door of a university scientist with a big smile on his face. The man shows a worn magazine photograph of a priceless diamond, then proudly announces that he has found the diamond's "sister," pulling from his pocket a sizable, flawless crystal of quartz. Despite detailed explanations of the differences between quartz and diamond, no argument can dissuade him and the man returns home, convinced he has struck it rich, if only he can find an honest person who will pay him for his "diamond." For the scientist, determining the identity of the quartz crystal at a glance without performing any tests was simply a matter of noting the external form of the crystal. Quartz and diamond occur in very different forms due to their differences in chemistry and associated arrangements of atoms. An elongate crystal with a hexagonal shape, that is six-sided when viewed from the point at the top of the crystal (color plate 7a), is most definitely not a diamond, which has eight flat surfaces or *faces*, only four of which appear when viewed from the point at the top of the crystal.

For many minerals that have crystallized in excellent to perfect forms, the visible shape reflects the internal arrangement of atoms and serves as a useful property for identification (color plate 17a). In fact many of the properties of solid materials are a function of how their atoms or molecules are arranged, including the manner in which they break, their resistance to scratching or other deformation, and even the temperature at which they melt. It follows, then, that understanding how atoms are arranged in crystals and how they become so arranged can be a valuable tool for identifying minerals as well as for gaining a more general understanding of mineral structure and stability. It is not possible in a single chapter (or even a single book) to cover the breadth and depth of these topics; so here we provide a basis for further study and a background we hope will be sufficient for using mineral symmetry as an aid in mineral identification.

Order, Disorder, Crystallinity, Nucleation, Crystal Growth, and Morphology

Human beings have an innate appreciation for order. Ordered patterns can be purely decorative, as in wallpaper patterns, or structurally important as in the arrangement of bricks in a wall. Like the repeated pattern of bricks that give strength and stability to a wall, an ordered arrangement of atoms gives some strength and stability to the solid chemical elements and compounds we call minerals.

The word **crystal** has a variety of meanings in everyday life, but to a geologist it implies some very specific things. A geologist might refer to a specific, single piece of material bounded by smooth surfaces as a crystal, but he or she might also refer to an irregular piece of that material as a crystal or as a crystalline substance. When something is described as being "crystalline" it does not necessarily mean that it is transparent, or forms a particular shape, or even that it is particularly pleasing in its appearance. It means that the atoms making up the material are arranged in an ordered, three-dimensional pattern repeated over and over again (this is called *long-range order*).[1] Liquids and gases, which do not have a long-range order of atoms, do not qualify as crystalline; there are also numerous solid materials that do not meet this criterion. Window glass and volcanic glass (obsidian), for example, have some order but it is only localized (i.e., the order is *short-ranged*). Liquid crystals, such as those found in LCD

(Liquid Crystal Display) panels, have long-range order but only in one or two dimensions.[2] These are intermediate examples. Something with a true lack of order is referred to as **amorphous** (from the Greek, meaning without form), but we can refer to all partially ordered and amorphous materials simply as noncrystalline.

In most cases within the Earth's crust, an ordered state is more stable than a disordered state. In thermodynamic terms, the reaction of a disordered phase to its ordered equivalent has a negative ΔG_r so it proceeds spontaneously, like the rock falling to the valley bottom, discussed in chapter 2. It may not be immediately clear why atoms should "prefer" an ordered state to a disordered one. Considering that it is easier to throw items into a drawer rather than organizing them within the drawer, it would seem, superficially, that it would be "easier" for atoms to collect together in a random arrangement than for the system to expend the energy to become organized. Yet, recall from chapter 2 that when the electrons of atoms in close proximity form bonding orbitals, *they lower their combined energies*. This is critically important. Recall too that the spatial orientations of the orbitals of any given atom control the orientations of the surrounding chemical bonds in which those orbitals participate. Thus, the shared bonds determine the positions of atoms that surround a core atom and the resulting arrangement is an ordered one. Even after a glass is formed, over time the atoms in the glass rearrange themselves into a crystalline form in order to lower the energy of the arrangement.

How does a mineral form as a crystalline material? As noted in chapter 3, ions may be dissolved in crustal fluids, such as water or a magma, but how do such ions end up with an ordered arrangement within a crystal? Consider a small beaker that contains 18 grams of pure water.[3] If we add one atom each of sodium and chlorine to this beaker, both of the atoms would readily ionize and form Van der Waals bonds with the surrounding water molecules (figure 4.1a). Obviously, with only two ions, the likelihood of these two ions coming into contact (to form a NaCl mole-

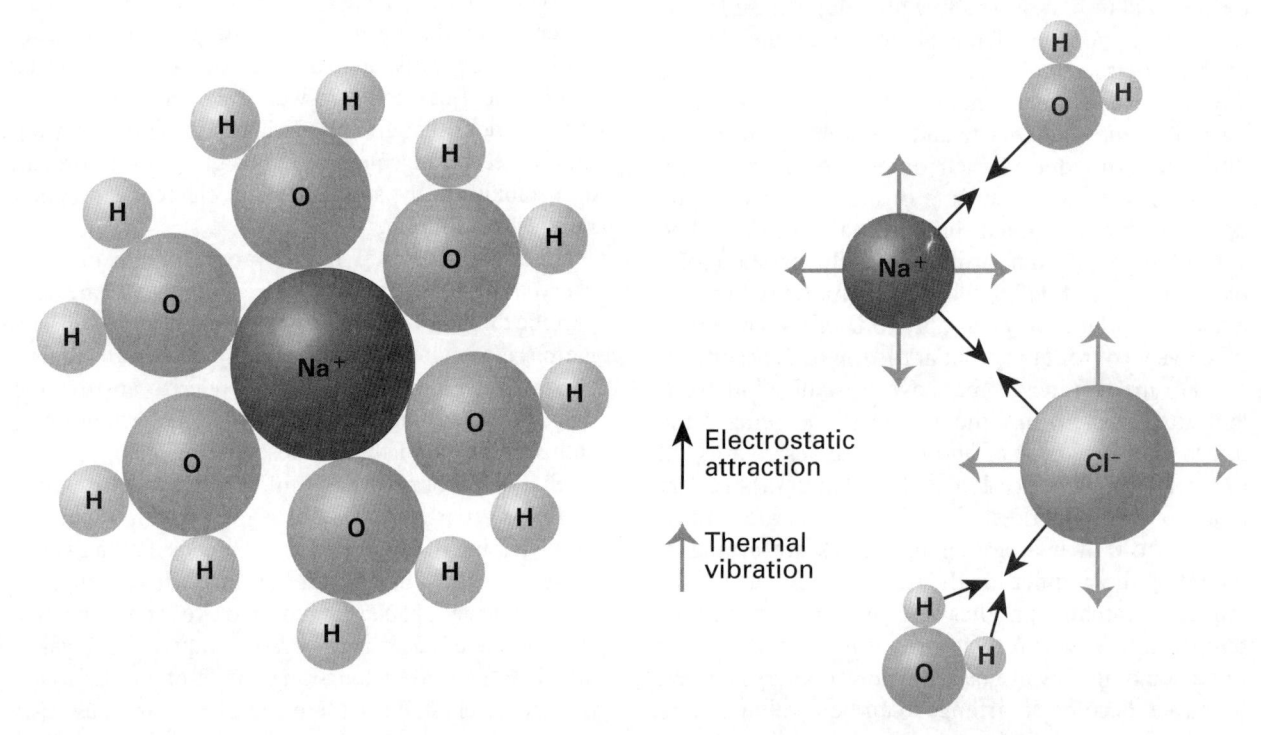

(a) A sodium ion, dissolved in water, surrounds itself with water molecules by attracting the negatively charged side of the oxygen of the H₂O molecule.

(b) Sodium and chlorine ions attract each other, but may be separated by the attractions of surrounding water molecules or by their own vibrations.

Figure 4.1
Formation of a crystal nucleus.

cule) is vanishingly small. Even if several hundred million more ions of sodium and chlorine are added, the possibility of them connecting to form a coherent mass of NaCl is effectively zero, because there are just not enough ions. We describe this situation by saying that the fluid in the beaker is **unsaturated** with respect to NaCl. It should be obvious that we *could* add enough NaCl to the fluid to make it **saturated**, a condition in which no more sodium and chlorine ions will dissolve under the existing conditions of pressure and temperature. In this saturated state, it would seem that with so many ions available, crystals of halite (NaCl) would begin to appear (color plate 19a). This frequently is not the case. To understand why, consider (qualitatively) the forces acting upon the ions in the fluid.

As shown in figure 4.1b, the Na^+ and Cl^- ions undergo a mutual electrostatic attraction which, if the ions join, creates an extremely small mass of halite. This process is referred to as **nucleation**, and the tiny mass is called a **nucleus**.[4] The electrostatic attractions of the water molecules, as well as the thermal vibrations of the two ions, oppose the nucleating (attractive) force. So, in order for the ions to remain joined, a balance of forces must be attained. When individual nuclei begin to form at the point of saturation and they form a nucleus, the nucleus may increase in size by attracting more ions from the fluid. Yet until the nucleus reaches a critical size, the dissociational forces opposing the attraction of the ions are strong enough to separate at least some of the ions from the forming nucleus. In order to overcome these forces—in order for the nuclei to become large enough to remain stable—a fluid generally requires some degree of **supersaturation:** a somewhat forced, temporary, and nonequilibrium condition in which the fluid contains more than the number of ions necessary for saturation.

The newly formed nuclei may remain suspended, they may sink to the bottom of the fluid container, or they may attach to whatever "wall" is bounding the fluid (a beaker in the lab or the walls of fracture, or magma chamber underground). The nuclei may also link with other nuclei and continue to grow. If the fluid increases in supersaturation slowly, then relatively few nuclei tend to form, yielding a few large crystals. A fluid that rapidly increases in supersaturation creates many nuclei and the result is a solid consisting of more numerous, smaller crystals. Considering the crystallization of a nucleus and a subsequent incipient crystal of halite, for every positively charged sodium ion there clearly must be a matching, negatively charged chlorine ion to form the growing nucleus. The concept of a balance of charges within an electrically neutral crystal is called **stoichiometry**. Stoichiometry applies to covalently bonded substances and molecules, as well as ionic crystals. There can be exceptions, however, due to crystal defects, impurities, or variable oxidation states and materials with unbalanced charges may have nonstoichiometric formulas.[5]

In general, ions are added to a growing crystal nucleus in an ordered fashion, controlled by the layer of ions on the outermost surfaces of the nucleus. The crystal shape (or as we call it, the **morphology)** that results from growth depends both on the symmetry of the crystal, which depends on the atoms involved, and on the rate at which ions can be added to the external surface. Since there are thousands of ways that small sets of the 92 natural elements can combine in different proportions to make crystals, crystals of different compositions have many different morphologies. It is less than clearly intuitive that two specimens of identical composition might have very different forms, as is the case with such minerals as calcite ($CaCO_3$) and aragonite ($CaCO_3$), or galena (PbS) (figure 4.2). Similarly, it may not be intuitively obvious that materials composed of entirely different elements might form crystals of the same morphology, as is the case with galena (PbS) and halite (NaCl).

Consider a cross-section at the atomic level though a tiny crystal of halite (figure 4.3). The cross-section has several external flat surfaces called **crystal faces** (depicted as lines in 2D) that are produced by the growth of the crystal. Comparing the bonding within the crystal with that at these surfaces or faces, we find significant differences. A sodium ion that occurs several Å below any surface is surrounded (in six directions in 3D) by chlorine ions that bond to it (in the 2D figure 4.3, the interior Na^+ ion with four short arrows showing bonds with adjoining Cl^- is totally surrounded). This Na^+ ion has all of its bonds fulfilled. In contrast, sodium ions at various locations *along the surface* have some unfulfilled bonds. Depending on the orientation of the surface relative to the structural arrangement of the ions in the material, there will be greater or fewer unfulfilled bonds (in figure 4.3, unfulfilled bonds = short arrows directed toward the surface from interiors of Na^+ ion located along the crystal edge). The more unfulfilled bonds on the surface, the greater the possibility that an appropriate ion will be attracted from the fluid. In general, corners (or kinks) will attract the most ions, edges or steps will attract the next most, and flat surfaces (like the one across the top of the crystal) will attract the least ions (because nearly all the bonding possibilities are fulfilled in an ion within a flat surface). Thus, a crystal face that is underlain by kinks will tend to grow faster versus one underlain by flat planes of atoms, which will grow more slowly (figure 4.4a).

This control on the rate of growth helps explain a pair of observed phenomena about the external shapes of crystals. The first is seen if we compare the morphologies of large and small crystals of the same mineral. In general, the smaller crystals have more numerous and a wider variety of faces than do larger crystals (figures 4.4b; 4.4c). As a crystal gets larger, the faces growing the fastest end up disappearing as they grow themselves into extinction (figure 4.4d). The second phenomenon is found in the observation that certain types of crystal faces are markedly more common than others. This phenomenon, first noted in the 17th century, is explained by the **Law of Bravais**, which states that the most common crystal faces are those that contain

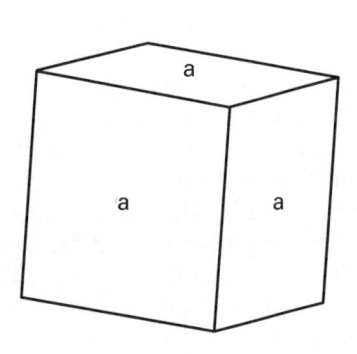

(a) A cubic morphology with six faces.

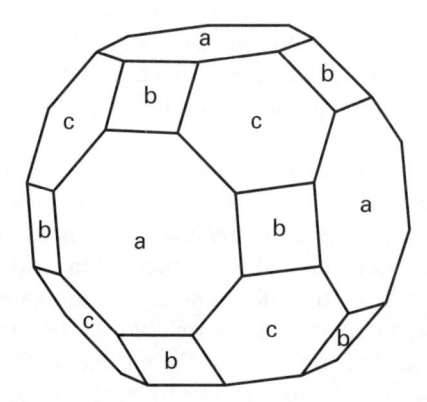

(b) A crystal with a combination of cubic (faces a), octahedral (faces c), and dodecahedral (faces b) forms.

Figure 4.2
Different crystal shapes (morphologies) in the same mineral. Sketches of two galena (PbS) crystals showing different morphologies. The letters correspond to unique crystal faces.

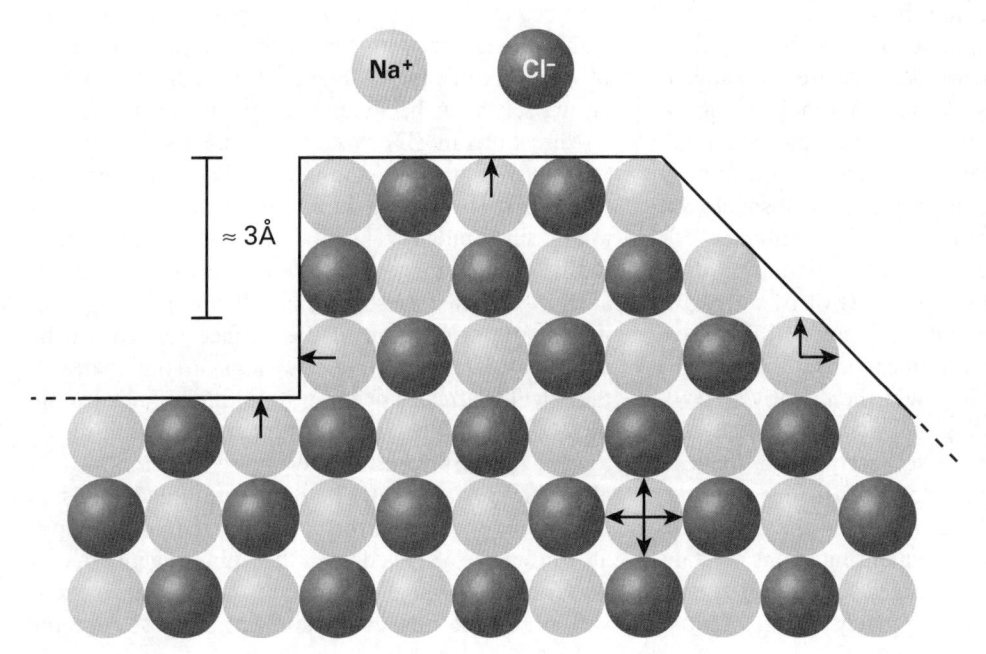

Figure 4.3
Cross-sectional view of atomic structure near the surface of a halite crystal. Sodium ions deep below the surface are entirely surrounded by chlorine ions, so all bonds are fulfilled. Sodium ions at the crystal surface have one or more unfulfilled bonds, depending on the orientation of the surface. For example, note that sodium atoms along edges (marked by straight lines) have unfulfilled bonds marked by the arrows.

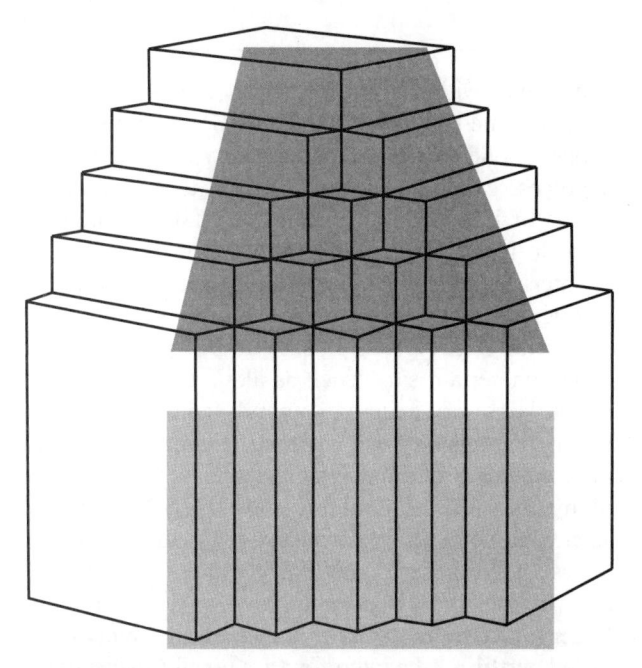

(a) Kinked (dark gray) faces will grow more quickly than stepped faces (light gray), which will grow more quickly than the flat faces (white).

(b) Example: a crystal of garnet with faces showing stepped growth.

(c) Example: a crystal of sphalerite with faces showing stepped growth.

(d) Diagram showing the changing shape of a crystal face with growth. Although the original crystal has six faces, the three larger faces grow more rapidly than the three smaller faces. Eventually, this more rapid growth toward the three corners of the outermost crystal outline eliminates the three smaller faces.

Figure 4.4
Growth of crystals to form faces.
[*Source:* (a) Modified from Lasaga, 1990; (d) after Perkins, 2002.]

the greatest number of lattice points (described later in this chapter). While it has been shown that the Law requires modifications to apply to a few minerals,[6] it remains an excellent first approximation for explaining the existence of some common crystal faces.

SOLID SOLUTION

Crystals grow *in* melts or *from* melts or solutions that contain the specific atoms that comprise the mineral that is crystallizing. From the discussion above, it might seem that a given crystal forms solely from specifically attracted atoms or ions, and it might also seem that once those atoms or ions are in place, they are locked down until the crystal melts, dissolves, or is otherwise broken down. In truth, a pure crystal is an idealized state that can only be approached in a laboratory. In reality, melts and solutions contain a wide variety of elements and in many cases crystals incorporate multiple elements within their atomic arrangements. In some cases, additional ions fit nicely into the structure because of their size and charge, whereas in others, they enter the structure in a sort of forced or awkward way.

The binary solid solution diagrams discussed in chapter 2 (e.g., figure 2.14) represent minerals of the former type, in which different ions fit nicely into a structure. As an example of solid solution, begin by recalling the halite crystal, crystallizing from a solution of sodium ion and chlorine ion in water, described above. Now, if potassium chloride (which becomes potassium ion and chloride ion in the solution) is added to the solution, an interesting phenomenon occurs. Solid potassium chloride, which occurs naturally as the mineral sylvite, has an arrangement of ions identical to halite, and if there is an appreciable amount of potassium ion dissolved in the water, some of the K^+ ions will be incorporated into the crystallizing halite. The crystal can be considered to be mostly halite (NaCl) with some sylvite (KCl) dissolved in it, and it has the formula (Na, K) Cl. Therefore, with two (or more) cations occupying the same kinds of positions within the structure of this crystalline substance (essentially, one ion substituting for the another) the material is or has become a *solid solution*. Many other examples of solid solutions exist in the realm of natural minerals.

If the chemical environment surrounding such a solid solution crystal changes, a number of reactions can occur within the crystal in response to those changes. These include (1) the breakdown of the homogeneous solid solution mineral into two separate solids, a process called **exsolution**, and (2) the exchange of atoms or ions with surrounding crystals or fluids, as they do in solid solution crystallization (described in chapter 2). Under certain conditions, elements are able to migrate (that is, **diffuse**) from atomic site to atomic site within solid crystals. In exsolution, ions diffuse and collect in certain sites (such as along twin planes), so that a larger crystal predominantly composed of one phase (one mineral) will include small areas composed predominantly of another phase (another mineral). In many cases, the resultant intergrowth is too small to be seen with the naked eye, but can be seen in microscopic views. In a few cases, most notably in the case of albite (Na-plagioclase feldspar) intergrown as small areas in a larger potassium feldspar crystal—an intergrowth called **perthite**—the intergrowth is readily visible to the unaided eye, commonly as tiny, straight to sinuous lines of different color within the host crystal (color plate 8a, b).

Problem 4.1

Expensive stemware is often referred to as "lead crystal." Is this an accurate description? Why or why not?

THE PROPERTIES OF SYMMETRY

What is symmetry? Symmetry was described by Aristotle[7] simply as the relationship between parts of an entity, but we here define **symmetry** as the property of a material in which *essentially identical* parts of that material are related in some way to one another (for example by reflection or rotation), but occupy a different position or orientation. The term is generally used to describe some type of order, either among groups of objects or within a single object. Although the topic of symmetry can become quite involved, the basic details can easily be understood by beginning with the simplest case.

Symmetry in One Dimension

The advantage of beginning a discussion of symmetry in one-dimensional space is twofold. First, it is easy to grasp. Second, it allows introduction of nearly all of the topics needed to understand symmetry in higher dimensions. So, consider the symmetry along a line. Along a line, there are only two directions of possible movement of an object or image. Since one direction is simply the opposite of the other, it is clear there is really only one independent spatial dimension. If an object is placed on the line, duplicated, and the copy is moved some distance away (referred to as **translating** the object), the resulting arrangement has a symmetry that did not previously exist (figure 4.5a). This process of adding symmetry to a collection of objects is called a **symmetry operation**. In the case just described, the operations are duplication and translation and the process is **translational symmetry**. When the operation is repeated a number of times, the object occurs again and again, each copy of the object translated from its neighbors by an identical distance (figure 4.5b). Clearly, this is more orderly than would be a translation with random sizes of repeat distances (figure 4.5c). The constant separation distance between the objects, or repeat translation, is what makes the arrangement orderly. Replace each of the physical objects with a mathematically idealized point and the physical translation by a vector (a vector is a line segment that has a specified length and direction) and the ordered arrangement of points is called a **lattice**.[8] The lattice is generated by repeating the vector a (termed a **basis vector**), in both directions from the starting point or **origin** (figure 4.5d). In one-dimensional space, there is only one possible type of lattice, although the length of a may vary between individual arrangements.

One-dimensional symmetry has other possibilities. Imagine that for every duplication and translation of the object it is also reflected, as if by an invisible mirror that only works on adjoining objects (figure 4.6). Each object along the line then becomes a mirror image of its two neighbors, with the invisible mirror located exactly half way between each pair of objects. Repeating the mirror along the line replicates the object, creating a third object that appears in the same orientation as the original, so the operation is closed (meaning no

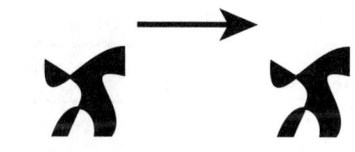

(a) Duplicating and translating an object creates symmetry.

(b) Symmetry is added by repeating the same translation multiple times.

(c) Non-identical translations do not create symmetry.

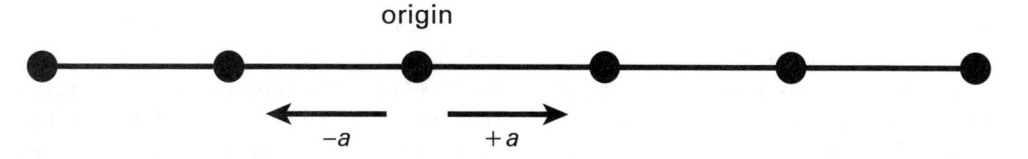

(d) A one-dimensional lattice.

Figure 4.5
Linear symmetry.

Figure 4.6
Mirror points along a line cause objects to be reflected as well as translated (see text for explanation).

new operation is created by using it). A point midway between each of the translated objects that causes each object (and in fact all of the one-dimensional space) to become a mirror image of its predecessor is a *mirror point*. The only part that is not reflected is the mirror point, which gives this second type of symmetry its name—**point symmetry**. Whereas translational symmetry operations move objects (and thus the space itself) in a linear fashion, *point symmetry* operations move objects in more involved ways, yet leave at least one single point in space invariant. The symmetry possibilities for one dimension are now exhausted.

Now, we classify the groupings of symmetry operations. A closed assemblage of point symmetry operations is referred to as a **point group**, and in one dimension, there are two (called 1 and m). A closed assemblage of point and one-dimensional translation operations is a **line group**, and once again, there are two (p1 and p1m). The term "group" is used with a very specific intention, as it derives from the mathematical underpinnings of symmetry, a branch of mathematics referred to as group theory. We do not address this subject in any detail and refer those interested to *Mathematical Crystallography* by Boisen and Gibbs (1990).

Symmetry in Two Dimensions

The rationale of a lattice as a group of regularly arranged points can be extended to two dimensions (figure 4.7). Adding a new dimension means that a new basis vector is required, and by definition the vector must be independent (i.e., it cannot be created by combinations) of the other basis vectors.[9] Whereas the one-dimensional space was filled by a line created by repeating the vector *a*, two-dimensional space must be filled by plane figures created by repeating vector *a* and a new vector, *b*. This vector need not be of the same length as *a* and is rotated from *a* by some angle (labeled as γ).

How will plane figures of two-dimensions fill the available space? Imagine laying tile on a kitchen floor and doing so with a single shape of tile. If the tile

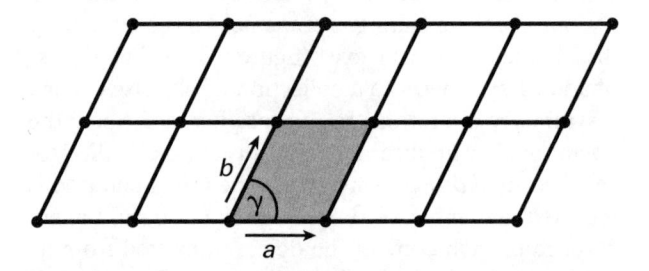

Figure 4.7
A two-dimensional lattice. The angle between the basis vector *a* and vector *b* is defined as γ.

Problem 4.2

Describe the point symmetry displayed by human beings.

---

serves to protect the wooden floorboards, the tiling must be continuous. There can be no gaps between tiles and not every possible shape of tile can be used to fill space. These requirements also apply to atomic symmetries in minerals. The reason there can be no gaps and only certain tile shapes can be used lies in the point symmetries found in two-dimensional space. The problem of filling space, referred to mathematically as **tessellation**, is not just a matter for home improvement. The arrangements of atoms that make up every bit of solid matter must also tessellate in the same way.

If we take a rectangular arrangement of lattice points ($a \neq b$, γ = 90°) (figure 4.8a), mirror points would sit halfway along both of the *a* vectors. Since all of the space within the rectangle would be reflected in the same fashion, however, the true mirror in two dimensions is not simply a pair of points but the line that connects them, a **mirror line**. Similarly, there is another mirror line perpendicular to the first that connects the mirror points halfway along the *b* vectors. A new point symmetry element then arises at the point of intersection of these two mirrors, a **rotation point**, as shown in figure 4.8b. In the case of two perpendicular mirrors, the rotation operation about the point takes the given object (or space) and rotates it by 180°.

Although there are an infinite number of possible rotations about points, the requirement to fill space with one shape results in only five possibilities: 360° (termed a 1-fold rotation because the operation completes in one step); 180° (2-fold); 120° (3-fold); 90° (4-fold); and 60° (6-fold) (figure 4.9). [Note that other rotational symmetries can occur if the restrictions are relaxed; that is, if there is no requirement to completely fill space or if more than one type of tile can be used (figure 4.9f). Flowers and leaf patterns in plants may have a variety of rotation angles, some virus particles have 5-fold symmetry, and a wide variety of tilings can be made of mixed tile shapes (figure 4.10 on p. 59). These latter types of symmetries are considered noncrystallographic because they cannot occur across long ranges in crystals.]

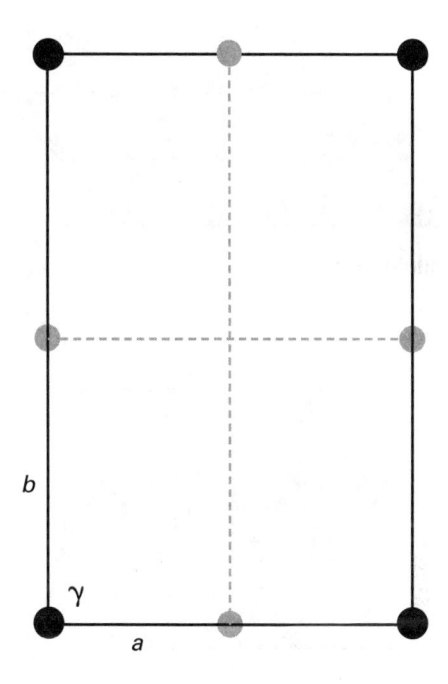

(a) Connecting parallel mirror points creates mirror lines.

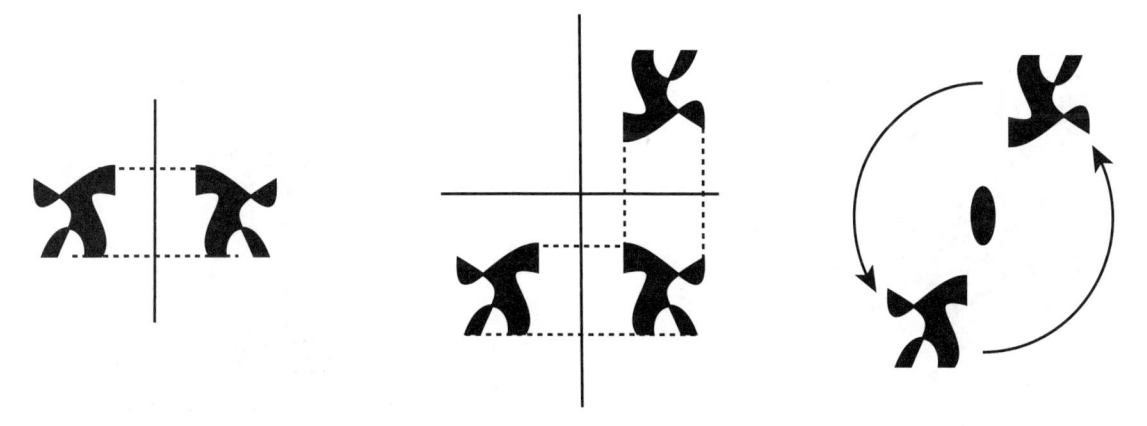

(b) Two mirror lines intersecting at right angles produce the same result as a 2-fold rotation point.

Figure 4.8
Mirror lines of symmetry.

58 Chapter Four

As noted earlier, two dimensions require two basis vectors (*a* and *b*) and some angle (γ) between them that allows complete tessellation of the plane. The individual plane figures that fill the space (analogous to the floor tiles) are called **unit cells**, and the basis vectors are the **unit cell axes**. Dividing space into unit cells provides a very useful simplification in understanding crystals. Because all the symmetry elements and details of the atomic arrangements must be the same within every unit cell, a complete description of a unit cell provides a complete description of the crystal.[10]

The unit cells illustrated as numbers 1 and 2 in figure 4.11 have one lattice point at each of their four corners, and each of these lattice points is shared by the three adjacent unit cells. This type of unit cell is considered **primitive**, as each corner contributes 1/4 of a

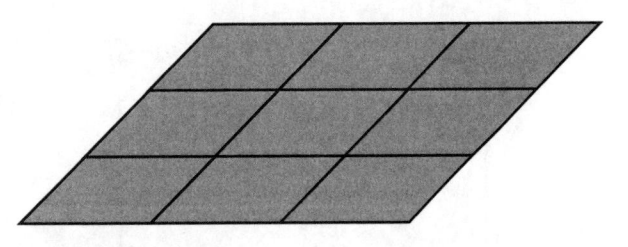

(a) Parallelograms.

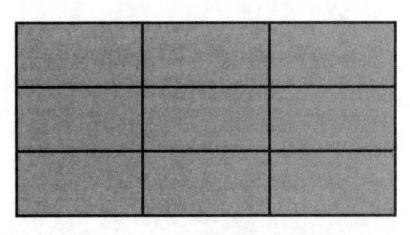

(b) Rectangles.

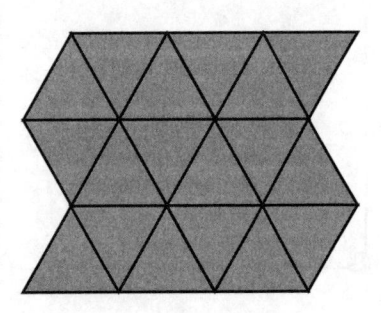

(c) Triangles.

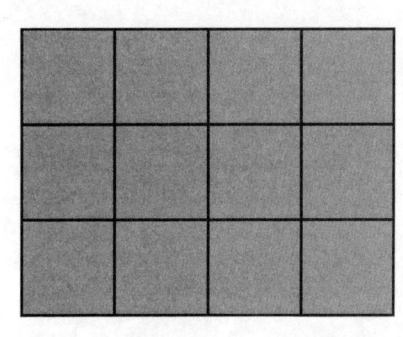

(d) Squares.

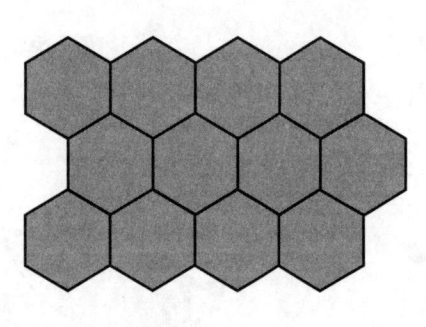

(e) Hexagons.

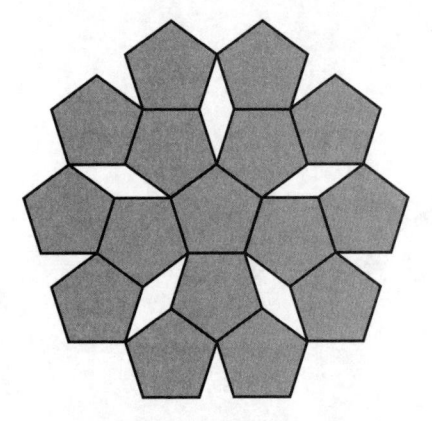

(f) Pentagons (nor any other shape) by themselves cannot tessellate two-dimensional space.

Figure 4.9
Tessellating two-dimensional space.

lattice point to the cell, adding up to one lattice point per cell. It is always possible to select a nonprimitive cell (figure 4.11, cell 3 and figure 4.12) to describe the symmetry, but in most cases, the choice of a primitive unit cell is simpler. An exception is made to this rule if the nonprimitive unit cell will more clearly illustrate the overall lattice symmetry. Unit cells that have more than one lattice point are described as **centered**.

With a variety of different types of unit cells, it is useful to group them into **crystal systems** (table 4.1).

Definition of the systems is accomplished by allowing the unit cell vectors a and b to be equal or unequal by turns and setting the angle γ between them to the different permissible values. This is directly analogous to classifying floor tiles by their shapes, and the result is that for two-dimensional space, there are five types of

> **Problem 4.3**
>
> In figure 4.12, the hexagonal crystal system is shown as having a primitive rhombic cell. Could a hexagonal cell also be chosen? Would this cell also be primitive? Illustrate.

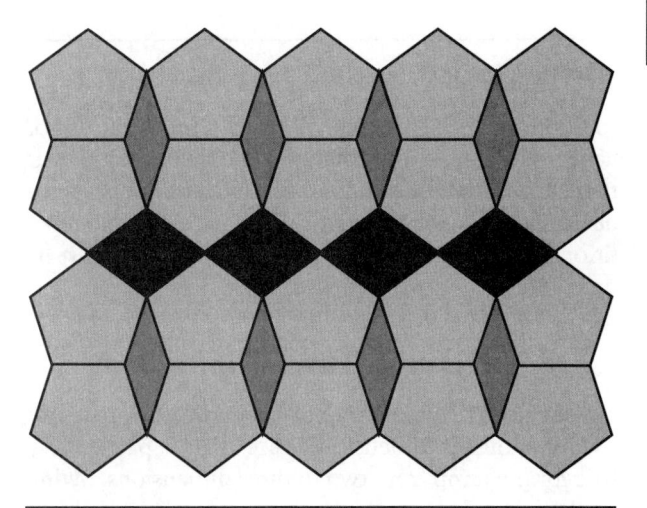

Figure 4.10
Two-dimensional space can be tessellated with multiple shapes of tile, as in this example with regular pentagons and two different types of regular parallelograms.

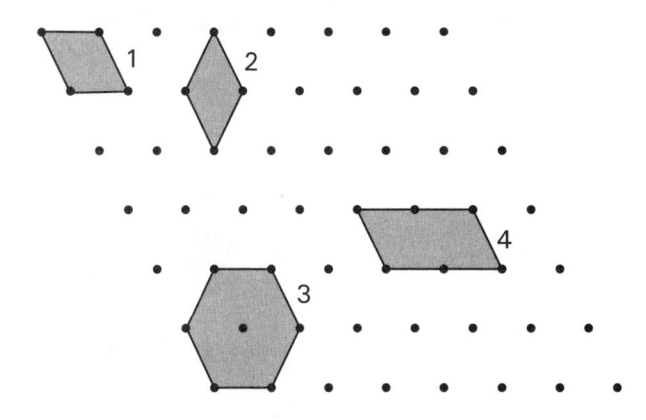

Figure 4.11
A variety of cells can be chosen for any given lattice. Unit cells 1 and 2 are primitive and cells 3 and 4 are nonprimitive. Primitive cells always enclose a smaller area than nonprimitive cells.

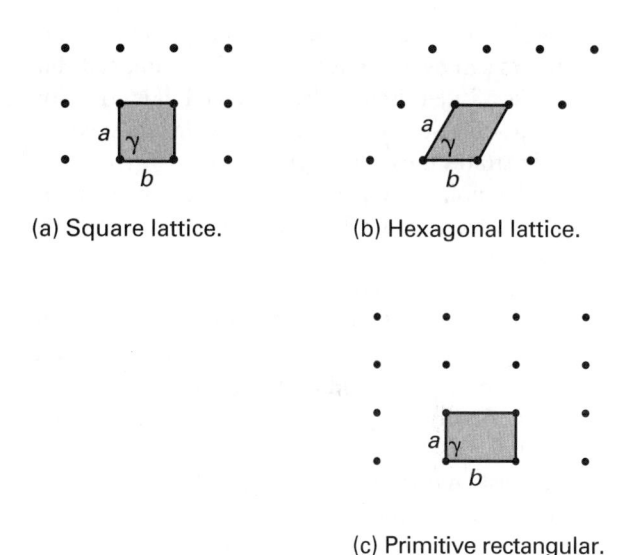

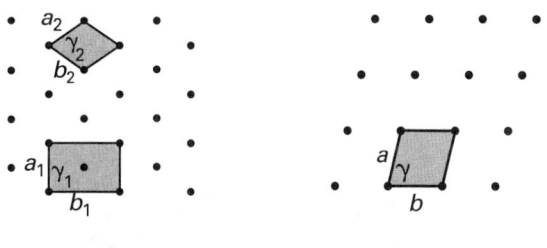

(a) Square lattice.

(b) Hexagonal lattice.

(c) Primitive rectangular.

(d) Centered rectangular cell and primitive cell alternative.

(e) Oblique.

Figure 4.12
Constraints on the types of lattice in two dimensions with examples of each. For the centered rectangular cell, a primitive diamond cell is also shown. The centered cell is preferred, as it more clearly illustrates the lattice symmetry.
(*Sources:* Modified from Putnis, 1992; Klein, 2002; and Perkins, 2002.)

Table 4.1 Constraints on the Types of Lattice in Two Dimensions Corresponding to Illustrations in Figure 4.12

	Crystal system	Vector constraints	Angle constraints	Unit cell shape and type
a	Square	a = b	γ = 90°	Square (primitive)
b	Hexagonal	a = b	γ = 60°*	Rhombic (primitive)
c	Rectangular	a ≠ b	γ = 90°	Rectangular (primitive)
d	Rectangular	a ≠ b	γ = 90°	Rectangular (centered)
		a = b	γ ≠ 60°, 90° or 120°	Diamond (primitive)
e	Oblique	a ≠ b	γ ≠ 60°, 90° or 120°	Parallelogram (primitive)

≠ indicates that equivalence is not required, but may occur by chance.

* In this example, the acute angle (60°) has been chosen for γ. By redefining the origin, γ can be set as the obtuse angle (120°).

Source: Modified from Putnis (1992), Klein (2002), and Perkins (2002).

plane lattice (figure 4.12) in four crystal systems. All of the lattices can be generated by primitive unit cells but in the case of the primitive diamond cell, it is common instead to choose a centered rectangular cell, which better illustrates the overall symmetry.

At this point, the possible combinations of point symmetry elements can be enumerated and the plane point groups listed. The simplest are the six single element groups, five composed just of one rotation axis (1, 2, 3, 4, and 6) and the sixth (m) of just a mirror. By combining four of the five rotation groups with the mirror element, four additional groups (2mm, 3m, 4mm, and 6mm) are generated (figure 4.13), resulting in a total of ten.

The last step is to consider how these ten groups can combine with the five plane lattice types to produce the plane groups. Not every plane point group can match up with every lattice type; the square lattice is incompatible with any group that contains a 6-fold rotation, and the hexagonal lattice similarly cannot accommodate a 4-fold rotation. The final result is that there are 17 unique configurations (figure 4.14). It can be demonstrated mathematically that all possible repetitions of an object in two-dimensional space must be one of these 17 possibilities.[11]

Symmetry in Three Dimensions

Features of symmetry of a lower dimension can be used in a higher dimensional case. This is particularly true for the jump from two to three dimensions, owing to the need for artistic perspective to display three-dimensional features on a two-dimensional page.

As noted previously, when a new dimension is added, another basis vector is needed. The third vector required for three-dimensional space (*c*) extends away from the *ab* plane and thereby requires two more angles (α and β) to define a **space lattice** (figure

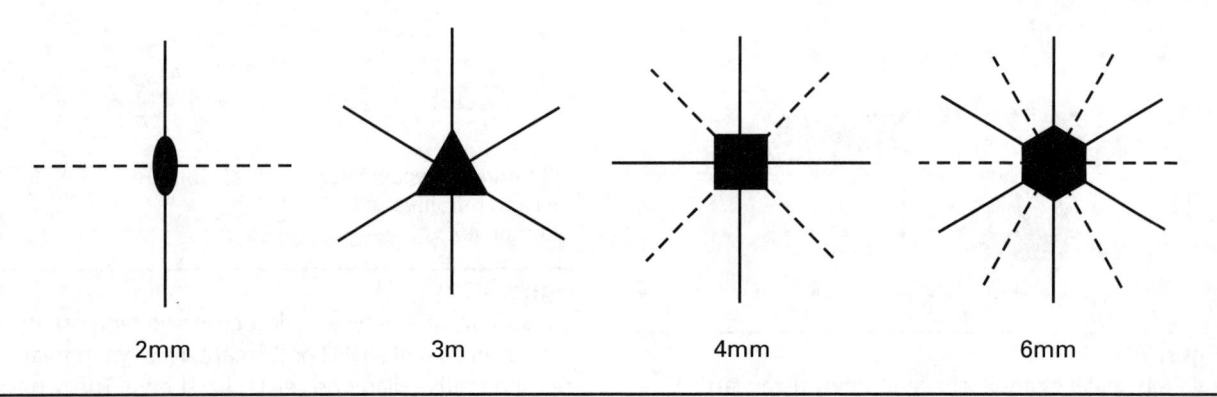

Figure 4.13
Combining mirror lines (solid) with the possible rotations around rotation points generates new plane point groups. Some of these combinations produce additional mirror lines (dashed). See text for additional discussion.

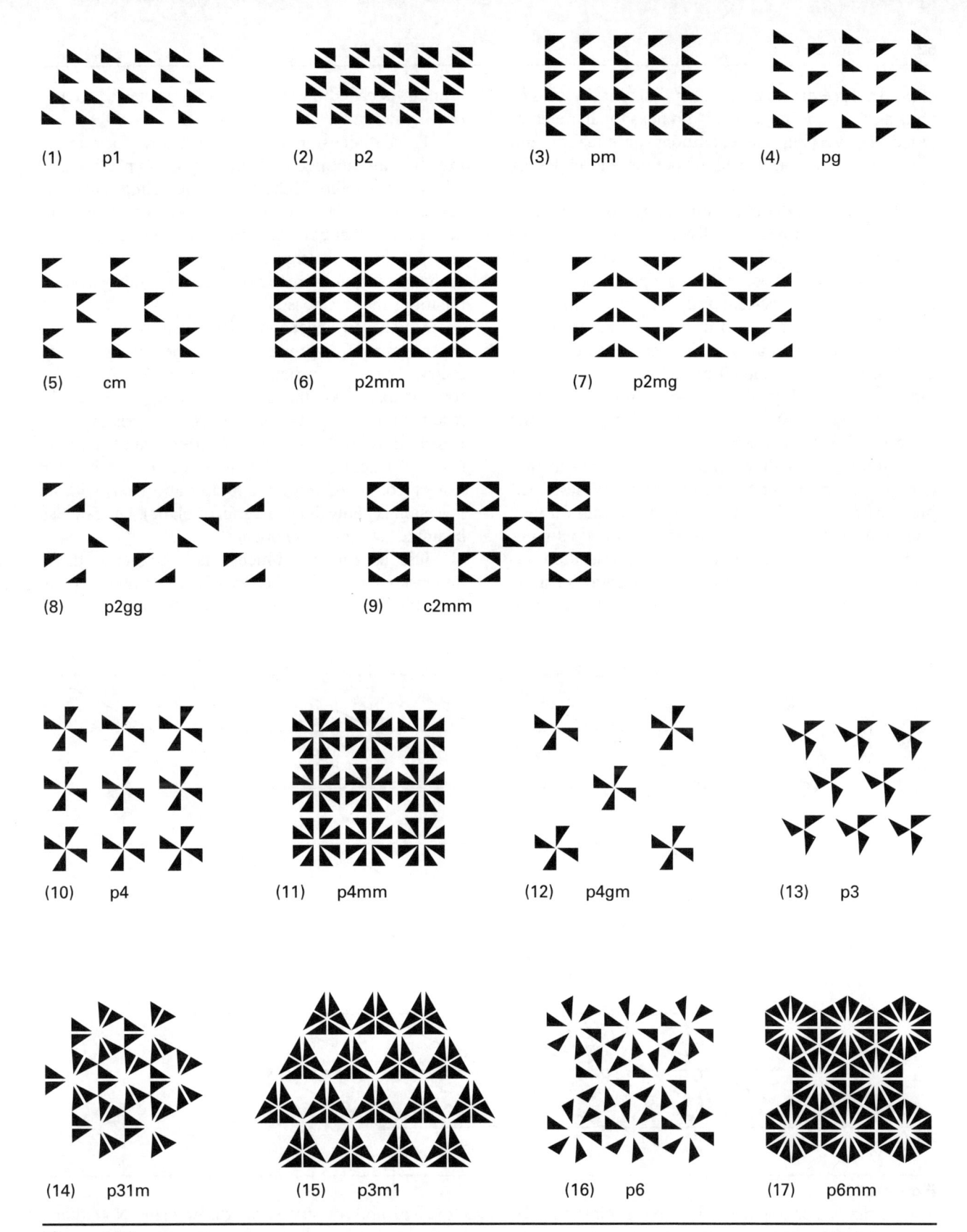

(1) p1
(2) p2
(3) pm
(4) pg
(5) cm
(6) p2mm
(7) p2mg
(8) p2gg
(9) c2mm
(10) p4
(11) p4mm
(12) p4gm
(13) p3
(14) p31m
(15) p3m1
(16) p6
(17) p6mm

Figure 4.14
The 17 plane groups. The numbering and shorthand symbols for each are those assigned in the *International Tables for Crystallography*.

4.15). In three dimensions, seven different crystal systems arise (six, if the trigonal and hexagonal are combined) by varying the conditions of equality and inequality for *a*, *b*, and *c* and the permissible angles for α, β, and γ.

The additional dimension allows for additional types of nonprimitive unit cells. Unit cells with lattice points in two opposing faces are called **end-centered** (or side-centered), those with lattice points in all faces are referred to as **face-centered**, and those with a lattice point entirely enclosed in the unit cell are **body-centered**. The total number of lattices, both primitive and centered, is 14, and all crystalline substances fall into one or another of the seven symmetry classes (table 4.2 on p. 64), and have one of the 14 Bravais lattices[12] (figure 4.16).

Addition of a third dimension also means that changes in the point symmetry elements described previously are required, as well as inclusion of a new one (figure 4.17 on p. 64). A mirror line in two dimensions becomes a **mirror plane** in three, and the rotation point becomes a rotation line or **rotation axis**. The new symmetry element, which operates through a point, is called an **inversion**, often referred to as a *center of symmetry*.

Three-dimensional space also allows for rotation axes that are not perpendicular to the *ab* plane, along with the possibility of these axes interacting. There are restrictions on how rotation axes can combine. One limitation is that two rotation axes always combine to create a third axis (figure 4.18 on p. 64). The second limitation is that there are only certain polyaxial combinations that are possible.[13]

Finally, an inversion center can combine with other symmetry elements to create new types of symmetry. As shown in figure 4.19 (on p. 65), if a 4-fold rotation axis is combined with an inversion center, the result is a new type of rotation axis, an axis of *rotoinversion*. Rotoinversion axes may be equivalent to other symmetry elements (a 2-fold axis of rotoinversion that results from combining a 2-fold rotation axis with an inversion is equivalent to a mirror plane), but can also be unique symmetry elements.

Just as for two-dimensional symmetry, three-dimensional point symmetry elements combine to produce additional symmetry elements. When all of these

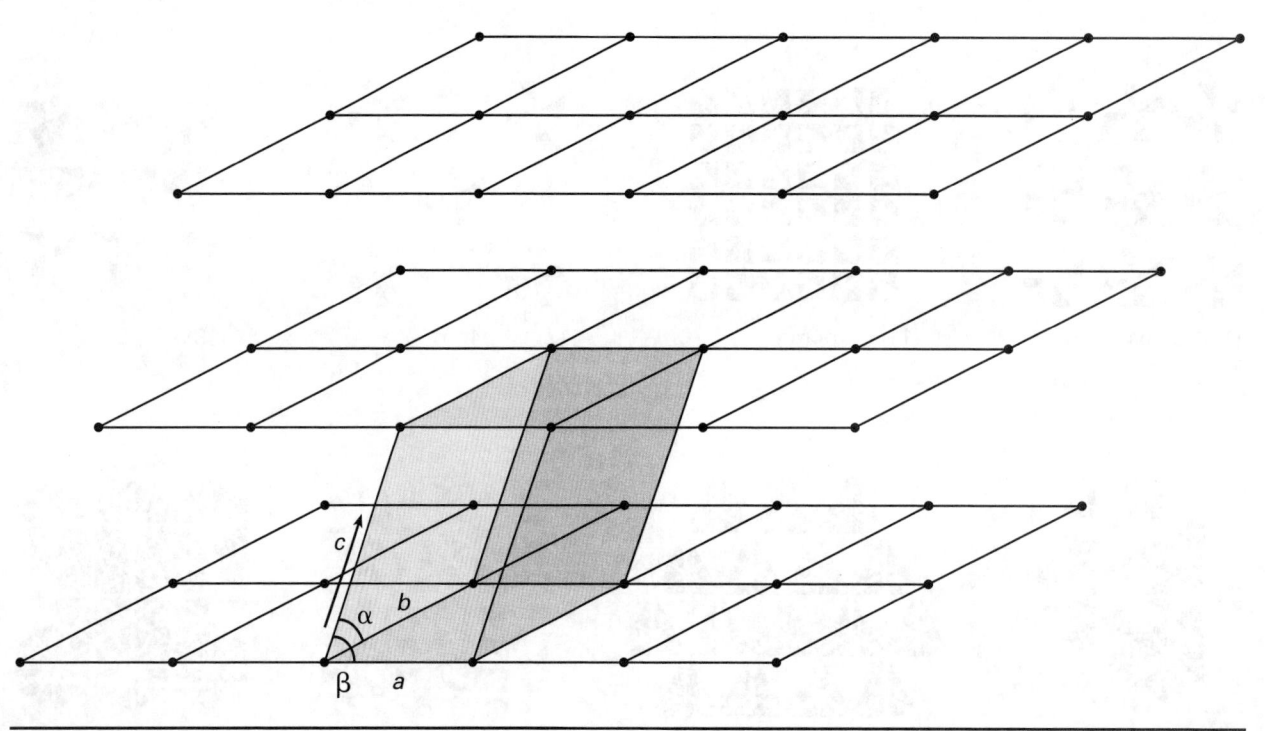

Figure 4.15
A three-dimensional space lattice. The required third basis vector (*c*) extends outward from the *ab* plane. Note that adding a third vector requires that two additional angles be defined: α, the angle between *b* and *c*; and β, the angle between *a* and *c*.

Crystallization and Crystallography 63

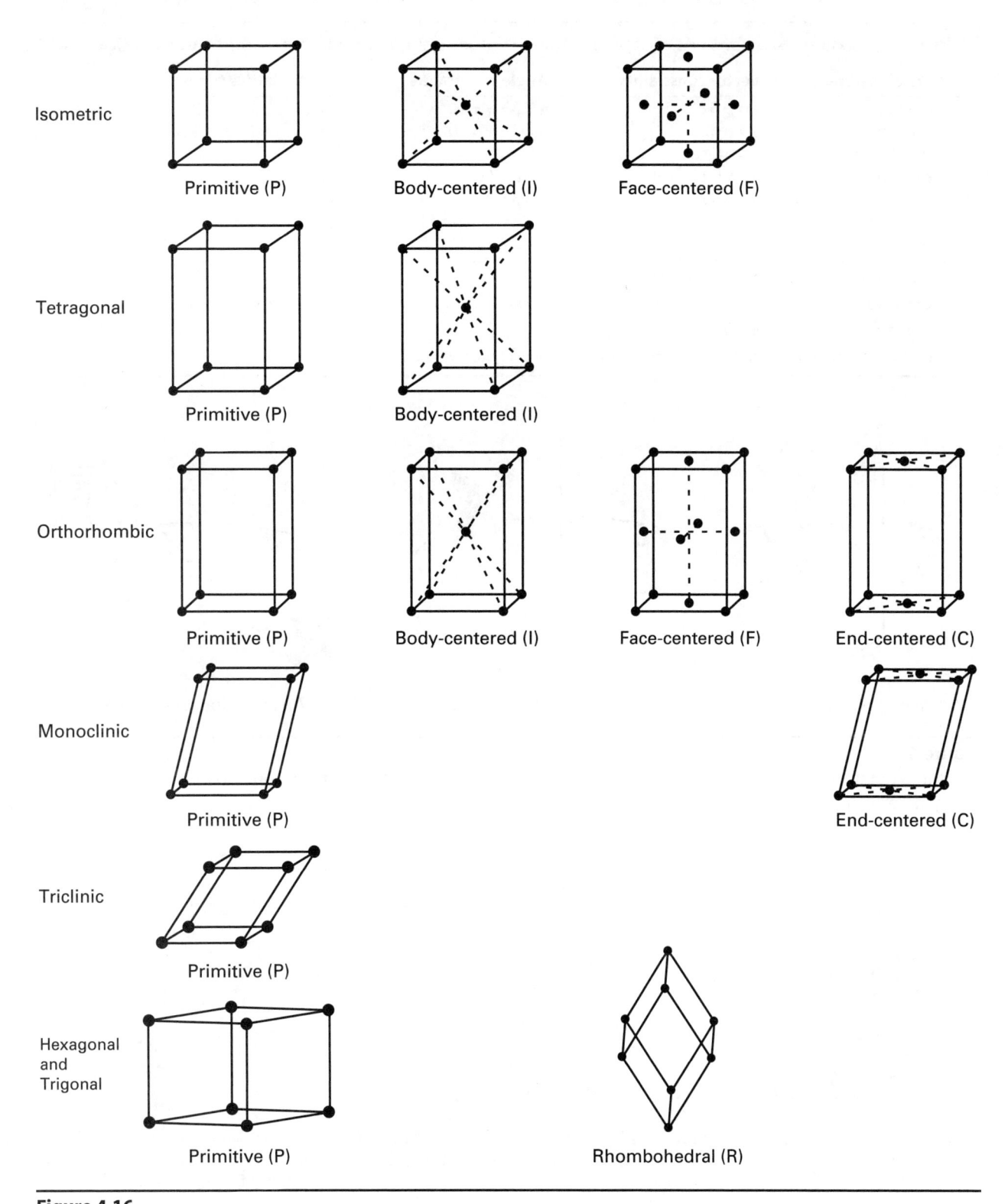

Figure 4.16
The 14 Bravais lattices, including end-centered (or side-centered), face-centered, and body-centered lattices.

Table 4.2 Constraints on the Type of Lattice in Three Dimensions Corresponding to Illustrations in Figure 4.16

	Crystal system	Vector constraints	Angle constraints	Bravais lattice types[1]
1	Isometric	$a = b = c$	$\alpha = \beta = \gamma = 90°$	P, F, I
2	Tetragonal	$a = b \neq c$	$\alpha = \beta = \gamma = 90°$	P, I
3	Orthorhombic	$a \neq b \neq c$	$\alpha = \beta = \gamma = 90°$	P, F, I, C[2]
4	Hexagonal	$a = b \neq c$	$\alpha = \beta = 90°, \gamma = 60°$	P
5	Trigonal[3]	$a = b = c$	$\alpha = \beta = \gamma \neq 90°$	R
6	Monoclinic	$a = b \neq c$	$\alpha = \gamma = 0° \neq \beta > 90°$	P, C
7	Triclinic	$a \neq b \neq c$	$\alpha \neq \beta \neq \gamma \neq 90°$	P

$\neq$ indicates that equivalence is not required, but may occur by chance.

[1] Lattice types are designated by capital letters: P, primitive; F, face-centered; I, body-centered; C, end-centered; R, rhombic.

[2] The most commonly chosen end-centering.

[3] The trigonal system can be considered as a subclass of the hexagonal system.

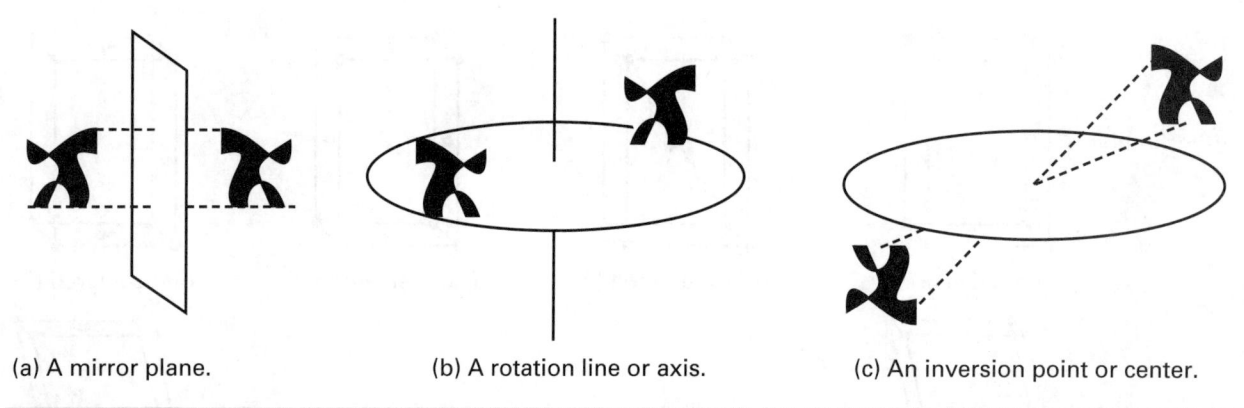

(a) A mirror plane. (b) A rotation line or axis. (c) An inversion point or center.

Figure 4.17
The three point symmetry operations in three-dimensional space. In each case at least one point is invariant under the symmetry operation.

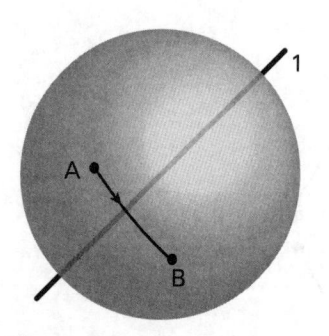

(a) Rotation from point A to point B occurs about the axis labeled 1.

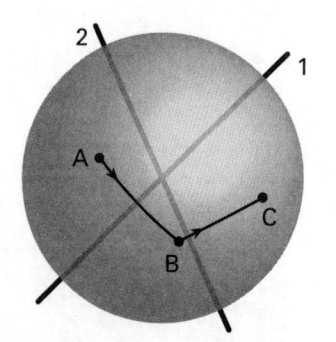

(b) Rotation from point B to point C occurs about the axis labeled 2.

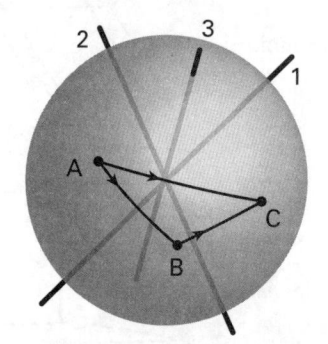

(c) This is identical to a rotation from point A to point C about a suitably located axis 3.

Figure 4.18
Rotations in 3D.

possible combinations are exhausted, the results are the 32 point groups (table 4.3). These represent all of the ways that an object can be repeated (and atoms or molecules arranged) in three dimensions, while leaving at least one point invariant. Combining these point groups with the 14 Bravais lattices results in the 230 space groups—the total number of ways in which an object can be repeated in three-dimensional space. Few of the typical methods for identifying minerals in hand specimens can distinguish between different space groups that are **isogonal** (have corresponding symmetry elements) with a particular point group. The symmetry classes and point symmetry are usually sufficient for hand-specimen identifications and descriptions, but more sophisticated analyses require differentiation among space groups.

CRYSTAL FEATURES

When examining minerals in hand specimens or under magnification using a petrographic microscope, the significance of the underlying symmetrical arrangement of atoms may be apparent. In hand specimens, some will display flat surfaces that are crystal faces, which developed as the mineral grew; in many cases, these faces are characteristic enough to provide a considerable aid in identification. Through the microscope, these crystal faces appear as straight boundaries outlining individual mineral grains. Some specimens in hand samples may show flat surfaces, the **cleavage faces** or **cleavage surfaces** that result from breaking of the mineral (e.g., color plates 10a; 18c). Cleavage faces are also characteristic features of certain minerals and they, too, reflect a long-range atomic order. (Cleavage will be discussed in greater detail in chapter 5.) In microscopic analyses, the relationships between lines representing both crystal faces or cleavage directions and the optical properties of the sample are diagnostic in many minerals.

Simply noting the presence of crystal or cleavage faces in hand specimens is less useful than being able to relate them in some quantitative way to the internal arrangement of the constituent atoms. The concepts of symmetry developed previously can provide such a link, because the basis vectors provide a convenient framework for describing lines and planes within crystals. As before, an examination of these topics in two dimensions will allow for easier conceptualizations that can be generalized to three dimensions.

Lines and Planes within Crystals: Miller Indices

Figure 4.20a illustrates a two-dimensional lattice with a point of origin in the center and the unit cell axes *a* and *b* extending in both positive and negative directions. From the origin, a separate vector extends to a lattice point three translations along the *a* axis and two translations along the *b* axis, or at the coordinates of (3, 2). A useful shorthand notation for vectors such as this encloses the components in square brackets without a comma: [32], or in general, [uv]. As can be

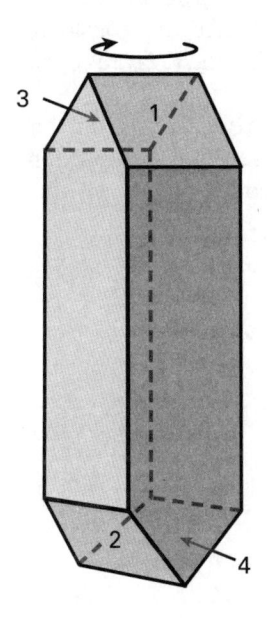

Figure 4.19
A crystal with a 4-fold rotoinversion axis. Each of the numbered faces is related to the next numbered face via a counterclockwise rotation of 90° followed by an inversion.

Table 4.3   The 32 Three-Dimensional Point Groups*

	Crystal system	Point Groups
1	Isometric	23, 2/m$\bar{3}$, 432, $\bar{4}$3m, 4/m, $\bar{3}$2/m[1]
2	Tetragonal	4, $\bar{4}$, 4/m, 422, 4mm, $\bar{4}$2m, 4/m 2/m 2/m
3	Orthorhombic	222, 2mm, 2/m 2/m 2/m
4	Hexagonal	6, $\bar{6}$, 6/m, 622, $\bar{6}$m2, 6/m 2/m 2/m
5	Trigonal[2]	3, $\bar{3}$, 32, 3m, $\bar{3}$2/m
6	Monoclinic	2, m, 2/m
7	Triclinic	1, $\bar{1}$

[1] Pronounced "two three," "two over m bar three," "four three two," "bar four three m," "four over m," "bar three two over m."

[2] The trigonal system can be considered as a subclass of the hexagonal system.

*The symbols used are part of the Hermann-Mauguin system of notation.

seen, there are several other vectors related to the first by a symmetry operation; the notation for *all these vectors taken as a group* is <32>, or more generally, <uv>. Some of these vectors extend in the direction of –*a* or –*b* or both. To distinguish negative from positive directions, a line is placed over the negative components (e.g., $\bar{2}$), and the number is pronounced "bar two."

This methodology can also be used for lines in a lattice, but with some modifications. Figure 4.20b is the same lattice with a line passing through a number of lattice points. This line intersects the *a*-axis three translations from the origin, and intersects the *b*-axis two translations from the origin. How should such a line be designated? The obvious choice would be to refer to the intercepts directly, e.g. (32),[14] but this can become clumsy in some circumstances. For example, consider a second line that intersects the *a*-axis two translations away from the origin and is parallel to the *b*-axis. If we were to refer to the intercepts, the result would be (2, ∞), because one of the precepts of Euclidian geometry is that parallel lines can only intersect at infinity. An alternative to dealing with infinities would be to "invert the intercepts," creating fractions that can be converted to whole numbers when any common factors are reduced. In this case the intercepts of 3 and 2 "invert" to become 1/3 and 1/2. These can be multiplied by 6 (becoming 6/3 and 6/2), which can be reduced to whole numbers (23). Using this procedure, the line parallel to *b* then becomes (10) (i.e., 1/2 and 1/∞ → 2/2 and 2/∞ → 1 and 0). These modified intercepts are referred to as Miller indices.[15] In the case of negative components, a line is placed over a negative Miller index and pronounced with the prefix "bar." Note that due to the clearing of common factors, all parallel lines in the lattice are considered equivalent, i.e., the lines (44), (33) and (22) are considered as equivalent to the line (11).

This terminology is readily extended into the third dimension. Vectors are given three indices, [uvw], and lines in two dimensions become planes in three

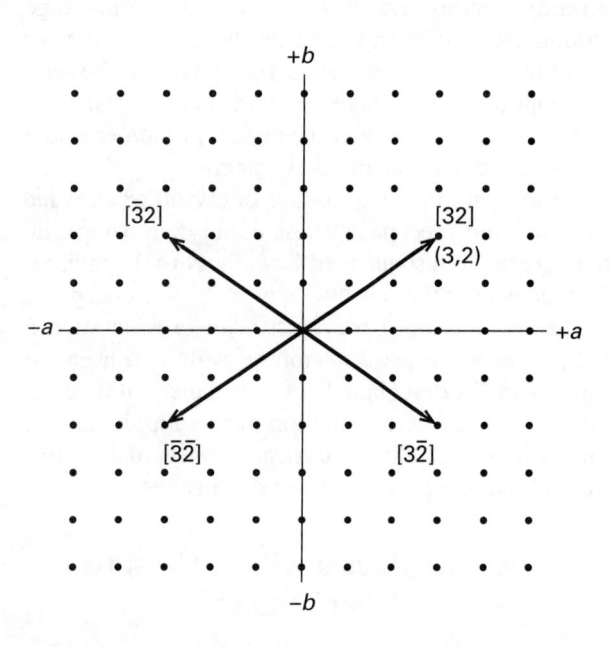

(a) A two-dimensional lattice with four symmetry-related vectors, showing lattice intercepts. The vectors are described as "three-two" rather than "thirty-two."

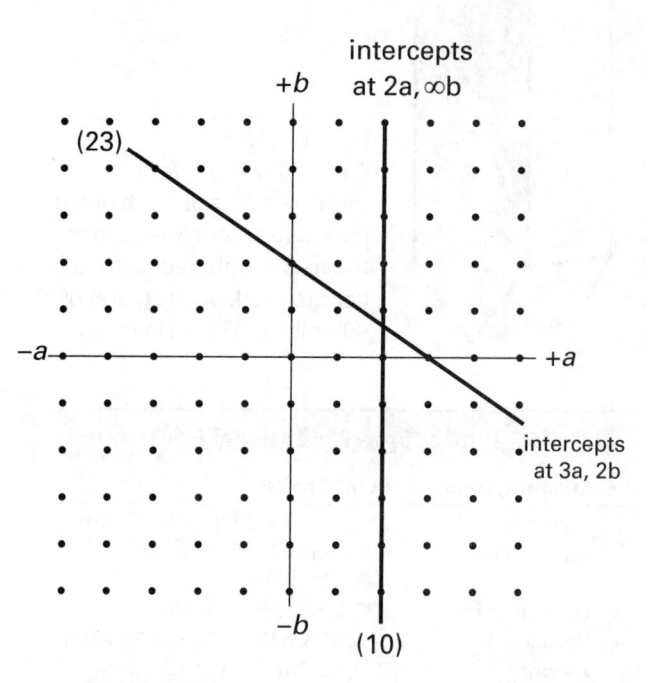

(b) Miller indices in a two-dimensional lattice with two lines illustrated, a line with intercepts of 2 on *b* and 3 on *a*, and a line with intercepts of ∞ on *b* and 2 on *a*. The Miller indices of these lines are (23) and (10), referred to as "two-three" and "one-oh." See text for explanation.

Figure 4.20
Lattice intercepts and Miller indices.

dimensions which are also described by three indices (hkl). This methodology is completely general and may be used with all symmetry classes; but for historical reasons, Miller indices for hexagonal crystals commonly have four parts (hkil). The fourth index, i, is derived from the first two by the relation i = – (h + k).

A simple example of Miller indices is provided in figure 4.21 by a labeled chromite crystal. In this case, each of the eight faces intersects each of the axes at one unit length, so that the indices all have numerical values of 111. The differences among the Miller indices of the faces in this case are only differences involving whether a particular face intersects the crystallographic axes in the positive or negative direction. In the latter case, "bar 1" is applied to the intercept. The figure shows the four faces on the front of the crystal labeled with Miller indices; labeling of the remaining four is left as an exercise for the student. Another simple example can be visualized for a face on a galena crystal—the crystal depicted in figure 4.2b. If the three crystallographic axes each extends from and is perpendicular to one of the crystal faces marked a, then the face marked b on the front right of the crystal would intersect the axes projecting to the right and to the front left at equal distances from the origin, but would not intersect the axis projecting up from the face a on the top of the crystal. Hence, the face b on the right middle would have a Miller index of (110). The face a on the top of the crystal, on the other hand, would intersect only the axis extending upward through this face and would have the Miller index of (001).

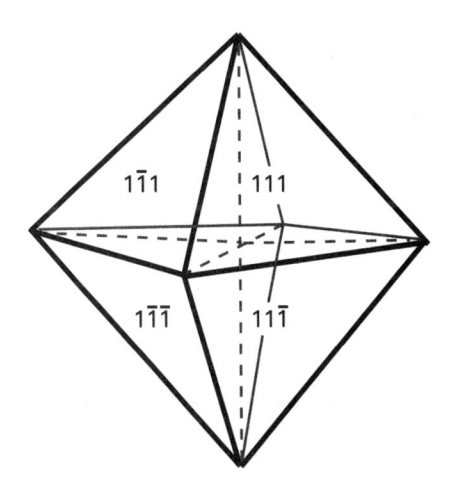

Figure 4.21
An octahedral crystal of chromite with Miller indices applied to the front four faces.

Problem 4.4
Determine the Miller indices for planes that intercept the a, b, and c axes at the following points:
(a) 1, 3, 2
(b) 3, 3, 3
(c) 2, ∞, 4
(d) 3, 1, 3

Problem 4.5
Consider an isometric mineral that crystallizes in cubes. Each of the cube faces is perpendicular to one of the unit cell axes, both positive and negative. What are the Miller indices for all six cube faces?

Problem 4.6
In figure 4.16, the nonprimitive orthorhombic lattices are body-centered, face-centered, and end-centered, whereas the nonprimitive isometric lattices are just body- and face-centered. Why is there no end-centered lattice?

Faces, Morphologies, and Forms of Crystals

The earliest well-known principle in the study of the morphology of crystals is the Law of Constancy of Interfacial Angles published in 1669 by Nicholas Steno. It formally states that *the angles between equivalent crystal faces on different specimens of the same material are identical*. The Law clearly implies a relationship between the internal symmetry and the external morphology of crystals.

If we examine crystals of various minerals, we can observe that well-formed crystals of many minerals generally have distinctive and specific shapes (morphologies) and faces (figure 4.22a). The distinctive external morphology commonly exhibited by a mineral is referred to as the **habit** of the mineral. (Habits are discussed further in the context of mineral identification in chapter 5.) The habit is descriptive and is not specifically related to the internal symmetry of a crystal. In some cases, these morphologies are simple and recognizably familiar; for example, cubic crystals, rectangular box-like forms, or eight-sided diamond

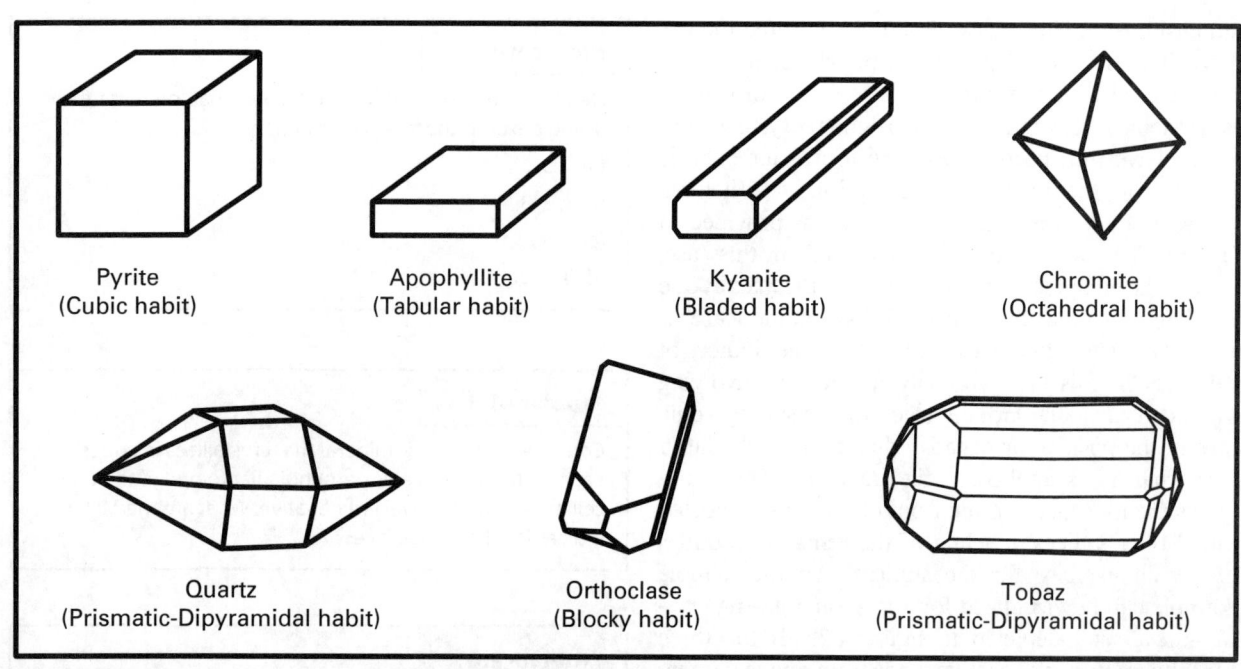

(a) Some habits common in a few selected minerals.

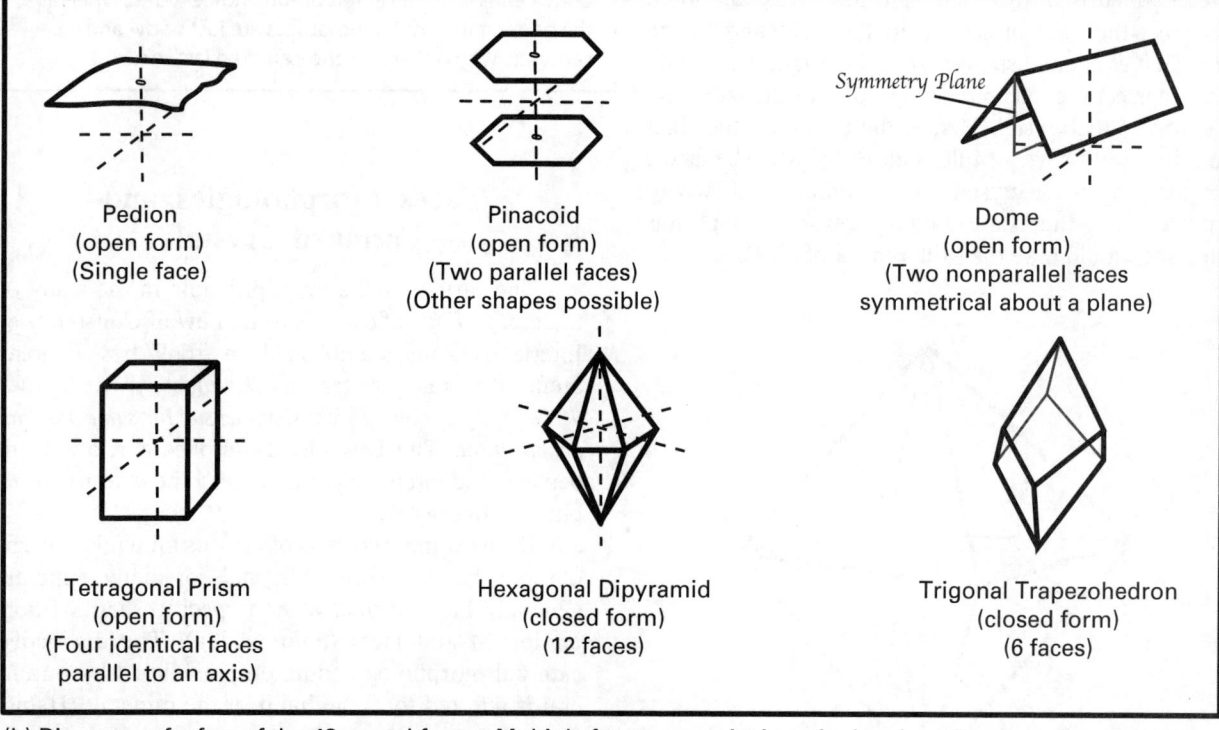

Pyrite (Cubic habit)
Apophyllite (Tabular habit)
Kyanite (Bladed habit)
Chromite (Octahedral habit)
Quartz (Prismatic-Dipyramidal habit)
Orthoclase (Blocky habit)
Topaz (Prismatic-Dipyramidal habit)

Pedion (open form) (Single face)
Pinacoid (open form) (Two parallel faces) (Other shapes possible)
Dome (open form) (Two nonparallel faces symmetrical about a plane)
Symmetry Plane

Tetragonal Prism (open form) (Four identical faces parallel to an axis)
Hexagonal Dipyramid (closed form) (12 faces)
Trigonal Trapezohedron (closed form) (6 faces)

(b) Diagrams of a few of the 48 crystal forms. Multiple forms can exist in a single mineral sample. For example, multiple prisms and dipyramids could be identified on the topaz crystal in (a).

Figure 4.22
Habits versus forms in minerals.
(*Source:* (b) based on and, in part, modified from Hurlbut, 1961.)

shaped crystals. In other cases, crystals can have unusual shapes or a bewildering array of crystal faces. In addition, particular faces may be distorted by the conditions of growth.

In crystallography, we want to describe more than the morphology when we describe crystals. One thing we are able to describe is referred to as the crystal form. The word "form" is used in crystallography in a special way: A **crystal form** can be one face, but is typically a group of essentially equivalent crystal faces, each of which has the same relationship to the symmetry elements as the others—that is, the faces are all related by a symmetry operation.[16] Thus in contrast to the habit, the *forms* displayed by a crystal can be identified and related specifically to internal symmetry. Some common forms are shown in figure 4.22b, including forms such as domes, prisms, and dipyramids, and the relationship of these forms to the crystallographic axis is also shown. The relationship of the faces to symmetry elements is illustrated by reference to the crystal in figure 4.23. In the crystal, each of the rectangular and equivalent crystal faces on the sides of the crystal has the same relationship to a rotation axis and is related to its neighbors by a rotation of 60°. The individual faces, of course, are designated using the Miller indices described above, and in this case, for a crystal of the hexagonal system, the crystal face designation has the form (hkil) (if we consider that the crystallographic axes go from edge to edge between faces of the crystal, the front face on the crystal would be designated [10$\bar{1}$0]). Together, the six essentially equivalent faces (the hexagonal prism faces) constitute a form and the method of notation of that form is one in which the indices are listed in brackets; for example, {hkl} or {hkil}. The specific form that includes all the vertical faces of the crystal in figure 4.23 is {10$\bar{1}$0}. In this case, this crystallographic designation indicates both the group of faces and their symmetrical relationships.

Crystals that belong to higher symmetry classes produce forms that have more faces than do forms occurring on crystals of lower symmetry classes. The number of faces belonging to a particular form is called the *multiplicity factor* of the form.[17] A crystal of low symmetry (triclinic or monoclinic, for example) may show forms with a multiplicity factor of one.

Figure 4.23
The six faces on the sides of this beryl crystal are all related by the symmetry operations of a 6-fold rotation and therefore all belong to the form {10$\bar{1}$0}.

Because a single plane cannot enclose any region of space, such forms are referred to as **open** (figure 4.22b). The hexagonal prism of the crystal in figure 4.23 is open. In contrast, crystal forms that enclose space are referred to as **closed** forms (figure 4.22b). The six faces of a cubic crystal (figure 4.2a), which has high symmetry (and is isometric), enclose space and define a closed form.

A number of minerals display crystal forms that are characteristic enough to allow easy identification. Pyrite, for example, occurs frequently enough in crystals with 12 faces of the form {210} that the crystals are called pyritohedra (figure 4.24b; color plate 17a). Other forms that can be diagnostic include the dodecahedral {110} form of garnets (figure 4.24a; color plate 14e) and the scalenohedral {211} form of calcite (figure 4.24c).

Cleavages may also have characteristic forms. Calcite displays rhombohedral {10$\bar{1}$1} cleavage (figure 4.25a), fluorite shows octahedral {111} cleavage (figure 4.25b), and both halite and galena have cubic cleavages of form {100} (figure 4.25c; color plates 17b; 19a).

(a) Garnet showing the {110} and {211} forms.

(b) Pyrite showing the {210} form.

(c) Calcite showing the {211} form.

Figure 4.24
Minerals with characteristic crystal forms.

(a) Calcite cleavages have a {10$\bar{1}$1} form.

(b) Fluorite cleavages have a {111} form.

(c) Galena and halite have a {100} form.

Figure 4.25
Minerals with characteristic cleavage forms.

Summary

Crystallography is the study of the atomic arrangements and the resulting crystal forms that develop in crystalline solids. Atoms in crystals are arranged in symmetrical arrays to produce lattices. Symmetry, the repetition of elements of a material or image, is of two types, point symmetry and translational symmetry. Translational symmetry describes objects duplicated at a distance from an origin, whereas point symmetry involves duplication involving at least one duplicate point. Point symmetry involves reflection across a mirror plane, rotation about an axis, or inversion through a point, or a combination of rotation and inversion. Point and translational symmetries yield arrangements of atoms (structures) that are grouped in point groups and space groups, respectively. Individual arrangements of atoms that define a filled space are called unit cells, the smallest arrangement of atoms that can define the structure of a particular crystal.

Crystals commonly have flat surfaces—the faces—produced by growth of the crystal. Faces are bounded by edges, which end in apices. Cleavage surfaces (or cleavage faces), flat surfaces produced by breaking, are similar in appearance to crystal faces but are a result of repeated planar zones of weakness within crystals rather than arrangements that bound the outer surfaces of crystals. Special directions related to the internal lattices of crystals are called crystal axes. The intercepts of crystal faces on crystal axes can be assigned numbers called Miller indices that define the three-dimensional position and orientation of faces in relation to those axes. Crystal forms are groups of crystal faces that are related via symmetry operations. Closed forms enclose space (such as a cube), whereas open forms do not fully enclose space (e.g., an open ended rectangle called a prism).

Selected References

Boisen, M. B., and Gibbs, G. V. 1990. *Mathematical Crystallography* (revised), in *Reviews in Mineralogy* 15, Mineralogical Society of America, Washington, DC. 460 p.

Buerger, M. J. 1978. *Elementary Crystallography: An Introduction to the Fundamental Geometrical Features of Crystals* (revised). MIT Press, Cambridge, MA. 528 p.

Cullity, B. D. 1967. *Elements of X-Ray Diffraction*. Addison-Wesley, Reading, MA. 514 p.

Dove, M. 2003. *Structure and Dynamics: An Atomic View of Materials*. Oxford University Press, Oxford, UK. 334 p.

Hammond, C. 1997. *The Basics of Crystallography and Diffraction*. International Union of Crystallography, Oxford University Press, Oxford, UK. 249 p.

Klein, C. K. 2002. *Mineral Science* (22nd ed.). John Wiley, New York. 641 p.

Perkins, D. 2002. *Mineralogy* (2nd ed.). Prentice-Hall, Upper Saddle River, NJ. 483 p.

Putnis, A. 1992. *Introduction to Mineral Sciences*. Cambridge Press, Cambridge, UK. 457 p.

Notes

1. There is no fixed distance that distinguishes long-range order from short-range order or even disorder. Ordering on the scale of 2Å is decidedly disordered and ordering on the scale of 1000Å is certainly long-range in crystals. See Cullity (1967).
2. Dove (2003).
3. Note that 18 grams of water, which is 18 cm^3 at 25°C, represents one mole of water. This contains 6.023×10^{23} water molecules.
4. In this context, the word "nucleus" has nothing to do with the central mass within an atom consisting of protons and neutrons.
5. Pyrrhotite, for example, contains both ferrous and ferric iron and has a formula of $Fe_{1-x}S$.
6. Donnay and Harker (1937) have shown that a more general formulation of the Law of Bravais must consider aspects of space group symmetry (Nesse, 2000).
7. Perkins (2002).
8. It is common for the term lattice to be misused, for example, in referring to a "crystal lattice of atoms." Strictly speaking, a lattice can only be an array of points (Boisen and Gibbs, 1990).
9. Boisen and Gibbs (1990).
10. This makes the assumption that the crystal is perfect with no defects in all directions and that the atoms are absolutely stationary, neither of which is true for real crystals. See Cullity (1967).
11. Buerger (1978), Hammond (1997).
12. Named for the crystallographer Auguste Bravais who first demonstrated the 14 unique lattices (Hammond, 1997).
13. A simple demonstration using spherical trigonometry can be found in Buerger (1978, pp. 36–45), and a more rigorous proof using formal group theory is in chapter 5 of Boisen and Gibbs (1990).
14. The notation for lines in a two-dimensional lattice encloses the components in parentheses and the generalized form of this is (hk). A group of symmetry related lines is denoted as {hk}. Bloss (1971).
15. After British crystallographer William Miller, who popularized them. Hammond (1997).
16. Bloss (1971).
17. Bloss (1971).

Minerals
Their Classification and Identification in Hand Specimens

INTRODUCTION

Mineralogy, the branch of geology devoted to the study of minerals, is closely allied with other areas of geology, such as geochemistry (the study of the chemical composition and associated reactions in rocks, minerals, and other Earth materials) and petrology (the study of rocks). Minerals are naturally occurring chemical compounds, so many aspects of the chemistry of Earth necessarily require an understanding of minerals. Inasmuch as rocks are made up of minerals, an understanding of the formation, properties, and classification of rocks are all dependent on knowledge of mineral content.

Numerous analytical techniques have been developed to study minerals, but the traditional core of mineralogy is the examination of mineral samples that can be held in one's hand (i.e., hand specimens). Hand-specimen identification is particularly useful in routine field studies. For example, in mapping sandstones in the Coast Ranges of California, rock units can sometimes be quickly distinguished by determining whether veins in the rocks are composed of quartz or calcite. Chapter 6 discusses advanced techniques used in more detailed, analytical evaluations of minerals. Such advanced methods are essential to a variety of research areas, including petrologic studies, mineral and ore formation studies, and structural studies.

Although identification of minerals in hand specimens usually seems like a daunting task to beginners, in fact, numerous common minerals can be identified readily by making use of a few easily observed physical properties. Once these minerals are familiar, recognition of their occurrence in rocks, generally in smaller grains, allows for the identification of those rocks in hand specimens. To understand why minerals have such distinct and characteristic properties, we first look at how minerals are classified.

MINERAL CLASSIFICATION

The definition of a mineral was given at the beginning of chapter 1: a **mineral** is considered to be an element or chemical compound that, *in general*, but with some exceptions, is a solid, inorganic, naturally occurring, and crystalline material produced by geological processes. Given this definition, how many materials meet these criteria? As of early 2014, there were roughly 4900 named minerals, enough to fill a book devoted solely to listing mineral names, formulas, and symmetry classes (Back and Mandarino, 2008). This book must be updated periodically to include the new minerals being discovered and named each year. There is even a book, for reference, that lists discredited, inappropriate, and obsolete mineral names (Bayliss, 2000). With so many different mineral possibilities, the need for a framework for discussing minerals systematically is obvious, something broadly analogous to the Linnaean system of classifying living organisms.

The simplest means of classification is to categorize minerals by their fundamental characteristics: what atoms they contain, and how these atoms are arranged (figure 5.1). Two minerals with the same atoms in the same arrangement must necessarily be the same, and two minerals with different atoms in different arrangements are unique and different. The term **isostructural** is used to describe minerals that share an atomic arrangement but are composed of different kinds of atoms. In contrast, two minerals with the same atoms, but in different arrangements, are referred to as **polymorphs**. Although based on fundamental characteristics, two problems arise from this type of classification: (1) The crystal structures of all minerals must be determined before they can be classified, and (2) a very wide variety of minerals fit into the unique category, making the scheme only slightly more helpful than no scheme at all. So why discuss such a frame-

Chapter Five

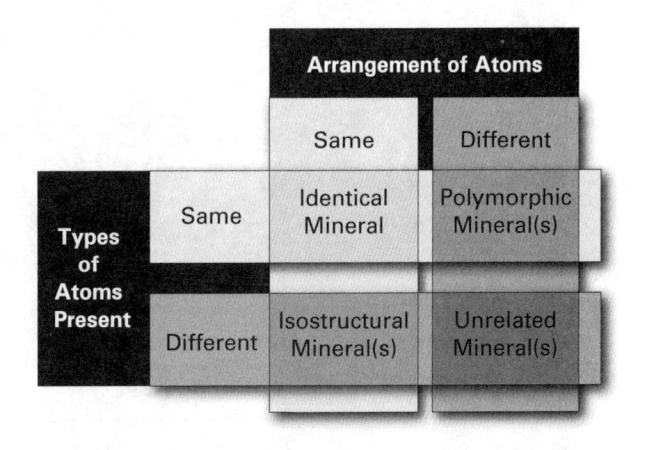

Figure 5.1
Classification of minerals by types and arrangements of atoms.

work? A discussion is useful, because isostructural and polymorphic characteristics are important.

Compare two pairs of minerals, one pair isostructural and the other polymorphic. Halite (NaCl) and galena (PbS) (figure 5.2a) represent an isostructural pair, sharing the same arrangement of atoms, with each type of atom alternating with the other type in a three-dimensional checkerboard pattern. Halite is transparent and lightweight, whereas galena is opaque and heavy, so at first glance they seem to share no similarities. As shown in figure 4.25c, however, both minerals break in an identical manner, strongly suggesting this particular property is controlled by the atomic arrangement. Calcite and aragonite (both $CaCO_3$), in contrast, have very dissimilar atomic arrangements

Problem 5.1

The International Mineralogical Association Commission on New Minerals, Nomenclature and Classification (IMA-CNMNC) recognized 4859 unique mineral species as of early 2014. (a) Locate the website for the commission and determine how many proposals for new minerals were approved in the last year. (b) On average, how many new minerals are approved every year?

Problem 5.2

Why can't two minerals be isostructural and polymorphic at the same time?

(figure 5.2b; color plate 1c), and once again, they seem to share few similarities, as they break differently, their crystal shapes differ (figure 5.2c), and they have different densities and hardnesses. A drop of dilute acid on each, however, will produce an immediate reaction, releasing CO_2 as the acid bubbles and the mineral dissolves. In this case, the chemical similarity results in a useful property. These two examples suggest that it is the combination of atomic arrangements and atomic contents that give minerals the unique macroscopic properties that permit their identification.

A more traditional approach to mineral classification derives from systematic attempts at a chemically based classification. The Swedish chemist Berzelius developed such a classification, which was later modified by the American mineralogist Dana (Gaines et al., 1997) (table 5.1). In this classification approach, minerals are first grouped by their principal anion or polyanionic unit, such as SO_4 or CO_3, and then are classified structurally into subgroups. The chemical data alone ensure that a mineral is assigned to a proper category, but this scheme also requires structural information for complete classification. Since virtually all but the most obscure and rare minerals have had their crystal structures determined, complete classification is generally possible.

The structural details of individual mineral groups are of interest mostly to specialists and will not be described here. For laboratory work in which such information is needed, books such as Mottana et al.

Table 5.1 Mineral Classification Based upon Atomic Arrangements and Content

Native Elements

Sulfides & Sulfosalts
Oxides & Hydroxides
Halides
"ates": Carbonates, Sulfates, Phosphates, Arsenates, Nitrates, etc.
Silicates
    Nesosilicates
    Sorosilicates
    Cyclosilicates
    Inosilicates
        Single chain
        Double chain
        Other (triple, mixed)
    Phyllosilicates
    Tectosilicates

(*Sources:* Gaines et al., 1997, after Berzelius, 1818, and Dana, 1837.)

Minerals **75**

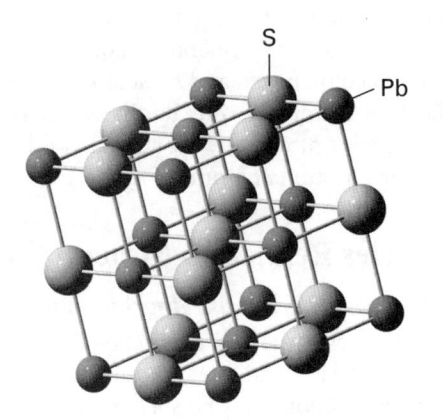

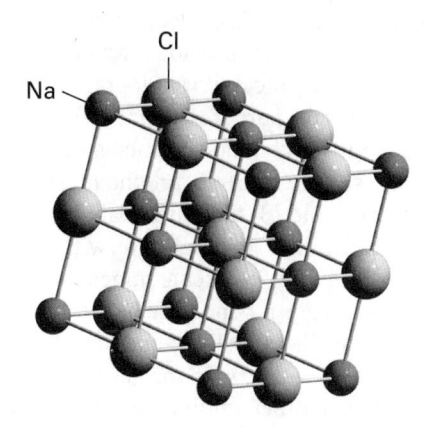

(a) The lead (Pb) and sulfur (S) atoms in the crystal structures of galena (left) and the sodium (Na) and chlorine (Cl) atoms in the crystal structure of halite (right) are arranged in an identical fashion, making them isostructural.

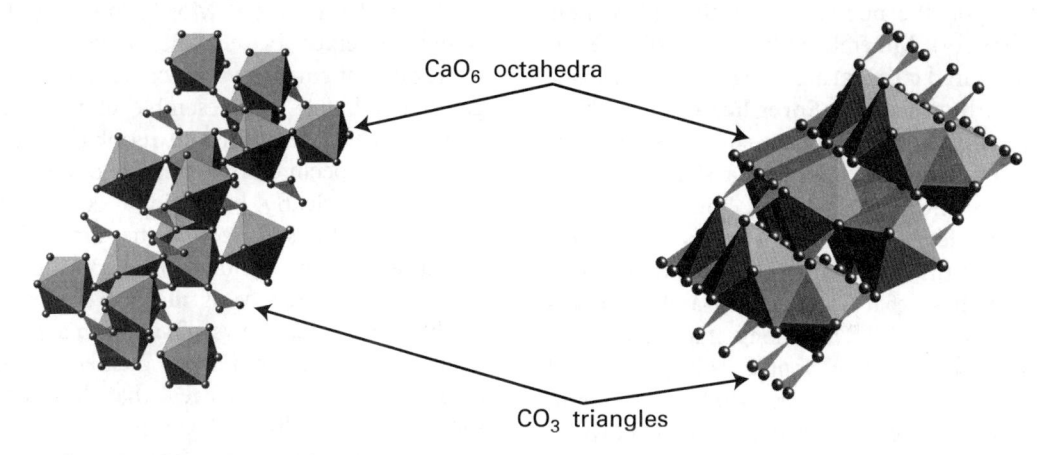

(b) The arrangements of CaO_6 octahedra and CO_3 triangles in the structures of calcite (left) and aragonite (right) $CaCO_3$ polymorphs result in very different crystal forms for the two minerals.

Figure 5.2
Isostructural versus polymorphic crystal structures.

(c) Very different crystal forms of the $CaCO_3$ polymorphs displayed by mineral samples of calcite (left) and aragonite (right).

(1978), Deer et al. (1992), Dyar et al. (2008), and Klein and Dutrow (2008) provide a wealth of information of a crystal chemical, crystallographic, and optical nature. For our purposes, since (1) oxygen and silicon are the most common elements in the crust of Earth, (2) the silicate minerals are the most important crustal minerals, and (3) these minerals are the fundamental "rock-forming" minerals, we will discuss and illustrate the structures of silicate minerals.

Silicon and oxygen bond together to form a tetrahedron, with the silicon at the center and four oxygen atoms at the four corners of the tetrahedron (figure 5.3a). Unlike most other anionic units found in minerals, the silicate tetrahedra frequently link together by sharing oxygens at the corners of the tetrahedra (figure 5.3b). The structural subgroups of silicate minerals are then based upon the number of tetrahedral corners shared (figure 5.4). Minerals with isolated silicate tetrahedra (no shared corners) are designated as **nesosilicates** (or orthosilicates). **Sorosilicates** are those minerals in which two tetrahedra share a single corner. In the **cyclosilicates**, tetrahedra share two corners, forming short chains that link back upon themselves to form rings. **Inosilicates** also form chains (and are therefore known informally as chain silicates), but these chains extend out and are effectively infinite in length.[1] These long chains may also be cross-linked to produce double, triple, or mixed chains and in such cases, the cross-linkage results in tetrahedra sharing between two and three corners. **Phyllosilicates** are the extreme case of this kind of sharing, with an extremely large number of chains cross-linked so that each tetrahedron shares three corners with adjacent tetrahedra, producing a sheet structure. The final possibility, wherein all four corners of all tetrahedra are shared in a three-dimensional framework, characterizes the **tectosilicates** (known informally as framework silicates).

IDENTIFYING MINERALS IN HAND SPECIMENS

The unique combination of chemical composition and physical arrangement of the atoms within a mineral, as noted above, gives rise to a set of unique physical properties for each mineral. The structural origins of certain of these physical properties are related, allowing the properties to be grouped into broad categories. For example, the way that light interacts with either the bulk or the surface of a mineral specimen generates the properties of color, luster, and streak that can be used in identification. Similarly, the hardness, cleavage, and fracture are all controlled by the strength of the bonding between the constituent atoms of a mineral.

Properties Related to Mineral Growth

Crystal Forms, Habits, and Aggregational States

The flawless, multifaceted crystal viewed in a museum showcase and the irregular hunk retrieved from a backyard both may be specimens of quartz, but they have very different outward appearances. In the best cases, mineral samples consist of large, single crystals that clearly display those properties needed for their identification. Most commonly, however, a sample observed in the field or lab may consist of aggregates of small crystals or partial crystals, aggregates that in no way resemble either perfect single crystals in a museum or a textbook description of the mineral. A specimen that consists of a single grain or crystal may clearly break in a consistent fashion, whereas another specimen of the same mineral might appear as a shapeless lump of microscopic crystals, a state often referred to as **massive** (color plate 17c). Slightly magnified views of such a massive specimen using a hand magnifier, or greatly enlarged views through a microscope, reveal that the specimen consists of small, individual grains and may reveal that each of these grains has the consistent pattern of breakage characteristic of the single larger crystal. This is a common enough occurrence that it is worth emphasizing: *It is the individual crystals or fragments that show the specific mineral properties; aggregations of smaller grains may display them inconsistently or not at all.*

Despite such potentially wide variations in appearance, many minerals do tend to take on particular shapes, either as individual crystals or in polycrystalline masses. In such cases, the particulars of the appearance can be helpful in describing and identifying minerals. The quality of a single crystal specimen of a mineral is typically assigned one of three descriptors: **euhedral**, which means that the specimen is entirely bounded by crystal faces; **subhedral**, where the specimen is partly bounded by crystal faces and partly by other surfaces;[2] or **anhedral**, meaning no crystal faces are present. In general, euhedral crystals are rare and highly prized as mineral specimens.

Euhedral and subhedral crystals will show some aspects of their internal symmetry, their forms, in the

Minerals 77

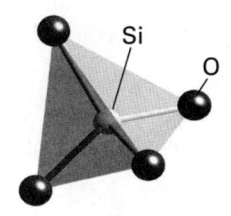

(a) Silicate structural unit, with silicon atom at center surrounded by tetrahedrally coordinated oxygen atoms.

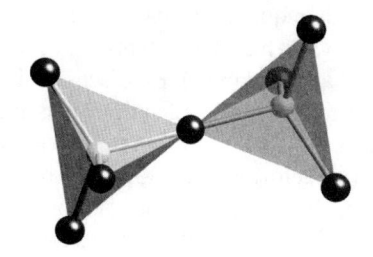

b) A pair of silicate tetrahedra sharing a single corner oxygen atom.

Figure 5.3
Silicate structures.

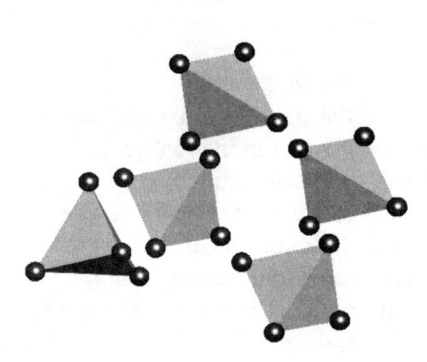

(a) Olivine, a nesosilicate.

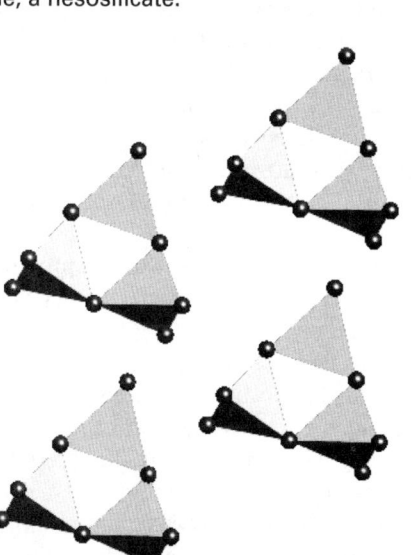

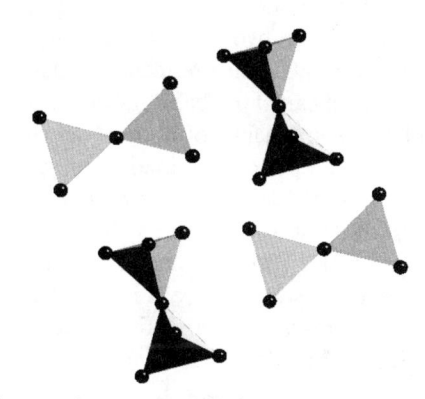

(b) Åckermanite, a sorosilicate.

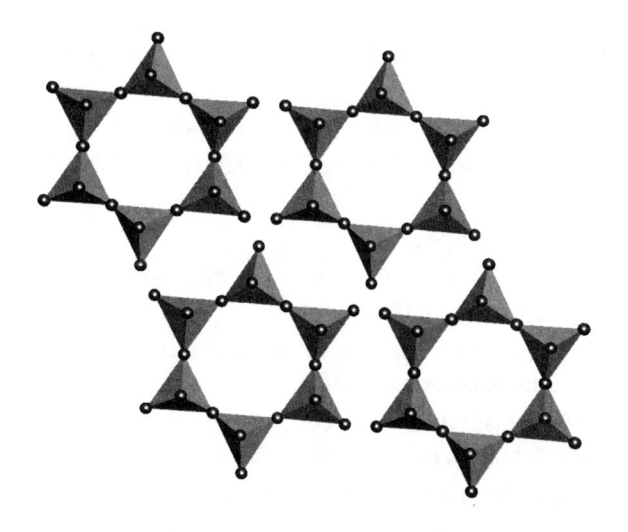

(c) Benitoite, a 3-member ring cyclosilicate.

(d) Tourmaline, a 6-member ring cyclosilicate.

Figure 5.4
Some major silicate tetrahedral structures, with mineral examples of various arrangements of silicate tetrahedra.

repetition and frequency of particular faces. For such crystals, recognizable point symmetry elements like rotation axes and mirror planes (discussed in chapter 4) can be used to aid in identification. It is important to realize, however, that variable rates of growth during formation (or deformation after growth) can camouflage the actual symmetry. Euhedral does not necessarily mean perfect! The ability to distinguish between faces that are different for symmetrical reasons and faces that appear different due to growth or deformation is an acquired skill. Anhedral crystals usually will not show any obvious symmetry. More advanced methods of identification, such as the X-ray diffraction or optical techniques described in chapter 6, are required for the determination of the symmetry of anhedral specimens.

Terms used to describe the overall shape of a specimen are divided into two groups. Individual crystals of particular minerals have a tendency to develop into a particular shape called the **habit**, as noted in chapter 4. Examples of some habits of mineral specimens are shown in figure 5.5, and examination of color plates 7–20 reveal typical habits of many of the common minerals. In contrast to the habit, a clustered group of crystals typical of a specific mineral is referred to as the **aggregational state** of the mineral. Samples of aggregational states are shown in figure 5.6 (on p. 80). Common habits include:

- acicular (needle-like, figure 5.5e)
- bladed (slender and flattened, like a knife blade, figure 5.5d; color plate 14b)
- capillary or fibrous (hair-like, figure 5.5f; color plate 10c)
- cubic (in the form of a cube, color plate 17a)
- dendritic (like the branches of a plant, color plate 15d)
- equant (of more or less equal length, width, and thickness; figure 5.5b)
- platy (flat like a plate or board with two subequal dimensions and one thinner dimension, figure 5.5h)
- sphenoidal (wedge-shaped, honey brown crystals of titanite in color plate 22d)
- tabular (similar to bladed, but thicker in the intermediate dimension)

Some of the more common aggregational states include:

- botryoidal (tiny to microscopic clusters with an outer bumpy surface like the outer surface of a bunch of grapes, figure 5.6b)
- coxcomb (jagged aggregates of platy or tabular crystals)
- granular (clusters of small anhedral, typically equant crystals, figure 5.6a; color plate 14f)
- massive (microscopic or nearly microscopic clusters without specific or regular form)
- radiating (fibrous to acicular crystals growing away from a point, figure 5.6g)
- stellate (star-like radiating acicular to tabular crystals)

In many cases, the selection of a term for the shape of a given mineral can be a matter of personal preference, where the character is transitional or has multiple possible descriptors (e.g., capillary versus fibrous). For example, millerite commonly is found in masses described as radiating *acicular* or radiating *capillary* aggregates, whereas the individual *botryoidal* masses of goethite may appear *radiating* on broken edges. Terms may also be combined.

Crystallographic Intergrowths

Many of the aggregation states described above form even though the individual crystals have little or no apparent crystallographic relationship to each other. It is possible, however, for the orientations of two or more adjoining crystals to be strictly controlled by the underlying atomic arrangements. Such crystallographically oriented intergrowths are referred to as parallel, epitaxial, or twinned, depending on the particular crystals and how they are related.

Parallel intergrowths occur when crystals of the same mineral are aligned so that mutual symmetry elements are parallel. The atomic arrangements match up from crystal to crystal, so that the intergrowth might be considered to be a single crystal with subparts. One well-known example of this is the scepter overgrowth of one quartz crystal onto another (figure 5.7 on p. 81).

Another possibility of growth is represented by the overgrowth of two different minerals that share a similar plane of atoms. Hematite (Fe_2O_3) and rutile (TiO_2) have very different symmetries, as well as different compositions, but in both minerals, the metal atoms are surrounded by six oxygen atoms in an octahedral coordination. This means that the {0001} plane of hematite and the {010} plane of rutile share a virtually identical arrangement of oxygen atoms, and crystals of rutile can grow on the surface of hematite crystals (figure 5.8 on p. 81). Intergrowths of this type are called **epitaxial intergrowths**. Many mineral examples of epitaxy are mere curiosities, but the principle has tremendous industrial importance—for example, modern computer chips are created by growing chemically

(a) Massive aggregations show no distinguishing features, as in this sample of massive kaolinite.

(b) Equant habit exhibits equal lengths in all directions, as shown here by garnet.

(c) Prismatic habit, here shown in topaz, is exemplified by crystals elongated in one direction.

(d) Bladed habit has the form of a knife blade, as shown here in kyanite.

(e) Acicular (needle-like) habit appears in amphiboles, such as the dark gedrite shown here, and in some other minerals.

(f) Capillary (or fibrous) habit is like fine hairs, as in this sample of millerite.

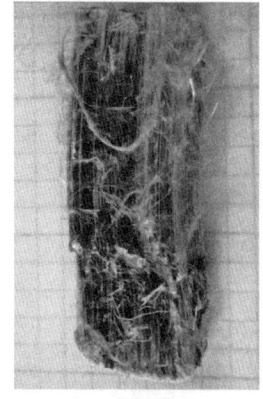

(g) Fibrous habit characterizes the asbestiform minerals, such as this chrysotile.

(h) Tabular habit, as in this hematite, is a habit in which crystals are thin in one direction and longer in the other two directions (like a table).

Figure 5.5
Examples of mineral habits and an aggregation.

(a) Granular aggregations are made up of individual somewhat equant grains, as shown here in olivine, with included granular, black chromite.

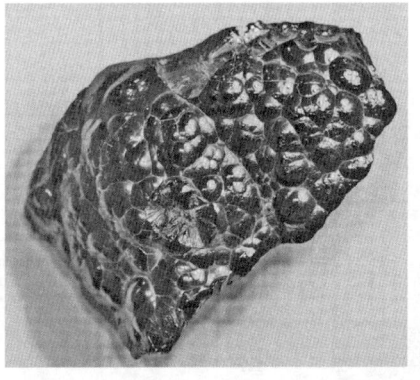

(b) Botryoidal aggregations have a surface resembling the surface of a bunch of grapes, represented here by a sample of hematite.

(c) Pisolitic aggregates are made up of pea-sized, concentrically banded grains, shown here in bauxite.

(d) Banded aggregations show distinct layers, as in the chalcedony and opal of agate.

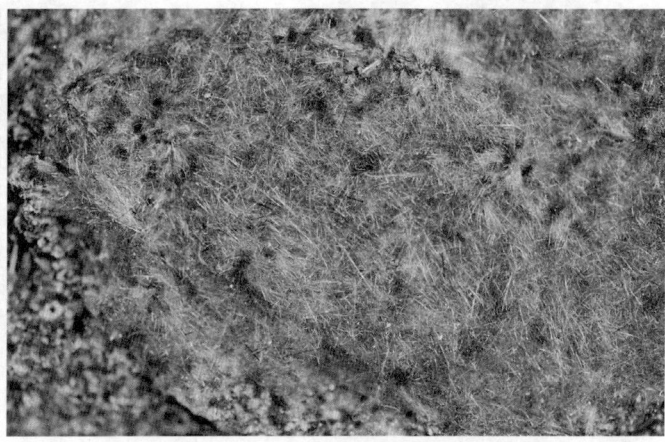

(e) Fibrous aggregates are made up of tiny fibers, here represented in a sample of actinolite.

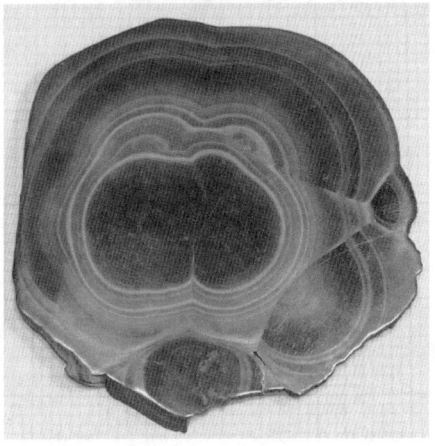

(f) Concentric aggregations have bands arranged in a roughly circular shape, as is the case here in malachite.

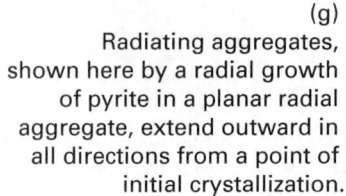

(g) Radiating aggregates, shown here by a radial growth of pyrite in a planar radial aggregate, extend outward in all directions from a point of initial crystallization.

Figure 5.6
Examples of mineral aggregations.

doped silicon or silicon oxide epitaxially on a pure silicon surface.

Twins occur where intergrown crystals of the same kind are related by a symmetry element (termed a twin element) that is not present in either of the untwinned crystals. These can develop during original growth of minerals, can result from deformation of mineral grains, or can form by other transformations (Buerger, 1945). Viewed in this context, twinning is interpreted as a crystal "mistake" or "defect" (figure 5.9). Twin elements are analogous to the point symmetry elements of a plane, rotation axis, or symmetry center and can be described by Miller indices and zone axis symbols just as their symmetry element ana-

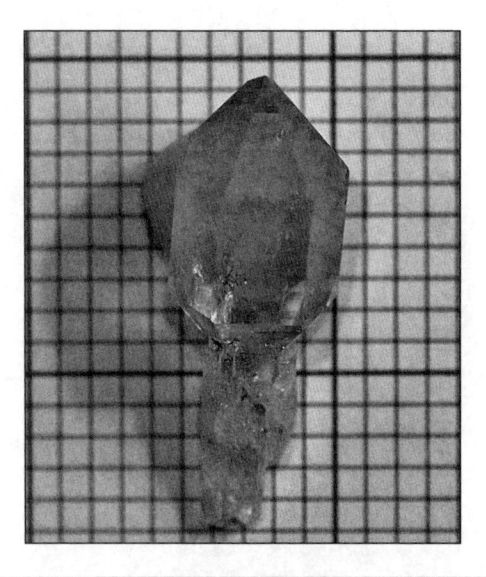

Figure 5.7
Overgrowth of a larger quartz crystal on a smaller one, referred to as a scepter overgrowth.

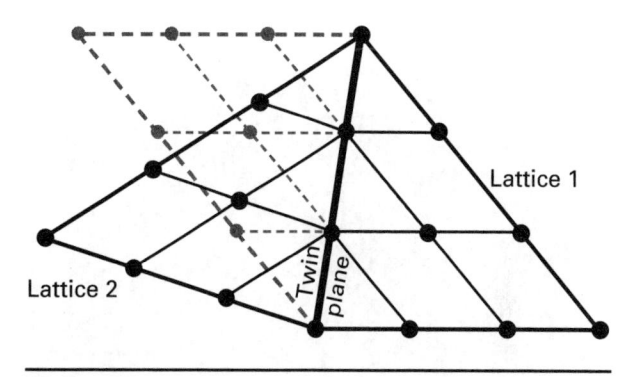

Figure 5.9
Schematic view of a twinned two-dimensional monoclinic crystalline lattice. Two subcrystals are represented by the lattices 1 and 2. A "mistake" in the atomic alignment results in the two separate crystals being related by a mirror (twin) plane that does not exist in either of the untwinned crystals. The gray lattice points and dashed lines represent the untwinned continuation of lattice 1.

(a) A natural example of epitaxial intergrowth, with darker rutile crystals on a larger hematite crystal.

(b) The structure of the intergrowth in (a), showing the TiO_6 octahedra of rutile (white) overlying, nearly perfectly, the FeO_6 octahedra of hematite (gray).

Figure 5.8
Epitaxial intergrowth.

82 Chapter Five

logs. Types of twinning that are common in specific minerals may also be described by a twin law that is given a specific name, such as the common Carlsbad Law twins of orthoclase feldspar.

Simple twins occur when there are only two crystals involved. These include *contact* twins, where two crystals share a single plane of contact, and *penetration* twins, where the crystals appear to be interpenetrating and are related by a twin axis (figure 5.10; color plate 14d). *Complex* twinning links multiple crystals. If the successive twin planes are parallel to each other and there are several or many of them, the twins are called **polysynthetic**. This kind of twinning characterizes plagioclase feldspars (figure 5.11b; color plate 8c, d). If the twin planes are not parallel, the twins are called **cyclic** (figure 5.11).

(a) Quartz twinned on the {11$\bar{2}$2} plane (Japan law twin).

(b) Gypsum twinned on the {100} plane (swallowtail twin).

(c) Orthoclase twinned along the {001} axis (Carlsbad twinning).

(d) Staurolite twinned on the {231} plane.

Figure 5.10
Examples of simple twins.

(a)
Chrysoberyl twinned on the {110} plane. The three twinned crystals together appear as a hexagon.

(b)
Oligoclase twinned on the {010} plane. The joins (twin planes) between the many repeated crystals produces the striations (straight lines).

Figure 5.11
Examples of complex twins.

Properties Related to Mineral Interactions with Light

Color and Coloring Phenomena

Electromagnetic radiation may be visualized as a series of particles (photons) that travel through space in waves. These waves vary greatly in length, from over a kilometer in the case of some radio waves to less than a picometer (10^{12} meter) for high energy gamma rays.[3] Those electromagnetic waves with wavelengths between approximately 390 and 770 nanometers represent visible light;[4] that is, those limits designate the fraction of the spectrum that can be perceived by human eyes. This visible spectrum (figure 5.12) can be subdivided into a series of bands that correspond to colors ranging from red to violet and known by the acronym ROYGBIV, with each letter representing a color (red, orange, etc.). Excluding some exceptions to be described later, no minerals generate their own light, and they are therefore observed in light coming from an external source.

When a photon strikes an atom, the energy it carries is transferred to one of the orbital electrons, raising the electron to a higher energy level. This level is typically not stable, so this electron rapidly returns to its original level, either by producing a photon of the same energy as the original photon that is re-emitted, or by absorbing the energy and passing it to the rest of the atom in the form of a vibration.[5] If a re-emitted photon travels in the same direction as the original photon was traveling, the result is *transmitted* light.[6] If

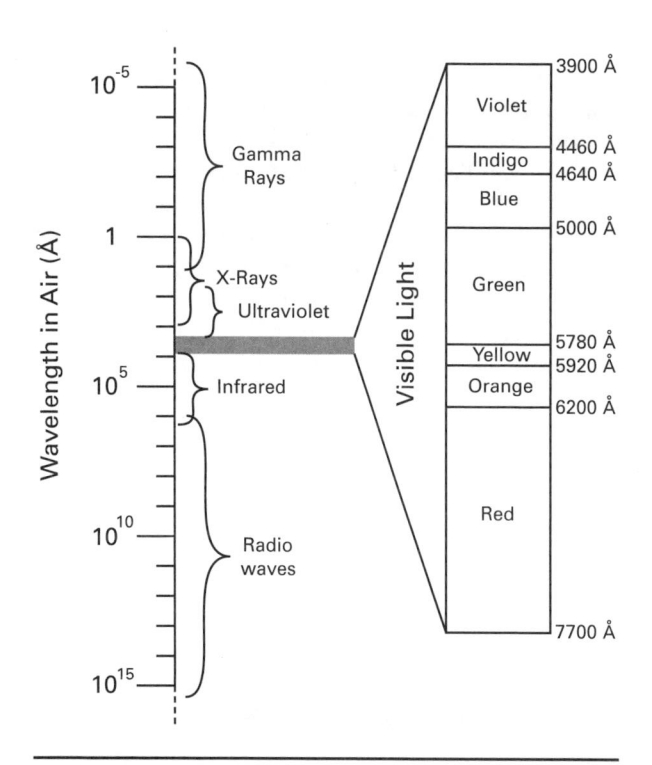

Figure 5.12
The electromagnetic spectrum. Note the very narrow range of wavelengths visible to humans and the continuous change in colors from 7700 Å to 3900 Å (770 to 390 nm).
(*Source:* after Bloss, 1971. Used with permission.)

the photon travels back in the direction of the original photon, it is *reflected* light, and any other direction of light travel is considered *scattered* light. A *transparent* material is one in which the bulk of the light is transmitted,[7] whereas in an *opaque* material, most of the light is absorbed, reflected, or scattered. Any material with behavior between these two possibilities is considered to be *translucent*.

Assuming that absorption and re-emission are independent of wavelength, if most of the incident photons are re-emitted, the reflected or scattered light is the same color as the incident light and the reflecting surface appears white. If most of the incident photons are absorbed, the surface will appear black (figure 5.13a).[8] When some wavelengths are preferentially either absorbed or re-emitted, as illustrated in figure 5.13b, the resultant light is enriched in certain wavelengths and the mineral appears to have a color. Azurite, for example, preferentially absorbs the red through green and indigo through violet light, giving the mineral its distinctive blue color.

The selective absorption of light may be caused by a variety of factors. The presence of transition metals, either as essential components or trace impurities, is one common cause. Transition metal atoms, bonded in mineral structures, may have multiple oxidation states or unfilled *d*-orbitals containing unpaired electrons. The electron energy levels that result frequently correspond to the energies of visible light, allowing them to absorb specific wavelengths. Light absorption can also result from defects in the crystal structure. Such defects, or color centers, can occur during crystallization if an atom of the wrong valence fits into the structure, or if a "correct atom" is "misplaced" in the structure. Color centers can also be created after mineral formation, if high-energy radiation disturbs the atoms in the structure.[9]

Another common source of mineral color is the presence of large-scale impurities, called inclusions. Inclusions may be small mineral crystals incorporated into larger crystals during the growth of the larger crystal (i.e., oikocrysts), or they may form within an existing crystal via *exsolution*, resulting from changes in the temperature, pressure, or chemical state. Microscopic or submicroscopic crystals of hematite are common examples of included crystals that impart a pinkish color to some potassium feldspars and a red color to the variety of quartz called jasper. Trapped fluids in tiny bubbles (fluid inclusions) may also produce color directly or by scattering light that has been transmitted through part of the crystal.

A few minerals have colors that are characteristic, such as the common olive-green color of olivine, the deep blue of azurite, or the golden-yellow of gold (e.g., color plate 14f; see additional images in the online resources). In most minerals, however, the effects that create the colors differ from specimen to specimen. Common minerals such as quartz, fluorite, and calcite may take on a wide variety of colors (color plates 7a; 18c; 19c). This variability makes the color of a mineral one of the least useful properties for identification.

Colored effects, especially those referred to as a "play of colors" as is found in precious opal or labradorite, are due to interference as light interacts with the mineral surface. The effect is very similar to the colored interference caused by oil on water. Such

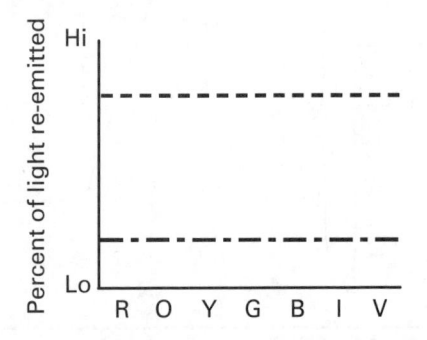

(a) A white mineral (upper dashed line) and a black mineral (lower dot-dash line) re-emitted all colors equally.

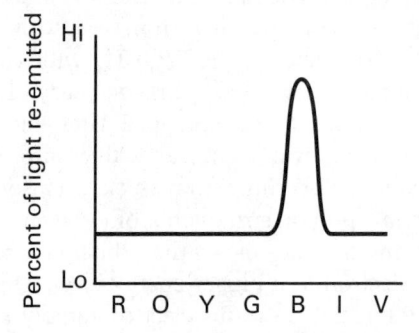

(b) A mineral re-emitting a narrow band of color, in this case, blue.

Figure 5.13
Schematic plots of the re-emission of absorbed light as a function of color.

effects are caused by microstructural details on the surface that are separated by distances approximately equal to the wavelength of the light that strikes them. Other effects in minerals that are not due to absorption, but to physical factors, include (1) star-like rays of light in a mineral, an effect called *asterism*, exhibited by star rubies and sapphires; and (2) *chatoyancy* or fibrous sheen in the semi-precious stone called "tiger's eye." Both of these effects are caused by fibrous inclusions within the mineral grains.

Streak

Many of the phenomena that contribute to variations in color of a bulk mineral sample are of little importance if the sample is powdered. The tiny fragments that result from crushing a mineral grain transmit a greater fraction of the incident light and thereby minimize the effects of impurities and inclusions. The color of a powdered mineral is therefore more characteristic, and potentially more useful, in hand-specimen identification. One easy way to powder a small quantity of a mineral specimen is to rub it on an unglazed porcelain tile, leaving a line or streak of the mineral powder on the tile. The test is called a **streak** test, and the tile is normally called a streak plate.[10] The mineral to be tested must be softer than the streak plate, otherwise the streak of powder formed by scratching is that of the streak plate and not that of the mineral.

For most minerals with a nonmetallic luster, the streak is white or has only a pale color. In these cases, the results of the streak test are no more diagnostic than the overall sample color. Minerals with metallic or submetallic lusters, however, such as native elements, sulfides, sulfosalts, and some oxide minerals, tend to have either darker or more distinctly colored streaks, or both. Hematite is a particularly good example of the utility of the streak test, as it may occur in earthy, fine-grained, brick-red masses or in very different looking masses of small, silver, metallic crystals that give the specimen a "metal-flake" appearance. In both cases, the aggregations will leave a diagnostic red-brown streak on the streak plate, providing an easy and reliable means of determining the mineral identity.

Luster

Luster is a subjective property, unlike the more objective optical properties of color and reflectance, both of which can be instrumentally measured. It refers to the quality of light traveling from the mineral to the eye of the observer. The luster includes light reflected or scattered from the surface, as well as light that is transmitted through the mineral, and is usually described in comparative terms such as metallic, glassy, earthy, or resinous.

If the incident light is nearly completely reflected from the surface, the mineral is described as having a metallic luster. A nonmetallic luster is one in which a larger proportion of the light is absorbed, transmitted or scattered. Distinguishing between these two possibilities would seem to be a trivial task, but it can be more difficult than it sounds. It is common for beginning students to confuse dark-colored, nonmetallic minerals with metallic minerals. To understand why, recall that valence electrons in metals are very delocalized, creating a sea of electrons with a variety of energy levels between atoms. When light strikes this electron sea, all energies are readily absorbed and re-emitted in all directions. A metallic material is therefore "shiny," regardless of the angle at which light strikes the surface. The bonding electrons in nonmetallic materials are much more localized, with fewer electron energy levels. This allows greater light penetration and more chances for scattering, absorption, or transmission; unless the light arrives at a particular angle (called the *critical angle*). Light that arrives at this angle is entirely reflected away at an identical angle[11](figure 5.14). In a typical classroom with multiple overhead lights, almost any angle is a critical angle

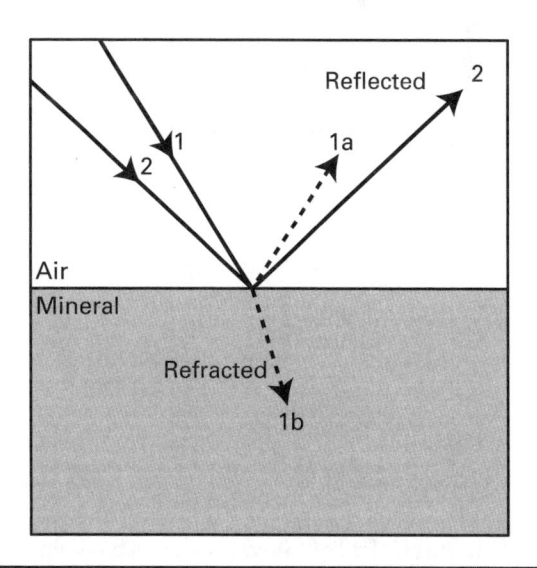

Figure 5.14
Light rays striking mineral surfaces at most angles (one of which is shown as ray 1) are partly reflected (ray 1a), partly absorbed, and partly refracted or bent (ray 1b). Light that strikes at a critical angle (ray 2) is entirely reflected away. The exact angle is dependent on the refractive index of the mineral (see chapter 6).

for one or more lights, so dark-colored, nonmetallic minerals can appear to be reflecting enough light to be metallic. A simple way to check for a metallic luster is to hold the hand specimen in one hand and shield it from the overhead lights with the other. A truly metallic mineral will still show bright reflections, whereas a nonmetallic one will appear dull. In cases where a mineral appears only faintly metallic, the term "submetallic" is often used.

Differentiating between types of nonmetallic lusters is even more subjective and based on experience. The most common type of nonmetallic luster is a **vitreous** luster, meaning that the mineral surface resembles a piece of glass (color plate 7a). A related but less common luster is *adamantine luster*, like very reflective glass, which occurs in minerals such as diamond or rutile that have a characteristic called a "high refractive index."[12] Other possible nonmetallic lusters include *silky* (having a sheen created by many narrow parallel fibers, color plate 10c); *waxy* (like candle wax, color plate 7b; 10d); *earthy* (dull and grainy, like a dirt clod, color plate 16c, d); *dull, greasy* (as if coated with a lubricant, color plate 7d); or *resinous* (like resin or amber).

Luminescence

As noted previously, the majority of mineral specimens produce no visible light of their own (unless they are heated to incandescence). Some specimens, however, will emit visible light if they are irradiated with higher energy light (ultraviolet light or X-rays). Such a phenomenon is called **luminescence**.[13] Luminescence occurs primarily due to the presence of trace chemical impurities in a mineral, particularly transition metals, which act as activators.

The interaction of the UV light or X-rays with a typical mineral surface is generally the same as with visible light (figure 5.15a). The incident photons are re-emitted at nearly the same wavelength or absorbed and converted into vibrations by the electrons they affect. In luminescent specimens, however, activator elements create intermediate energy levels that act as "traps" for the excited electrons (figure 5.15b). The result is a multi-stage process in which excited electrons drop down to the level of a "trap" by giving up part of the incident energy absorbed, and they remain at the trapped level for a short period, before releasing the rest of the energy as a photon. If the difference

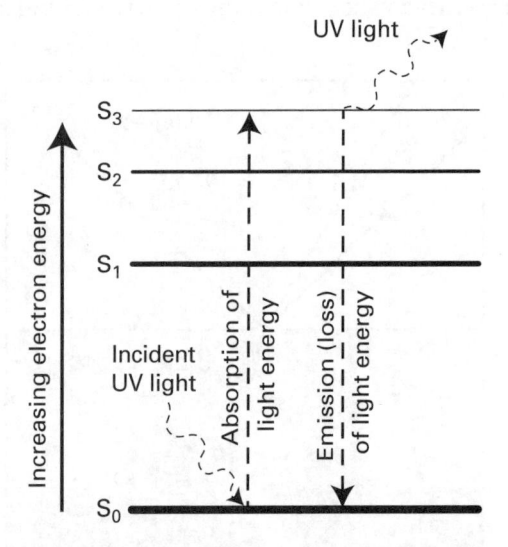

(a) Non-fluorescent minerals absorb the energy of incident ultraviolet light (lower wave-like UV line), then immediately re-emit the same energy (upper wave-like UV line).

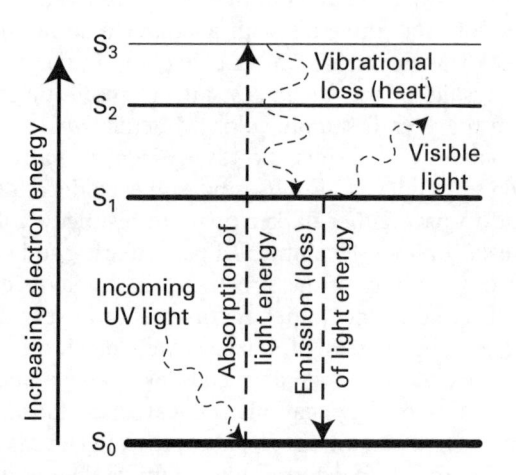

(b) Fluorescent minerals dissipate some of the absorbed energy of incoming UV light (lower wave-like UV line) as vibrations (heat) then emit the remaining energy (represented by the difference between S_1 and S_0) as visible light (upper wave-like visible light line).

Figure 5.15
Schematic Jablonski diagrams depicting issues of fluorescence.

between the original energy of the electron and the energy level of the trap matches that of visible light, the specimen luminesces. If this occurs immediately after absorption (and therefore ceases when irradiation ceases), the phenomenon is named **fluorescence**. If there is a time lag between absorption and visible re-emission, it is called **phosphorescence**. Normally fluorescence and phosphorescence are very low in intensity and can only be seen under low light conditions.

As is the case with color, luminescent phenomena vary between mineral specimens and can be unpredictable. An excess of activator elements can actually cause the luminescence to decrease or disappear entirely, a result referred to as "quenching." Commonly fluorescent minerals include fluorite (for which fluorescence was named), calcite, scheelite, and willemite. One of the most famous localities for fluorescent minerals is the Franklin-Sterling Hill district in New Jersey, where manganese acts as an activator in calcite and willemite. Perhaps the most famous example of mineral phosphorescence can be found in the Hope Diamond, which glows a brilliant red after illumination with short-wave ultraviolet light.

Properties Related to Mineral Strength

Tenacity

The resistance of a mineral to deformation—such as breaking, bending, or tearing—is referred to as its tenacity. Most minerals can be described as **brittle**, meaning that they will break without obvious deformation, a result of their dominantly covalent bonding. Because such bonds are mostly directional and not easily displaced, the minerals respond to an applied force by breaking. The exceptions to this general tendency are unusual enough to be useful in some mineral identification.

Phyllosilicate minerals tend to be flexible or elastic in nature due to differences in the bonding of the silicate layers that comprise the mineral structures. Within the layers, the bonding is strongly covalent, but interlayer bonding varies in character and is typically much weaker. In chlorite and talc, for example, interlayer bonding is largely of the Van der Waals or hydrogen bond types, so that when a fragment of one of these minerals is bent, the interlayer bonds reconfigure to the new shape and the fragment stays bent. Chlorite and talc are thus described as having **flexible** sheets. Muscovite and biotite are ionically bonded between the silicate sheets, and these bonds provide a restoring force that causes a mineral fragment to snap back after bending. This is referred to as **elastic** behavior.

Ductility (capable of being drawn into wire), *malleability* (capable of being hammered into thin sheets), and *sectility* (capable of being cut into shavings) are all properties that are associated with metallically bonded minerals. The electron sea of the metallic bonds permits the atoms to move past each other with relative ease, which permits a large amount of deformation to take place without the sample actually breaking.

Cleavage and fracture

In order for a mineral specimen to break into pieces, the chemical bonds within the structure must be broken. Because not all bonds in all crystals have the same strength, nor do all directions in a crystal have the same number of bonds, it is understandable that in some minerals the crystals break in some directions (those with weak and fewer bonds) more readily than in others. Careful observation of how a mineral breaks, therefore, provides a unique window into the atomic arrangement—a means of macroscopically observing one of the fundamental characteristics of the mineral—and also provides useful information for mineral identification.

Cleavage and fracture represent the opposite ends of the spectrum for how a mineral breaks. When a mineral is described as having **cleavage**, it means that there is a series of parallel planes in the mineral structure that have fewer or weaker bonds between them, resulting in a repeated series of parallel surfaces along which the mineral may be broken yielding flat cleavage faces (color plates 17b; 18c). In contrast, **fracture**, which occurs where there are no systematic differences in how the atoms are bonded, produces breaks that are random in nature (right end of sample in color plate 8a). Figure 5.16 shows a pair of classic examples of the differences between the two phenomena. On the left are fragments produced by breaking a drinking glass. In the original sample, the way in which the glass breaks is more a function of preexisting microscopic damage and the nature of the force applied, than it is a function of any repeated internal structures, so no two glass fragments will have the same shape except by chance. On the right in the figure are fragments of calcite, which has three planes of cleavage that are not at right angles to each other. The surfaces of the fragments are more or less flat cleavage faces and the fragments are dominantly rhombus shaped, so the resultant three-dimensional forms are referred to as rhombohedra. Regardless of how often a

Figure 5.16
Fracture fragments of a glass plate (left) and cleavage fragments of calcite (right).

specimen of calcite is broken, these three planes define the shapes of the cleavage fragments.

Because the cleavages are related to the atomic arrangement within a mineral, they can be precisely characterized using crystallographic terminology, like Miller indices (as described in chapter 4). For most general descriptions, however, the cleavage is described by the number of unique directions of parallel planes of breakage that can occur (1 for muscovite, 2 for plagioclase, 3 for halite, etc.) and by the angles between the cleavage directions, plus the quality of the break (perfect, good, imperfect, poor). There are specific terms for cleavage fragments, such as cubic for halite, octahedral for fluorite, or rhombohedral for calcite. It is often the recognition of characteristic cleavages that makes certain minerals instantly identifiable to experienced individuals. This skill is one of the principal means of identifying small mineral grains within a rock. As noted earlier, however, students commonly have difficulty in recognizing that cleavage is a property only of an individual mineral grain or crystal, and not of an aggregate of mineral grains that generally will not show cleavage.

Fracture is normally a less useful property than cleavage, except in that it shows evidence of a lack of cleavage. Nevertheless, fracture is useful in identifying minerals in rocks. There are some specific terms used for fracture, such as hackly (jagged like the edge of a torn metal surface), splintery, fibrous, and conchoidal (curved like a conch shell). There is some overlap between these terms and those used for aggregational states.

Problem 5.3

Miller indices can be used to describe cleavage surfaces. Can they also be used to describe fracture surfaces? Why or why not?

Parting

It is possible for a mineral to break along apparently planar surfaces, even if the mineral has no cleavage in a given direction or has no cleavage at all. These *pseudocleavage* planes can be imposed on a mineral due to twinning, exsolution, or stress-induced deformation. In cases where a mineral displays such flat surfaces, they are called **partings**. Partings can generate a fair amount of confusion for students, who may read that a mineral has no cleavage yet clearly see something resembling cleavage in a specimen before them. Some garnet samples used in laboratories notoriously exhibit partings and some feldspars have partings that suggest more than the two normal cleavages of this mineral.

The key to distinguishing partings from cleavages is to recognize that cleavage is inherent to all areas of all specimens of a mineral, whereas parting is imposed only upon some areas of some specimens. The result is that parting surfaces may be less smooth than cleavages and may have a slight curvature. If parting is caused by twinning, the twin plane may appear regular, but there will not be matching parallel planes in the individual crystals. Spacing between partings created by the exsolution of another mineral often has variable widths.

Hardness

The hardness of a mineral is defined as its resistance to being scratched on a clean surface, or its "scratchability" (Klein, 2002). Scratching a mineral surface requires the breaking of chemical bonds, as do cleaving or fracturing of a mineral. But unlike cleavage or fracture, which depend on all types of bonds in the mineral, hardness is more a function of the weakest bonds present, much like the strength of a chain is dependent on its weakest links.[14]

The classic scale for hardness determination, published by Friedrich Mohs in 1824 and subsequently named for him (figure 5.17), is comparative, using commonly occurring minerals (Hurlbut and Klein, 1977). Typical modern usage of the scale is based on everyday items such as a fingernail (2.5), a copper penny (3.5), and a knife blade or glass plate (5.5). When quantitative hardness measures were developed, Mohs' choices of minerals for the scale were realized to be "exceedingly canny" (Bloss, 1971). As shown in figure 5.17, plotting the Mohs hardness versus quantitative values produces a nearly exponential curve.

The hardness test, actually a scratch test rather than a hardness test strictly speaking, is simple enough to perform but may be subject to some pitfalls in interpretation. One simply scratches, with firm pressure, one mineral using the corner of another. Minerals with the same or nearly the same Mohs hardness will generally scratch each other, which illustrates the need to test both minerals against one another. Accurate determination of Mohs hardness (scratchability) requires selection of a fresh surface for testing, as alteration products are generally softer than the original minerals. Scratching a soft mineral like gypsum on a glass plate will produce a deceptive line of powdered gypsum, but will not leave a permanent mark on the glass. Hence, the gypsum is said to be "softer." By the same token, a granular aggregation of olivine (harder than glass) may separate when rubbed on the glass, leaving a trail of grains and crushed oliv-

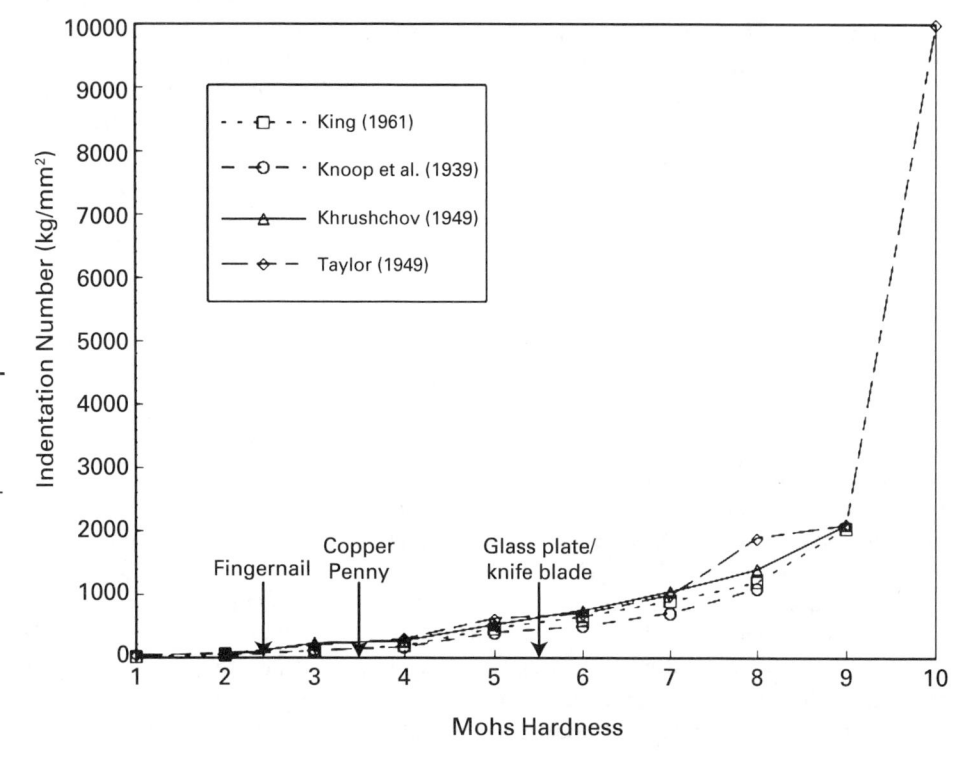

Figure 5.17
Mohs scale of hardness, including approximate hardness values for household items, plotted vs. quantitative hardness measures.
(*Sources:* After Bloss, 1971, and Plendle and Gielisse, 1963, from data by King, Knoop et al., Khrushchov, and Taylor. Used with permission.)

ine that hides the marks on the glass. In testing Mohs hardness, it is also possible to use the ease of scratching as an additional guide. For example, a specimen of augite will scratch a glass plate only with difficulty, meaning it has a hardness close to the glass, whereas a corundum specimen will easily leave a mark with only the slightest pressure, owing to corundum's much greater hardness.

In some cases, hardness may differ appreciably in different directions on a mineral surface, depending on the bonds that are broken, but for most minerals, these differences are small and can only be detected by instrumental methods. A notable mineral with differential hardness is kyanite, which has a Mohs hardness of 5 along the length of its crystals, and a Mohs hardness of 7 across the width.

> **Problem 5.4**
>
> You are given samples of two minerals, potassium feldspar and topaz, and told that these are the two minerals you have been given. Both are in clear, anhedral masses showing no cleavage faces. Using only a glass plate, describe how you could identify which is which.

A Property Related to Mineral Mass: Specific Gravity

The density of a solid material is defined as its mass divided by its volume, and is generally given in units of grams per cm^3 (g/cc). Determining the volume of an irregularly shaped mineral specimen can be difficult so historically, instead of density in g/cm^3, the densities of minerals have been presented as a **specific gravity**, the ratio of the mass of a mineral to the mass of an equal volume of water at 4°C. At this temperature, water has a density of 1 g/cm^3, so the specific gravity is simply the density expressed as a unitless number.

From its definition, density (and therefore specific gravity) is a function of both the mass of the atoms within a mineral and the closeness of packing of the atoms. Galena (PbS) and halite (NaCl) are isostructural, so the large difference between their specific gravities (7.5 vs. 2.2, respectively) is due mostly to the presence of the more massive lead atoms in galena. In contrast, the differing specific gravities of the three polymorphic forms of Al$_2$SiO$_5$—andalusite (3.16–3.20), kyanite (3.55–3.66), and sillimanite (3.23)—arise from each forming under different pressure and temperature conditions that compress or expand the structures.

The various specific gravities of individual minerals seemingly should be as useful as cleavage in identifying hand specimens, but this is not the case in general practice. Many of the most common minerals have similar specific gravities, falling in the range 2.6–2.8. Although instruments can be used to *measure* these specific gravities, methods of *estimating* specific gravity of hand specimens do not have a precision better than roughly 0.5 units. Hence, specific gravity is typically used only to distinguish minerals that seem unusually heavy or light relative to the average values of common minerals. A simple and quick method of estimation and comparison called "hefting" is useful. One takes two specimens of the same size, places one in each hand, and gently bounces the samples up and down several times to feel the difference in perceived weight. Then, one switches the specimens between hands and repeats the bouncing procedure. Specific gravity differences greater than 1 become readily apparent, and differences as low as 0.5 can be felt with several iterations, if the specimens are identically sized.

Specific gravity values can be determined in a variety of ways using instruments and fluids. One instrumental method is the balance method, which involves determining the weight of a specimen in air, then the weight of the specimen in water. Dividing the weight in air by the difference between the two weights yields the specific gravity. The balance used can be either a spring (Jolly) or torsion (Berman) type. Pycnometric methods make use of the fact that water has a density very close to 1 g/cm^3 at room temperature. The specimen is first weighed, then submerged in a known volume of water, and the volume of water displaced by the specimen is determined. The specific gravity is the specimen weight divided by the specimen displacement. For precise determinations, a bottle of accurately known volume called a pycnometer is used. Lower precision measurements can be made using a graduated cylinder. Heavy liquid methods attempt to match mineral and liquid densities. The specimen is submerged and the liquid composition is altered until the specimen begins to float, at which point the liquid density is determined. Liquids used in the past for this technique included bromoform (2.89) or methylene iodide (3.33), which were mixed with acetone (0.79) to adjust the density. A newer, safer, and easier to use alternative is sodium polytungstate (2.89) mixed with water. Each of the measurement methods described above can provide precise measurement, but they are time consuming, lab-based procedures.

Other Mineral Properties

Magnetism

In chapter 2 it was noted that the spin quantum number used to describe orbital electrons could only take on one of two values. These are interpreted as indicating the orientation of the axis around which an electron spins: the axis can point up (+1/2) or it can point down (-1/2). The spinning electric charge of the electron generates a small magnetic field. If all of the electrons in an atom or ion are paired, the individual fields cancel each other out. Such a material is referred to as **diamagnetic**. Diamagnetic materials are not only not attracted to magnetic fields, they actually repel them, although in most materials this repulsion is too weak to be easily observed.[15] If some electrons remain unpaired, the material is **paramagnetic**. Unpaired electrons are common in cases where transition metal atoms or ions occur in the structure, because in these elements, the process of filling the *d* (or *f*) orbitals can result in multiple unpaired electrons. The overall strength of the magnetic field created by unpaired electrons depends on other factors (the orientations and magnitudes of magnetic domains), and the character of the magnetic field determines whether a mineral is designated as *ferrimagnetic*, *antiferromagnetic* or *ferromagnetic* (the discussion of which is beyond the scope of this text).

Although a number of minerals are affected by strong magnetic fields, like those generated by large electromagnets,[16] only a few respond to the weaker field generated by a small permanent magnet like that used in hand-specimen identification. Magnetite is strongly attracted to such a small magnet, making it one of the easiest minerals to identify. Ilmenite and pyrrhotite show a weak attraction, best observed by suspending the magnet on a string, slowly bringing the mineral specimen near one end or edge of the magnet, and noting a very slight torsion of the string.

Acid Reaction

Carbonate minerals such as calcite and aragonite [$CaCO_3$], dolomite [$CaMg(CO_3)_2$], siderite [$FeCO_3$], magnesite [$MgCO_3$], and azurite [$Cu_3(CO_3)_2(OH)_2$] will react with acids, producing gaseous carbon dioxide in the process. This reaction can be strong enough that bubbles of CO_2 quickly develop and burst when a drop of acid is placed on the mineral surface. The phenomenon is called *effervescence*, but is commonly referred to as "fizzing."

The release of CO_2 is caused by the acid attacking the bonds between the metal cations and the CO_3 polyanion, as well as the C–O bonds within the CO_3 group. Because the strength of the metal-carbonate bonds varies with the metal(s) present, the intensity of the reaction to a dilute acid will vary, depending on the mechanical state of the mineral and the temperature of the acid. Calcite and aragonite react to dilute, room temperature HCl with immediate effervescence, whereas dolomite reacts weakly, if at all. If, however, some bonds in dolomite are broken mechanically by grinding the mineral into a powder before applying acid, effervescence occurs readily. Magnesite and siderite, in contrast, only effervesce if the dilute HCl is heated.

Odor, Water Solubility, and Feel

For a very few minerals, it is possible to produce a characteristic odor. Sulfide minerals, if scratched, produce a rotten egg odor caused by the formation of hydrogen sulfide. Arsenic-bearing minerals, like arsenopyrite, similarly yield a garlic-like odor. Clay minerals, like kaolinite, will produce an earthy smell akin to that of a wet basement if the surface is dampened with water or heavy breath.

Another test involves assessing the water solubility of samples using the natural perspiration of your hands. Since the advent of hand-specimen identification, chloride minerals like halite or sylvite were identified by their tastes, with halite being salty and sylvite being bitter. Safety concerns have now rendered such tests inadvisable. So now, equivalents can be found by testing the water solubility. Repeatedly rubbing soluble specimens between one's (perspiring) fingers will cause some of the soluble minerals to dissolve from the surface and re-precipitate on the skin. After a few moments, the salty film becomes noticeable, much like the salt film that forms on your skin after swimming in the ocean. The film yields a sticky feel. The film is easily removed with running water and the test for solubility is positive.

Some minerals have a distinctive feel. Minerals that have appreciable amounts of weak bonding (like Van der Waals bonds) can break off tiny fragments to create a thin coating on the fingers. Particularly for minerals with layered structures like graphite, molybdenite, or talc, the coating can feel greasy or soapy. Because of this, graphite and molybdenite are used as industrial lubricants, and talc is ground finely to produce talcum powder.

Using various sets of the tests and observations described here, common minerals of hand-sample size can generally be identified with little difficulty. Identifications of this type are useful as a first assessment in field mapping and in routine laboratory studies of rocks. For more complete and accurate identifications of minerals, detailed assessments of mineral character, or petrologic studies of rocks, however, laboratory analyses using petrographic microscopes, X-ray diffractometers, electron microscopes, and other instruments are necessary (chapter 6).

> **Problem 5.5**
>
> You are handed samples of calcite, dolomite, and magnesite that have an identical appearance (white color, rhombohedral cleavage), along with a bottle of dilute HCl containing only 3 drops of acid. Outline a procedure to identify all three samples with the acid.

SUMMARY

The ability to identify and interpret the rocks of Earth is dependent on the ability to identify minerals, commonly beginning with identifications in hand specimens. This requires familiarity with some of the characteristic physical properties of minerals, as well as an understanding of how minerals are classified. The most widely used classification assigns minerals to chemical groups based on the principal anion or polyanionic unit, then subdivides groups into subgroups based on structural features. Of particular importance in this scheme is the classification of the different silicate minerals, which are grouped by the number of corners of their SiO_4 tetrahedral building blocks that are shared.

Some properties useful for identifying and describing minerals derive from growth patterns. Specimens are labeled euhedral, subhedral, or anhedral depending on whether they are completely, partially, or not at all bound by crystal faces. Individual mineral crystals may take on characteristic habits, or groups of mineral crystals may occur in characteristic aggregational states. Occasionally, the individual crystals within aggregations are related by their crystallography—for example, where dissimilar minerals are grown together along a similar plane of atoms to form an epitaxial intergrowth. Twins represent identical crystals joined along a symmetry element not present in the individual crystals and can be paired or in groups of multiple crystals.

Other properties useful for identification include mineral interactions with light, styles of breaking, and characteristics such as density, magnetism, and solubility in acid. For example, minerals that are black or white absorb, reflect, and scatter light equally across different wavelengths; whereas minerals that preferentially absorb or re-emit some wavelengths appear as colored. Inclusions and chemical impurities can impart colors to minerals, making color of limited use in identification; however, if a mineral is powdered, the color of the resultant streak may be characteristic. How a mineral reflects light—the luster—is described subjectively by comparing the reflected light to reflected light in other common materials. Properties that originate from the strength of the mineral include the resistance to breakage (tenacity) and patterns of breaking—for example, fracture (or random breaking) versus cleavage (breaking in parallel planes). The degree to which a mineral resists scratching is called the hardness, and the classic Mohs scale of hardness lists a series of common minerals in order of increasing hardness, ranging from talc at 1 to diamond at 10. The property related to the mass of the mineral is the specific gravity, which is the ratio of a mineral's mass to the mass of an equivalent volume of water. Although more accurate methods of measuring density are available, in many cases, simple comparison of two mineral samples by hefting them in opposite hands provides sufficient information for identification. Additional properties that may assist in the identification of a mineral include determining a mineral's magnetism, its reaction to dilute hydrochloric acid, its odor, its water solubility, and its feel when handled.

Selected References

Back, M. E., and Mandarino, J. A. 2008. *Fleischer's Glossary of Mineral Species* (10th ed.). Mineralogical Record, Tucson, AZ. 346 p.

Bayliss, P. 2000. *Glossary of Obsolete Mineral Names.* Mineralogical Record, Tucson, AZ. 235 p.

Bloss, F. D. 1971. *Crystallography and Crystal Chemistry.* Holt, Rinehart and Winston, New York. 545 p.

Buerger, M. J. 1945. The genesis of twin crystals. *American Mineralogist,* v. 30, pp. 469–482.

Klein, C. K. 2002. *Mineral Science* (22nd ed.). John Wiley, New York. 641 p.

Knoop, F., Peters, C. G., and Emerson, W. B. 1939. A sensitive pyramidal diamond tool for indentation measurements. *Journal of Research of the National Bureau of Standards,* v. 23, pp. 39–61.

Palache, C., Berman, H., and Frondel, C. 1944. *The system of Mineralogy of James Dwight Dana and Edward Salisbury: Dana* (7th ed.). John Wiley, New York. 834 p.

Notes

1. The chains are, of course, not infinite in the mathematical sense, but compared to a tetrahedron approximately 2Å in width even a chain 1 cm long is effectively infinite.
2. These surfaces may be fracture surfaces, or they may be cleavages or partings that can be mistaken for crystal faces.
3. In electromagnetic radiation, wavelength and energy are correlated; long wavelength waves are low energy and short wavelength waves are high energy.
4. Bloss (1971).
5. Just as photons are the quanta of light, vibrational waves can also be considered as having quantized equivalents called phonons.
6. In general, the re-emission direction differs from the incident direction by a slight angle, resulting in an apparent bending (refraction) of the incident light. The angle is a function of how much the re-emitted light has been slowed relative to the incident light, and the ratio of these two velocities—the index of refraction—is an important property of the material in question.
7. Virtually all transparent materials reflect or scatter some light; a perfectly transparent object would be invisible!
8. This assumes that the incident light is white; if the incident light is colored, the reflected light will be the same color.
9. This effect is demonstrated by placing powdered NaCl or KCl into the sample holder of an X-ray powder diffractometer and illuminating an area of the powder for several minutes.
10. Although most commonly white, some black streak plates are available and are particularly useful for specific color distinctions.
11. See chapter 6.
12. See chapter 6.
13. Blacklight posters and glow-in-the-dark paints are everyday examples of luminescent materials.
14. In silicate minerals, the hardness is dependent not on the Si-O bonds in the silicate tetrahedra, but on other bonds in the structure. This accounts for the variability in silicate mineral hardness from 1 in talc to 8 in topaz (Klein, 2002).
15. The most spectacular exception to this can be found in superconductors, which repel magnetic fields so strongly that they can levitate magnets, a phenomenon called the Meissner effect.
16. Instruments such as the Franz magnetic separator make use of this fact to separate the minerals for analyses in crushed rock samples.

Advanced Methods of Studying Minerals and Rocks

6

INTRODUCTION

The unique combination of mineral chemistries and crystal structures lead to properties that make mineral identification possible. The methods by which minerals can be identified in hand specimens by many characteristic properties were introduced in chapter 5. There are important properties not discussed there, however, that vary only slightly from mineral to mineral and cannot be distinguished in hand specimens or by visual observation. In addition, in some samples minerals occur as grains that are too small to be evaluated using hand-specimen methods. For this and other reasons, more sophisticated techniques have been developed for mineral description and identification. One such technique useful beyond the standard hand-specimen analysis methods was provided in chapter 5 in the discussion of specific gravity, where we note that the "hefting" method is useful for larger specimens with significant differences in density, but describe balance and pycnometric techniques that provide useful results for small samples or those with subtly differing specific gravities.

A complete understanding of the nature and origin of minerals and rocks requires information that cannot be obtained from hand-specimen analysis. Analyses beyond the hand-specimen level typically require some specialized training and use of expensive and sophisticated equipment. A thorough discussion of these more sophisticated methods of analysis is well beyond the scope of this book, but a basic introduction to a few techniques (1) will help students begin to understand how these techniques are applied; and (2) will inform discussions of chemical properties and other characteristics of Earth materials in subsequent chapters. Note that advanced methods are simply additional ways of characterizing the chemical and structural aspects of minerals and the chemistry of other Earth materials. More thorough descriptions of advanced techniques are available in textbooks, some of which are listed in the references.

HISTORICAL METHODS

Some of the advanced analytical techniques date back more than a century. Such early methods can be readily divided into chemical and physical methods of analysis. For example, in past centuries relatively simple chemical tests, devised for qualitatively determining mineral chemistry, used a device called a blowpipe and an alcohol lamp (e.g., Brush and Penfield, 1911). A horizontally directed flame was used to create characteristic light emission spectra generated from powdered minerals by some elements heated in the flame. The blowpipe was also used to test how easily mineral fragments would melt—determining their fusibility—and to produce a characteristic residue from volatilization of mineral fragments on blocks of charcoal.

Perhaps the predominant, pre-electronic physical analysis technique involved measurement of interfacial angles of crystal faces, a technique known as **goniometry** (e.g., Phillips, 1963). Recall that Steno's Law of Constancy of Interfacial Angles states that the angles between particular faces for a given mineral crystal are always the same. Simple calculations based on goniometric measurements allowed the determination of the **axial ratios**—ratios of the lengths of the unit cell axes a, b, and c. For all minerals of symmetry lower than isometric, these ratios are unique and can be used as an aid in identification.

These historical techniques have been replaced by far more sophisticated instrumentation that allows the collection of far more accurate data. Yet, the earlier methods set science on the track of collecting both details of chemistry and information on crystallography.

Optical Mineralogy

Optical mineralogical analyses are physical analyses that make use of the properties of light. These analyses are commonly extended to the study of rocks, as well. Light consists of electromagnetic waves, with each color having a somewhat limited wavelength (λ) range (figures 5.12; 6.1). Light may travel either through empty space or through some material called a *medium* (chapter 5). In a vacuum, light waves move at a very high speed.[1] Normally, as light waves travel, they vibrate in a very complex way involving both electric and magnetic vectors. Each of the atoms in a light source, which is a very large number of atoms, rapidly radiates waves (1) that are not in phase (meaning that the peaks and valleys of the waves are not coordinated) and (2) that have varying vibration directions. Consequently, the large number of individual waves arriving simultaneously at any given spot from all the atoms in a white light source collectively (and for our purposes here) presents the *appearance* of a ray of light that vibrates in all possible directions perpendicular to the line of travel (see the left side of figure 6.2).

When light strikes the surface of a material, part of the light energy is absorbed by the orbital electrons of the atoms that make up the material, producing increases in the vibrations of the atoms of the material. Some light is re-emitted, contributing to the processes of reflection, scattering, and transmission described in chapter 5. In the case of light transmission, both the speed of the light and nature of its vibrations can be changed.

Light Interactions with Matter

When light is transmitted through any medium—whether solid, liquid, or gas—it is slowed down when it is absorbed and re-emitted by the orbital electrons of the atoms making up the medium. The amount of speed reduction is a function of how many electrons are present in these atoms, a characteristic described as the *electron density*. The directions in which light waves vibrate may also be restricted, a phenomenon called **polarization** (figure 6.2).

Rather than attempting to determine the absolute speed of light within any given material, it is instead customary to refer to the **index of refraction** (designated by the letter *n*), defined as the ratio of the speed of light in a vacuum to the speed of light in the medium. For example, if a material slows light to 50% of its speed in a vacuum, *n* for the material is 2.0

Figure 6.1
Travel direction and wavelength of light waves. This two-dimensional view shows that light wave vibrational vectors vary in size in a sinusoidal pattern and are oriented perpendicular to the direction of travel of the wave (in this case, toward the right).

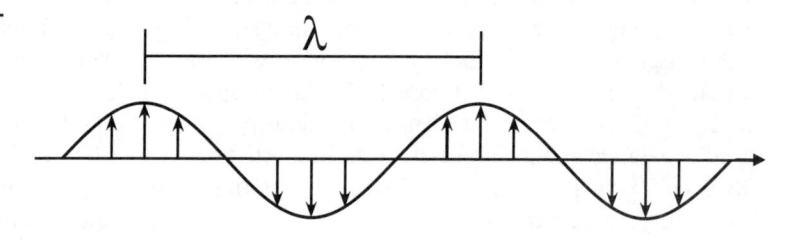

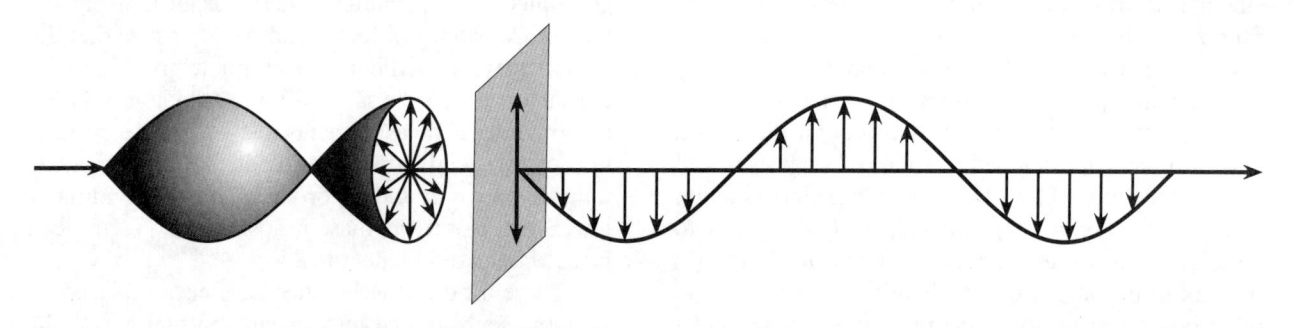

Figure 6.2
Polarization of waves. A polarizing filter (shown here as a gray plane surface) polarizes the incoming light by only permitting light that is vibrating in a single direction to pass through. Note that a second polarizing filter oriented at right angles to the first and allowing light vibrating only sideways (versus up and down) to pass through, would block all the remaining light from passing.

(speed of light divided by ½ speed of light = 2).[2] If the electron density, on average, is the same in every direction in a material, then the index of refraction is also the same in every direction, as is the nature of the light vibration. This is true for gases, liquids, amorphous solids such as glass, and crystalline solids that have isometric symmetry. Materials in which the index of refraction is the same in all directions are called **isotropic**. In contrast, **anisotropic** materials are those in which the electron density differs significantly with direction. As a result, in such materials, the velocity of light and the index of refraction differ in various directions, and materials will have more than one index of refraction. One consequence of this is that light may be strongly polarized on passing through such materials. Some minerals do not transmit light, especially many metallic minerals, and therefore are **opaque**. They do not have indices of refraction measurable using a standard petrographic microscope.

The multiple indices of refraction in anisotropic materials depend on the direction light travels through the crystal. In addition, the symmetry of the crystal limits the number of unique or *principal* refraction indices that can exist. Crystals that have tetragonal, hexagonal, and trigonal symmetry have two principal indices of refraction, one *parallel* to the unique c-axis and designated as n_e, the other *perpendicular* to the c-axis and designated as n_o. These two can be used to define an ellipsoid called the *uniaxial indicatrix* (figure 6.3a). All other indices of refraction have values that lie between the two principal values of n_e and n_o. Crystals that have tetragonal, hexagonal, and trigonal symmetry are called **uniaxial** because there is one unique optic axis—that is, one axis perpendicular to which all refractive indices are equal (figure 6.3a). In figure 6.3a, the n_e is larger than the n_o index and the mineral is said to be positive in sign. If $n_o > n_e$, then the mineral is said to be negative in sign. These signs are measured using the accessory plates inserted above the stage, as described below.

Crystals with orthorhombic, monoclinic, or triclinic symmetry have three principal indices of refraction (n_α, n_β, and n_γ), which do not necessarily

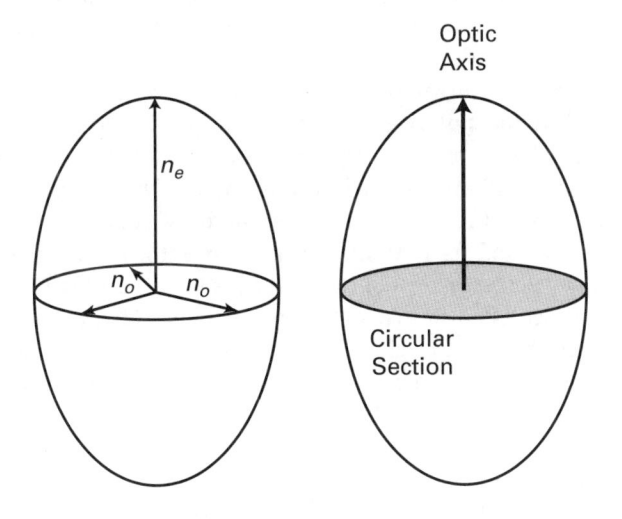

(a) Schematic diagram of principle optical directions in uniaxial crystals (the uniaxial indicatrix). The refactive index along the c-axis is n_e; perpendicular to this, all directions in the circular section have the same refractive index, n_o. Any direction other than these varies between n_e and n_o, depending on the geometry. Note that for this example, $n_e > n_o$, but the opposite case will also occur, with $n_o > n_e$. The single optic axis in uniaxial minerals is perpendicular to the circular section.

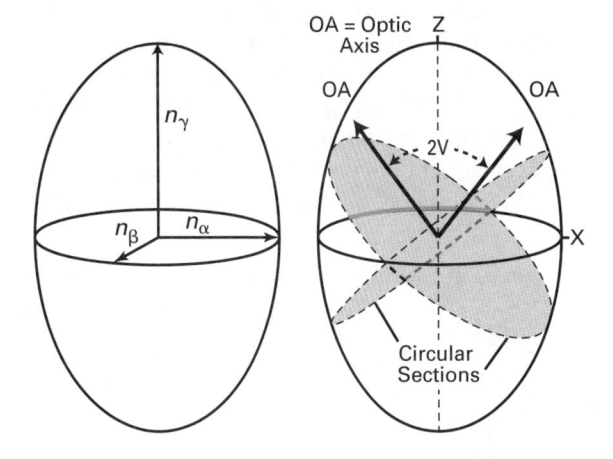

(b) Schematic diagram of principle optical directions in biaxial crystals (the biaxial indicatrix). None of the refractive indices are constrained to lie along the unit cell axes. Indices are labeled so that $n_\gamma > n_\beta > n_\alpha$. In this case there are two separate circular sections and two separate optic axes perpendicular to each of them.

Figure 6.3
Optical indicatrices.

correspond with any unit cell axes (figure 6.3b). These three indices define the shape of the biaxial indicatrix. Such crystals have two circular sections in which all refractive indices are equal, and therefore, have two optic axes (perpendicular to these circular sections). As a result, such crystals are called biaxial (figure 6.3b). The angle between the two axes is measurable and is called the 2V. If the acute angle of the 2V is centered on the Z axis of the indicatrix, the mineral has a positive sign, whereas if the acute angle is centered on the X axis, the mineral has a negative sign.

The effects of light passing through anisotropic crystalline materials are normally too subtle to be observed in hand samples of minerals. These effects, however, can be observed easily using a clear cleavage fragment of calcite, commonly called "Iceland Spar" (figure 6.4). When this fragment is placed over an image on a piece of paper, two versions of the image will appear on the upper surface of the crystal, representing the two rays of light that follow the two directions of the principal values of n in this uniaxial material. In rotating the crystal, two observations can be made: (1) one version of the image appears higher than the other—the one representing the faster ray of light; and (2) one image will appear to revolve around the other—reflecting the fact that the light rays do not follow the same paths. The two end-member rays of light are called the *ordinary ray* and the *extraordinary ray*, corresponding to n_o and n_e (figure 6.3a). Both of these rays of light are polarized by the calcite; that is, they are constrained to vibrate in only one direction.

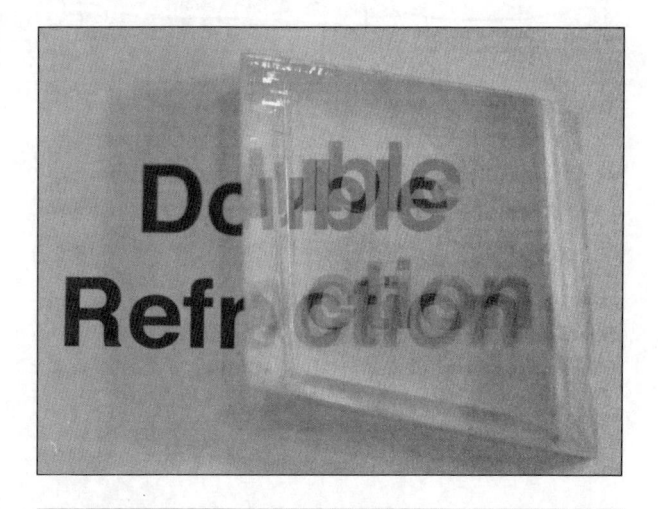

Figure 6.4
Double refraction through a cleavage rhombohedron of high-quality, transparent calcite, called "Iceland Spar."

Design of the Petrographic Microscope

In order to observe and measure the special properties of minerals as they interact with light, a special kind of microscope has been designed—the petrographic microscope. The petrographic microscope is the "workhorse" of geologic studies, offering insights into not only the identity of minerals but also the relationships between or among minerals combined in rocks. As with typical biological microscopes, light passes up through a hole in the stage and through the transparent materials of interest that rest upon the stage. The light is then focused by *objective lenses* of different magnifying ability, so that the object on the stage can be observed through the eyepieces or *oculars*. Petrographic microscopes differ from traditional microscopes in a number of important ways (figure 6.5). The most obvious difference is that petrographic microscopes have a circular stage that can be rotated through 360°. A *vernier* along one edge of the stage allows for the precise determination of the amount of rotation, which helps in mineral identification. Less obvious are the placement of a pair of perpendicularly oriented polarizing filters: one below the stage and fixed in the *optical path* (the line that the light follows from the mirror or light source to the ocular lens) called the **polarizer**; the other above the stage and movable in or out of the optical path, called the **analyzer**. The process of inserting the analyzer is called "crossing the polars" or "crossing the Nicols," the latter derived from William Nicol, inventor of the first analyzer used in such microscopes (crossed polars or crossed Nicols is usually abbreviated as XN). A polarizer absorbs all light except that vibrating in a single direction. If two polarizers are oriented so that one allows only "east-west" vibrating light to pass and the other allows only "north-south" vibrating light to pass, all light under crossed polars (XN) is absorbed and the field of view goes dark or **extinct**. An isotropic mineral grain in the optic path, under examination on the rotating stage, is concealed from view because such a grain does not alter the vibration directions of the light. Aniostropic grains, however, alter the vibration directions, polarizing the light and appearing in a variety of colors.

In addition to the polarizing filters, there are also two specialized lenses, one below the stage (the *condensing* or *converging* lens) and one above (the *Bertrand-Amici* lens). These can be moved into the optical path to aid in the determination of the uniaxial or biaxial nature of a mineral grain in question and to determine

Figure 6.5
The petrographic microscope.

(a) Basic features of a petrographic microscope indicated by labeled parts.

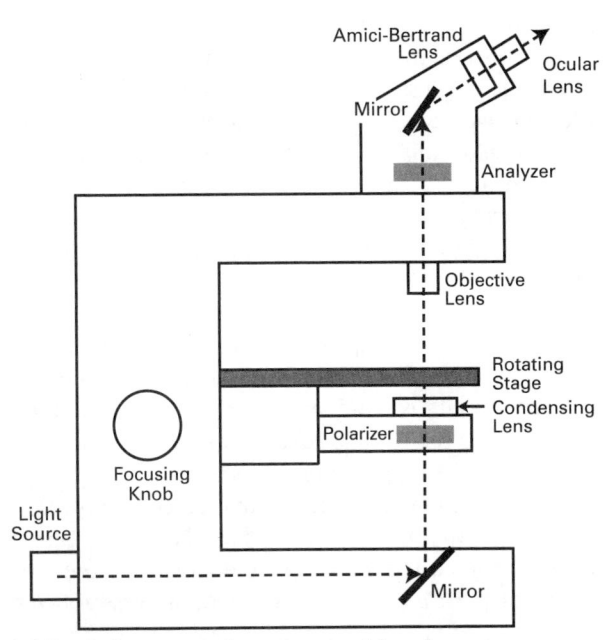

(b) An example of a petrographic microscope (in this case a Leitz™ brand).

other optical properties. One additional feature of the petrographic microscope is a slot in the tube above the stage into which small plates of mineral mounted in metal, called accessory plates, may be inserted. These are used to make determinations of specific optical properties of minerals.

Use of the Petrographic Microscope

Routine studies using the petrographic microscope are of two types. First, but not common, is the examination of a sample of crushed mineral grains scattered on a glass slide in special oils of fixed refractive index (a *grain mount*) for the purpose of determining the optical character and the specific refractive indices of minerals. Since the grains are crushed, they have no relationship to one another. For grain-mount studies, the refractive index (or indices) of the mineral grains is compared to that of a chosen oil, which has a known refractive index. Additional mounts are made with oils of different indices until the mineral and oil indices are closely matched, whereupon the results are compared to measured values of known minerals.

The second, more common, type of analysis with the petrographic microscope involves the examination of a largely translucent slice of a mineral or rock, a *thin section*, typically cut and ground down to 30 thousandths of a millimeter (30 μm = 30 microns) (figure 6.6). In such thin sections, the thickness and positions of grains relative to one another are fixed. Grain orientations can be changed in two dimensions by rotating the stage of the microscope. Specific optical effects, representing the particular optical properties of the minerals in the thin section, are observed, measured, and compared to mental or book images and data of specific effects—standardized for 30 μm. Thus, mineral properties and the minerals themselves can be identified, some information on mineral compositions may be obtained, and—in sections of rock—inter-grain relationships can be observed. In addition, the optical properties of different mineral grains can be compared.

Thin sections of a mineral or rock are made by cutting off, with a diamond-studded saw blade, a small (20 × 40 mm) slab of the sample. Each slab is ground smooth and glued to a glass slide. Most of the slab is then sliced away with another similar saw. The thin layer remaining is carefully ground to a thickness of 30 microns.

The standard procedure for a routine examination of a sample using a petrographic microscope is provided in Appendix C, which also includes visual tools for preliminary mineral identification. Some details of the procedure are also described here, because the technique and properties observed reinforce an understanding of the optical characteristics of minerals. Generally, the minerals in a thin section are first examined in light that only passes through the polarizer, called *plane polarized light*. Several observations may be made. First, relative estimates of the indices of refraction for grains of specific minerals in the thin section are obtained by visual comparison with neighboring grains. The higher the index of refraction of a mineral, the bolder are the outlines of the mineral and the greater the contrast with lower index minerals. This relative difference in refractive indices is called *relief*, and provides a key property for mineral identification. High-relief minerals seem to rise up, whereas low-relief minerals appear flat. Second, under plane polarized light (PL), many minerals are colorless (or nearly so), whereas other minerals display characteristic colors. In some minerals the colors change in intensity or shade when the stage is rotated—a phenomenon known as **pleochroism**. Specific color changes are characteristic of particular minerals. Third, crystal morphologies, revealed by straight edges reflecting the crystal faces, and cleavages, straight lines within the grains, plus the relationships between morphology and cleavage, can be observed. Alternatively, the absence of cleavage may be noted. In addition, minerals such as quartz seldom show alteration, whereas minerals such as the feldspars, which may look superficially similar, are commonly altered and appear cloudy. All of these observations in PL help identify and characterize the minerals in the thin section.

The section is next viewed in crossed polars (XN). By viewing the thin section under XN (i.e., by inserting the analyzer), it is possible to characterize the symmetry of the minerals. As described above, isotropic

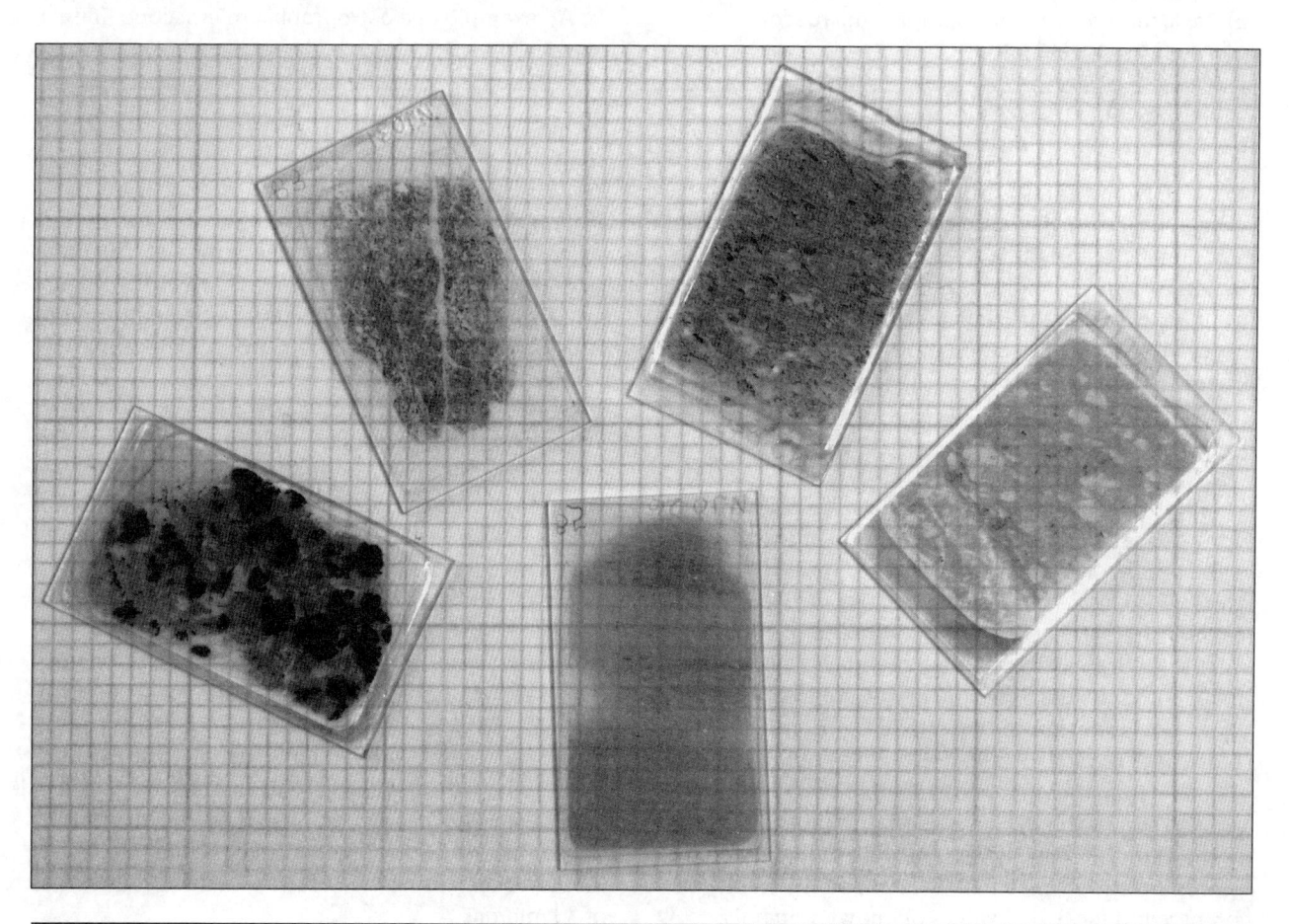

Figure 6.6
Examples of petrographic thin sections of different rock types.

minerals do not alter the vibration directions of the light passing through them, so under XN they become extinct (go completely dark) and remain extinct when the stage is rotated though 360°. Since anisotropic minerals do change the vibration direction of transmitted light, as the stage is rotated under XN through 360°, these minerals become extinct at four positions 90° from one another (unless viewed directly down an optic axis, in which case they appear to be isotropic). For uniaxial minerals, the extinction positions are parallel to cleavage directions and crystal faces, whereas for biaxial minerals the extinction positions are commonly, but not always, at an angle to cleavage directions and crystal faces. The angles are generally characteristic of individual minerals and may be measured by rotating the stage and using the vernier. Between the points of extinction, minerals will display colors resulting from the interference of the different light rays. The characteristics of these colors, which may appear in rainbow–like patterns, are indicative of the differences in velocities and refractive indices of the light rays for the various directions of travel and vibration in the crystal. Specific values of index difference (called birefringence and designated by the Greek letter β) are characteristic of particular minerals. These and related topics are discussed in more detail by Wood (1977), Wahlstrom (1979), and Bloss (1999), and optical data are tabulated in Phillips and Griffen (1981), Deer et al. (1992), and Raymond (2009).

Thin-section analysis is standard practice in petrologic studies. In thin sections, internal features of mineral grains, such as exsolution lamellae, as well as inter-grain relationships (textures), can reveal significant information about the history of formation (and later change) within a rock.

> **Problem 6.1**
> Can an isotropic mineral be pleochroic? Why or why not?

> **Problem 6.2**
> An isotropic mineral (which has only one refractive index), can be identified when the index of the immersion oil matches that of the mineral. Should immersion oils with multiple refraction indices be used to assist in the identification of an anisotropic mineral? Why or why not?

POWDER X-RAY DIFFRACTION (XRD)

Following the discovery of X-rays by Roentgen in 1895, a debate ensued over whether X-rays should be considered as waves or particles.[3] In 1912, German physicist von Laue realized that if X-rays were waves, their wavelength was nearly equal to spacings in crystals, and that they might be diffracted in direct analogy to the diffraction of visible light by a ruled grating.[4] Thereafter, he reported the capture of a diffraction pattern on photographic film, leading to a technique of single crystal and powder diffraction analyses of mineral specimens.[5]

X-Rays and Diffraction

X-rays are short wavelength electromagnetic waves (0.02–100 Å or 0.002–10 nm) (figure 5.12). Geological instrumental analyses using X-rays begin by directing a high voltage beam of electrons into a solid target, normally made of a metal such as copper or tungsten. When the electrons in this beam collide with the atoms of the target, they slow down and give up their excess energy as X-ray photons. This produces a spectrum of X-rays of varying wavelengths analogous to a source of white light. Some of the high voltage electrons from the beam also knock off electrons in lower orbitals of the atoms of the target. Higher orbital electrons then drop down to fill the lower orbital levels, and in doing so also give up their excess energy in the form of X-rays. These X-rays have very specific wavelengths, analogous to a monochromatic (single wavelength) light source. The exact wavelength depends on the differences in energy between the orbital levels that lose electrons and those that supply new electrons. In copper, for example, an electron that drops from an L_3 orbital to a K orbital produces an X-ray of wavelength 1.54056Å. These monochromatic X-rays are used for the study of minerals.

The new X-rays generated by the target material are directed as a beam towards a sample. When X-rays strike the electrons surrounding an atom in the sample, they are absorbed and re-emitted, as is the case with light. When they are re-emitted from the closely spaced atoms within a crystal, the waves can interfere with each other due to their short wavelengths. When the waves are out of phase with each other—that is, when the peak of one wave matches a trough in another wave—they cancel out (figure 6.7a), a process referred to as destructive interference. When the waves are in phase with each other—that is, when the peaks of two waves match—they add together in constructive interference (figure 6.7b). It is this constructive interference that produced the diffraction patterns observed by von Laue.

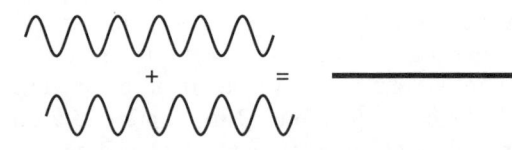

(a) Diagram showing destructive interference. When the crests of one wave are aligned with the troughs of another, the waves are out of phase and interfere destructively, so that no light wave results.

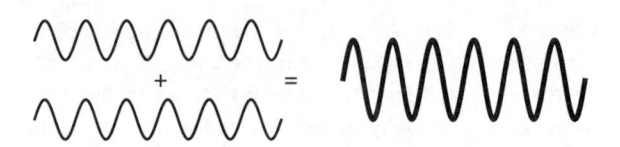

(b) Diagram showing constructive interference. When the crests and troughs of two waves are aligned, they interfere constructively, enhancing the waves.

Figure 6.7
Light wave interactions.

The creation of constructive interference requires that certain geometrical constraints be met. As shown in Figure 6.8, the re-emitted waves will only be in phase if the path length xyz is equal to an integral number of wavelengths. The mathematical formulation of this constraint was developed by William Bragg, and is therefore known as the Bragg equation, which is

$$n\lambda = 2d \sin \theta$$

In the expression, λ is the wavelength of the radiation used, θ is one-half the measured diffraction angle, and d is the spacing between particular planes of atoms within the crystal. The X-rays diffracting from each plane in the crystal produce a single beam and a discrete spot on a piece of film, which is used as a recording device. The beams together produce a pattern of discrete spots collected on the film. The arrangement of the spots is a function of the arrangement of the

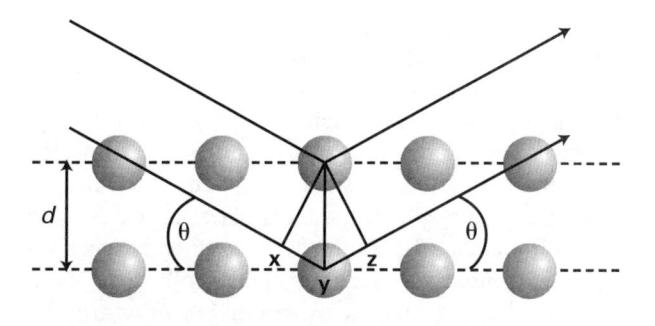

Figure 6.8
The geometry of diffraction. Two incoming waves arrive from the upper left. For the two waves to remain in phase, the distance x-y-z must equal an integral number of wavelengths. d is the distance between layers of atoms and θ is the angle of incidence.

Problem 6.3

On a powder diffractometer using Cu Kα_1 radiation (l = 1.54059Å), a peak is measured at 25° 2U. What is the d-spacing of the planes in the crystal that produced this peak?

atoms within the crystal. The intensity of the X-rays at each spot is a function of the electron density of the diffracting plane and is thereby a function of the types of atoms in the crystal.

XRD Analysis Procedures and Applications

The single crystal method, which produces a von Laue pattern on film, is still the standard for determining the structures of new materials. There are difficulties in conducting such analyses and the method is not used for routine study of geological materials. In routine modern analyses, powdered mineral is used as a target and diffraction creates a series of nested cones of diffracted X-rays, the angles and intensity of which are recorded electronically.

The procedure for XRD powder pattern analysis is as follows. The material to be examined is crushed to a fine powder (< 10 μm), packed into a sample holder ("mount") or sprinkled on a glass slide, and loaded into the diffractometer (figure 6.9a). The X-rays diffract from the specimen according to Bragg's law (figure 6.9b) and the intensity of diffracted X-rays is measured at different angles by a detector similar in basic concept to a Geiger counter. The result is a series of peaks on a plot of intensity versus diffraction angle (figure 6.10). Because each peak represents the diffraction from a given set of planes in all of the crystals in the powder, the peaks can be *indexed* or labeled with the Miller indices of the diffracting planes.

The dominant use of the powder diffraction method is for the identification of unknown materials. Once the measured 2θ values of the peaks have been converted into *d*-spacings via the Bragg equation, the list of peaks can be compared with those of known materials previously measured. Modern computer databases contain several tens of thousands of powder patterns, and a search and match procedure can often identify unknown patterns in a few seconds. Materials within a mixture can also be identified, since the diffraction from one component of a mixture has no impact on the diffraction of another. With careful calibration, it is also possible to determine the relative percentages of each component.

Indexing the diffraction peaks also permits the direct determination of the unit cell dimensions for the material. This can provide information about the composition of the material and the degree of solid solution, as the substitution of one atom for another in a mineral structure almost invariably changes one or more of the unit cell dimensions. Much more information about this useful method of analysis can be found in Bloss (1971) and Moore and Reynolds (1989).

OTHER COMMON ANALYTICAL METHODS

Many other instruments are used for mineral analysis, including the scanning electron microscope and the electron- and ion-beam microprobes. These instruments facilitate various chemical and physical analyses.

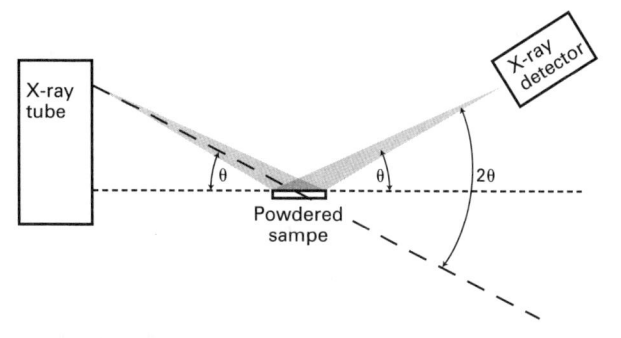

(a) Photograph of a modern (Rigaku™) X-ray powder diffractometer.

(b) The Bragg-Brentano focusing geometry of most modern powder diffractometers is depicted in this diagram, which shows the positions of the X-ray tube and X-ray detector and the relationship between θ and 2θ. The X-ray detector is movable and may be rotated through an angle of 2θ relative to the undiffracted X-ray beam.

Figure 6.9
X-ray diffraction.

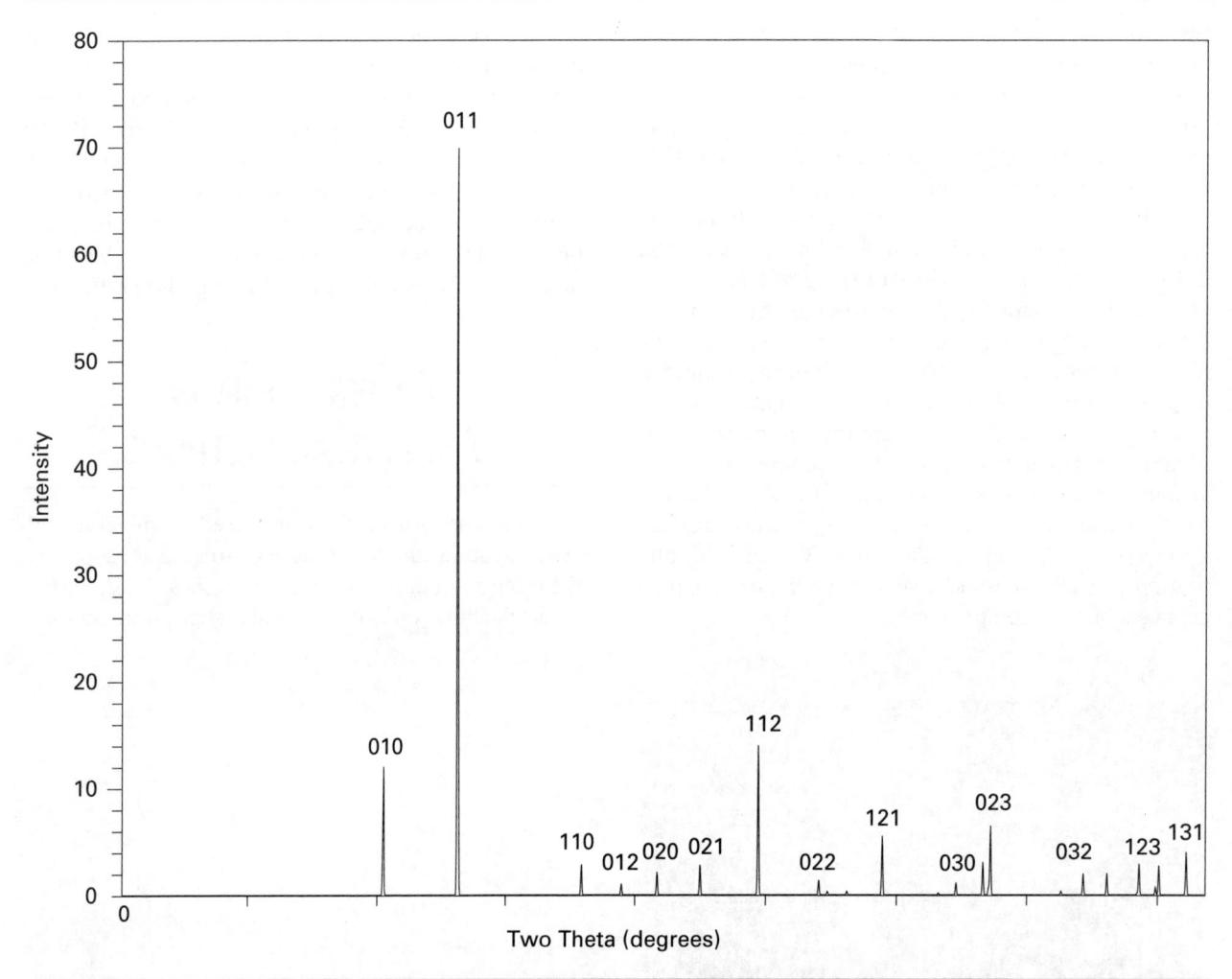

Figure 6.10
A calculated powder X-ray pattern for quartz, using Cu Kα radiation. The peaks represent diffraction from individual atomic planes and are labeled with the appropriate Miller indices.

Electron-Beam Microprobe (EM)

The electron-beam microprobe is an instrument used for chemical analyses of extremely small volumes of minerals, on the order of 10–30 cubic micrometers. In fact, several spots on a single mineral grain can be analyzed for mineral growth history evident as compositional zoning. As discussed above, a beam of high-energy electrons striking a target of a *known* metal generates characteristic X-rays of known wavelength. If the beam of electrons instead strikes a target of unknown composition, characteristic X-rays of unknown wavelength are generated. Measurements of the wavelengths of the X-rays emitted allows identification of the elements in the mineral grain, and measurements of the intensities of the X-rays provides a means of determining the relative proportions of those elements.

An electron-beam microprobe (EM) (figure 6.11) is an instrument designed for just this purpose. It is essentially a modification of a scanning electron microscope (SEM). A normal SEM generates and focuses an electron beam onto the surface of a specimen in order to view the features of that surface at magnifications well beyond that possible with light microscopes. If outfitted with appropriate instrumentation, qualitative chemical analyses that show the approximate amounts of various elements in a sample can also be carried out.

The modifications required to convert an SEM to an EM can be made as part of the original instrument design or by adding specialized equipment to an SEM. The processes of SEM and EM analyses are basically the same. The beam of electrons strikes the surface of the specimen under investigation, generating X-rays. To analyze via wavelength, called wavelength dispersive spectroscopy or WDS, the generated X-rays are focused by crystals with known interplanar *d*-spacings that diffract them into detectors similar to those in diffractometers. These detectors measure the intensities produced. Energy dispersive spectroscopy (EDS) collects the X-rays in detectors of a different design, which sort the X-rays by the energies of the photons collected.[6] The analyses either may be qualitative, with the results simply noting the presence of different elements, or quantitative in a table reporting the specific amounts of each element. Quantitative analysis requires periodic analysis of standards of known composition and subsequent comparison of the data from the unknown with the data from the standard materials of known elemental composition.

Figure 6.11
A sketch of the basic features of an electron-beam microprobe (right) and a photo of it (below).

> **Problem 6.4**
>
> The usefulness of quantitative analysis results from an electron microprobe is fairly self-evident; of what use might qualitative results be?

X-Ray Fluorescence (XRF)

X-ray fluorescence analysis is another technique of chemical analysis. In general, larger volumes of sample are required for XRF. Unlike the electron beam in a microprobe, the original X-rays are not as readily focused on a small area for the analysis of individual grains, which makes XRF useful mostly for bulk chemical analyses of rocks (rather than minerals, although both mineral and rock analyses are possible). The major advantage of the method is that the sensitivity is high, allowing for accurate analyses of elements that may occur at very low levels, such as rare-earth elements. As with the microprobe, analyses can be conducted either by measuring the wavelength or the energy of the X-rays that are produced, and may be qualitative or quantitative in nature.

As with the previously described techniques, X-rays and their behaviors are at the core of XRF analysis, but one major difference exists. In the same way that high-energy electrons generate characteristic X-rays as they interact with the atoms of a target material, other types of radiation can also generate characteristic X-rays. Rather than electrons knocking out the lower orbital electrons in the target material atoms, in XRF, high-energy photons can be used to produce the same result.

To use an X-ray source to produce this interaction, X-rays must be of an energy level higher than those that would be produced by the target. This is directly analogous to the process of using UV light to generate visible light in fluorescent minerals, and hence, the analytical method is referred to as X-ray fluorescence.

The XRF procedure begins with grinding of the sample to a powder. This improves the homogeneity of the sample and the reproducability of the resultant analyses. The powder is commonly fused into a glass with a fluxing agent, added to minimize any effects caused by individual grains. Then, as in the other techniques described here, the sample is targeted with a beam and X-rays generated are measured with electronic detectors.

> **Problem 6.5**
>
> Figure 5.15b shows (schematically) how visible light is emitted when a fluorescent mineral is illuminated with ultraviolet light. Sketch this diagram, then add an additional sketch at the same scale showing how X-ray fluorescence compares to this.

> **Problem 6.6**
>
> If an acquaintance were to bring you a sample from his "rock" collection and tell you he thinks it is a solid native platinum nugget, what steps might you take to provide your friend with information about his sample? Consider the procedures described in chapters 5 and 6 when forming your answer.

Summary

Hand-specimen study cannot be used successfully in all cases to identify a mineral specimen and cannot reveal details of optical, structural, chemical, and textural properties. Unlike properties that can be discerned in hand specimens, other properties may differ subtly between minerals. In other cases, mineral grains may be too small for testing or observation of properties. For cases such as these, specialized equipment or techniques may be required for description and identification of minerals.

Historically, chemical tests were performed using devices called blowpipes, while physical tests included measurement of the angles between the faces of unknown crystals using a goniometer. Modern analyses are conducted using equipment, commonly computerized, that ranges from relatively simple to significantly complex in use, design, and theoretical underpinnings. Optical studies are conducted with the specially designed petrographic microscope. Although similar to a conventional light microscope, a petrographic microscope differs in having a circular, rotating stage and polarizers above and below the stage. Light transmitted through a mineral grain is reduced in velocity due to the interaction between the photons of light and the orbital electrons of the atoms in the mineral. It may also be polarized, having its vibrations restricted to particular directions, and the directions of the vibration rotated. The measure of the velocity reduction is called the index of refraction of the mineral. In minerals with isometric symmetry, light travels in all directions with the same velocity, and the minerals, called isotropic, have only one refractive index. Minerals with lower symmetry are anisotropic and have multiple refractive indices, because light travels in different directions with different velocities. Some minerals do not transmit light, especially many metallic minerals, and therefore are opaque. Use of the petrographic microscope reveals these optical properties of minerals and often facilitates identification of individual minerals. Microscopes also allow study of the relationships between minerals, as well as details of intragrain features like exsolution lamellae and twins.

X-ray diffractometers are used for identification and structural studies of minerals. In X-ray diffraction (XRD) studies, a beam of electrons is directed at a known target material and X-rays—short wavelength electromagnetic waves—are produced. The electrons dislodge inner orbital electrons from atoms of the target, which then emit characteristic X-rays of specific wavelength. The wavelength-specific X-rays are directed in a beam at a study sample. Because X-ray wavelengths are comparable in length to the spacing between planes of atoms in crystals, they are diffracted from crystals where the geometry allows for constructive interference to occur. The resulting, diffracted X-rays appear in distinctive, measurable patterns. Although analysis of single crystals is possible, most geological analyses involve powdered sample diffraction, used to identify unknown materials, determine the components in mixtures, and measure unit cell dimensions.

Among the other instruments used for mineral analysis are the scanning electron microscope and the electron- and ion-beam microprobes. The electron-beam microprobe generates characteristic X-rays, again by sending a beam of electrons to a target material. In the case of the microprobe, however, the target is an unknown material and characteristic X-rays emitted are measured in order to provide a chemical analysis of the material. X-ray fluorescence analysis (XRF) is a broadly similar method of analysis that uses a diffractometer X-ray generator to produce high energy X-rays to stimulate emission of characteristic X-rays needed for chemical analysis of an unknown material. Scanning electron microscopes are used to create highly magnified images of crystals and crystal surfaces, as well as to conduct semi-quantitative chemical analyses.

SELECTED REFERENCES

Bloss, F. D. 1971. *Crystallography and Crystal Chemistry.* Holt, Rinehart and Winston, New York. 545 p.

Bloss, F. D. 1999. *Optical Crystallography.* Mineralogical Society of America, Washington. 239 p.

Brush, G. J., and Penfield, S. L. 1911. *Manual of Determinative Mineralogy with an Introduction on Blowpipe Analysis* (16th ed.). John Wiley, New York. 312 p.

Farmelo, G. (ed.). 2002. *It Must Be Beautiful: Great Equations of Modern Science.* Granta, London. 283 p.

Johnson, N. E. 2001. X-Ray Diffraction Simulation using Laser Pointers and Printers. *Journal of Geoscience Education*, v. 49, pp. 346–350.

Moore, D. M., and Reynolds, R. C., Jr. 1989. *X-Ray Diffraction and the Identification and Analysis of Clay Minerals.* Oxford University Press, Oxford. 332 p.

Perkins, D. 2002. *Mineralogy* (2nd ed.). Prentice-Hall, Upper Saddle River, NJ. 483 p.

Phillips, F. C. 1963. *An Introduction to Crystallography.* John Wiley, New York. 340 p.

Wolfe, C. W. 1953. *Manual for Geometrical Crystallography.* Edwards Brothers, Ann Arbor. 263 p.

Wood, E. A. 1977. *Crystals and Light: An Introduction to Optical Crystallography* (2nd ed.). Dover, New York. 156 p.

NOTES

1. Approximately 3×10^{10} cm/sec^2.
2. The index of refraction (or indices if the material has more than one) is a characteristic property of a material. Identifications of unknown minerals can be made solely by the determination of these refractive indices.
3. The question of waves vs. particles was not resolved until 1923 by de Broglie, who showed that both interpretations were equivalent (see Farmelo, 2002).
4. This phenomenon can be easily demonstrated with a laser pointer and either a fine mesh sieve or an array of dots printed on transparency by a high-resolution printer. See Johnson (2001).
5. The actual experiments were performed by Walther Friedrich and Paul Knipping. See Moore and Reynolds (1989).
6. For electromagnetic radiation, wavelength and energy are inversely related, as shown by the formula $E = hc/\lambda$, where h is Planck's constant, c is the speed of light, and λ is the wavelength of the radiation.

Igneous Rocks

INTRODUCTION

Igneous rocks are rocks formed by the solidification or crystallization of melted rock materials. Recall from chapter 3 that these melts are called *magmas*, but an aspect of these magmas not emphasized in chapter 3 is that magmas may not be entirely liquid. Most magmas contain substantial amounts of crystals, rock fragments, and gases. Where magmas crystallize well below the surface of Earth, they form **plutonic rocks** characterized by visible and generally interlocking mineral grains (color plate 21). Most magmas crystallize (solidify) at depth, but somewhat less than 30% of magmas erupt or crystallize near or on the surface of the Earth to form **volcanic rocks** (White et al., 2006; Cashman and Sparks, 2013) (color plates 24b–e, 25a, b). Volcanic rocks, in hand specimens, are distinguished by one or more of the following features:

- Glasses that largely lack crystals
- Minerals that are too small to see or identify in hand specimens
- Mixes of a few minerals large enough to see in a hand specimen in a matrix called *groundmass*, composed of minerals that are too small to see or identify (color plates 24e, 25a, b)
- Ash composed of highly fragmented volcanic materials with grains the size of sand or smaller
- Breccia, composed of small to large fragments of volcanic rocks that formed previously and were fragmented by explosive eruption or flow of lavas

Igneous rocks are recognized and classified on the basis of the constituent minerals and the textures of the rocks. The **texture** is the combination of the arrangement, sizes, and shapes of mineral grains (and fragments), and the relationships between the grains and fragments. Larger features of the rocks, called *structures*, also aid in recognition of igneous rocks. The minerals, chemical compositions, textures, and structures of igneous rocks all provide clues to the history of the rocks. This chapter describes the features of igneous rocks and the history learned from these features.

THE ORIGIN OF MAGMAS

Because magmas are melted rock materials, they must be formed by melting of rocks somewhere in the Earth. As we consider magma formation, it is useful to distinguish magmas of four types. **Primitive magmas** form by melting of unaltered mantle rock. It is doubtful that any primitive magmas form today, but early in the history of Earth such magmas did develop. **Primary magmas** are melts of any rock type, but melts that are largely unmodified after they form. These magmas contrast with **derivative magmas**, which are significantly modified magmas that have been changed in composition through one or more of several magma-modification processes. **Parental magmas** are magmas that give rise to derivative magmas via modification. Thus, primary magmas may become the parental magmas of derivative magmas.

Primary magmas form by way of three processes—flux melting, decompression melting, and thermally induced melting. **Flux melting** occurs when fluids mix with hot rocks and lower the melting point, resulting in melting of the rocks. This process is important above subduction zones (figure 7.1). Here, where the subducting plate descends into the mantle, the heat at depth in the mantle causes the rocks of the subducting plate to heat up, driving water from them and freeing water from water- and hydroxl-bearing minerals. Thus, this water is of two types: (1) water between grains (intergranular water) and (2) structural water, the water that is part of the composition of the minerals in the rocks. Intergranular water is driven off at relatively lower temperatures than the structural water found in minerals. Minerals such as micas, hornblende, and serpentine—all of which contain, in their formulas, combined water in the form of (OH)⁻—will be transformed by the heat of the Earth into anhydrous (dry) minerals such as feldspar, pyroxene, and olivine. The water driven from the rocks during both heating and mineral transformations will enter surrounding rocks (Peacock, 2003; Brown, 2013).[1] If water enters the hot mantle rocks above the subduction zone, it lowers the melting point of these rocks and they partially melt to form a magma.[2] Once enough magma has formed, it will tend to collect or "pool." Pools of magma tend to rise toward the surface, under pressure, due to the low density of the magma relative to the residual surrounding rocks.

Decompression melting occurs in situations where the pressure is substantially reduced on already hot rocks. Convection in the mantle—the same convection that drives plate movements—and rising masses of hot mantle rock called **diapirs**, will move mantle rocks towards the surface by way of plastic flow. As the rocks rise, the pressure on them is reduced, while the temperatures remain high. Reduction of pressure (or decompression) causes melting at high temperatures, especially in mantle rocks, which contain the chemical components of the magma type called *basaltic* magma.

Experiments in the laboratory have revealed that the melting curve for dry mantle rocks slopes from lower pressures at lower temperatures toward higher pressures at higher temperatures (figure 7.2). As the mantle rock diapir moves plastically toward the surface (toward lower pressure), it will lose relatively little heat (the temperature will not fall much) and it will cross the *mantle solidus* (remember the phase diagrams in chapter 2). Recall that the solidus is the boundary that separates conditions of totally solid rock (in this case, at greater depth) from conditions of partial melting (at shallower depth). As a result, the mantle rock will partially melt, as the pressure falls and the diapir enters the crystal + liquid field (revisit the phase diagrams in chapter 2). These melts will separate from the unmelted (residual) rock and rise more quickly towards the surface, due to their relatively lower density (that is, their greater buoyancy).

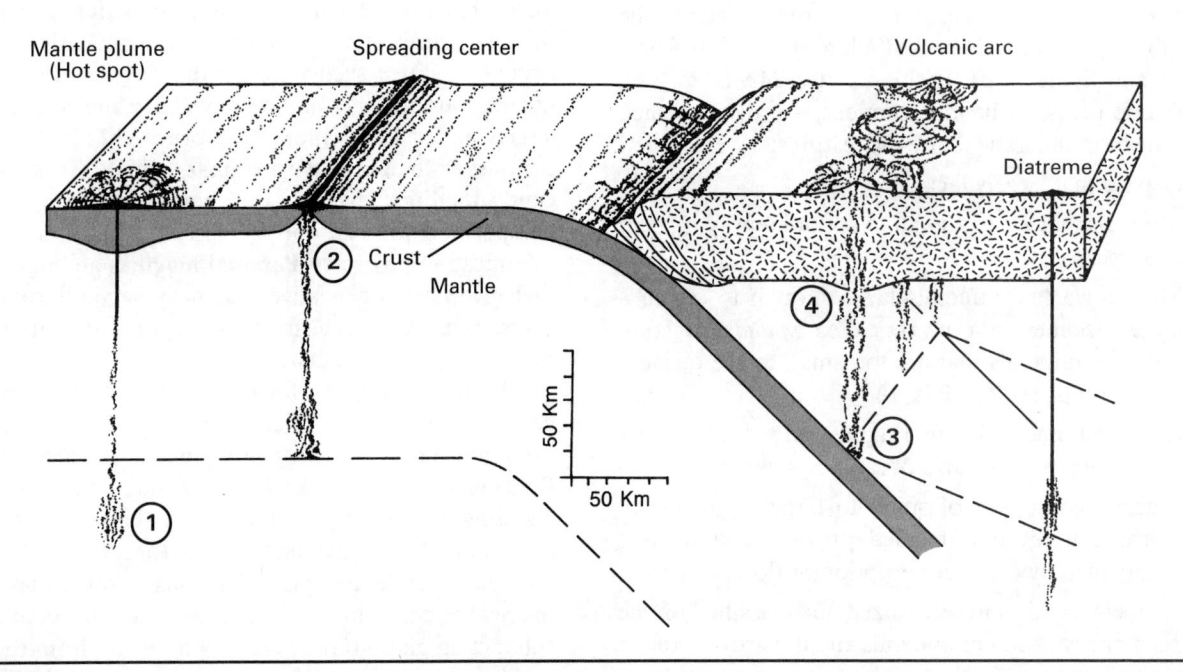

Figure 7.1
Sketch showing selected sites of magma generation. ① Deep mantle plume site (forming a hot-spot at the surface). ② Mid-ocean ridge spreading-center site. ③ Base of mantle wedge above a subduction zone beneath a continental volcanic arc (on an active margin). ④ Base of the crust beneath a volcanic arc. The diatreme on the far right forms from magmas generated deep in the mantle and can be a source of diamonds (see chapter 11).
(*Source:* Modified from Raymond, 2007, p. 88.)

Igneous Rocks 111

Both flux melting and decompression melting of mantle rocks typically yield basaltic/gabbroic magmas—magmas with relatively low silica and relatively high magnesia, iron, and calcium. These magmas are typical parents of the mid-ocean ridge basaltic rocks (MORB) formed at plate spreading centers and many of the rocks formed in volcanic arcs (figure 7.1).

Thermally induced melting primarily occurs at the base of the crust. It also occurs below volcanoes within some subduction zones, where increases in temperature due to friction may contribute somewhat to the overall temperature increase, and hence, to melting.[3] Beneath the crust, basaltic magmas rising quickly toward the surface will move farther and farther from the solidus as they rise (i.e., they "move" figuratively on the temperature-depth graph of figure 7.2, away from the mantle solidus). As they rise, they move well into the field of partial melting. As a result, they have substantial amounts of heat that can be given up before the melts crystallize (solidify or "freeze"). Such a condition, one in which a magma has extra heat, results in the magma being *superheated*. As these superheated basaltic magmas rise beneath continents, they encounter the base of the continent and may accumulate or "pond" there.[4] The heat of the magma pools may be enhanced by heat from radioactive decay of

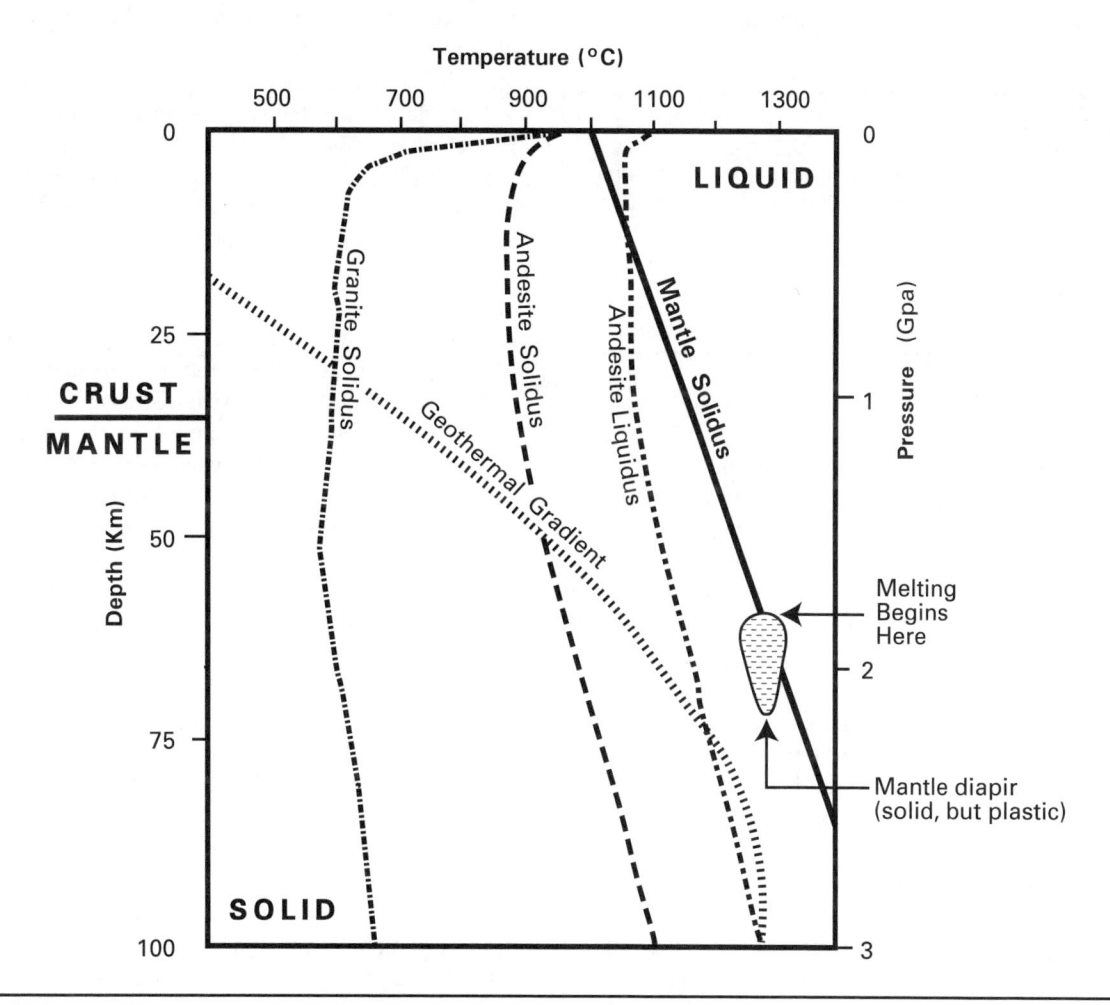

Figure 7.2
P,T grid showing curves for the beginning of water-fluxed melting of granite, melting of andesite, and melting of dry mantle (ultramafic) rock; the curve for complete melting of andesite; and a typical geothermal gradient curve. Melts (liquids) occur to the right of the curves and solid materials occur to the left. A schematic rising mantle diapir (teardrop shaped blob at lower right) is shown crossing the mantle solidus, a point at which melting would begin.
(*Sources:* Modified from Marsh, 1979; with additions from Stern, Huang, and Wyllie, 1975; and the authors.)

minerals, from shearing during deformation, and from exothermic mineral reactions (e.g., Clark et al., 2011; Bea, 2012). Continental crustal rocks, such as granite and muscovite schist, are generally richer in silica, alumina, potash (K_2O), and soda (Na_2O) than other types of rocks, and have a lower melting temperature (a lower solidus) than mantle rocks or basaltic rocks (figure 7.2). As a result, if basaltic magmas pond at the base of the crust, they can give up enough heat to partially melt the crust, particularly if other sources of heat contribute to overall heating of the basal crust. Such heating yields high-silica magma of the *rhyolitic/ granitic* type. Both the rhyolitic/granitic magmas and the basaltic/gabbroic magmas will be primary magmas as long as they are not chemically modified.

> **Problem 7.1**
>
> On a copy of figure 7.2, sketch the path of a granite magma diapir that forms from melting at 750°C at the base of a 40 km thick crust; moves up to 20 km, where it is at 700°C; and moves up to 4 km, where it is at a temperature of 690°C. If the rate of movement is relatively slow, what likely will happen to the magma at this point?

What evidence is there that magmas form in the mantle and at the base of the crust? Mantle and lower crustal sources for magmas are indicated by magma chemistry, including isotope chemistry, **xenoliths** (foreign inclusions of mantle rock) in the magmas, and seismic (earthquake) evidence. For example, under Hawaii, volcanic eruptions may be preceded by a series of earthquakes that occur first at greater mantle depths, and then at progressively shallower crustal depths prior to an eruption.[5] Clearly, magma moving from the mantle towards the surface breaks the rocks along the way, generating earthquakes, and these reveal the movement of the magma from the mantle to the shallow levels of the crust. The earthquakes sequences also can be linked, as they have been at mid-ocean ridges, to changes in minerals that accompany the migration and temporary pooling of the magma, further revealing details of the magma history and pathway on its way to the surface (Viccaro et al., 2016).

The chemical and xenolithic evidence of a mantle source for magmas is of several types. Laboratory studies of mineral stability fields—the P,T conditions under which particular minerals or combinations of minerals are stable—reveal that the upper mantle rocks contain different minerals at different depths. Plagioclase, for example, occurs in mantle rocks at shallower depths, whereas garnet reflects considerably greater depths. Pieces of mantle rock containing garnet, plus many other kinds of rock, occur locally in basalts. These fragments of pre-existing rocks included in magmatic rocks (called either *xenoliths* or *nodules*) reveal the kinds of rocks that exist below the volcanoes, both at the depths of magma formation and the depths along the path through the mantle taken by the magmas on their way to the surface.

In general, the upper mantle consists of rocks dominated by the minerals olivine, orthopyroxene, and clinopyroxene. Rocks containing both olivine and pyroxene are called *peridotites*. Peridotites containing olivine, orthopyroxene, and clinopyroxene (called lherzolite) at the shallowest depths of the mantle may contain plagioclase feldspar as an aluminum-bearing mineral phase (figure 7.3). At these shallow depths, some petrologists also consider that the mantle will contain an omphacite-pyrope (clinopyroxene-garnet) rock called *eclogite*. Magmas and the derivative magmatic rocks that contain xenoliths or nodules of eclogite, plagioclase peridotite, and various crustal metamorphic rocks indicate that the magmas have passed from the uppermost mantle through the crust on their way to the surface (MacGregor, 1968).[6] Inasmuch as they contain no xenoliths of rocks from deeper levels of the mantle, we assume that these magmas formed in the plagioclase peridotite zone of the mantle. Below the plagioclase peridotite zone is a spinel peridotite zone in which the mineral spinel is the Al-bearing mineral phase. Magmatic rocks containing xenoliths of spinel peridotite as well as xenoliths of crustal metamorphic rocks, eclogite, and plagioclase peridotite, likely formed in the spinel peridotite zone of the mantle and picked up pieces of plagioclase peridotite, eclogite, and crustal metamorphic rocks as they rose to the surface. Similarly, garnet is an aluminum-bearing phase at even greater depths in the mantle. Thus, xenoliths of garnet peridotite, along with other types of xenoliths derived from rocks at shallower depths, reflect formation of magma at the greater depths at which garnet is the stable Al-bearing mineral in peridotite.

Chemical evidence of a mantle origin for magmas is based on an understanding of the relationships between various elements and the minerals and rocks in which they occur. Generally, rocks react to attain equilibrium with their surroundings. They do so by exchanging elements with the surrounding rocks. If a

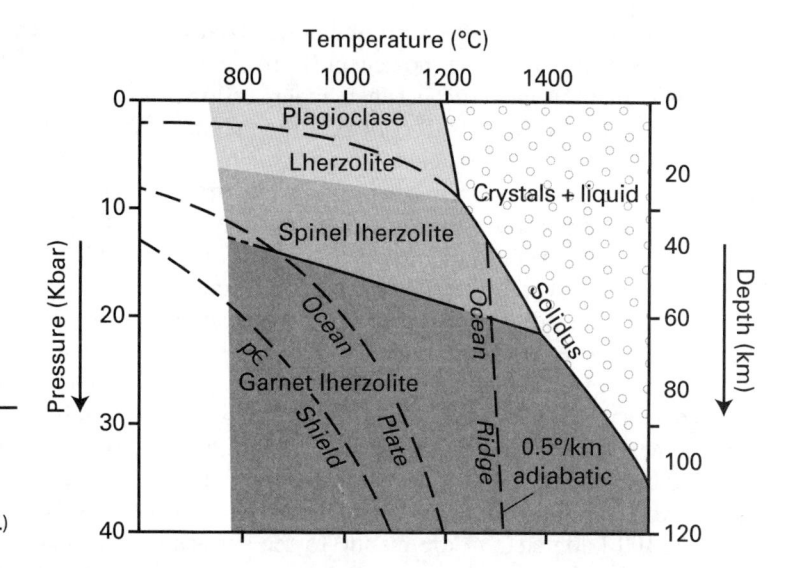

Figure 7.3
Stability fields of upper mantle rock types, with the peridotite solidus.
(*Source:* From Morse, 1980, p. 407. Used with permission.)

rock (or magma) is rich in an element such as Mg but the surrounding rocks are poor in that element, there will be a tendency for Mg to migrate from the rock enriched in this element into the rock that is poor in the element, so that the two rock types have a more balanced distribution, an equilibrium, relative to Mg. In chemical terms, there is a *chemical potential gradient* between the two masses of rock that the attainment of equilibrium will eliminate. Through movement of ions along chemical potential gradients and resulting exchanges of elements between rocks (or magma and rock) of different composition, rocks tend toward a homogeneity of composition. In the mantle, the ratio of $Mg^{++}/(Mg^{++} + Fe^{++})$, a number called the **magnesium number**, is about 0.7. Magmas that have formed in equilibrium with mantle rocks tend to have similar magnesium numbers (about 0.7). In contrast, some crustal rocks may have Mg numbers as low as 0.10 or 0.20 and magmas that might form from such rocks would have similarly low Mg numbers. The Mg numbers of the mid-ocean ridge basalts (MORBs) that make up the seafloor are typically 0.30 to 0.76.[7] The higher values indicate equilibration between the MORB magmas and the mantle rocks. The lower values suggest that low Mg rocks were involved in the history of the magma. Similarly, Sr isotope ratios (Sr^{87}/Sr^{86}) for MORBs are generally 0.702 to 0.705, values typical of the mantle rocks, and are unlike the higher values of 0.710 to 0.730 typical of the rocks of the continental crust.[8] Chemical data such as these indicate that many magmas originate in the mantle. Different values for these chemical indicators suggest that magmas originated in the crust or had crustal rocks somehow involved in their history. If two magmas mix or magmas are derived from more than one source, the evaluation of the chemical evidence becomes more difficult.

In general, melting occurs where two or more minerals (phases) are in contact. (Refer to figure 2.12a showing a eutectic binary system and note that in a cold system, as the temperature rises, the first melt to form will have a eutectic composition because the system has two components, A and B, mixed together. The eutectic melt forms at lower temperature than would a melt of either A or B alone.) The melt initially forms at grain boundaries. As melting progresses, the melt accumulates as pockets, lenses, or layers of melt that join to form dikes, sills, or bodies of melt, particularly along shear zones and foliations in metamorphic rocks (Sawyer, 2001; Brown, 2013).[9] If melts crystallize at depth, they form plutonic rocks. The available arguments and evidence, including experiments and calculations, suggest that only about 5% to 10% of a rock must be melted in order for there to be extraction and migration of melt away from the zone of melting.[10]

MOVEMENT OF MAGMAS

Once magmas form, they move toward the surface, generally due to their low densities and fluid mobility. The temperature of the magma and the surrounding rocks, the size of the magma body, and the physical characteristics of the surrounding rocks also may exert control on magma movement (Burov et al., 2003). Magma movements occur where magmas flow: (1) through cracks created by the force of the magma pushing upwards (a process called *dike propagation*), (2) through zones of shear or conduits (faults and fractures) resulting from strain, (3) through porous mantle rock, or (4) in masses as upside-down tear-drop shaped blobs called *melt diapirs*. In general, movement along cracks and pores is thought to be more rapid than magma movements in diapirs.

The rates at which magmas move have been estimated or calculated using a variety of methods.[11] Rates of movement through magma-induced cracks are estimated to be between 0.83 cm/second and 400 cm/second.[12] At these rates, magmas could move from a depth of 100 kilometers in the mantle to the surface over a period of time ranging from 139 days to 1/3 of a day. Hot, fluid basaltic magmas (magmas low in silica) typically move relatively rapidly, especially where they generate or follow cracks within the mantle and crust.

Diapirs are considered to move more slowly, at rates of 10^{-3} to 10^{-8} cm/second.[13] At such rates, the diapirs would take between 316 years and 32 million years to reach the surface—if they did not solidify. Rhyolitic/granitic magmas that move up in diapirs likely will solidify before reaching the surface, unless their rate of ascent is rapid and their size is large. This is so particularly because the solidus curve for melts of rhyolite (or granite) composition slopes in a peculiar way, so that the silicic magmas moving toward the surface cross the boundary from the zone of partial melting into the zone of solidification as the pressure is lowered near the surface (see the curve of the granite solidus in the upper left of figure 7.2). Furthermore, in small diapirs, the strength of the roof rocks above the diapir may inhibit upward movement (Burov et al., 2003). Therefore, relatively cool, viscous (sticky) rhyolitic or granitic magmas that move relatively slowly through cracks and in small diapirs generally solidify before they reach the surface. Large diapirs, however, may erupt at the surface in volcanic eruptions, as they did at Yellowstone less than a million years ago.[14] In part, this may be due to a combination of buoyancy, rapid accumulation of rhyolitic melt (segregated from crystal mushes), brittle failure of roof rocks, formation of dikes or conduits, and the production of bubbles that exsolve from the magma at low pressures (near the surface) and may help to drive the magma toward eruption (Eichelberger et al., 2000; Burov et al., 2003; Cashman and Sparks, 2013).

MODIFICATION OF MAGMAS

Primary magmas are commonly modified as they move toward the surface. The main processes of modification are magma mixing, assimilation of rock, and fractional crystallization. **Magma mixing** occurs where two magmas of different composition come into contact and mix to form a derivative magma of a new composition. For example, the (low-silica) basaltic magma that ponds at the base of the crust may later mix with the (high-silica) rhyolitic magma formed by melting of that crust, to form a new derivative magma of intermediate composition (a dacitic or andesitic magma). In some cases, conditions of temperature and chemistry preclude magmas from mixing, so they **mingle** to become interspersed as blobs or masses of the two magma types.

Assimilation is a process in which magmas dissolve or absorb surrounding rocks, changing the magma composition. As magmas flow rapidly through fractures, they may create stresses that cause fracturing of the wall rock to yield pieces of rock that may then be assimilated. In diapirs, stresses may similarly contribute to failure of the more brittle enclosing rocks, especially in the roof above the diapir, yielding fragments that are assimilated. Assimilation generally requires that the magma be superheated. If not, the amount of heat required to melt or absorb rock fragments would represent a large enough heat (energy) loss to cause the magma to solidify. Where solidification occurs prior to complete melting or assimilation, foreign rock fragments called **xenoliths** remain unmelted and appear in the solidified rock (figure 7.4a; color plate 2a).

Fractional crystallization is a process in which part of a magma (a fraction) is crystallized and the

---

Problem 7.2

Explain why xenoliths of basalt composition would not tend to melt in a granite magma at a temperature of 680°C intruded to a depth of 5 km.

Problem 7.3

Examine the photograph in figure 7.4a. Considering cross-cutting relationships, describe the geologic history revealed by the rocks and structures (e.g., inclusions) in this photograph.

Igneous Rocks 115

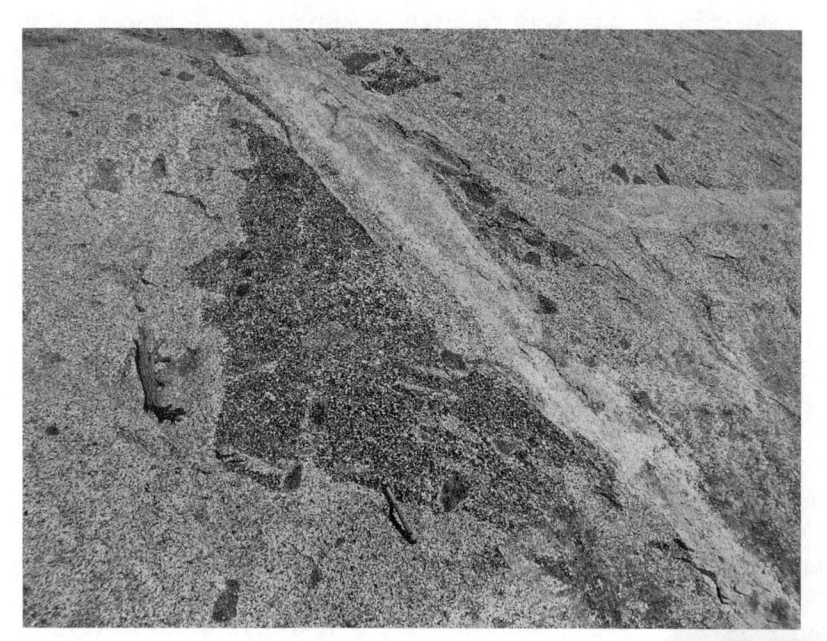

(a) Xenoliths of mafic igneous rock (dark masses of hornblendite in center) in granodiorite of the Sierra Nevada batholith, Wilson Reservoir, CA. Granitoid dikes, including white granite aplite, intrude the granodiorite and xenoliths. Pen (13.5 cm) resting on xenolith-granodiorite contact at lower center provides scale.

(b) Photograph of glacially smoothed, aplitic granite dike in granodiorite, Olmsted Point, Yosemite National Park.

Figure 7.4
Photographs of some plutonic structures.
(Photos by Loren A. Raymond.)

remaining melt is separated from the crystallized rock. This process may occur during magma movements but more commonly occurs while magmas are stored in fractures or magma chambers beneath the surface. The first crystals to form as a magma begins to crystallize (usually crystals of one or two minerals) have a composition different from that of the magma as a whole (remember the crystallization process in the discussion of phase diagrams in chapter 2). Hence, crystallization removes elements from the magma in proportions different from those that existed in the original magma. Preferential removal of certain elements from the magma changes the magma composition.

In general, fractional crystallization leads from magmas generally poor in Si, Na, and K (basaltic/gabbroic magmas) to those richer in Si, Na, and K (rhyolitic/granitic magmas). Early crystallization of Ca-rich plagioclase and the low-silica minerals olivine and pyroxene removes Ca, Mg, and Fe from the magma and ultimately tends to enrich it in Al, Si, Na, and K. The famous petrologist N. L. Bowen noted that during crystallization experiments, early crystallization of Ca-rich plagioclase was followed by successive crystallization of Ca,Na- plagioclase and Na-rich plagioclase in what he called a continuous reaction series (a solid solution change involving no basic change in structure of the minerals). He also noted in his experiments that early-formed olivine changed to pyroxene in a reaction that was more abrupt than gradual and involved a change in crystal structure from olivine, a mineral with isolated silica tetrahedra, to pyroxene, a chain silicate. Combining this information with observations in rocks, he envisioned a discontinuous reaction series involving olivine, pyroxene, amphibole, and biotite. Together, the two series and a high-silica residue are known as Bowen's Reaction Series (figure 7.5). One value of the concept of the reaction series is that it emphasizes the fact that fractional crystallization of a basic (low-silica) magma leads from Mg, Fe-richer, low-silica rocks composed of early-forming minerals to Si, Al, Na, and K-richer, high-silica rocks composed of late-forming minerals. Other reaction series formed from different compositions and under different conditions are possible.

Assimilation and fractional crystallization often occur in the same magma. The combined process, occurring commonly under volcanic arcs, is referred to as **AFC** (assimilation/fractional crystallization).

Some processes of modification are processes that contribute to differentiation of the Earth. In its early history, because Earth formed from accumulated planetesimals (small masses of matter), it was rather heterogeneous in terms of the distribution of elements. Various processes, including partial melting of mantle rocks that concentrates elements such as Si and Al in upward-migrating melt, resulted in concentration of those elements in the crust. Similarly, magma modification processes such as fractional crystallization contribute to differentiation by further concentrating elements such as Si, Na, and K in melts.

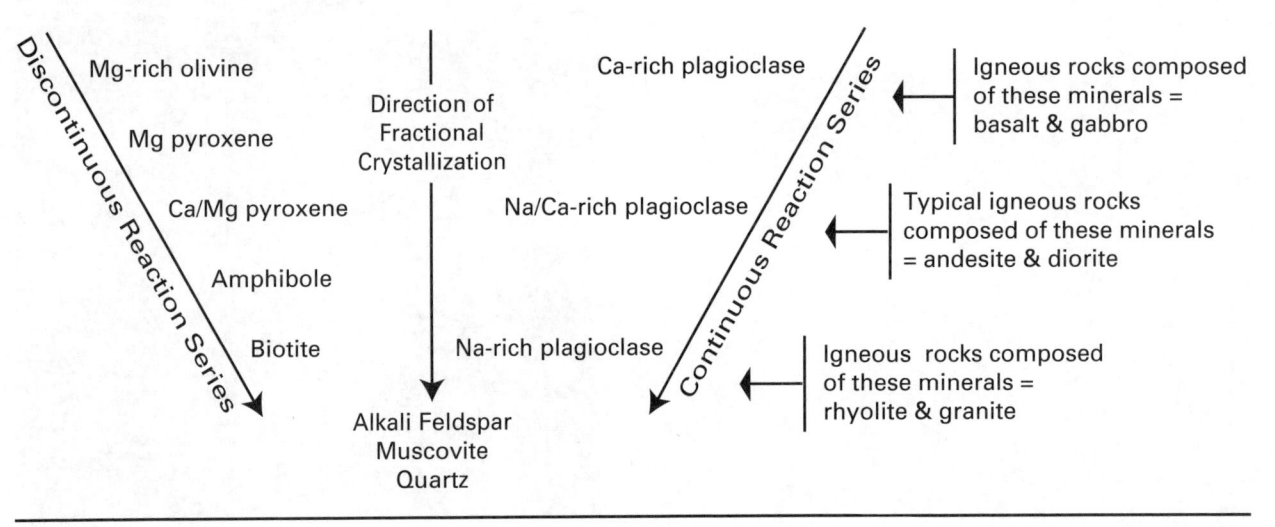

Figure 7.5
Bowen's Reaction Series showing discontinuous and continuous sides. On the right is a list of rocks that would typically form from evolving magmas experiencing fractional crystallization following Bowen's idealized reaction series.

Emplacement and Eruption of Magmas and the Formation of Igneous Structures

Magmas are said to be *emplaced* if they crystallize at depth in the Earth to form the bodies of rock called *plutons*. Plutons are *structures* that have a variety of shapes and sizes and these characteristics reveal something of the emplacement process (figure 7.6). In terms of size, plutons that have a surface area of exposure greater than 100 square kilometers are called **batholiths**. Many batholiths are lens shaped, so that they are longer and wider than they are thick.[15] Large batholiths are usually composite, made up of several smaller plutons. Growing evidence suggests that batholiths and smaller plutons are formed by incremental additions of batches of melt (Glazner et al., 2004; Anderson et al., 2013).[16] Large plutons that are dish shaped and layered are called **lopoliths**. If smaller than 100 km^2 in area of exposure, plutons are called **stocks** or simply **plutons**, provided they are somewhat equant in shape. If they are tabular and cut across surrounding rock structures, the plutons are called **dikes** (figure 7.4b). If they parallel the layers into which they are emplaced (intruded), the plutons are called **sills**. Thick sills with convex-up tops are called **laccoliths** (figure 7.6).

Considerable controversy exists with regard to the nature of the processes most common in batholith and stock emplacement.[17] One aspect of the controversy involves the process of magma movement, whether it is by diapirism or propagation through cracks. The possibilities are that: (1) magmas move through fractures into zones of tension or shear, where space is created for them; (2) magmas move through the crust

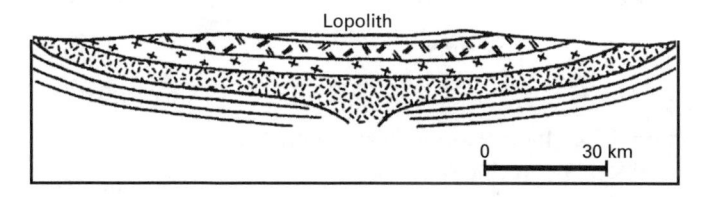

Figure 7.6
Diagrams of selected intrusive (plutonic) structures.
(*Sources:* (a) From Raymond, 2007, p. 19; (b) modified from Gilbert, 1880; (c) modified from Hunt et al., 1953; (d) modified from Balk, 1937; (e) from Raymond, 2007, p. 19.)

(a) Lopolith (note scale).

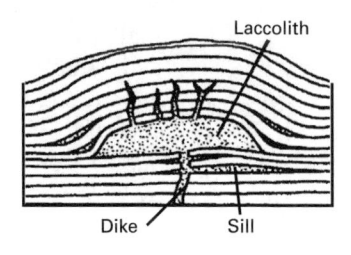

(b) Laccolith (variable scale).

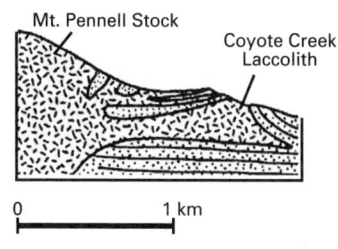

(c) Stock and laccolith complex.

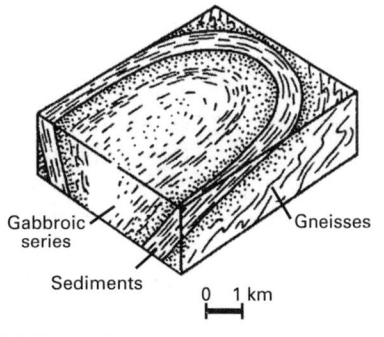

(d) Funnel.

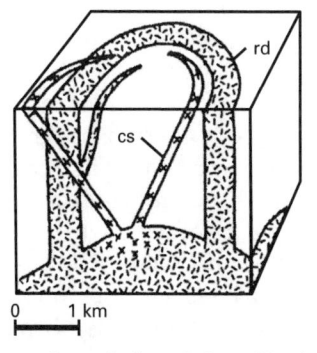

(e) Cone sheet (cs) and ring dike (rd).

by assimilating the rocks in their paths; (3) magmas **stope** their way through the crust (i.e., they break off and engulf blocks of the country rock along their routes toward the surface); or (4) magmas emplace themselves forcefully by diapir "ballooning" and pushing country rocks aside (figure 7.7). While there is evidence suggesting that many batholiths and smaller plutons are formed by incremental additions of batches of melt, evidence supporting the view that each of these processes may take place in one pluton or another suggests that no one process is responsible for the emplacement of all plutons.

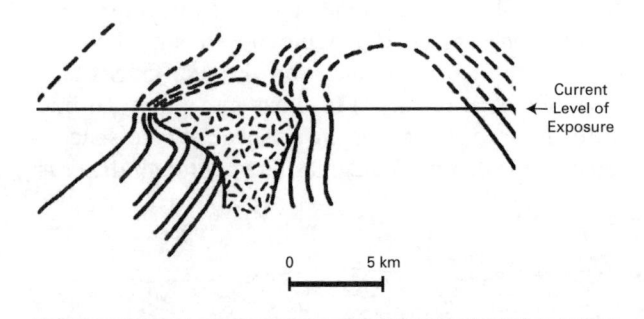

Figure 7.7
Cross-section through the Papoose Flat pluton, Inyo Range, SE California, showing deformation in surrounding country rocks that suggests forceful emplacement.
(*Source:* From Sylvester et al., 1978. Used with permission.)

Some magmas migrate all the way to the surface, where they escape the subsurface in volcanic eruptions. These may be relatively quiet eruptions of lava, like those at Kilauea in Hawaii, or they may be explosive, like the eruption of Mount St. Helens in the State of Washington in 1980. As magmas approach the surface, they may become multiphase mixtures of crystals, gas, and melt.[18] Rapid rise of the magma and resulting loss of pressure (decompression) cause frothing, fragmentation, and explosive eruption of the magma (Sugioka and Bursik, 1995; Cashman and Sparks, 2013). Alternatively, new melt entering a magma body at depth may trigger an eruption, as may stresses that develop within Earth.[19]

The structures produced by large, explosive eruptions are **pyroclastic sheets**. These are tabular masses usually composed of fine-grained material called **ash** that is blown into the atmosphere and spread over hundreds or thousands of square kilometers. Ash is lithified to form the rock called **tuff**. Small-scale eruptions yield smaller layers of pyroclastic material, including ash and breccia layers. Quiet eruptions generally result in the formation of crudely tabular to elliptical bodies of lava called **flows**, but in the case of siliceous lavas, mounds of lava called **domes**, as well as flows, may form.

Where eruptions of lava come from a series of fractures, a lava plateau may form, whereas eruptions from single vents at a point tend to construct the conical mountains—the volcanoes that we usually associate with volcanic rocks. **Lava plateaus** are large, tabular masses of lava flows that typically are hundreds to thousands of meters thick and hundreds to thousands of square kilometers in area. Similar accumulations of flows that erupt from a large number of point sources, such as shield volcanoes, yield a **lava plain**. If lava erupts from a single point source, lava forms volcanoes of a variety of sizes and shapes, each shape depending on the styles of eruptions, the material erupted, and the volume of material erupted. Large volcanoes with relatively gentle slopes are called **shield volcanoes** or **shield cones** and are composed predominantly of low-silica (basaltic) lava flows (figure 7.8a). In contrast, the stereotypical volcanic cone, the stratovolcano or **composite cone**, is smaller in volume and consists of alternating flows and pyroclastic layers ranging from low- to high-silica types of rocks (basalts to rhyolites) (figure 7.8b; 7.9e). Small pyroclastic eruptions of short duration may yield small cones of cinder called **cinder cones,** whereas small cones of lava produced on a flow are called **hornitos**. Mound-like extrusions of highly viscous lava produce **domes** that typically reach only a few hundred meters in height (figure 7.8b).

The surfaces and internal parts of lava flows also have distinctive structures. Pencil-like cooling fractures are called **columnar joints** (figure 7.9a on p. 120). Lavas with ropy, relatively smooth flow surfaces are called **pahoehoe lava**, whereas lavas with flow surfaces characterized by broken fragments of volcanic rock are called **aa lava** (figure 7.9b,c). Rounded to cylindrical structures in lava flows, called **pillows**, form where lavas flow into or are erupted within significant bodies of water (figure 7.9d). Within hand-specimen sized pieces of volcanic rock, gas holes called **vesicles** are common and in older rocks, where these have been filled with minerals, they are called **amygdules**.

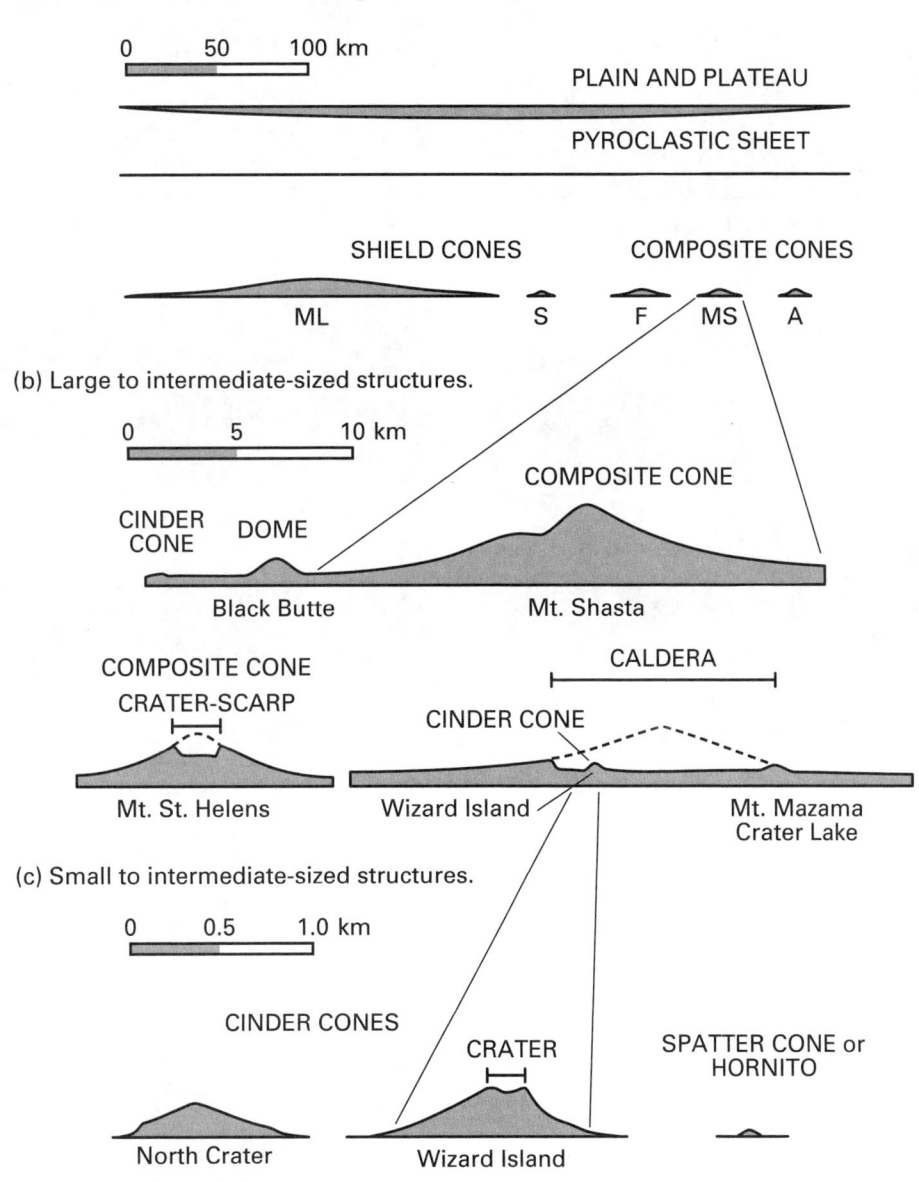

Figure 7.8
Sizes and shapes of various volcanic features. Note the differences in scales among the three panels.
(*Source:* Diagram from Raymond, 2007, p. 10; with panel (a) based on Cook, 1968 and Williams and McBirney, 1979.)

120   Chapter Seven

(a) Columnar joints in andesite, Lost Creek Dam, OR.

(b) Ropy pahoehoe lava, Craters of the Moon National Monument, ID.

(c) Aa (blocky) lava flow, Homestead lava flow, Lava Beds National Monument, CA.

(d) Cross-sectional view of rounded to tubular shaped Jurassic pillow lavas, Golden Gate National Recreation Area, CA. Scale provided by 6-cm lens cap at left center. Pillows, with point on lower side, show top to upper left.

(e) Composite volcano, Mt. Shasta, CA, with smaller cone, Shastina, on right, and cinder cones on lower flank.

Figure 7.9
Photographs of some volcanic structures.
(*Sources:* (a) to (c) Raymond, 2007, figure 2.9, figure 2.8b, and 2.8c, respectively; (d) and (e) photos by Loren A. Raymond.)

Crystallization and Solidification of Magmas and the Resulting Textures

Crystallization of magmas takes place in both shallow and deep locations. As they rise toward the surface, magmas that slow down at depths greater than 2–3 km and then cease to move upward crystallize to form plutonic rocks. These rocks generally have **phaneritic textures**, *in which nearly all the grains are large enough to see and identify with the unaided eye* or a low-power hand magnifier (color plates 21; 22a, b). Magmas that make their way to very shallow depths or erupt at the surface form volcanic rocks characterized by *aphanitic* and related textures. **Aphanitic textures** are those *in which the grains are predominantly too small to see with the unaided eye* or a low-power hand magnifier. **Glassy textures** are those with smooth and curved rock surfaces characteristic of glass. In some cases, rocks composed mostly of glass are characterized by glass fibers separated by abundant, typically elongate *vesicles* (holes). Such textures are called *pumiceous* rather than glassy (color plate 24b).

Magmas typically pass through three stages of crystallization during cooling (figure 7.10). In general, at the highest temperatures a melt exists in which atoms do not form any lasting bonds with other atoms (stage 1). The melt has the characteristics of a high-temperature fluid. Many melts may never exist in this state, because they are not hot enough and because they contain crystal fragments and nuclei. As a high-temperature melt cools, crystal nuclei form and crystals begin to grow at a temperature that allows bonds to persist. This temperature varies depending on the composition of the melt. The variation in temperature means that on a phase diagram showing the stable phases present in the melt or magma, the temperature at which crystals start to form is a curve. Recall from chapter 2 that this is called the *liquidus* (figure 7.10, curve l_1 - l_2 - l_3). Recall, too, that for any given composition, as the temperature falls, the melt of that composition will cool and intersect the liquidus. When it does so, crystals begin to form (i.e., crystals nucleate and grow). The result is that the melt, or magma, then consists of a mixture of crystals and melt (figure 7.10). This mixture represents the second stage (stage 2). When all the melt has crystallized (stage 3), the material has become totally solid (e.g., below point C_3 on figure 7.10). For geologic systems, the magma has crystallized and is now a rock. The crystallization shown in figure 7.10 is a simple one for the solid solution of plagioclase feldspars. Magma compositions are far more diverse and the crystallization history more complex. Still, the general principles apply.

The process of crystallization of a mineral is one in which atoms move from the melt to a solid collection of atoms constituting a nucleus or a crystal surface. Recall from chapter 4 that **nuclei** are small ordered arrangements of atoms and that crystal surfaces consist of planar arrangements of atoms that are typically ordered in the structure of specific minerals.

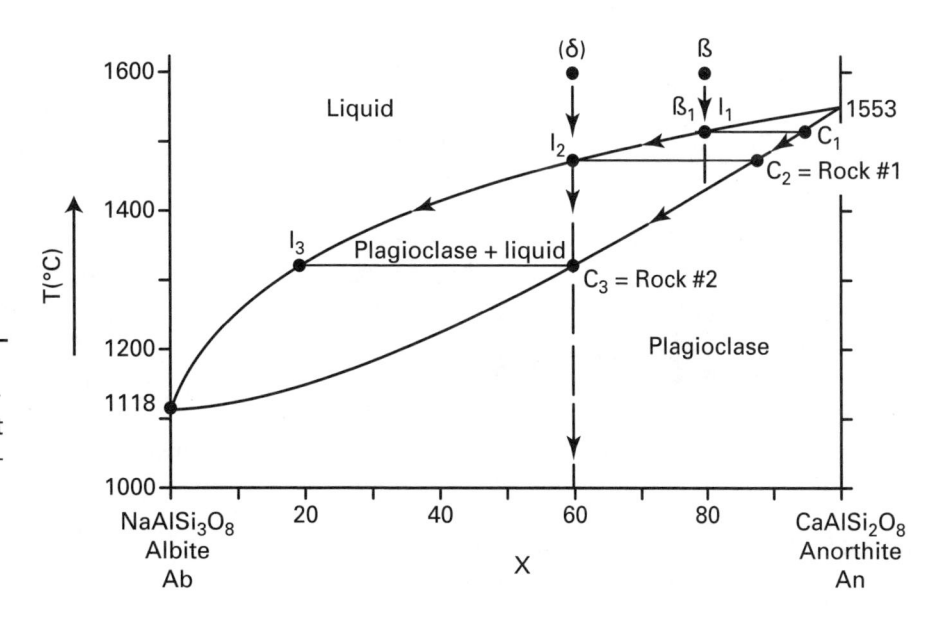

Figure 7.10
The system NaAlSi$_3$O$_8$ (albite)–CaAl$_2$Si$_2$O$_8$ (anorthite). See text for a discussion of the crystallization path shown.
(Source: After Bowen, 1913; Kushiro, 1973; and modified from Raymond, 2007, p. 72.)

Atoms are attracted to this surface as a mineral grows. In rocks, after early crystallization of the first mineral begins, other minerals also begin to crystallize so that at the atomic level during crystallization, a kind of frenetic movement exists in which many different ions migrate toward appropriate crystal faces of the many different kinds of minerals that are forming.

The crystallization of a rock composed of many minerals can be viewed as a simultaneously occurring, two-stage process for each of the many minerals of the future rock. The process for each mineral first involves the *nucleation* of crystals and then involves the **growth** of those crystals (Kirkpatrick, 1975; Swanson, 1977).[20] If no nuclei exist in a melt, the nuclei must first form. For each mineral, recall from chapter 4 that this comes about when energies in the melt are such that the rate at which nuclei can form and retain bonds between atoms exceeds the rate at which the bonds are broken and the incipient or beginning arrangements of atoms disperse back into the melt. Nucleation from a melt (or magma) containing virtually no remnants of former crystals is referred to as *homogeneous nucleation*. Where atoms simply attach themselves to existing surfaces or nuclei to form incipient, growing crystals of a mineral grain, the nucleation is referred to as *heterogeneous nucleation*. Heterogeneous nucleation is more common, both (1) because preexisting nuclei and surfaces often exist in melts, as unmelted crystal fragments or as crystals at the walls of the fractures or chamber containing the melt; and (2) because the energy required is less, inasmuch as fewer bonds must be created.

Once nuclei form, the growth of the mineral grains can occur by the addition of atoms to the ordered arrangement of atoms that formed the nucleus. Growth is influenced by a number of factors including: (1) the composition of the melt, (2) the temperature of the melt, (3) the kind of nuclei present, (4) the number of nuclei present, (5) the rate of cooling during growth, and (6) the rate of diffusion of chemical species through the melt to the growing crystal.[21] For example, recall that high-silica magmas have high viscosity (a property of being "thick" relative to water), which reduces the rate at which atoms can diffuse through the magma. Hence, in such magmas crystals grow more slowly than in others, if all other factors are equal. Conversely, low-silica magmas generally have lower viscosities allowing more rapid ion migration. Viscosity is affected by the presence of crystals as well as bubbles, which form when reduction in pressure during magma ascent allows volatile phases such as H_2O to exsolve from the magma (Cashman and Sparks, 2013).

The factors that control the nucleation and growth of crystals in magma control the texture of the particular rock formed from that magma. Since *texture* is a composite of the size of grains, the arrangement of grains, and the shapes of inter-grain boundaries (the relationship of one grain to another), variations in these features yield numerous different textures. Some of the more common textures seen in hand specimens are depicted in the sketches of figure 7.11. Hand-specimen and microscopic (thin section) views of two plu-

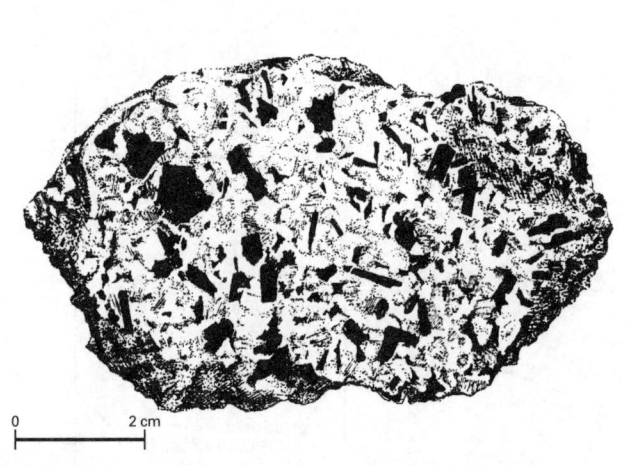

(a) Hypidiomorphic-granular texture in biotite-hornblende granodiorite, Sierra Nevada batholith, CA.

(b) Trachytoidal texture in gabbro. Locality unknown.

Figure 7.11
Sketches of some plutonic textures in hand specimens.
(*Sources:* Raymond, 1984b; 2009, Figs. 10, 13a, and 13c.)

Figure 7.11 *(cont'd.)*

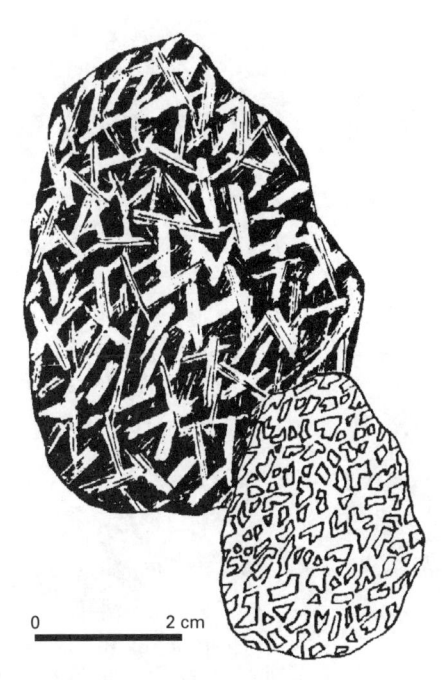

(c) Diabasic texture in gabbro (schematic) showing distribution of plagioclase (white, lath-shaped grains) and pyroxene (black grains). Small diagram shows distribution of pyroxene. Compare (c) to (d).

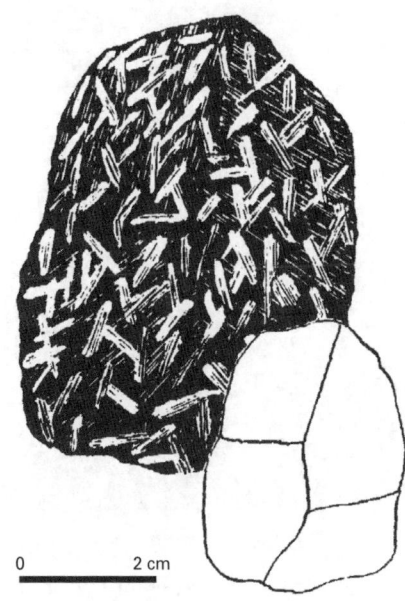

(d) Ophitic texture in gabbro (schematic) showing distribution of plagioclase (white, lath-shaped grains) and pyroxene (black grains). Small diagram shows the size and shape of the large pyroxene grains that enclose the plagioclase laths.

tonic textures are shown in figure 7.12. Some textures in volcanic rocks are depicted in figures 7.13 (on p. 126) and 7.14 (on p. 127).

Among the common phaneritic textures are the hypidiomorphic-granular, diabasic, pegmatitic, and porphyritic textures.[22] **Hypidiomorphic-granular texture**, one of the most common textures of plutonic rocks, is a texture in which, *on average*, grains are equant and subhedral. Typically this texture features a mixture of euhedral, subhedral, and anhedral grains (figure 7.12a, c; color plate 2b, c). **Diabasic texture** consists of randomly oriented, rectangular (lath-shaped) plagioclase crystals with intergranular grains of pyroxene, opaque minerals, or other minerals (figures 7.11c; 7.12d; color plate 2d). **Trachytoidal texture** also typically occurs in feldspar-rich rocks and consists of crudely aligned feldspar crystals surrounded by other mineral grains (figure 7.11b). Some plutonic rocks have very coarse-grained textures with dominant grains larger than three centimeters in length. These are the **pegmatitic textures** (figures 7.12b; 11.5b; color plate 21a). If there are two dominant sizes of visible crystals in a rock, some large and others rather small, the texture is porphyritic and properly should be called **phaneritic-porphyritic**. The few larger crystals in such a texture—the **phenocrysts**—crystallize from a few scattered nuclei. Hence, there is a low nucleation density for that mineral. In rocks in which there is a range of grain sizes from large to small but the sizes are dispersed more or less evenly over that range, the texture is referred to as **seriate**. If a larger grain encloses smaller mineral grains, the texture of the large grain is said to be **poikilitic**, but this is a mineral texture rather than a rock texture (figure 7.12c). In plagioclase-pyroxene rocks, large pyroxenes may include smaller plagioclase grains to form the special poikilitic texture called **ophitic texture** (figure 7.11d). Diabasic texture contrasts with ophitic texture, because in diabasic texture the plagioclases are relatively larger than the pyroxenes and the pyroxenes occur between the plagioclase grains (figure 7.11c). In some cases, a grain will have a core of one mineral and a rim of another, resulting in a **corona texture**. Corona texture in granitic rocks, where the rim is albite and the core is alkali feldspar, is called **rapikivi texture**.

(a) Hypidiomorphic-granular texture in biotite-hornblende granodiorite (slightly weathered), Carson-Iceberg Wilderness near Mosquito Lake, CA. Note the large, nearly euhedral zoned pagioclase grain near the lower center. Scale has 1 cm divisions.

(b) Graphic texture in pegmatitic granite, Pink Glaze Mine near Custer, SD. Dark linear grains are quartz, and the white to light gray material is a single, large pink grain of potassium feldspar. Scale has 1 cm divisions.

Figure 7.12
Photographs and photomicrographs of some plutonic textures.
(*Source:* Photos by Loren A. Raymond.)

Commonly, volcanic rocks are either aphanitic or aphanitic-porphyritic in texture. In **aphanitic-porphyritic texture,** two main sizes of grains exist (a bimodal population of grains)—a set of larger grains and a set of typically aphanitic surrounding grains referred to as the groundmass (figures 7.13a; 7.14a; color plates 24c, e; 25a, b). Such rocks are sometimes called *phyric* rocks. In cases where volcanic magmas solidify under conditions that prohibit the formation of crystals, such as extremely rapid cooling, the resulting texture is **glassy** and the resulting rock, *obsidian*, is composed of glass. Frothing of a flowing magma produces closely spaced tubular holes, giving the rock a **pumiceous texture** (color plate 24b). Where glass is erupted explosively, froths, and is fragmented, it forms glass ash. The texture resulting from the accumulated fragments of the ash, which in some cases are cemented together, is a **pyroclastic texture** (figure 7.14d). In contrast, if a magma erupts more "quietly" to produce flows, crystals may be aligned to form the volcanic equivalent of trachytoidal texture, a **trachytic texture** (figure 7.14b; color plate 24c). The aphanitic equivalent of diabasic texture is **intergranular texture** (figure 7.14c). If feldspar grains or other minerals are surrounded by glass rather than by crystals of other minerals, the texture is called **vitrophyric**.

Grain-size categories in igneous rocks are limited. Very coarse-grained pegmatitic rocks have grains greater than 3 cm, whereas coarse-grained rocks have grains of 3 cm to 5 mm. Medium-grained rocks have grains that on average fall in the 5 mm to 1 mm range. Fine-grain size is less than 1 mm, but below the range of visibility without a microscope the texture is considered to be aphanitic.

Chemical Composition and Classification of Igneous Rocks

Chemically, igneous rocks usually contain between 40 to 77% silica (SiO_2). Chemical analyses of rocks are usually reported in percentages of the

Igneous Rocks **125**

(c) Photomicrograph of Alpine Quartz Diorite, near Alpine, CA, showing hypidiomorphic-granular texture.
B = Biotite   H = Hornblende   Q = Quartz
P = Plagioclase feldspar   Px = Pyroxene (Augite)
Note the small oikocrysts (included small crystals) in some larger crystals, making a poikilitic texture for those minerals.

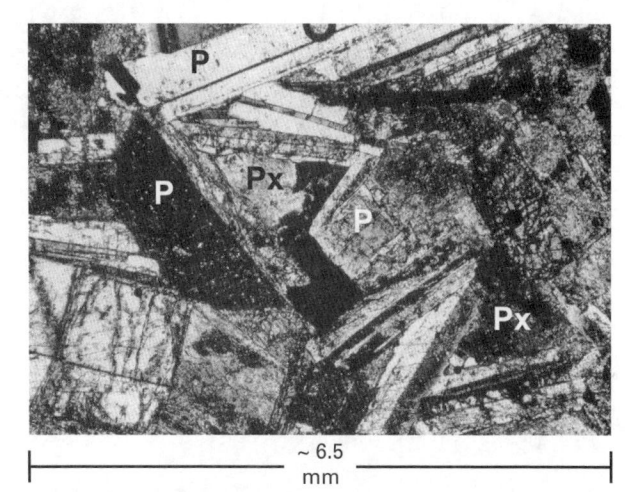

(d) Photomicrograph of diabasic texture in gabbro from Coast Range Ophiolite, Carbona Quadrangle, CA.
P = Plagioclase   Px = Pyroxene
Note how the pyroxene occurs between the somewhat larger plagioclase grains.

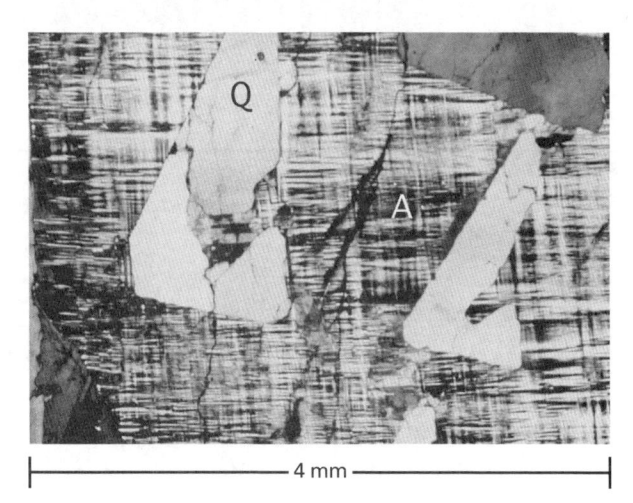

(e) Photomicrograph of graphic texture in pegmatitic granite, McKinney Mine, Spruce Pine District, NC.
A = Alkali (K-rich) feldspar   Q = Quartz

Figure 7.12 *(cont'd.)*

oxides. Examples of representative rock analyses (presented in table 7.1 on p. 126) reveal some of the variations in chemistry. Rocks may be oversaturated in silica, in which case they typically have quartz as a significant mineral. Alternatively, they may be saturated and contain either feldspar without significant quartz or feldspathoid minerals, or they may be undersaturated and silica deficient, containing olivine, feldspathoid minerals, or both. Quartz and feldspathoids do not occur in equilibrium in the same rock. Rocks may be rich in alumina (peraluminous rocks); have moderate amounts of alumina relative to lime (CaO), soda (Na_2O), and potash (K_2O), in which case they are called metaluminous; or they may be relatively poor in alumina but rich in alkalis (Na_2O and K_2O), in which case they are called peralkaline rocks. Peralkaline rocks typically contain distinctive sodic minerals such as the blue amphibole, riebeckite.

All rocks, including igneous rocks, are usually classified either on the basis of a combination of texture and mineral abundance, or on the basis of rock chemistry. Phaneritic igneous rocks, composed of

126   Chapter Seven

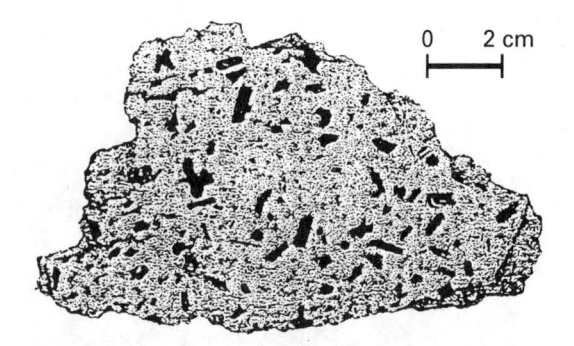

(a) Aphanitic-porphyritic texture in hornblende andesite, from Helena, MT area.

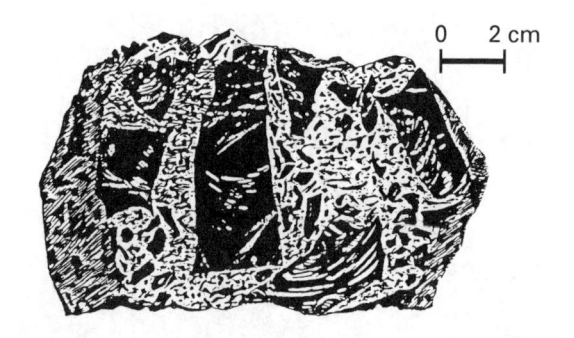

(b) Breccia texture in rhyolite obsidian breccia, Glass Mountain, CA.

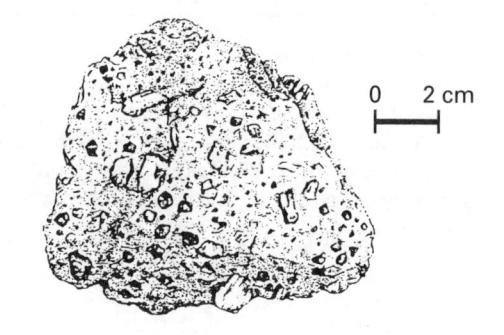

Figure 7.13
Sketches of some volcanic textures in hand specimens.
(*Source:* Raymond, 1984b, p. 42.)

(c) Seriate texture, a texture with a gradation in grain sizes from small to large, in porphyritic rhyodacite from Castle Crags, CA.

Table 7.1 Chemistries of Some Common Igneous Rock Types

Oxide	Dunite	Basalt	Gabbro	Andesite	Granodiorite	Dacite	Granite	Rhyolite
SiO_2	40.08*	49.2	50.12	61.3	63.47	67.6	73.93	77.26
TiO_2	0.01	2.03	0.19	0.8	0.72	0.48	0.18	0.18
Al_2O_3	0.29	16.09	20.01	16.9	15.81	16.1	12.09	11.54
Fe_2O_3	0.31	2.72	0.8	2.26	2.14	3.3	2.91	0.85
FeO	7.62	7.77	4.29	3.08	3.03	0.1	1.55	0.13
MnO	0.11	0.18	0.09	0.08	0.09	0.07	0	0.03
MgO	49.69	6.44	7.91	2.1	2.28	0.9	0.08	0.2
CaO	0.11	10.46	13.97	5.95	4.72	3.2	0.31	0.58
Na_2O	0.05	3.01	1.74	4.36	3.32	4.5	4.66	2.96
K_2O	0.01	0.14	0.05	1.44	3.22	1.9	4.63	4.65
P_2O_5	0.0	0.23	tr	0.18	0.17	0.14	0.0	tr
Other	0.58	1.65	0.73	0.0	1.32	1.36	0.71	1.99
Total	98.86	99.92	99.9	98.45	100.29	99.65	101.05	100.37

* = Values in weight percent. Tr = trace amount
Sources: **Dunite**: Sample DTS-1 (Flanagan, 1976); **Gabbro**: Stillwater Complex, Montana, sample EB43 (Hess, 1960); **Basalt**: "Oceanic tholeiite" (MORB = Mid-Ocean Ridge Basalt), Sample AD-2, Atlantic Ocean (Engle et al., 1965); **Andesite**: Hornblende Andesite, Mt. Hood, Oregon (Wise, 1968); **Granodiorite**: Hornblende-biotite granodiorite, Half Dome Granodiorite, Yosemite National Park, California, Sample 31 (Bateman and Chappell, 1979); **Dacite**: Sample MJG-73, near Bear Point, Central Oregon Cascade Range (Greene, 1968); **Granite**: Quincy Granite, Quincy, Massachusetts (Warren, 1913); **Rhyolite**: "Lithoidal welded tuff" rhyolite, Beatty, Nevada (Cornwall, 1962).

Figure 7.14
Photomicrographs of some volcanic textures. H = hornblende, P = plagioclase feldspar, Px = clinopyroxene.
(*Source:* (a) Raymond, 2007, p. 139; (b) Raymond, 2007, p. 99; (c) Raymond, 2007, pp. 110,126; (d) photo by Loren A. Raymond.)

(a) Aphanitic-porphyritic texture in hornblende andesite, Black Butte, Shasta County, CA.

(b) Trachytic texture in olivine basalt, from east of Barstow, CA. Large olivine phenocryst is surrounded by matrix with aligned laths of plagioclase feldspar.

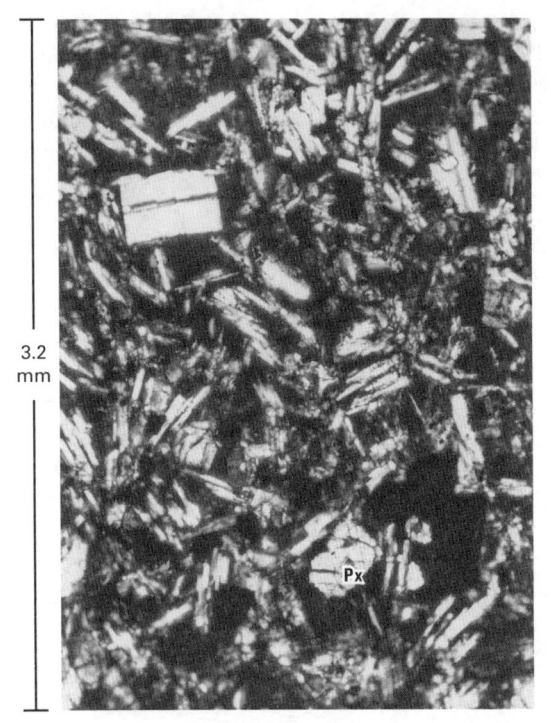

(c) Intergranular texture in basalt, with microphenocrysts of plagioclase feldspar and clinopyroxene (px) in a matrix of lath shaped plagioclase, pyroxene, and opaque minerals (magnetite?). Columbia River Plateau, WA.

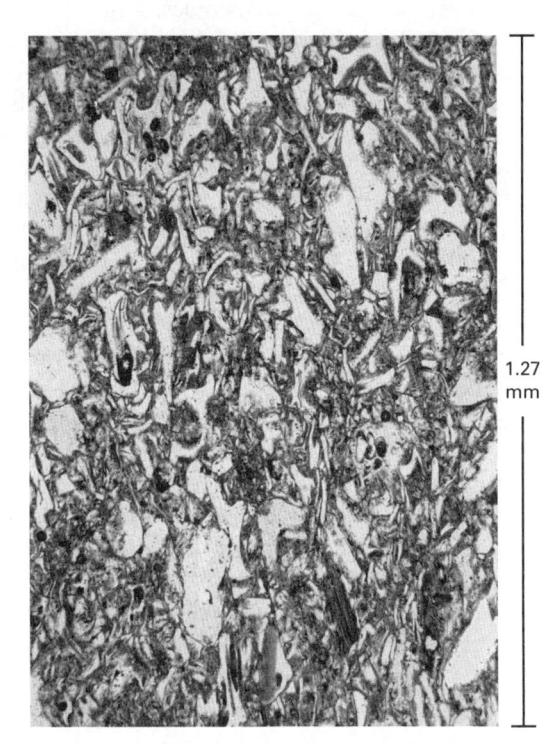

(d) Pyroclastic texture in vitric tuff rhyolite, Valley Springs Rhyolite, near Columbia, CA. Irregular grains are shards of glass.

grains large enough to see and identify using the unaided eye or the eye and a hand lens, are most easily classified based on observed minerals and textures. In examining such rocks, mineral percentages are estimated or measured visually and subdivided into:

- essential minerals, which are used to classify the rock
- characterizing accessory minerals, which occur in abundances of 5% or more but are not essential
- minor accessory minerals, which occur in abundances of less than 5 percent and are not essential

The essential minerals for the common plutonic rocks are quartz, plagioclase feldspar, K-feldspar (alkali feldspar), and the feldspathoids. These minerals, together with other common silicate minerals such as micas, amphiboles, pyroxenes, and olivine are important minerals of the igneous rocks (table 7.2). Additional characterizing and accessory minerals are listed in table 7.2.

Figures 7.15 (on p. 131) and 7.16 (on p. 132) show two classifications for phaneritic rocks with moderate to abundant amounts of quartz, feldspathoid minerals (foids), and feldspars. The RSO classification (figure 7.15), based on a system used in the United States in the last century, is easily remembered and used in the field. The classification system in figure 7.16, the International Union of Geological Sciences (IUGS) classification of phaneritic feldspathic rocks, is the most widely used, although in the field it is far more difficult to employ because one must: (1) sum the essential minerals at the corners of a triangle, (2) divide the percentage of each observed essential mineral by the sum, in order to obtain a normalized percentage, and (3) plot a point based on the three normalized percentages on a triangular graph—all in order to arrive at a name. This procedure is demonstrated using sample data in Box 7.1 (on p. 135). The IUGS provides a rectangular classification that approximates the fields of the triangular classification (Appendix B, figure B.4). This classification has some similarities to the RSO classification but uses the somewhat different European and IUGS fields for such names as granite and quartz monzonite. One important aspect of the IUGS scheme is that the dual-

> **Problem 7.4**
>
> Using the data in table 7.1, construct a line graph showing the changes in CaO, Na_2O, and K_2O across this table. What conclusion might you draw from the pattern you see?

triangle structure of the classification emphasizes the incompatibility of quartz and feldspathoid minerals in the same rock. Special names are used for rocks poor in quartz but rich in plagioclase and ferromagnesian minerals (the gabbros and related rocks) (figure 7.17 on p. 133) and rocks dominated by ferromagnesian minerals (the ultramafic rocks) (figure 7.18 on p. 134).[23] In petrologic studies, microscopic analyses—point counts—involve counting several hundred grains in a rock to determine mineral percentages and are used to classify rocks.

Volcanic rocks, which are generally aphanitic or aphanitic-porphyritic, are difficult to impossible to identify using the same techniques as those used for classifying phaneritic (plutonic) rocks. Two approaches are taken in classifying and naming volcanic rocks. One is to use visible minerals to derive a name, while ignoring the minerals and glass of the aphanitic part of the rock. The name is derived from classification charts such as that shown in figure 7.19 (on p. 136). The second and more commonly used method of classifying volcanic rocks is to rely on chemical analyses and chemical parameters as a basis for deriving a name (figure 7.20 on p. 137).

Pyroclastic rocks, those volcanic rocks composed of fragments produced by explosive eruptions, require additional designators. If a rock is composed of particles smaller than 2 mm in diameter, recall that the rock is called a tuff. Crystal tuffs are dominated by crystal fragments. Lithic tuffs are dominated by rock fragments. Vitric tuffs are dominated by glass shards and fragments. A tuff is said to be *welded* if the materials have annealed (i.e., welded together by heat) to form a hard mass. **Lapilli tuffs** contain scattered fragments of rock larger than 2 mm in diameter. **Volcanic breccias** are dominated by large, angular clasts.

Table 7.2 Characteristic Minerals of Igneous Rocks

ESSENTIAL MINERALS — All Silicate Minerals

Mineral	Chemical / Structural Group	Distinctive Properties	Commonly Occurs in
Quartz	Tectosilicate	hexagonal; any color; H = 7; vitreous; conchoidal fracture; optically positive	Granitoid rocks; Rhyolites to quartz basalts
Potassium feldspars	Tectosilicate	white, pink or green; H = 6; 2 cleavages @ 90°; Carlsbad twinning; low birefringence	Granitoid rocks including granite, granodiorite, syenite; Trachyte and latite; Foid syenite to foid monzosyenite
Plagioclase feldspars	Tectosilicate	white or colored gray, green, brown, or black by impurities; H = 6; polysynthetic twinning (shows striations)	Granitoid rocks; Monzonite to gabbro; Foid monzodiorite to foid gabbro; Feldspathoidal latite to feldspathoidal gabbro (tephritoids and some foiditoids)
Nepheline	Tectosilicate	hexagonal; white to gray or colorless; H = 5.5–6; greasy luster; optically negative	Feldspathoid (foid)-bearing rocks of all kinds; Usually occurs with abundant K-feldspar or Na-rich minerals
Leucite	Tectosilicate	tetragonal; white to gray; H = 5.5–6; occurring as twinned and equant trapezohedral crystals in volcanic rocks; low birefringence	Rare mineral in K-rich volcanic rocks
Sodalite	Tectosilicate	cubic (isometric and isotropic); pale yellow, blue, green, pink, or gray; H = 5.5–6; poor cleavage	Rare mineral in Na-rich, silica-poor igneous rocks
Enstatite	Inosilicate (single chain)	gray, green, brown, or bronze; H = 5.5–6; blocky shape with 2 cleavages @ 90°	Gabbros, including various norites; Basalts; Ultrabasic rocks such as orthopyroxenite, websterite, harzburgite, and lherzolite
Pigeonite	Inosilicate (single chain)	brown, greenish-brown, or black; H = 6; blocky shape with 2 cleavages @ 90°	Basalt flows and dikes
Augite	Inosilicate (single chain)	brown, greenish-brown, or black; H = 5–6; blocky shape with 2 cleavages @ 90°	Gabbro to granodiorite (plutonic rocks); Basalts and andesite; Some peridotites (lherzolite) and clinopyroxenite
Olivine	Nesosilicate	dark to light green, yellow; H = 6.5–7; typically lacks cleavage	Mg-rich olivine in some gabbros and basalts, dunite and peridotites; Fe-rich olivine in acidic alkaline rocks, including some granites and rhyolites

ACCESSORY MINERALS

Silicate Minerals

Mineral	Chemical / Structural Group	Distinctive Properties	Commonly Occurs in
Muscovite	Phyllosilicate (sheet silicate)	colorless to white and light green, brown, and yellow; H = 2.5–3; one perfect cleavage	Granite and other granitoid rocks including pegmatitic-textured rocks
Biotite	Phyllosilicate (sheet silicate)	brown to black; H = 2–3; one perfect cleavage	A wide range of igneous rocks of both plutonic and volcanic nature
Phlogopite	Phyllosilicate (sheet silicate)	brown to bronze; H = 2–3; one perfect cleavage	K-rich, leucite-bearing rocks and some carbonatites
Tourmaline	Cyclosilicate (ring silicate)	black (variety schorl) and various colors including red, green, and yellow (variety elbaite); H = 7; habit = trigonal (3-sided) prisms; marked pleochroism optically	Granite and especially pegmatitic-textured granite

(continued)

Mineral	Chemical / Structural Group	Distinctive Properties	Commonly Occurs in
Silicate Minerals			
Garnet	Nesosilicate	any color; H = 6–7.5; vitreous to resinous luster; no cleavage, but can have parting; optically isotropic	Metamorphic rocks of many types; Granites and some rhyolites
Hornblende	Inosilicate (double-chain silicate)	black to dark green; H = 5–6; 2 cleavages @~56° and 124°; acicular (needle-shaped) to rectangular prismatic crystals	A wide range of plutonic and volcanic rocks
Hastingsite	Inosilicate (double-chain silicate)	black to dark green; H = 5–6; 2 cleavages @~56° and 124°; acicular to rectangular prismatic crystals	Na-rich foid-bearing rocks, especially nepheline syenite
Riebeckite	Inosilicate (double-chain silicate)	black to dark blue; H = 5; 2 cleavages @~56° and 124°; acicular to rectangular prismatic crystals	Na-rich granite, syenite, and rhyolites; Fe-rich metamorphic rocks
Aegirine	Inosilicate (single-chain silicate)	green to greenish black; H = 6; 2 cleavages @ ~90°; commonly in prismatic crystals	Na-rich granite, syenite, nepheline syenite and other foid-bearing rocks
Allanite	Sorosilicate	black to light brown; H ≈ 5–6; elongate crystals with relatively poor cleavage; parallel extinction optically	Granite to granodiorite, syenite, and monzonite; Some metamorphic rocks such as skarn
Zircon	Nesosilicate	various colors from red-brown to colorless; typically in elongate crystals; adamantine; high relief and high birefringence optically	A wide range of granitoid igneous rocks
Titanite (Sphene)	Nesosilicate	gray, brown, yellow, or black; resinous to adamantine luster; wedge-shaped to equant crystals	A wide range of igneous rocks; Some metamorphic rocks such as gneiss and schist
Nonsilicate Minerals			
Ilmenite	Oxide	black; submetallic to metallic luster; slightly magnetic; equant to elongate crystals; optically opaque	A wide range of igneous rocks, especially mafic rocks such as gabbro and basalt; Metamorphic rocks such as high-temperature schist
Magnetite	Oxide	black with black streak; submetallic to metallic luster; magnetic; irregular to octahedral crystals; optically opaque	A wide range of igneous rocks
Chromite	Oxide	black to brownish black; submetallic to metallic luster; nonmagnetic; octahedral crystals or granular, anhedral grains; optically opaque	Dunite, peridotites, and pyroxenites
Spinel	Oxide	variously colored (e.g., black, brown, pink, green); vitreous luster; octahedral crystals	Some mafic to ultramafic igneous rocks
Hematite	Oxide	silver, black, to maroon and red-brown; red-brown streak; metallic to dull earthy luster	Some K- and Na-richer plutonic and volcanic rocks
Rutile	Oxide	light yellow to red or black; typically in acicular crystals; red needles with adamantine luster are notable; tetragonal	A variety of igneous rocks, particularly plutonic rocks
Pyrite	Sulfide	yellow-gold; metallic luster; H = 6–6.5; in cubes and pyritohedrons; optically opaque	A variety of igneous rocks
Apatite	Phosphate	brown, green, yellow, blue, and colorless; vitreous luster; H = 5; hexagonal prisms common	A wide variety of igneous rocks as tiny accessory grains

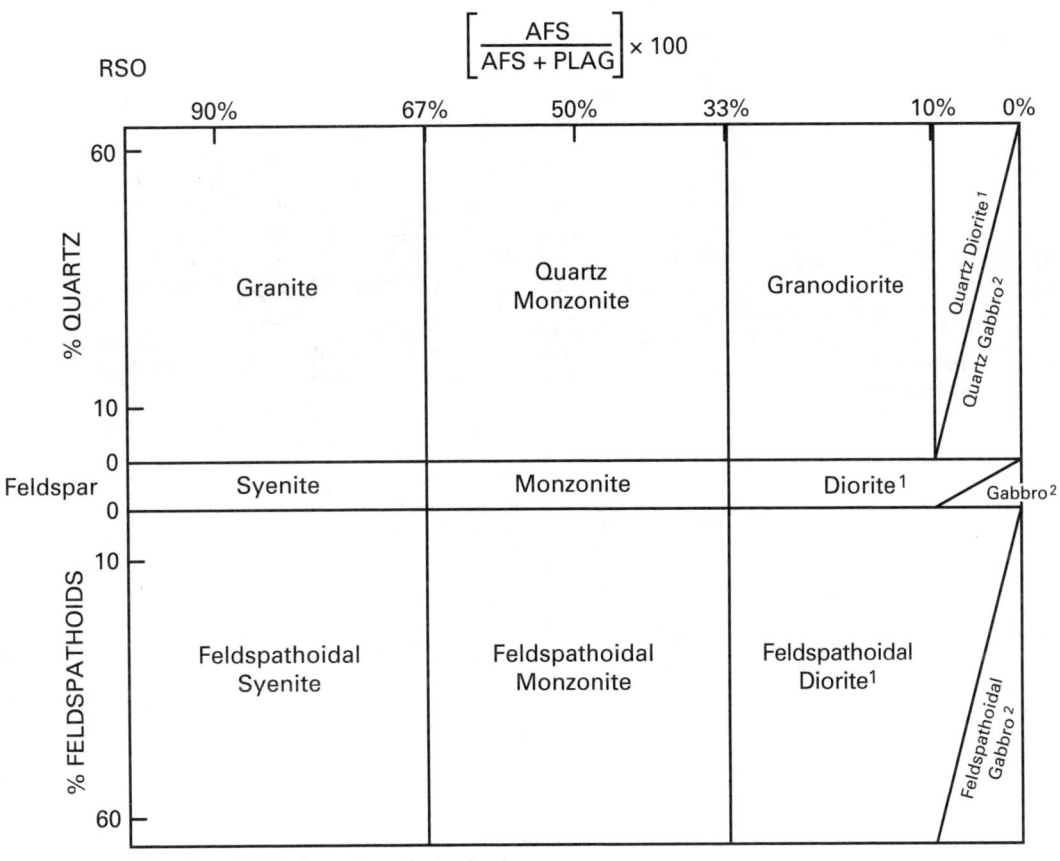

Figure 7.15
The RSO classification of phaneritic plutonic rocks.
(*Source:* Raymond, 2009.)

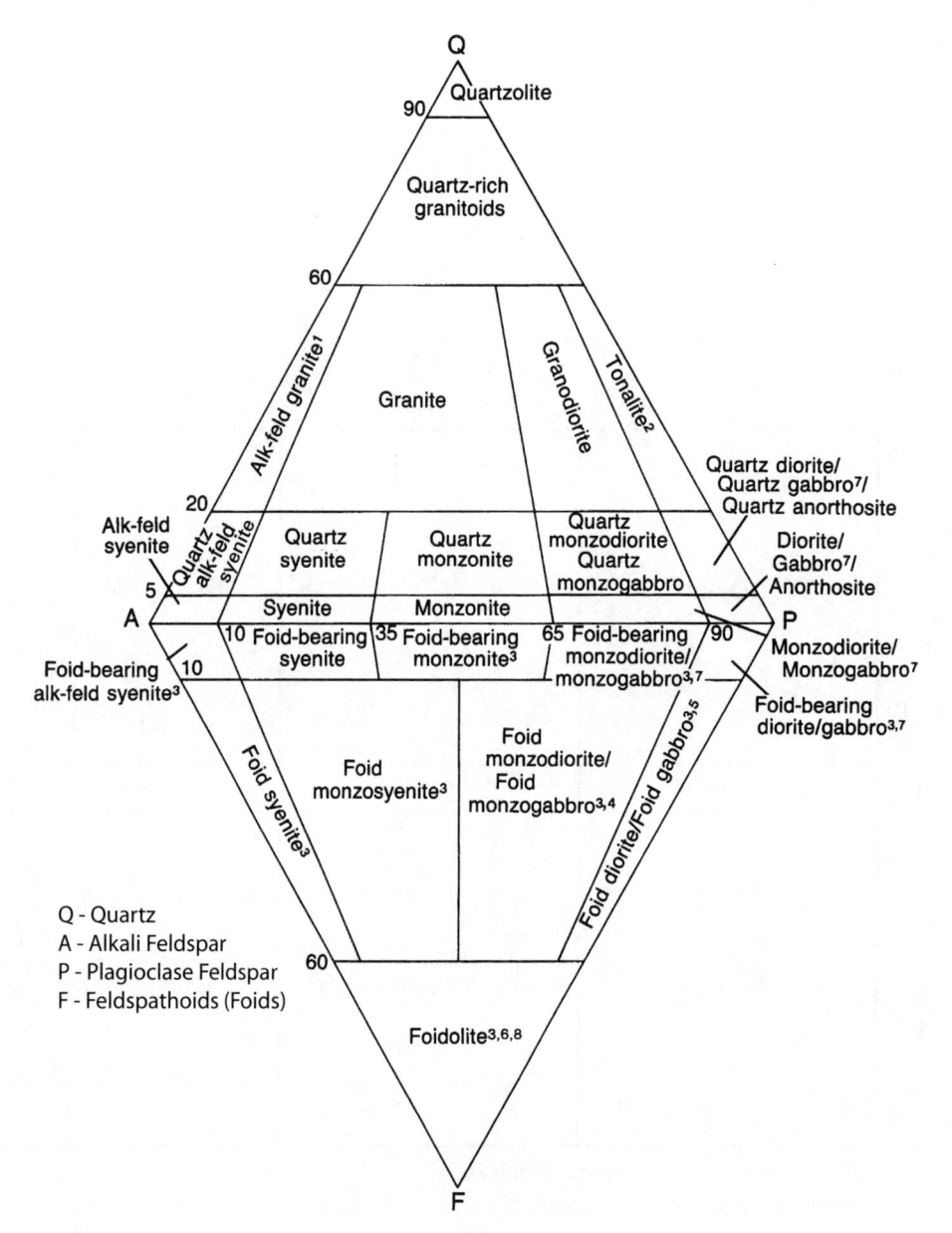

[1] Specify alkali feldspar(s) present in each case (e.g., orthoclase granite).
[2] Alaskite may be used for very light-colored alkali feldspar granites.
[3] Trondhjemite may be used for light-colored tonalites containing oligoclase or andesine.
[4] Specify feldspathoid present in each case (e.g., sodalite-bearing syenite).
[5] Specify feldspathoids present in every case (e.g., nepheline syenite).
[6] Essexite may be used for nepheline monzodiorite/monzogabbro. Theralite = nepheline gabbro; Teschenite = analcite gabbro.
[7] The distinction between various diorites and gabbros is based on the anorthite (An) content of plagioclase.
   In diorite, plagioclase is An < 50 and the common accessory minerals are hornblende, biotite, and augite.
   In gabbro (e.g., gabbronorite), plagioclase is An > 50 and common accessory minerals are olivine, orthopyroxene, and clinopyroxene.
[8] Many special names exist for rocks of this composition, e.g., ijolite and melteigite.

Figure 7.16
The IUGS classification of phaneritic plutonic rocks.
(*Source:* Modified from Streckeisen et al., 1973; Streckeisen, 1974, 1976, following Raymond, 1993. Used with permission.)

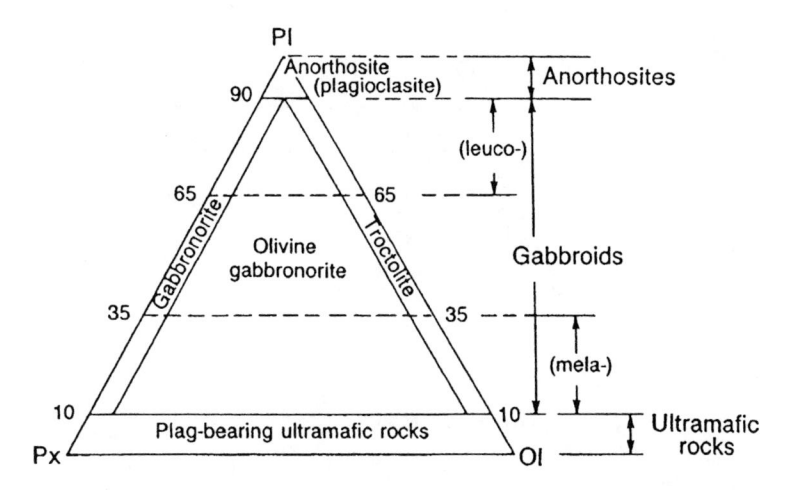

(a) For rocks composed of plagioclase, pyroxene, and olivine.

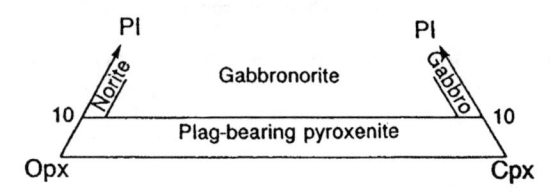

(b) For rocks composed of plagioclase, orthopyroxene, and clinopyroxene.

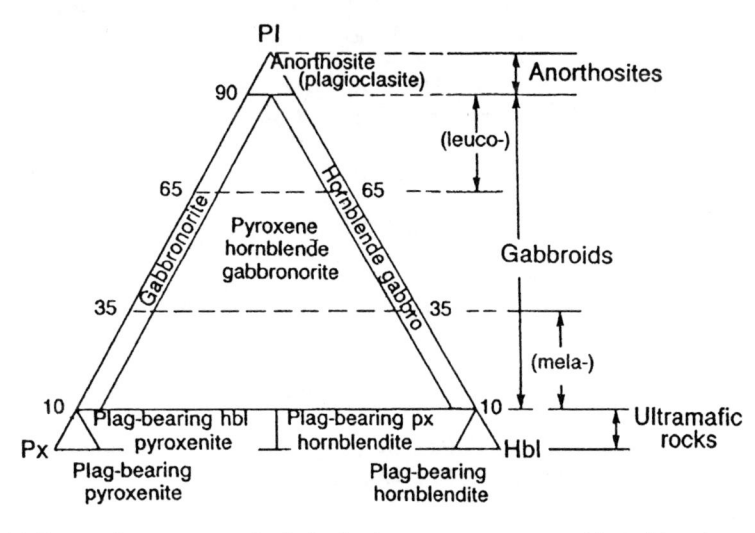

(c) For rocks composed of plagioclase, pyroxene, and hornblende.

Figure 7.17
The IUGS classification of gabbroic rocks.
(*Source:* Streckeisen, 1974, 1976. Used with permission.)

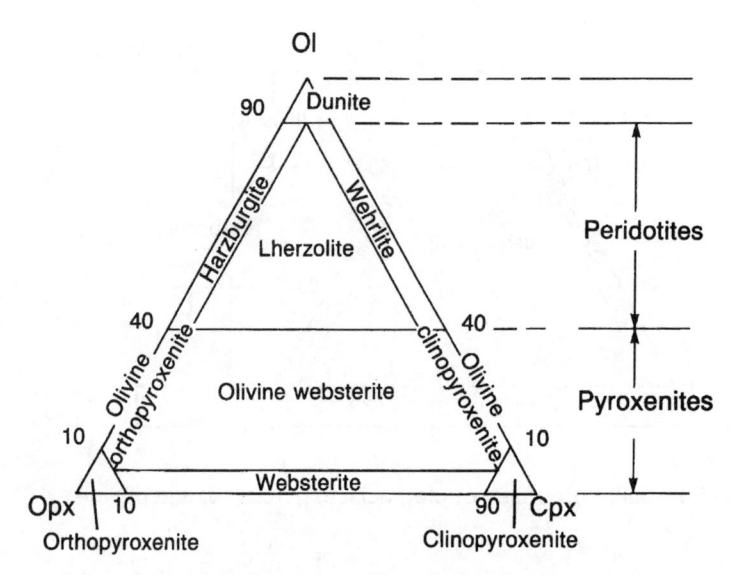

134 Chapter Seven

(a) Classification diagram for rocks with olivine and two pyroxenes.

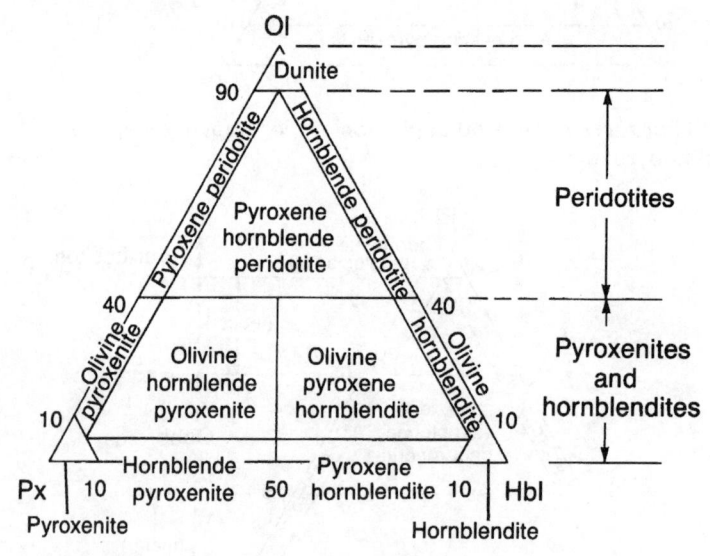

(b) Classification diagram for rocks with hornblende, olivine, and pyroxene.
OL = olivine   Px = pyroxene   Cpx = clinopyroxene
Opx = orthpyroxene   Hbl = hornblende

Figure 7.18
The IUGS classification of ultramafic (pyroxene-, olivine-, and hornblende-rich) igneous rocks.
(*Source:* Streckeisen, 1974, 1976. Used with permission.)

Box 7.1 — Plotting the Composition of a Rock on a Triangular Plot

Plotting a rock composition on a triangular plot is a three step process. Suppose we have a rock composed of:

40%	alkali feldspar
25%	quartz
17%	plagioclase
12%	biotite
4%	hornblende
1%	magnetite
1%	sphene
100%	TOTAL

The only minerals important for classification of the rock on the IUGS chart (figure 7.16) are quartz, alkali feldspar, and plagioclase feldspar. We obtain a total for these minerals, i.e.

40% AFS (alkali feldspar)   **[Step 1]**
25% Qtz (quartz)
17% Plag (plagioclase)
82%

To normalize each of the percentages of these essential minerals to 100%, we use proportional relationships, dividing the percentage of each by the total of the three minerals. Thus,

AFS: $\dfrac{40\%}{82\%} = \dfrac{AFS}{100\%}$; AFS = 49%   **[Step 2]**

Qtz: $\dfrac{25\%}{82\%} = \dfrac{Qtz}{100\%}$; Qtz = 30%

Plag: $\dfrac{17\%}{82\%} = \dfrac{Plag}{100\%}$; Plag = 21%

Note the new total is 100% for the normalized percentages of essential minerals.

[Step 3]

Step 3 is to plot the position of the composition and obtain the name from the IUGS triangular plot. Note that each corner of the plot represents 100% of the mineral marked at that corner. Hence, at the top of the triangle, the point of the triangle marked Qtz is 100% quartz. Likewise, the Plag corner is 100% plagioclase and the AFS corner is 100% alkali feldspar. The entire side opposite each apex represents 0% of the component at the corner. Percentage lines parallel those opposing sides across the triangle. The composition of the rock in this exercise plots at the black dot, where the percentage lines for 30% Qtz, 49% AFS, and 21% Plag intersect. The rock is a granite, as determined from figure 7.16.

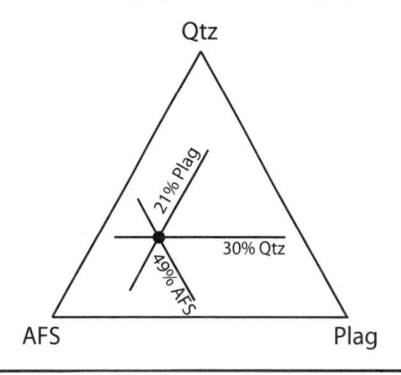

Problem 7.5

Using (a) the IUGS classification and (b) the RSO classification, classify a rock composed of 58% alkali feldspar, 20% plagioclase feldspar, 17% quartz, 4% hornblende, and 1% magnetite. What does the result suggest about simply reading the word *granite* in an article?

BASALTS AND THEIR SIGNIFICANCE

Basalts are aphanitic to aphanitic-porphyritic volcanic rocks with silica contents of 45–52%. There are two main types of basalt. Tholeiitic basalt, or **tholeiite**, typically contains plagioclase and Ca-poor pyroxene (pigeonite or orthopyroxene) and chemically has 48–52% silica. Olivine is usually absent from tholeiites but does occur in some rocks. **Alkali olivine basalt** contains plagioclase, olivine (commonly as phenocrysts), and Ca-richer pyroxene (usually the clinopyroxene, augite). Alkali olivine basalt typically contains 45–49% silica.

Basaltic magmas are the most common and important types of magmas. They are parental to many rock types and to other magmas. In general, basaltic magmas form through partial melting of silica-poor mantle rocks. As discussed earlier, phase diagrams and studies of melting reveal that *partial melts* of rocks do not generally have the same compositions as the parent rocks from which they are derived, but instead tend to be richer in silica and alkalis. Hence, basalts are richer in silica and alkali elements than the mantle rocks from which they are derived.

At the mantle sites of magma formation, magmas of tholeiitic character form at shallower depths, whereas alkali olivine basalt magmas form at greater depths. Melting of the mantle occurs beneath so-

			Other phenocrysts or grains					
			Alkali-feldspar ±biotite	Plagioclase ±alkali-feldspar ±biotite	Hornblende ±plagioclase	Pyroxene	Olivine	Hornblende ±plagioclase
Porphyritic texture	Essential phenocrysts or grains	Quartz feldspar	f – Rhyolite	f – Dacite			/////	/////
		Feldspar only	f – Trachyte	f – Andesite (Light feldspar)		f – Basalt (Dark feldspar)		
		Felspathoids ±feldspar	f – Phonolite	f – Feldspathoidal andesite (Light feldspar)		f – Feldspathoidal basalt (Dark feldspar)	f – Basanite	
		None	f – Biotite felsite	f – Andesite (Light feldspar)			f – Basalt	
Aphanitic texture	No visible grains Glass<50%		Felsite (Light colored)			Mafite (Dark colored)		
	Glass>50%		Obsidian*					

*Obsidian is actually a textural term and obsidians are dominantly rhyolites that plot at the left end of the obsidian space.

Figure 7.19
Raymond's classification of common aphanitic and aphanitic-porphyritic feldspathic volcanic rocks. Texture and visible essential minerals are used to determine the row. The most abundant other phenocrysts and the feldspar types are used to determine the column. A prefix f- is used for field and hand-specimen determinations and a prefix p- (for petrographic) is substituted for the f- for determinations using a microscope.
(*Source:* After figure 23, Raymond, 2009.)

called "hot spots" (at the tops of mantle plumes), beneath mid-ocean ridges at spreading centers, in and above subduction zones (beneath the arcs), and beneath leaky (trans-tensional) transform faults (figure 7.1). Tholeiites exist at the surface in all of these locations, whereas alkali olivine basalts occur primarily in arcs and above hot spots.

The hot-spot magmas, derived from partial melting of plumes of solid mantle rock, rise toward the surface due to lower densities (relative to surrounding rocks), as do the plumes themselves. Lower density in the plumes is caused by heating of rocks at depth. As the hot, plastic rock of the plume rises, decompression melting yields a basaltic melt of alkali olivine basalt character at deeper sites in the mantle. As the plume continues to rise, the melts that form at shallower depths (< 80 km), are tholeiitic in composition. The Hawaiian Islands were formed by plume-generated volcanism at a hot spot and include both types of basalt.

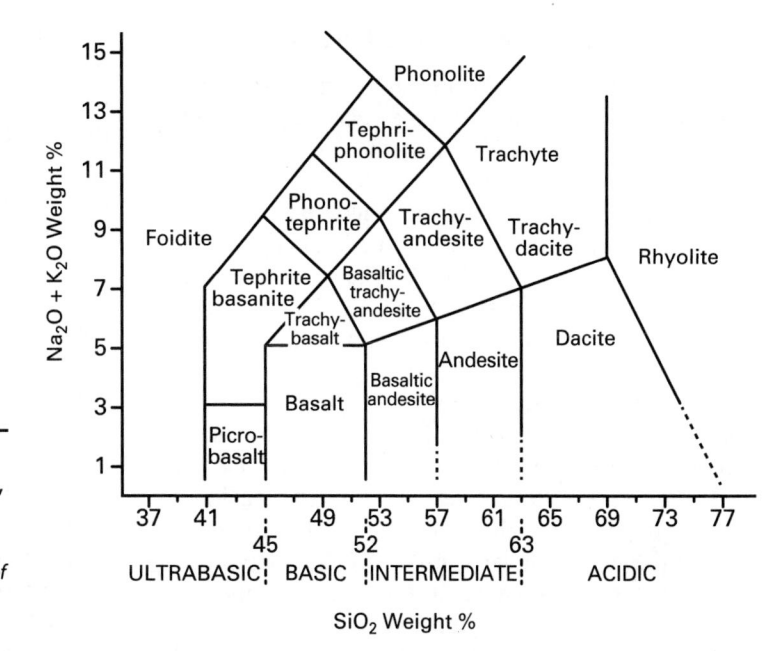

Figure 7.20
IUGS chemical classification of volcanic rocks, based on total alkalis and percent silica.
(*Source:* LeBas et al., "A Chemical Classification of Volcanic Rocks Based on the Total Alkali-Silica Diagram" in *Journal of Petrology*, 27:745–50, 1986. Used with permission of Oxford University Press.)

Decompression melting also yields the basaltic melts of the mid-ocean ridges that form mid-ocean ridge basalts (MORBs). All such melts are tholeiitic, although there are subtle variations in the MORBs that derive from several factors such as the location of melting relative to the spreading ridge structure. Debate continues with regard to whether MORB parent magmas are primary magmas, derivative magmas, or mixed magmas (Raymond, 2007, pp. 112–113). Regardless of their early history, crystallization of mid-ocean ridge magmas, both at depth and at the surface, produces the oceanic crust. MORB-like magmas also develop below continental rifts, such as the Rio Grande Rift of New Mexico and westernmost Texas.

On the edges of some continents as well as in ocean basins, the linear arrays of volcanoes called **arcs** contain abundant basalts. Hence, basaltic magmas must form beneath these sites. Here, the relative roles of decompression melting versus flux melting are not entirely resolved, but the dominant view is that flux melting plays a central role in arc magma genesis. H_2O, in particular, influences many aspects of melts formed beneath the arcs (Ringwood, 1974; Tatsumi, 1989; Gaetani and Grove, 2003; Kushiro, 1983, 2007; Kimura et al., 2009). As is the case for hot spots, the arcs may contain both tholeiitic and alkali olivine basalts.

A newly formed basaltic magma may take one of several "paths" before finally becoming a rock (i.e., magmas may have various possible histories). As basaltic magmas rise from their sites of origin, they may: (1) be modified by the various processes of modification to yield other magma types (later yielding a variety of rock types); (2) rise to the base of the crust and provide the heat that contributes to melting more siliceous crustal rocks to yield rhyolitic magmas; (3) crystallize at depth to form gabbros; or (4) erupt at the surface, where they crystallize to form basalts. By providing material and heat, basaltic magmas contribute to the formation of most other common magma types, and hence to most common igneous rock types (Hildreth, 1981; Raia and Spera, 1997).

IGNEOUS ROCK ASSOCIATIONS

In general, igneous rocks form at the various sites above the subsurface locations at which magmas form. Hence, many igneous rocks form at mid-ocean ridges and rifts (i.e., at spreading centers), in arcs above subduction zones, at hot spots, and along leaky transform faults. Rare igneous rocks also form in other settings, some of which are unusual. For example, hot, carbon-dioxide-rich, ultramafic (silica-poor, Fe-Mg-rich) magmas erupt explosively within continents to form volcanic vents called **diatremes** that contain the locally diamond-bearing **kimberlites** (as discussed in chapter 11). At these various sites, specific igneous rocks occur together with other rocks to form distinctive rock associations, the petrotectonic assemblages.

Mid-Ocean Ridge Rocks

If we examine mid-ocean ridges, which scientists have done with the aid of submersible diving vessels and ocean drilling, we find that the dominant rocks formed at these sites are igneous. The petrotectonic assemblages of the mid-ocean ridges also include some sedimentary rocks such as breccias, deep marine limestones, mudrocks, and some metamorphic rocks—particularly rocks altered by hydrothermal (hot water) fluids and heat. In this section we focus on the igneous rocks.

The mid-ocean ridges are zones of tension. Lithospheric plates are pulled apart here, probably by movements of convection currents in the mantle and the drag exerted by subducting plates. The tensional forces and splitting of the lithosphere lowers the pressure on the hot, underlying mantle rocks, the mantle rocks rise, and decompression melting occurs. As noted above, the magmas produced by this melting are predominantly tholeiitic, and they crystallize to yield tholeiitic rocks.

Igneous activity at the mid-ocean ridges produces a crudely layered sequence of silica-poor rocks, with local zones of silica-richer rocks (table 7.3) (Engel and Fisher, 1975; Dilek and Furnes, 2014).[24] The uppermost rocks in the sequence consist of (1) tholeiitic MORB flows. Commonly the flows are pillowed, but many are unpillowed, layered basalts. Beneath the lava flows the structure and rock types vary, depending on the spreading rate, which ranges from ultraslow to slow-intermediate to fast-spreading. Overall, going downward, we see (2) a basalt-gabbro dike complex with fine-grained gabbros typically called *diabase*, (3) an unlayered plutonic complex of gabbros to silica-richer rocks (e.g., diorite), and (4) a layered plutonic complex of layered gabbros and ultramafic rocks that overlie (5) tectonically deformed and metamorphosed mantle rocks (mantle *tectonites*) of ultramafic composition, including peridotites, such as harzburgite and lherzolite, and some dunite (figure 7.21). In slow-spreading ridges, the plutonic complexes are thought to be minor and rocks such as lherzolite, dunite, and plagioclase peridotite may make up the lower parts of the crust (e.g., Dijkstra et al., 2001; Muntener and Piccardo, 2003; Abily and Ceuleneer, 2013). In contrast, in fast-spreading ridges the dominant peridotite is thought to be harzburgite and both dike and plutonic complexes are considered to be thicker than those of slow-spreading ridges.[25]

Drilling has not penetrated the deepest levels of the oceanic crust and underlying mantle, so that the structures and compositions there are based on geo-

Table 7.3   Igneous Rock Types Typical of Various Protectonic Assemblages

Rock Type t	Spreading Center (Rift)	Transform Fault	Subduction Zone/Mtn. Belt	Plate Interior
Granite	—	—	x	—
Quartz Monzonite	—	—	x	—
Granodiorite	—	—	x	—
Quartz Diorite	x	—	x	—
Syenite	—	—	—	x
Monzonite	—	—	—	—
Diorite	x	—	x	—
Gabbro	x	—	x	x
Dunite	x	—	x	x
Peridotites	x	—	x	x
Pyroxenites	x	—	x	x
Feldspathoidal rocks	—	—	—	x
Rhyolite	O	T	x	x
Quartz Latite	—	—	—	—
Rhyodacite	—	—	x	—
Andesite	O	T	x	x
Basalt	x	T	x	x

x = typical of particular settings
O = typical only in continental settings
T = only in trans-tensional transform faults
— = not common or typical

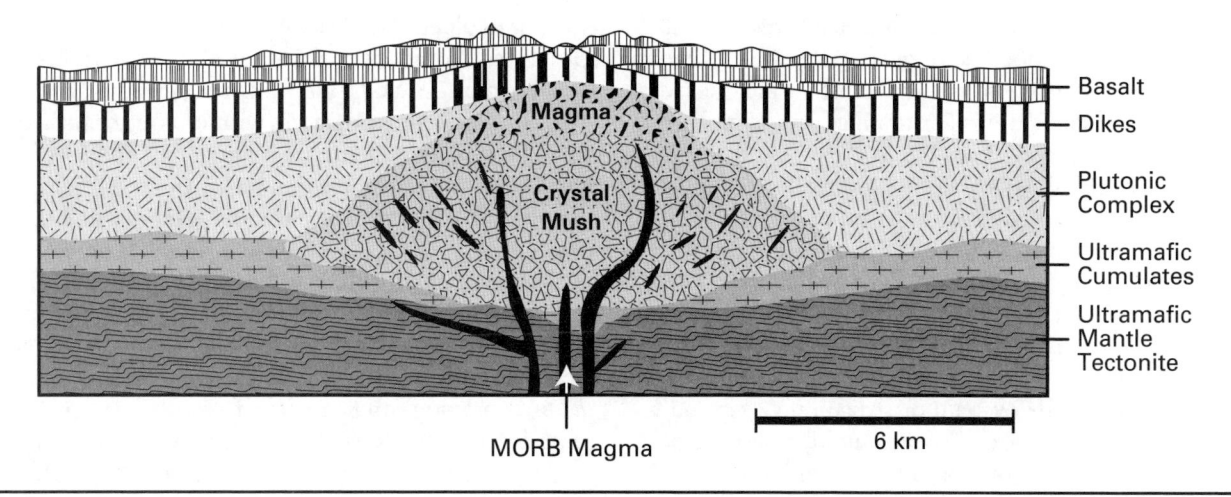

Figure 7.21
The simplified structure of a fast spreading mid-ocean ridge. The structures of very-slow to fast-spreading ridges are variable and more complex than shown. See Dilek and Furnes (2014) and Goodenough et al. (2014) for cross-sectional examples of complexities derived from ophiolite studies.
(*Source:* Modified from Raymond, 2007, p. 113, based on the ideas of Perfit et al., 1994 and Dilek et al., 1998.)

physical and petrologic considerations, as well as comparisons with rock sequences similar to those described above, found in mountain belts. The latter are called **ophiolites**. Yet, ophiolites and mid-ocean ridge rocks are not entirely analogous, creating difficulties and controversies of interpretation (Moores et al., 2000; Metcalf and Shervais, 2008; Dilek and Furnes, 2014).[26]

Rock Associations of Subduction Zones and Orogenic Belts

Where plates collide in regions above subduction zones, the variety of igneous rocks is greater than in the spreading center sites. In subduction zones, the subducting plate is carried down to depths where it is heated and dehydrated (figure 7.1). Hydrous fluids derived from this dehydration may contain a variety of elements and may rise into the overlying plate, causing both chemical changes (metasomatism) and flux melting of the mantle rocks in that plate. The melts that form rise through the mantle to the crust and are commonly modified before crystallizing to yield a variety of plutonic and volcanic rocks. Plate collisions along subduction zones also result in orogenic mountains, particularly where both plates contain continents. In orogenic mountains the igneous rocks cover a chemical spectrum, collectively referred to as the metaluminous calc-alkaline rocks, ranging from granite and rhyolite to gabbro and basalt (see Raymond, 2007, p. 34–35; 48–49). Locally, alkalic, calcic, potassic, and peraluminous types also occur.[27]

Volcanic rocks that constitute the arcs above the subduction zone typically include a range of rock types from basalt to rhyolite.[28] The intermediate rocks, *andesite* and *dacite*, are particularly characteristic. Whereas the basaltic magmas—which yield basalt and the plutonic equivalent, gabbro—are derived relatively unmodified from the mantle, the andesites and dacites form largely through a variety of magma modification processes. Rhyolite magmas develop where basalts melt the base of the crust and also, rarely, form by fractional crystallization of earlier-formed magmas. The rhyolite magmas may then mix with basalts to form andesitic, dacitic, rhyodacitic, and related magmas. Basaltic magmas may also assimilate surrounding rocks as they pass through the crust, yielding andesitic magmas. Thus, the combination of assimilation and fractional crystallization (AFC) is important in the formation of some andesites. Andesitic magmas rich in Mg also form locally, where melting is particularly extensive in the subduction zone.[29]

The roots of the volcanoes consist of batholiths and stocks composed of a range of rocks—the gabbros to granites—chemically similar to the volcanic rocks (Hildebrand et al., 2010). Beneath the arcs, quartz monzonites and granodiorites are the common rocks, and these likely form as magmas rising from the subduction zone are modified in the crust by AFC. Rhyolitic/granitic melts form by crustal melting or fractional crystallization. The granites crystallize from these melts (Clarke, 1992; Pitcher, 1997) and locally exhibit pegmatitic (very coarse-grained) and aplitic (sugary, fine-grained) textures. In ponded basaltic magmas in the batholithic belts, ultramafic rocks such as peridotites and pyroxenites form by fractional crystallization and crystal settling (table 7.3).

The igneous rocks of the orogenic belts are accompanied by a wide variety of metamorphic rocks and locally may be overlain or structurally associated with sedimentary rocks. Contact and dynamothermal metamorphic rocks (discussed in chapter 10) surround the plutons. The sedimentary rocks (chapter 9) may include the spectrum of continental to marine types. Together, all of these rocks constitute the petrotectonic assemblage of the orogenic belts.

Between continents and some offshore arcs, small ocean basins with spreading centers may form. These suprasubduction zone spreading centers yield rocks many consider to be similar in general, but with some chemical differences, to rocks formed at mid-ocean ridge spreading centers. These rocks are thought to be predecessors of many ophiolites in mountain belts.[30]

Anorogenic Igneous Rocks and LIPs

Anorogenic igneous rocks are those not associated with orogenic belts—that is, generally absent from the folded and thrust-faulted mountain systems. Strictly speaking, the rocks of the mid-ocean ridges are anorogenic. In addition, hot-spot igneous activity within plates—particularly within continents—and rifting within continents are anorogenic processes that produce a variety of igneous rocks, from granites and rhyolites to gabbros, feldspathoidal gabbros, and ultramafic rocks.

Commonly, the anorogenic rocks are alkalic (i.e., they are high in the alkalis Na and K relative to the amounts of Ca). Minerals such as the sodic pyroxene *acmite* and the alkali amphiboles *hastingsite* and *riebeckite* are common, and in some rocks the feldspathoid *nepheline* is characteristic. Typical anorogenic volcanic rocks are trachytes and latites. In some cases, where rifts and hot spots occur within continents, calc-alkaline, bimodal suites of rhyolite and basalt develop. The rocks of Yellowstone National Park and the Rio Grande Rift in the United States have this character.[31] In these settings, magma mixing and other modification processes may occur locally, yielding andesites and related rocks. Below the surface, silica-richer magmas may be modified through fractional crystallization and other processes to form layered magma chambers (figure 7.22) that erupt to form pyroclastic sheets of compositionally, reversely layered volcanic rock (higher-silica rocks on the bottom and lower-silica rocks on the top). Plutonic rocks that occur in fractionally crystallized alkali plutons include granites, gabbros, and nepheline syenites; plus the unusual nepheline-pyroxene rocks *ijolite* and *jacupirangite* as

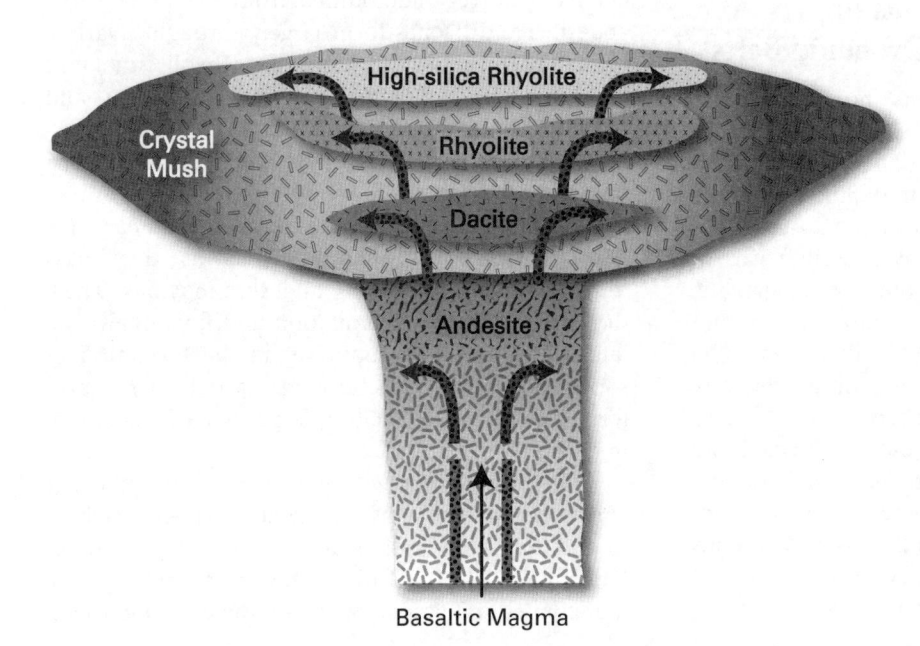

Figure 7.22
A diapiric, zoned magma chamber of fractionating magma. Heat and material may be provided by basaltic magmas that intrude the base of the column and feed the magma chamber. Final magma modification results from convection and fractional crystallization. (*Source:* Modified from Raymond, 2007, p. 131.)

> **Problem 7.6**
>
> Examine figure 7.22 carefully. Assume the magma chamber was 2/3 emptied by eruptions three times, refilling each time after an eruption, and that the eruption produced a pyroclastic sheet each time. Draw a cross section showing the layers of the pyroclastic sheets as they would appear after the third eruption.

well as the igneous carbonate rocks, the *carbonatites* (Erickson and Blade, 1963; Sorensen, 1974).

Within continents, explosive eruptions of fluid-rich ultramafic magma produce the *kimberlites*, which in some cases are diamond bearing. A wide array of other rare rocks of both volcanic and plutonic character may develop where alkaline magmas are modified during their ascent to the upper levels of the crust and the surface.

In the past, at a few locations on Earth, very large volumes of igneous rocks were formed in relatively short periods of time. Areas of such rocks, many of which are of basaltic chemistry, are called *large igneous provinces* or LIPs. While the details of LIP character and formation remain matters of considerable study, some certainly seem to have resulted from plume-induced magmatism.[32]

Some Final Comments

The igneous rocks are widely distributed on Earth. They still provide us with a number of problems to resolve: Are most ophiolites formed in supra-subduction zone settings? How important is stoping in the process of batholith formation? What are the primary triggers for rhyolite eruptions? Resolution of these problems often requires both field and laboratory data, as well as computational modeling studies.

As noted by Cashman and Sparks (2013), understanding igneous processes is of practical importance in geologic hazard assessment. Geologists have a role to play in clarifying—as much as is possible—how, why, and when volcanic eruptions will affect human populations. As we will see in chapter 11, igneous rocks also have practical value as resources used in human activities and as hosts for rare elements important to modern society.

Summary

Igneous rocks are formed by the solidification or crystallization of melted rock materials called magmas. Rocks crystallized at depth from magmas are called plutonic, whereas those formed at or very near the surface of Earth are called volcanic. Volcanic rocks may contain glass, minerals, or rock fragments and, in texture, may be composed of interlocking crystals (crystalline texture) or fragments of volcanic rock, glass, and minerals (pyroclastic texture). Plutonic rocks are crystalline in texture. The textures form via an interplay of nucleation and subsequent growth from initially formed nuclei that begin as small clusters of appropriately arranged atoms. Textures in igneous rocks may also have features formed by other processes, such as explosive eruptions. The textures, bulk chemical compositions, and the percentages of various minerals in igneous rocks serve as the basis for individual classification of these rocks.

Magmas may form as primitive melts of unaltered mantle rock or may be primary, having formed by melting of any preexisting rock. Magmas are parental if they are altered and give rise to another type of magma called a derivative magma. Primitive and primary magmas form by flux melting, decompression melting, or thermal melting of rocks in the mantle or at the base of the crust. Once formed, they rise into the crust or to the surface in periods ranging from a day or so for *basic* (low-silica) magmas to a calculated, but unrealistic 100+ million years for *acid* (high-silica) magmas. It is likely that less than 30% of magmas make it to the surface. Typical magmas (and the rocks formed from them) range from about 40% to 75% silica, with varying amounts of other major elements. Magma compositions are modified by assimilation of previously formed rocks, mixing of magmas, and processes of differentiation—especially fractional crystallization. The chemical compositions, particular minerals, and physical features of the rocks that form reveal the histories of the magmas and the rocks formed from them.

Intrusion and extrusion produce a range of structures ranging from batholiths exposed for areas greater than 100 km^2 to xenoliths that occur in hand specimens. Pyroclastic flows typically form sheets, and lava flows form sheets to lobate masses. Among the plutons, the dikes form cross-cutting, typically tabular structures, whereas the sills and laccoliths are concordant and lenticular to tabular in shape. Batholiths are large, variously shaped, but broadly lenticular masses of plutonic rock typically formed by multiple intrusions of magma. Many are composite, consisting of rocks of lower to higher silica content. Lopoliths are large dish-shaped batholiths that are typically dominated by low-silica rock types.

Igneous rocks are major components of most petrotectonic assemblages. Mid-ocean ridge assemblages are dominated by mafic and basic, low-silica rocks. Rift and hot-spot assemblages in the continents are characterized by bimodal mafic-felsic, low- to high-silica associations. Subduction zones yield a wide range of rocks of various compositions, as do some intracontinental, intraplate sites.

Selected References

Basaltic Volcanism Study Project (BVSP). 1981. *Basaltic Volcanism on the Terrestrial Planets*. Pergamon Press, New York. 1286 p.

Brown, M. 2013. Granite: From genesis to emplacement. *Geological Society of America Bulletin*, v. 125, pp. 1079–1113.

Cashman, K. V., and Sparks, R. S. J. 2013. How volcanoes work: A 25-year perspective. *Geological Society of America Bulletin*, v. 125, pp. 664–690.

Clarke, D. B. 1992. *Granitoid Rocks*. Chapman and Hall, London. 283 p.

Hildebrand, R. S., Hoffman, P. F., Housh, T., and Bowring, S. A. 2010. The nature of volcano-plutonic relations and the shapes of epizonal plutons of continental arcs as revealed in the Great Bear magmatic zone, northwestern Canada. *Geosphere*, v. 6, pp. 812–839.

Hildreth, W. 1981. Gradients in silicic magma chambers: Implications for lithospheric magmatism. *Journal of Geophysical Research*, v. 86, pp. 10153–10192.

Moores, E. M. 1973. Geotectonic significance of ultramafic rocks. *Earth–Science Reviews*, v. 9, pp. 241–258.

Raymond, L. A. 1993. *Petrography Laboratory Manual: Part 1, Handspecimen Petrography* (2nd ed.). GEOSI, Boone, NC. 154 p.

Raymond, L. A. 2007. *Petrology: The Study of Igneous, Sedimentary, and Metamorphic Rocks*. Waveland Press, Long Grove, IL. 720 p.

Sorensen, H. (ed.). 1974. *The Alkaline Rocks*. John Wiley, London. 622 p.

Viccaro, M., Giuffrida, M., Nicotra, E., and Cristofolini, R. 2016. Timescales of magma storage and migration recorded by olivine crystals in basalts of the March-April 2010 eruption at Eyjafjallajökull volcano, Iceland: *American Mineralogist*, v. 101, pp. 222–230, doi:10.2138/am-2016-5365.

Notes

1. The paths of fluid movement and the effects of fluid flow in subduction zones are subjects of considerable interest and research. The influence of water on a variety of compositional and other aspects of melting is also important (Gaetani and Grove, 2003; Brown, 2013). See, for example, studies by Ringwood (1974), Anderson et al. (1976), Kushiro (1983), Peacock (1987b, 1989, 1990a, 1990b, 1993), and Pearce and Peate (1995).
2. See note 1.
3. Turcotte and Schubert (1973), Peacock et al. (1994), and Peacock (1996).
4. Hildreth (1981) and Sisson et al. (1996). Also see Pichler and Zeil (1972) and Wyllie (1983) for discussions.
5. Eaton and Murata (1960), Koyanagi and Endo (1971). Also see Cashman and Sparks (2013, fig. 5).
6. MacGregor (1968, 1974), Basu (1975), Dawson (1981), Swanson et al. (1987), Blatter and Carmichael (1998), and many others discuss mantle xenoliths, whereas crustal xenoliths are discussed, for example, by Padovani and Carter (1977), Fodor and Vandermeyden (1988), and Parsons et al. (1995).
7. BVSP (1981, p. 139), Wilkinson (1982), and Davis and Clague (1987).
8. BVSP (1981).
9. Maaloe (1981), Nicolas (1986), LaPorte and Watson (1995), Brown et al. (1995), Brown and Rushmer (1997), and Sawyer (2001). Also see the calculations by Rabinowitz and Vigneresse (2004) and summary by Brown (2013) on melt segregation and migration.
10. See references in note 9, including Brown (2013) for a summary and additional references.
11. For example, see Marsh (1979, 1984), Petford et al. (1993), Kelley and Wartho (2000), Burov et al. (2003), Viccaro et al. (2016), and see Raymond (2007, p. 88–91) for a review.
12. Yoder (1976), Wright et al. (1979, in Wadge, 1980), Kelley and Wartho (2000).
13. Marsh (1984), Marsh and Kantha (1978), Burov et al. (2003).
14. Christiansen and Blank (1972), Hildreth et al. (1984), Gansecki et al. (1998), Miller and Wark (2008), Shervais et al. (2014).
15. For example, see Hamilton and Myers (1967, 1974), Richards and McTaggart (1976), McNulty et al. (2000), and the discussion of batholith shapes in Raymond (2007, p. 14–18, and note 16, p. 30).
16. Also see Harper et al. (2004), Glazner and Bartley (2006), Bartley et al. (2012), and Anderson et al. (2013).

17. See Paterson and Fowler (1993), Glazner et al. (2004), Glazner and Bartley (2006), Anderson et al. (2013) and also see the discussions of Bateman (1984), Paterson and Vernon (1995), and Petford (1996).
18. Eichelberger (1995), Vergniolle (1996), Wylie et al. (1999), Cashman and Sparks (2013).
19. Murphy et al. (2000), Cashman and Sparks (2013).
20. Also see Lofgren (1980), Kirkpatrick (1981), Brandeis and Jaupart (1987).
21. See the papers cited in note 20.
22. Raymond (1995), MacKenzie et al. (1982).
23. For IUGS rock classifications, see Streckeisen (1973, 1976, 1979, 1980) and LeBas et al. (1986).
24. Also see Moores and Jackson (1974), Hekinian et al. (1976), Coleman (1977), Ballard et al. (1982), Phipps Morgan et al. (1994), Perfit et al. (1994), Dilek et al. (1998), Robinson et al. (2000), and Dilek and Furnes (2011).
25. Because we have not seen the actual rocks deep in the oceanic crust and underlying mantle, analogies between ophiolites (found in mountain belts) and various aspects of oceanic crust in ocean basins and basins behind volcanic arcs (the suprasubduction zone back-arc basins) are used to infer the composition and structure of the lower parts of the ocean crust and underlying mantle. For references to the ophiolites and comparisons to ocean crust, see papers in Moores et al. (2000), and other reports, such as Boudier and Nicolas (1985), Girardeau et al. (1985), Girardeau and Mercier (1988), Dilek et al. (1998), Dijkstra et al. (2001), Muntener and Piccardo (2003), Nicolas and Boudier (2003), Thy and Dilek (2003), Robinson et al. (2000), and Abily and Ceuleneer (2013).
26. For example, see discussions and reviews in Moores (1969, 1982), Miyashiro (1973), Raymond (2007), Dilek and Furnes (2014), and Pilla et al. (2016).
27. For example, Ewart (1976, 1982), Dostal et al. (1977), McBirney and White (1982).
28. Refer to the references in note 27.
29. Tatsumi and Ishizaka (1981, 1982), and contrast with Myers and Johnston (1996).
30. See Bloomer et al. (1995) and such interpretations as those of Harper (1984), Monnier et al. (2000), Raymond et al. (2003), Metcalf and Shervais (2008), and Dilek and Furnes (2014).
31. Christiansen and Lipman (1972), Hildreth et al. (1991), Raymond (2007, p. 131ff.).
32. For example, see Sheth (2007) and Ernst, R. E., et al. (2005).

Weathering and Soils

8

INTRODUCTION

Soil is an important Earth material, and it is the parent of much sediment. Sedimentary rocks are derived from sediment formed through weathering and erosion of exposed rocks. Soil is, thus, an intermediate Earth material formed in the chain of materials: rock → soil → sediment → new sedimentary rock. **Soil** is defined as weathered and disaggregated rock and associated organic materials, *unmoved* from and overlying the bedrock.[1]

The importance of soil extends beyond its role in the formation of sediment. Soil is an important material from a variety of perspectives. From an ecological viewpoint, it is essential to the growth of plants and, therefore, to the functioning of complex ecosystems. Without soil, complex life systems of Earth would not have evolved as they did. From the practical point of view, soil provides the base for agricultural production of food plants and grains for animals, including humans. Historically, the transition from hunter-gatherer communities to agricultural communities occurred because crop production was possible along major rivers like the Nile, where alluvium derived from eroded soils was deposited to produce fertile floodplains. Furthermore, soil helps to retain and filter water, contributing to its value in agriculture and hydrogeologic systems. Where soils retain significant water, they can serve as aquifers that yield water for human needs. Soils also may develop into ore deposits in special situations where the soil-forming processes concentrate available, valuable elements (chapter 11).

We know the common elements present in the crust of Earth (chapter 1), and we recognize the major types of minerals that constitute that crust. Soils form by the weathering of those minerals at the *critical zone* of intersection of Earth system spheres—the lithosphere, hydrosphere, biosphere, and atmosphere. Because soils form and exist in that zone, they contain water, gases, and organisms—as many as nine billion microorganisms in three cubic centimeters of soil.[2]

WEATHERING PROCESSES

Weathering is a general term applied to the processes that change rock into soil. Two major types of weathering are recognized: physical weathering or **disintegration**, and chemical weathering or **decomposition**. The transformation of rock into soil involves both disintegration and decomposition and represents an overall change in the rock systems, including the minerals, towards a new state of thermodynamic equilibrium at a lower energy state. From a physical perspective, disintegration seems to work against this tendency towards lower energy, because in creating smaller particles it yields an increase in surface free energy. Nevertheless, from the chemical perspective, minerals that crystallize as components of rocks below the surface under higher energy conditions generally are no longer stable under surface and near-surface conditions and will undergo geochemical change to reach a lower energy state. Thus, under conditions at the surface of the lithosphere, various minerals react to form new minerals compatible with surface and near-surface conditions of pressure, temperature, and the chemical character of that setting. In detail, the reactions that occur take place at the nanoscale mineral-water interface (Putnis and Ruiz-Agudo, 2013). The geochemical changes are complex but clearly are facilitated by the fluid phase. As soils develop via complex geochemical changes, the kinds of solid-fluid interfaces available for reaction within the soil increase to include primary minerals, newly formed secondary minerals (such as clays), mineral-microbe complexes, organic matter, and oxide and carbonate surface coatings, creating opportunities for additional kinds of reactions (Chorover et al., 2007). In the end, the overall change in soil formation is towards a state of lower energy.

The two major types of weathering, disintegration and decomposition, work synergistically. Disintegration breaks large rock masses apart and reduces the grain sizes of rock and mineral particles while increasing the total surface area. The principal processes of disintegration include frost action (frost wedging and frost heaving), abrasion, unloading and exfoliation, and biological activities. Decomposition, especially acting on larger surface areas, changes the mineral compositions of the rock, mineral, and organic fragments, converting them to newly formed minerals and mineral assemblages that are stable.

Among the processes of disintegration, two types of frost action are extremely important processes in temperate to high-latitude climate zones. They are particularly effective in places such as the high mountains of temperate zones and in spring and fall months in colder climates, where daily freezing and thawing of water occur. **Frost wedging** occurs where water freezes in cracks. The water in contact with the air at the opening of the crack freezes first, as the temperature drops, and forms a seal on the fracture. Water expands upon freezing by nearly 11%,[3] and as a result the ice that develops within the sealed crack expands to create a pressure that breaks the rock apart. Experiments many years ago suggested that the force applied by the expansion of water, as it forms ice upon freezing in a rock fracture, could be several thousands of pounds per square inch (more than 1000 bars) (see note #3). Although now the exact forces are open to question, experience teaches that frost wedging is a reality. **Frost heaving** is a process that operates in previously developed soil and other grainy materials. Water freezes repeatedly in the soil, forming a wedge that increasingly forces soil materials apart. Both frost heaving and frost wedging create greater access of water to underlying materials, increasing opportunities for continued weathering.

Climate is less directly involved in the other processes of disintegration. **Unloading** is a process in which erosion removes both surficial materials and the overlying rock load from buried rock. This allows the buried rock to expand upward (and on steep slopes or cliffs, outward). Expansion creates tensional forces that break the rock along a set of fractures called **exfoliation joints**, which crudely parallel the ground or rock surface (figure 8.1a; color plate 3a). These joints become more closely spaced near the surface, where additional, more vertical joints in the rock cause slabs or sheets of exfoliated rock to break apart, yielding fragments that subsequently tend to slide downhill. **Biological activity** leading to disintegration includes such events as (1) the growth of tree roots in fractures (figure 8.1b; color plate 3b), which further widens the fractures and facilitates other processes of weathering; and (2) animal movements that cause rocks to fall or roll down slopes. **Abrasion**, the grinding of rock and mineral surfaces, occurs to a minor extent as a result of animal activities but occurs extensively where wind, running water, and ice tend to drag, bounce, or roll mineral and rock grains across a surface, grinding away both that surface and the surface of the grains.

Abrasion reduces grain sizes. Reduction in grain size by half doubles the surface area of a given volume of material. Thus, a single cubic meter of rock with a surface area of 6 m^2 (six sides of 1 m × 1 m is 6 × 1 m^2 = 6 m^2), if broken into eight equal blocks of 0.5 meters per side, would have double the surface area. Each of the blocks would have sides of 0.5 m by 0.5 m and a surface area of 0.25 m^2, but each block would have six sides, giving it a total surface area of 1.5 m^2 per block. Since there are eight half-meter blocks required to make the single cubic meter of rock, the total surface area is 8 blocks × 1.5 m^2/block = 12 m^2 of surface area—double the original amount.

If the cubic meter of rock were reduced to millimeter-sized cubic pieces, it would consist of 10^9 cubes with six sides, each of 1 mm by 1 mm = 6 mm^2 each. Thus, the resulting surface area would be six billion mm^2 or about 6000 m^2, one thousand times greater than the surface area of the original block.

$$10^9 \text{ blocks} \times 6 \text{ sides/block} \times (1 \times 10^{-6}) \text{ m}^2/\text{side} = 6 \times 10^3 \text{ m}^2$$

Inasmuch as chemical reactions proceed at rates that are proportional to the surface area, disintegration greatly facilitates decomposition by increasing surface area and thereby increasing chemical reaction rates.

Decomposition includes at least seven different processes: oxidation, reduction (recall redox in chapter 2), hydration, hydrolysis, dissolution, chelate formation, and colloid formation with base exchange. Commonly, as we will see below, the processes of oxidation, reduction, and hydration, in particular, yield new minerals—a process also called **authigenesis**. All are facilitated by the smaller grain sizes produced by disintegration.

Oxidation and reduction are opposites. Recall from chapter 2 that *oxidation* is the process in which the

Weathering and Soils 147

(a) Exfoliation joints dipping (tilted) left in foreground and middle ground parallel to the slope and dipping right in the background, parallel to the slope. Near Olmstead Point, Yosemite National Park.

(b) A tree has grown (and later died) in a crack in a glacially polished and grooved rock (lower left), the upper surface of which (mid-right foreground) is becoming etched by decompositional processes. June Lake, CA.

Figure 8.1
Aspects of weathering.
(Photos by Loren A. Raymond.)

valence or oxidation number of an ion is increased. A typical reaction of this type is the oxidation of magnetite to hematite, in which the ferrous iron in magnetite is oxidized by oxygen to form ferric iron in hematite.

$$4Fe_2O_3 \cdot 4FeO + O_2 \leftrightarrow 6Fe_2O_3$$
$$\text{magnetite} + \text{oxygen} \leftrightarrow \text{hematite}$$
$$8(Fe^{3+}) + 4(Fe^{2+}) \leftrightarrow 12(Fe^{3+})$$

Reduction takes the reverse course—that is, in reduction, ferric iron is reduced to ferrous iron, especially in the presence of acidic solutions. Hence, pyrite (FeS_2) is formed from hematite through reduction reactions, in some cases facilitated by sulfur-reducing bacteria. Reduction is far less important than oxidation as a weathering process.

Water is essential to reactions of the hydration, hydrolysis, and dissolution types. These reactions occur at the fluid-mineral interface and are initiated on the mineral surface (Putnis and Ruiz-Agudo, 2013). In **hydration** reactions, water *combines* with a preexisting compound in a mineral to yield a new hydrated mineral. For example, hematite combines at the atomic scale with water to become goethite in the following reaction:

$$Fe_2O_3 + H_2O \leftrightarrow 2FeOOH$$
$$\text{hematite} + \text{water} \leftrightarrow \text{goethite}$$

In contrast, in **hydrolysis**, an excess of H^+ or $(OH)^-$ is produced in solution by a reaction (i.e., the reaction may yield a weak acidic or weak basic solution).[4] Hydrolysis is important in the weathering of feldspars to form clays. A common reaction involves the two-stage breakdown below, beginning with hydrolysis and ending with hydration:

Hydrolysis

$$KAlSi_3O_8 + H_2O \leftrightarrow HAlSi_3O_8 +$$
$$\text{alkali feldspar} + \text{water} \leftrightarrow \text{unstable phase} +$$
$$K^+ + OH^-$$
$$\text{potassium ion} + \text{hydroxyl ion}$$

Hydration

$$2HAlSi_3O_8 + 9H_2O \leftrightarrow$$
$$\text{unstable phase} + \text{water} \leftrightarrow$$
$$Al_2Si_2O_5(OH)_4 + 4H_4SiO_4$$
$$\text{kaolinite} + \text{silicic acid molecule}$$

In this reaction, the unstable phase that develops by feldspar hydrolysis immediately hydrates.

Dissolution involves water in another way. Acids in aqueous solution and water itself are the two main agents of **dissolution**, a process in which those parts of a mineral at the grain-fluid interface are dissolved. For example, pyroxenes may dissolve by the following reaction:

$$(Fe, Mg, Ca)SiO_3 + 2H^+ + H_2O \leftrightarrow$$
$$\text{Ca-pyroxene} + \text{hydrogen ion} + \text{water} \leftrightarrow$$
$$Fe^{2+} + Mg^{2+} + Ca^{2+} + 4H_4SiO_4$$
$$\text{ions} + \text{silicic acid molecule}$$

(see endnote 4).

> **Problem 8.1**
>
> Consider the conversion of magnetite to limonite (FeO · nH_2O). Write a reaction and explain what processes of decomposition are involved in this reaction.

Water is also involved as a base for solutions in colloid formation and chelation. In the former process, very finely divided compounds called *colloids* develop in an aqueous solution. Colloids have negative charges associated with their structures, and they attract hydrogen ions to their surfaces. If they come in contact with minerals containing positive ions, such as potassium, these ions can be drawn out of the mineral structure and exchanged for the hydrogen ions of the colloid surface. Such an exchange destabilizes the mineral, just as was the case in the hydration example for feldspar presented above. As soil fluids move into parts of the soil with different chemical compositions and character (e.g., different pH), the K^+ ions may be re-exchanged for H^+ ions (Brady and Weil, 1999, p. 24). Thus, the combined reactions of **colloid formation and base exchange** represent a type of decomposition process. Similarly, in **chelation**, complex molecules called chelates that form in aqueous solutions (e.g., stream waters or groundwater) extract or draw out the metal cations from metallic mineral structures, causing those structures to break down. In general, the rates of weathering reactions and the exact details of dissolution and new mineral formation at the atomic level are undergoing reevaluation as a result of new nanoscale observations, but the critical role of fluid remains as an essential element of these processes (Putnis, 2002; Teng, 2013; Luttge et al., 2013).

Our understanding of the role of the biosphere in weathering is increasing.[5] Microbes, lichens, and plants produce organic acids that act to increase weathering rates and influence the nature of decomposition reactions. In addition, the decay of organic materials also yields acids and new organic particles. At the nanoscale, these organic products can alter mineral surface chemistry and affect reaction rates and ion mobility (Chadwick and Chorover, 2001; Chorover et al., 2007). Thus, the biosphere contributes in a significant way to weathering.

Together, the various weathering processes convert large monolithic masses of rock into decomposed rock and soil. The *chemical end products* of weathering, the various ions and molecules, enter into solution and are transported to other parts of the developing soil or are transported away in solution. The resulting *physical end products* of weathering, the rock and mineral grains, generally vary in size and stability, except in those settings where weathering over a long period of time has allowed the soil system to attain stability (equilibrium under the surface/near-surface conditions).

If we examine the minerals of crustal rocks in a variety of weathering environments, we find that the stability relations of common minerals vary. Several factors control the rates of weathering and kinds of products that become a part of the soil and groundwater (Velbel, 1999).[6] Earlier, Goldich (1938) thought that the relative stability of minerals was the reverse of Bowen's Reaction Series (figure 7.5); that is,

quartz → muscovite → K-feldspar → biotite and Na-feldspar → hornblende and NaCa-plagioclase → augite → Ca-plagioclase → olivine

In fact, climate is a major controlling factor in weathering and stability of minerals and may result in changes in this sequence. In humid, temperate regions, the general sequence of mineral stability is approximately that shown in figure 8.2. In arid climates, minerals that are less stable in humid climates may have greater stability than some of the more stable minerals of the humid climates. In a wide range of environments, the stability of quartz is high and olivine is low, but the detailed sequences of stability between these extremes are determined by local environmental conditions of climate and organic activity.

Ultimately, then, within decomposed rocks and soils, there exist relatively few minerals, plus molecules, ions, and an array of common inorganic products of weathering (table 8.1). As shown in table 8.1 and discussed below, these include residual quartz, feldspars, and other residual minerals; rock fragments; newly formed clays, iron oxides, and hydroxides (e.g., hematite, goethite); aluminum oxides; and precipitated materials derived from the ions and molecules that have gone into solution. In solution, the common products are silicic acid molecule (H_4SiO_4) and K, Na, Ca, Mg, Mn, and Fe ions. In addition, a variety of organic compounds are present. Soils and sediments are composed of the various solid materials and new minerals precipitated from the ions in solution.

Ease of Weathering ↑

Calcite
Magnetite
Olivine
Garnet
K-feldspar
Plagioclase, Chlorite
Hornblende
Biotite
Staurolite
Kyanite
Andalusite
Sillimanite
Quartz
Zircon

Figure 8.2
Relative stabilities of some minerals in a humid temperate climate.
(*Sources:* Based in part on Velbel, 1999; Nesbitt et al., 1997; Hyman et al., 1998; Murphy et al., 1998.)

Table 8.1 Common Weathering Products of Abundant Minerals of the Crust

Mineral in Rock	Weathering Products
Quartz	Quartz; dissolved silica as silicic acid molecule [H_4SiO_4]
White micas	Clays; Na, K ions; H_4SiO_4; gibbsite
Biotite	Clays; iron oxides; K, Mg, Fe ions; H_4SiO_4
Feldspars	Clays; Ca, Na, K ions; H_4SiO_4
Amphiboles	Fe oxides; Na, Ca, Mg, Mn, Fe ions; clays; H_4SiO_4
Garnet	Ca, Mg, Mn, Fe ions; clays; Fe oxides; H_4SiO_4; gibbsite
Al-silicates	Quartz; clays; gibbsite; H_4SiO_4
Epidote	Ca, Fe ions; clays; H_4SiO_4
Pyroxenes	Fe oxides; Ca, Mg, Mn, Fe ions; clays; H_4SiO_4
Magnetite	Hematite; goethite; limonites
Pyrite	Hematite; goethite; limonites; SO_4^{-2} ions
Dolomite	Ca, Mg, HCO_3 ions
Calcite	Ca, HCO_3 ions

The Structure, Chemistry, Mineralogy, and Physics of Soils

As soils develop, they generally develop a series of layers or *horizons*. The particular details of the soil structure, the soil mineral content, and the soil chemical composition depend on a variety of factors including the climate, the underlying rock type, the biota, the topography, and the degree of development of the soil. Degree of development, at least in part, is a function of time. Physical properties depend on mineral content, composition, and structure. In an idealized, well-developed soil formed in a temperate climate, the soil has three layers above the underlying rock (figures 8.3; 8.4; color plate 3c). The layer just above the rock is called the C-horizon. It is composed of partially decomposed and fractured bedrock. Clays and residual minerals—such as quartz, feldspar, micas, and other minerals, some of which may be partially decomposed—occur in this horizon. Above the C-horizon is the B-horizon, a layer sometimes referred to as the *zone of accumulation* because materials dissolved from the layer above (the A-horizon) re-precipitate and accumulate in this horizon. Depending on the climate and other factors, this horizon may contain abundant iron or aluminum oxides; some residual quartz, feldspar, and other minerals; and a variety of clays formed from the decomposing rocks below. If it is well-developed, the A-horizon, sometimes referred to as the *zone of leaching*, has sublayers within it. The uppermost part of the A-horizon, a layer separated out by some soil scientists as the O-horizon, consists of organic debris such as decaying leaves, branches, other plant parts, plus animal remains, all of which decompose into a variety of organic particles. Below that is a layer of mixed organic and mineral matter, underlain by a leached zone depleted in iron and other elements. The leached zone is designated as a separate E-horizon by some soil scientists (e.g., Brady and Weil, 1999, p. 66). The thicknesses of soil horizons vary and the boundaries are a bit diffuse and irregular.

Within the soil layers, the soil may have a *fabric* (Paton, 1978, ch. 5). Here, the term fabric refers to the combination of structures and textures that give the soil a physical character other than homogeneity. Of particular importance at the textural level are the

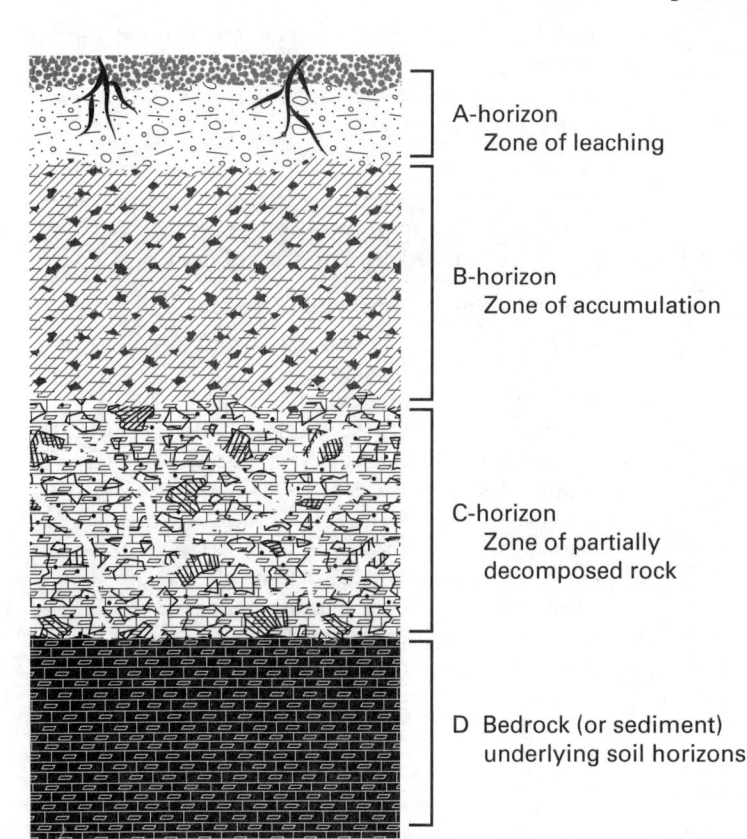

Figure 8.3
Major soil horizons. The A-horizon includes an upper zone of humus accumulation (the O-horizon). Roots may extend down into or through this horizon. The lower part of the A-horizon may be designated the E-horizon.

arrangements of soil particles and void spaces. In some soils, grains are packed together rather tightly and voids are minimized. In others, voids are abundant and comprise a major fraction of the soil volume. Superimposed on the textures, at a larger scale, are structures such as fractures that cut across the soil mass.[7] These and other physical properties are discussed next.

Chemistry of Soils

Reviewing the compositions of soils, we find four major physical components within the soil. These are the *mineral/rock fragments*—the solid, inorganic components of the soil; *water*, the liquid, inorganic component; *gases* present within voids; and solid, *organic*

Figure 8.4
Partially decomposed bedrock and overlying soil, with A, B, and C horizons marked. D is the bedrock of Mesoproterozoic Cranberry Mines Gneiss. The middle of the outcrop has a small fault that offsets the C-D contact. Notice the roots in the A-horizon near the right and left edges of the photo. The B-horizon is poorly developed and appears to contain some materials that may have been transported to the site prior to the latest soil development period. The 3-m exposure is located in Boone, NC.
(Photo by Loren A. Raymond.)

components. Each contributes to the *overall* physical properties of the soil, including soil textures and structures. The general chemical composition of a soil, in contrast, is considered in terms of the *individual* chemical components. The solid components of the soil, which typically make up about 50% of the soil volume,[8] are usually analyzed separately to determine their organic and inorganic chemical constituents, such as oxygen, iron, and carbon. Similarly, the colloidal substances (particularly clays) may be analyzed as a separate component; in some cases each colloidal substance is reported as a constituent with a particular chemistry.[9] Each fluid and gas, too, is usually analyzed separately as an individual component of the soil.

Chemically, as might be expected, the analyses of soils reveal that soils have a wide range of compositions. Overall, the major elements in soils are those typical of rocks and minerals—oxygen, silicon, aluminum, iron, calcium, magnesium, sodium, and potassium—but in organic soils carbon is abundant. Yet, there are wide variations in abundance in virtually all the major elements except oxygen. In general in soils in the United States, for example, Si ranges from 1.6% to 45%, C ranges from 0.06% to 57%; Ca ranges from 0.01% to 32%, Al ranges from 0.07% to 1%, and Fe ranges from 0.01% to >10% (Shacklette and Boerngen, 1984). Many of these variations are predictable. **Aridosols**—the soils of arid regions—commonly have high Ca concentrations; **histosols**—the soils of swamps and alpine tundra—have high C concentrations (up to 57%); and **oxisols**—the leached soils of the humid tropics—are high in Al.[10] Predictably, organic carbon is low in the aridosols (< 1%) (Brady and Weil, 1999, p. 448).

Like the solid component, soil water and soil gas—the aqueous and gaseous components of the soil—contribute to the variability in composition (Schwab, 2000; Tan, 1994). Soil fluids not only contain water, but also dissolved substances and colloids. Among the elements that are dissolved in soil fluids are Si, as silicic acid molecule (H_4SiO_4); ionic Al, Fe, Mg, Mn, Ca, Na, and K; N as a constituent of amino acid; inorganic ions, such as NH_4^+ and NO_3^-; and P as $H_3PO_4^-$ (table 8.2).[11] The gasses in the soil are typical of those of the atmosphere, although individual gas abundances may vary.[12] The most abundant components include N_2, O_2, CO_2, H_2O, and CH_4 (table 8.2).

The various elements and compounds in the soil thus occur as residual and new mineral grains, as colloids, as soil fluids and gases, and as living and nonliving organic materials, each of which has particular chemical makeup. As noted above, common residual minerals include quartz and feldspar; but also, depending on the bedrock source and degree of weathering, minerals such as muscovite, biotite, chlorite, amphiboles, garnet, epidote, aluminum silicate minerals, and zircon are present. Newly formed minerals—minerals that are stable at or very near the Earth surface and are produced by weathering—include opal; calcite; aragonite; a variety of clay minerals, including kaolinite, dickite, smectites such as nontronite and beidellite, illites, halloysite, and mixed-layer minerals; chlorites; gibbsite; zeolites; and the iron oxide and hydroxides hematite, goethite, and limonites. The colloids, which are finely divided materials dispersed in the aqueous fluid,[13] include both the clays and a variety of organic materials. Organic materials include a wide range of living and nonliving substances, including microbes

Table 8.2 The Abundant Elements in Soils

Element	Soil Solids	Soil Fluid*	Soil Gas
O	✓✓	✓✓	✓
H		✓✓	✓
N		✓	✓✓
C	±✓	✓	✓
Si	✓	✓	
Al	✓	✓	
Fe	✓	✓	
Mg	✓	✓	
Mn	✓	✓	
Ca	✓	✓	
Na	✓	✓	
K	✓	✓	
P		✓	

* Except for water, the concentrations of the abundant elements in fluids are far lower than are those of the solids.

✓ Present

✓✓ The most abundant elements

± Indicates that the element is only present in relative abundance in certain soils.

(*Sources:* Schwab, 2000; Tan, 1994, Ch. 4, 5; Tan, 1998, p. 51; Brady and Weil, 1999, p. 22–23.)

Problem 8.2

Examine figures 1.2, 1.3, and 1.4 and table 8.2. Using this information and information available on the Internet, for each element in the soil determine the soil solid, soil liquid, and soil gas phases, and whether the most likely source is the lithosphere, atmosphere, hydrosphere, or biosphere.

such as bacteria; fungi; worms, including nematodes; plant matter of various types, especially roots; and animal remains ranging from bones and decomposing, soft cellular materials to amoebae. The chemical compounds that constitute these materials are highly varied and include such materials as amino acids and other organic compounds called lipids and lignin.[14]

Physical Properties of Soils

The several physical properties of soils include the color, texture (especially the grain size distribution), structures, density, porosity, permeability, rheological (strength) properties, elasticity, and plasticity.[15] In terms of texture, soil scientists (in contrast to engineering geologists) focus on grain-size distributions in soils, paying less attention to grain shapes and intergrain relationships. The term *soil texture*, as used by soil scientists, generally excludes grains larger than 2 mm (e.g., Soil Survey Staff, 1975, p. 383). Grains larger than 2 mm include "gravels," cobbles, and boulders. Sand ranges from 2 mm to 0.02 mm (or 0.05 mm, depending on the classification), and silt ranges down to 0.002 mm (figure 8.5). In examining figure 8.5, notice that the Unified Soil Classification used in engineering practice differs substantially, in terms of grain-size designations, from the classifications used both by soil scientists (USDA, International Society of Soil Science) and geologists (Wentworth's Sediment Size Scale). In the Unified Soil Classification, the boundary between sand and gravel is much higher at 4.75 mm than is the same boundary used by USDA, ISSS, and Wentworth (2.0 mm). Similarly, the International Society of Soil Science uses a lower boundary—0.02 mm—for the sand/silt boundary than do Wentworth and the USDA (0.0625 and 0.05 mm,

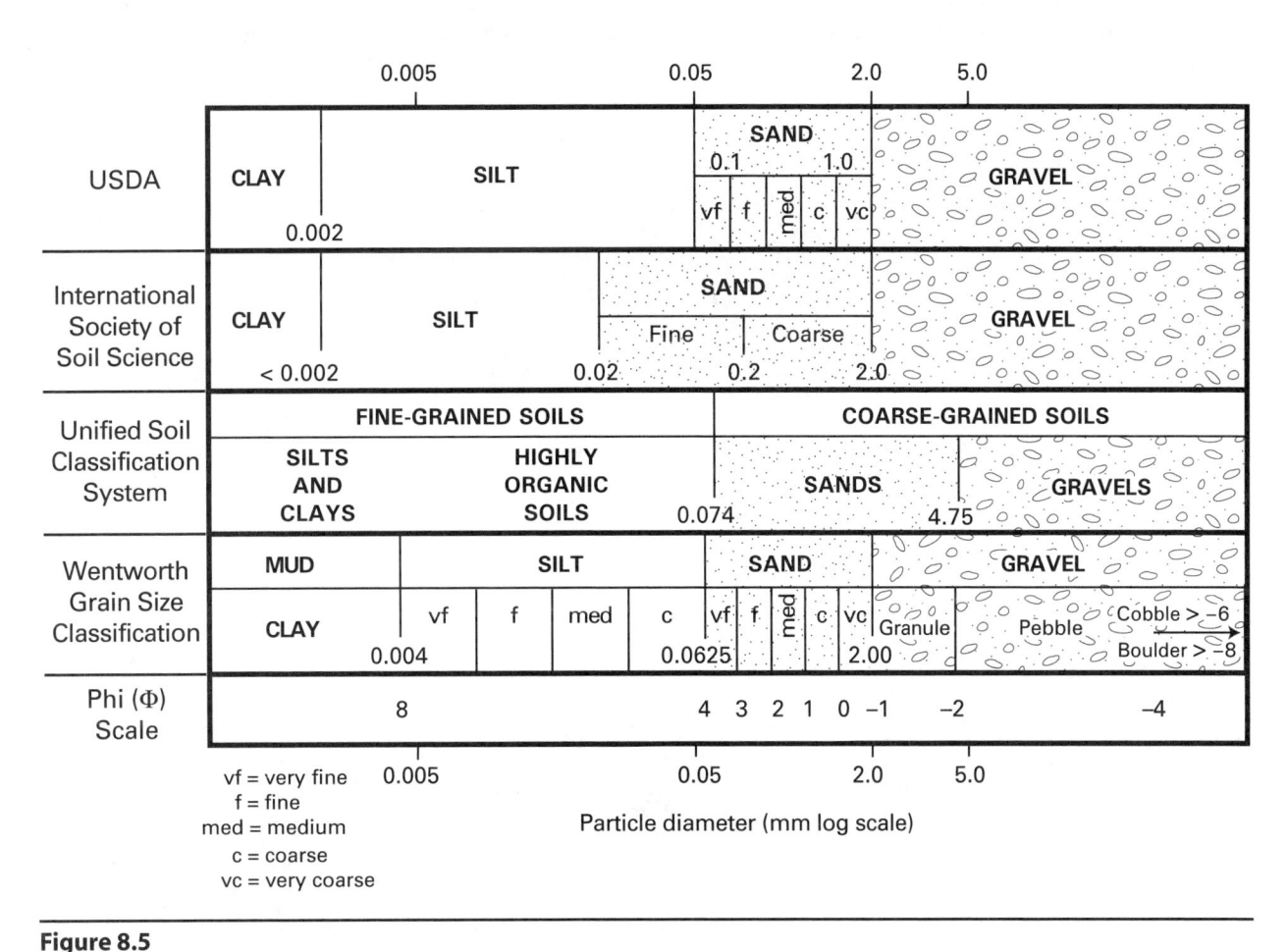

Figure 8.5
A comparison of grain-size categories used in different soil and sediment classifications.
(*Sources:* Wentworth, 1922; Krumbein, 1934; McManus, 1963; Soil Survey Staff, 1975; Merritt and Gardner, 1986, p. 6-4; Brady and Weil, 1999, p. 120.)

respectively).[16] Thus, different definitions are used in different fields of study. Regardless of how soils are classified or subdivided, it remains true that soil textures control the movement of water through soils, the strength of soils, and the physical behaviors of soils, particularly under stress.

Structures in soils are of two types: *primary structures* developed during soil formation and *secondary structures* developed in soils after they have formed. In soil science, the term structure refers to the arrangement of soil particles into aggregates called **peds** (Soil Survey Staff, 1975, p. 474; Brady and Weil, 1999, p. 130). Peds are described in terms of their sizes, shapes, and definition; and their relationships to intervening pore spaces is a critical aspect of soil structure.[17] Ped sizes can be defined by the size categories shown in figure 8.5. However, since a variety of particle sizes may clump together, the USDA also recommends use of the terms skeletal, sandy, loamy, and clayey for soils (and presumably, peds) dominated by particular grain sizes. *Skeletal* refers to particles of >2 mm, *sandy* has its usual meaning, *loamy* refers to soils dominated by very fine sand and silt, and *clayey* refers to clay-dominated soils.[18] Ped shapes include spherical (granular), platy, blocky, and prismatic types (figure 8.6).[19] Ped definition categories include weak, moderate, and strong. Where peds are absent, the soil is massive (Kay and Angers, 2000). Organic structures, such as roots, form both during and after soil formation. Secondary structures in soils include root lines, joints (which may be difficult to distinguish from the fractures that define peds), faults, and slickensides that develop along faults.

Density, porosity, and permeability in soil studies have their usual meanings. Soil densities vary from 0.1 in the organic soils called histosols to about 2.0 in vertisols, the soils rich in swelling clays (Brady and Weil, 1999, p. 136). Hard, cemented layers that develop in the soil, called *pans*, may have slightly higher densities. Densities are higher in poorly sorted materials

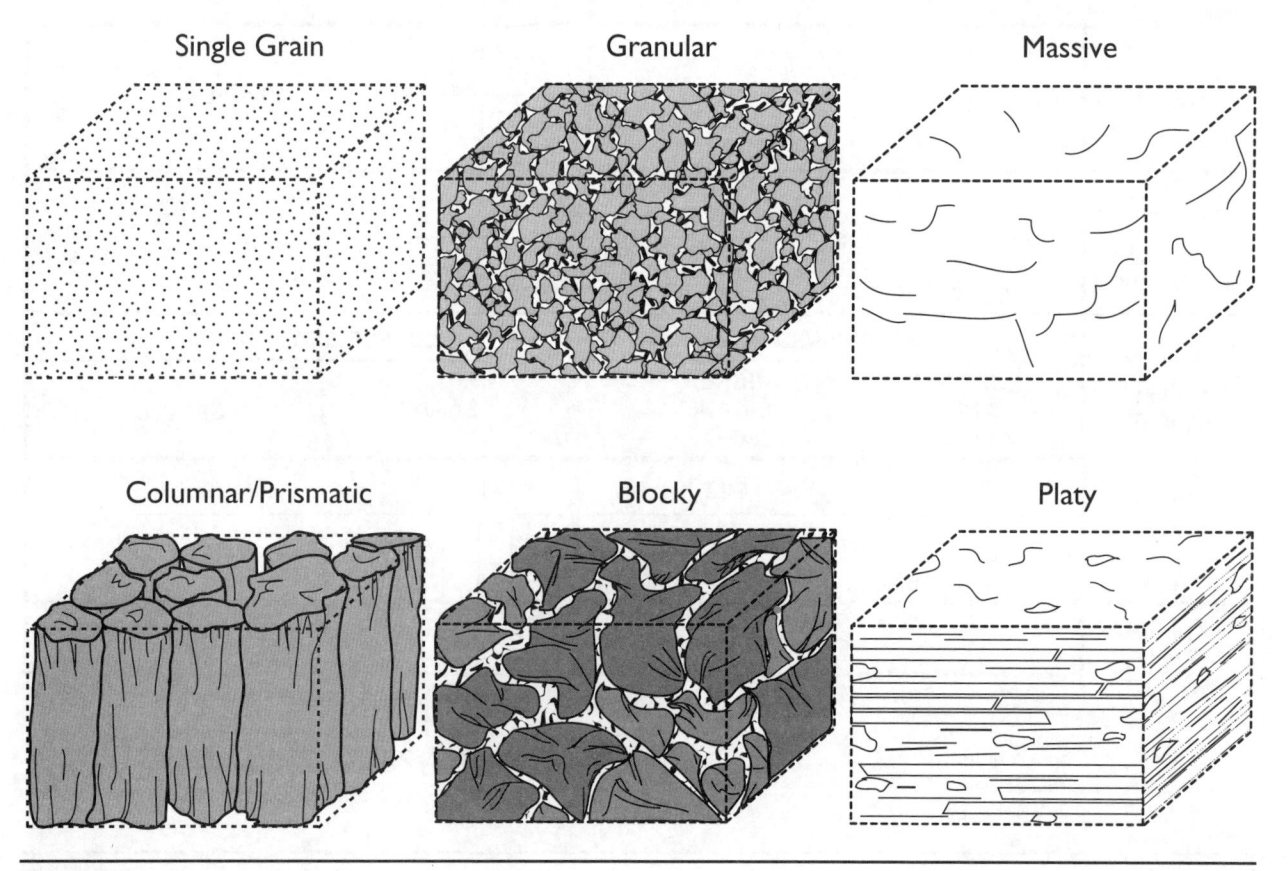

Figure 8.6
Ped structures. See text for discussion.
(*Sources:* Brady and Weil, 1999; Kay and Angers, 2000.)

versus well-sorted sandy soils. Permeability, in contrast, increases with degree of sorting. Porosities of typical uncultivated soils are in the range of 35–65%.[20] The pores themselves are divided into macropores (> 75 mm), mesopores (30–75 mm), micropores (5–30 mm), ultramicropores (0.1–5 mm), and cryptopores (< 0.1 mm).[21]

Other physical properties of soils—strength, elasticity, plasticity, and liquid limit—have particular importance to the practice of engineering and environmental geology. The strength of soils can be measured and described in relation to a variety of tests. Soil scientists refer to the degree to which a soil resists deformation as the *consistence* or *consistency*. In terms of consistence, soils are described as non-cemented, extremely weakly cemented, weakly cemented, moderately cemented, and strongly cemented. The field test for consistency involves an assessment of one's ability to drive a pencil (by hand) or a thumbnail into the soil. For classifying soils using the Unified Soil Classification (USC) and as a measure of physical properties, Atterberg limits (plastic limit and liquid limit) and measures of critical water contents are measured in soils (figure 8.7).[22] The Plasticity Index (PI), measured in percent, is the difference between the liquid limit and plastic limit. Hence, PI = LL – PL. The **Liquid Limit** (LL) is the percentage of water in a soil above which the soil flows under its own weight and below which the soil behaves as a solid. The **Plastic Limit** (PL) is the percentage of water in a soil marking the boundary between brittle, friable behavior and plastic behavior. Wetter soils behave plastically. Thus, in the USC, for example, soils with a high PI and a high LL are designated CH and are called *fat clays* or inorganic clays of high plasticity. Such clays or soils are easily deformed. The strength of cohesive soils is tested by squeezing a cylinder of soil between two plates using a device referred to as a *press*. Application of such external stresses (forces per unit area) using a press will cause the soil to undergo plastic deformation. If the sample is enclosed and subjected to fluid pressures on all sides, the test is a triaxial compression test. Materials that lack cohesion will deform permanently by shearing with shear stress being directly proportional to normal stress (i.e., the material follows Hooke's Law). Other materials will deform elastically, at least under low stresses, so that applied stresses yield temporary deformation that disappears when the stress is released. Still other materials behave plastically, retaining deformation after stress is released. All of these (and other) properties are particularly important to engineering structures built on soils.

Soil Classification

Because soils differ from place to place and in degree of development, a number of classifications have been created to distinguish different types of

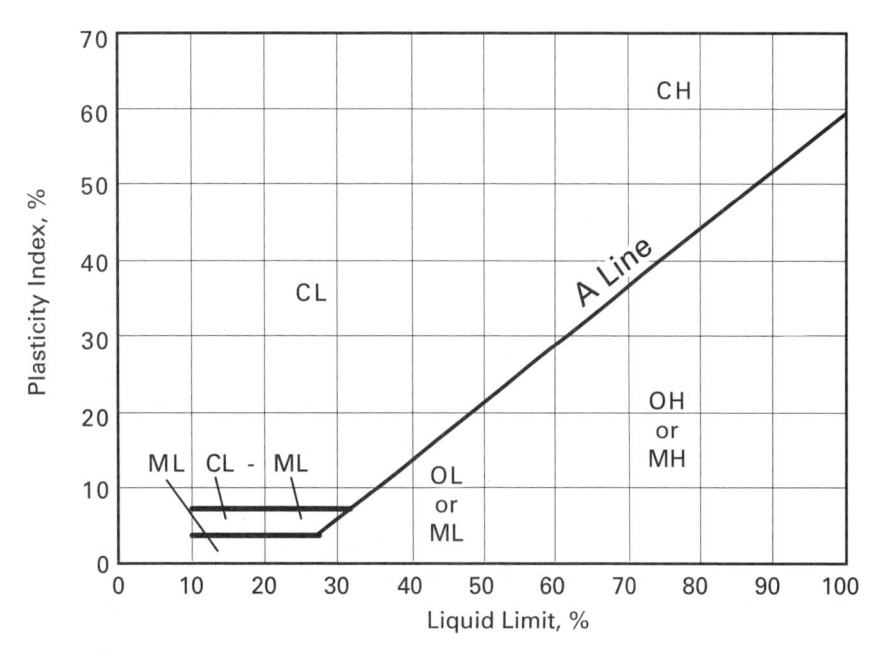

Figure 8.7
Diagram showing Atterberg limits of soils. Letters represent soil types listed in figure 8.8. See text for explanation.
(*Source:* Merritt, 1986. Used with permission.)

soils. Some older classifications were based on climate, with soils developed in areas of relatively low rainfall being designated as *pedocals* (because of accumulated calcium) and soils developed in areas of higher rainfall being designated as *pedalfers* (the *al* and *fe* in the word referring to concentrations of aluminum and iron that develop in such soils). *Laterite* is a term used for the hard, brick-like soils of tropical regions, soils that are enriched in iron and aluminum, and in unusual cases in other elements of economic value. In general, such older classifications and names are no longer used, although students may encounter anachronistic terms from these classifications in doing research—especially in older literature.

Modern soil classifications are based on particular features of the soil rather than on external controls. Many different soil classifications are used in the various countries around the world. Here, we introduce two classifications: the United States Department of Agriculture (USDA) classification used in agricultural fields and the Unified Soil Classification used in engineering practice.

The USDA Classification

The USDA classification divides soils into twelve soil types called *orders* (table 8.3) (Soil Survey Staff, 1999). Entisol, inceptisol, alfisol, ultisol, and oxisol represent soils of a sequence of progressive increases in leaching from Horizon A and accumulation in Horizon B. Entisols, for example, are very "immature" soils, with little leaching from Horizon A and no development of Horizon B (i.e., B is lacking). At the other end of this series, oxisols have a strongly leached A Horizon, significant accumulation of iron and aluminum hydroxides in layer B, and some evidence of breakdown of clays in Horizon B. The remaining soil orders have special characteristics. For example, histosols have a thick upper A Horizon composed of organic debris, whereas aridosols have thin A and B horizons and are characterized by calcium carbonate (*caliche*) accumulation in the upper layers of the soil.

The USDA classification is further subdivided into five additional sets or levels of subdivision. Below the order is the suborder, which includes 47 soil types. Below the suborders are the so-called "great groups," which include 185 variants. The lowest category of soil subdivision recognized is the *series*, of which there are 10,500 in the United States. The designation "soil type" is not used.

Unified Soil Classification

The Unified Soil Classification (USC) is used in engineering practice. As indicated in figure 8.8 (on p. 158), soils are separated into two categories: fine and coarse. Within these groups, the various subdivisions are based on texture, Atterberg limits, and other tests. Gravels (G) and sands (S) are each divided into four types, each of which has a symbol (e.g., GW represents well-graded = poorly sorted gravels). The first letter represents the grain size (e.g., gravel = G) and the second letter represents the distinguishing feature (e.g., W = well-graded). A third category of soils, the highly organic soils, is recognized based on composition rather than texture and is designated as PT.

Classification of soils using the USC *requires* that a variety of tests and analyses be completed in the laboratory. The details of performing these tests and the resulting values are beyond the scope of this text and are not shown in figure 8.8, but they are essential to proper classification of soils in the USC. This is especially true for the fine-grained soils, for which classification requires tests on strength, Atterberg limits, dilatancy (ability to change volume), and toughness. An example of defining laboratory criteria would be that GC soils must have Atterberg limits above the A Line (of figure 8.7) and have plasticity indices of greater than 7.

Problem 8.3

Using your best judgment, estimate (a) what USDA soil order might represent each of the following USC soil types: GW, SM, ML, and PT; and (b) what USC soil type best matches the vertisols of the USDA.

ORIGINS OF SOILS

Soils develop from parent materials through the various processes of weathering described earlier. The parent materials of soils include rocks of igneous, sedimentary, and metamorphic character, and unlithified sediments ranging from marine and glacial to desert types. Each of these brings a particular mineral, chemical, and textural character to the soil-forming process.

As weathering progresses over time, soils form through a series of processes that operate at rates dependent on climate, plant growth, and other organic activity, and time. Disintegration breaks underlying rock and available rock fragments into smaller pieces, increasing surface area and thereby facilitating chemical reactions. Decomposition reactions change unstable minerals into stable ones. Some materials are lost from the developing soil, having been removed by erosion from the surface or groundwater flow into the subsoil. Organic materials are added to the developing soil horizons by addition from above (e.g., the falling of leaves) and through organic activities by plants, animals, fungi, and microbes within the soil. Water is added from above through precipitation, is lost to the atmosphere through transpiration by plants and evaporation back to the atmosphere, and is lost to the groundwater reservoir as it moves to levels below the

Table 8.3 Simplified Version of the USDA Soil Classification

Soil Type	Soil Characteristics
Entisols	Very poorly developed soils, with little leaching from the A-Horizon and no development of a B-Horizon; Red, clay-rich (cambic), sulfuric, calcic, gypsic, or cemented horizons (petrocalcic or petrogypsic horizons), with little clay in B-Horizon, and soils with either a fragipan or an exchangeable sodium percentage of >15%.
Alfisols	Contains a thin A-Horizon and a clay-rich B-Horizon, with a fragipan, Na accumulation zone (natric horizon), or low CEC
Ultisols	Strongly leached A-Horizon with clay-rich (argillic or kandic) subsurface B-Horizon or fragipan and a base saturation of < 35% at a 2-cm depth or 75 cm below the fragipan
Oxisols	Strongly leached A-Horizon underlain by a subsurface B-Horizon composed of sandy silt to clay with little weatherable material (an oxic horizon) within the upper 150 cm, some accumulation of hydrated oxides of Al and Fe, and a low CEC
Spodosols	Acid soil with organic-rich A-Horizon and spodic horizon (a horizon with amorphous organic, Al, $\pm$ Fe materials) forming a "hardpan" at the B-Horizon level
Histosols	Organic materials extend down to the impermeable layer or more than 40 cm down, but soil lacks the properties of andisols
Inseptisol	Typically, highly colored (yellow, orange, red, or brown) thin mineral soil with dark humus-rich surface layer (epipedon). Inseptisols lack many features of other soil types, but commonly contain a cambic horizon and a fragipan.
Mollisols	Thick, dark, humus-rich, A-Horizon (mollic epipedon) and a Ca and Mg-bearing clay-rich B-Horizon; base saturation of >50% to an impermeable layer or at 1.8 m from the soil surface
Vertisols	Characterized by shrinking and swelling properties and 30% or more clay in upper 50 cm
Aridosols	Thin A- and B-Horizons with little organic material, and low soil-moisture regime, a salty (salic) horizon, or a weak B-horizon
Andisols	Andic properties (properties imparted by low-density materials, glass, pumice, and physically similar minerals)
Gelisols	Cryoturbation structures (or permafrost)

Definitions:
andic properties = the occurrence of low-density, glass, pumice, and like mineral materials
argillic horizon = clay-rich, subsurface horizon
cambic horizon = red, clay-rich subsurface horizon
CEC = cation exchange capacity
cryoturbation structures = structures produced by freezing
epipedon = diagnostic soil horizon formed at or near the surface
fragipan = uncemented subsurface soil that restricts entry of water and roots into soil matrix
kandic horizon = a particular kind of clay-rich, subsurface horizon with what are called "low activity clays" [low CEC].
mollic epipedon = thick, dark, humus-rich horizon
natric horizon = a subsurface horizon rich in clay and enriched in sodium
oxic horizon = subsurface horizon composed of sandy silt to clay and little weatherable material
salic horizon = a horizon with salts more soluble than gypsum
spodic horizon = horizon with amorphous materials composed of organic matter, Al, $\pm$ Fe materials

(*Sources:* Soil Survey Staff, 1975; Brady and Weil, 1999; Krauskopf and Bird, 1995; Arens and Arnold, 2000.)

soil horizons. Groundwater passing through laterally or vertically may bring in ions or remove them, depending on local conditions. As these processes operate, the various soil horizons form as materials move from one layer to another in response to local environmental conditions. Over time, if undisturbed, a mature soil develops that is characteristic of the particular region in which it is forming and the rock types from which it is derived.

Human activities commonly disturb soil-forming processes. Cultivation, excavation, and human induced erosion alter the paths and rates at which materials move during the extended times over which soils form.

PALEOSOLS

Paleosols are ancient soils developed in the past and preserved in the geologic record (Parnell, 1983; Retallack, 1990; Kraus, 1999). Paleosols are characterized by a variety of features including structures, tex-

MAJOR DIVISIONS			GROUP SYMBOLS	TYPICAL NAMES
COARSE-GRAINED SOILS More than half the material above No. 200 sieve size	**GRAVELS** More than half of coarse fraction above No. 4 sieve size	Clean gravels	GW	Well-graded gravels, gravel-sand mixtures, little or no fines
			GP	Poorly-graded gravels, gravel-sand mixtures, little or no fines
		Gravels with fines	GM	Silty gravels, poorly graded gravel-sand-silt mixtures
			GC	Clayey gravels, poorly graded gravel-sand-silt mixtures
FINE-GRAINED SOILS More than half the material less than No. 200 sieve size	**SANDS** More than half of coarse fraction less than No. 4 sieve size		SW	Well-graded sands, gravelly sands, little or no fines
			SP	Poorly graded sands, gravelly sands, little or no fines
			SM	Silty sands, sand-silt mixtures
			SC	Clayey sands, sand-clay mixtures
	SILTS AND CLAYS Liquid limit less than 50		ML	Inorganic silts and very fine sands, rock flour, silty and clayey fine sands with slight plasticity
			CL	Inorganic clays of low to medium plasticity, gravelly clays, sandy clays, silty clays, lean clays
			OL	Organic silts and organic silt-clays of low plasticity
	SILTS AND CLAYS Liquid limit greater than 50		MH	Inorganic silts, micaceous or diatomaceous fine sandy or silty soils, elastic silts
			CH	Inorganic clays or high plasticity, fat clays
			OH	Organic clays of medium to high plasticity
	HIGHLY ORGANIC SOILS		PT	Peat and other highly organic soils

Boundary classification: Soils with characteristics of two groups are designated by combinations of symbols (e.g., well graded gravel-sand mixtures with clay binder = GW-GC

Figure 8.8
The Unified Soil Classification (with values for various laboratory tests omitted). The letter symbols here, each representing a kind of soil, are determined using tests described in the text that reveal the properties reflected by the position of each soil as plotted on figure 8.7.
(*Source:* Merritt, 1986. Used with permission.)

tures, minerals, fossils, and colors; but they are difficult to recognize because they have undergone diagenesis, including compaction. Among the features most useful for recognizing paleosols are root traces (root fossils), soil horizon characteristics, and relict soil structures (Retallack, 1990, ch. 3).

Root traces consist of downward branching, irregular cylindrical structures.[23] Wispy networks of tubular structures are most likely root traces rather than burrows or other similar structures. In younger deposits, organic carbon traces of roots may remain, whereas in older deposits the root traces are infilled with minerals, such as clays.

Relict soil horizons with poorly developed structures are easily confused with sedimentary layers, especially where the paleosols lack root fossils. One distinguishing feature of paleosols is that paleosol horizons are sharply bounded at the top but grade gradually down through successive layers into underlying rock.[24] The gradational layers may be distinguished by gray-green layers above and red to purple layers below. At the paleosol surface, Fe-oxide/hydroxide stains may mark the contact with overlying layers. In terms of structures, where peds are preserved, they serve to indicate the presence of a paleosol; and typical platy, columnar, blocky, or granular structures reinforce the assessment that a layer is a paleosol. Another distinguishing feature of paleosols is the presence of soil nodules of calcite, anhydrite, or other minerals.

Paleosols provide valuable insights into geologic history. For example, they reveal the existence of a time gap (a hiatus) between layers that may help us understand episodic sedimentation or volcanic eruptions.

Summary

Soil is an important Earth material, essential to the growth of plants and the functioning of complex ecosystems, as well as being:

1. a precursor to many kinds of sediment
2. a cleanser of water
3. a foundation material in construction
4. a significant base for agricultural production of food plants and grains

Soil develops via weathering—the transformation of rock into soil—by processes that are physical (disintegration) and chemical (decomposition). The products of weathering include a variety of inorganic materials including quartz, feldspar, rock fragments, clays, iron and aluminum oxides and hydroxides, and precipitated minerals derived from the ions and molecules—including silicic acid molecule, and K, Na, Ca, Mg, Mn, and Fe ions—placed in solution by weathering. In addition to the inorganic components, soils contain gases, fluids, and living and nonliving organic materials. Chemical reactions that occur at the mineral-water interface are critical to decomposition.

Both disintegration and decomposition contribute to the formation of the physical properties of soils. These physical properties of soils are controlled by their compositions, textures, and structures and include color, density, porosity, permeability, strength, elasticity, and plasticity. The physical and chemical properties serve as the basis for many classifications, including the two common classifications used in the United States: the USDA classification used in agriculture and the Unified Soil Classification (USC) used in engineering practice. Versions of the USDA classification divide soils into either ten or twelve soil types called orders based on chemical and physical properties, whereas the USC is based primarily on physical properties.

Ancient soils can be preserved in the rock record as paleosols. Like contemporary soils, paleosols are characterized by a variety of features including structures, textures, minerals, fossils, and colors but they are difficult to recognize because they have undergone diagenesis (the post-depositional changes that include compaction) and authigenesis (the formation of new minerals). Paleosols provide valuable insights into geologic history.

Selected References

Brady, N. C., and Weil, R. R. 1999. *The Nature and Properties of Soils* (12th ed.). Prentice Hall, Upper Saddle River, NJ. 881 p.

Habib, P. 1982. *An Outline of Soil and Rock Mechanics.* Cambridge University Press, Cambridge. 149 p.

Retallack, G. J. 1990. *Soils of the Past.* Unwin Hyman, Boston. 520 p.

Soil Survey Staff. 1975. *Soil Taxonomy.* U.S. Department of Agriculture, Agriculture Handbook No. 436. 754 p.

Sumner, M. E. (ed.). 2000. *Handbook of Soil Science.* CRC Press, Boca Raton. 2112 p.

Tan, K. H. 1994. *Environmental Soil Science.* Marcel Dekker, New York. 304 p.

Notes

1. Important summary and reference works on soils include Brady and Weil (1999), Marshall et al. (1996), Soil Survey Staff (1975, 1999), Sumner (2000), and Tan (1998).
2. Singer (2003).
3. Increase in water volume during freezing is 10.9% (Verhoogen et al., 1970, p. 324) and was thought, in the past, to yield up to 30,000 lbs./in^2 (2000 bars or 200 million Newtons/m^2) of pressure (Leet and Judson, 1958, p. 80). The process of frost wedging may be more complex than simple lab examples suggest, but from practical experience, we know that the forces are large enough to crack steel engine blocks and break concrete sidewalks.
4. The roles of fluids, particularly water, and the nature of reactions in weathering are topics of considerable current research. Earlier, Krauskopf (1979, p. 36) and Krauskopf and Bird (1995, p. 36) discussed the general geochemical reactions that yield slightly acidic or basic solutions during hydrolysis; Huang and Kiang (1972) and Krauskopf (1979, p. 91–92) described the hydrolysis and hydration reactions involved in weathering of feldspar; and Koster van Groos (1988) described the weathering of pyroxene. New research is focused on the specifics of the reaction-controlling features of minerals, reaction rates, and the processes that occur at the nanoscale (see Putnis, 2002; Putnis and Ruiz-Agudo, 2013; Luttge et al., 2013; Teng, 2013, and the references therein).
5. Columbo and Violante (1997), Adamo et al. (1997), Moulton and Berner (1998), Chadwick and Chorover (2001), Amundson et al. (2007), Chorover et al. (2007).
6. Also see Velbel et al. (1996) and Wasklewicz (1994).
7. Some soil scientists use the word "structure" to refer the way soil particles are aggregated (e.g., see Tan, 1998).
8. Helmke (2000).
9. For example, see Tan (1998).
10. See the USDA soil classification (Soil Survey Staff, 1975) for soil names and see Brady and Weil (1999, ch. 12), Baldock and Nelson (2000), and Tan (1994, 1998) for information on organic chemistry of soils. Carbon values from Brady and Weil (1999, p. 448).
11. Tan (1994, ch. 5).
12. Tan (1994, ch. 4; 1998, p. 51), Brady and Weil (1999, p. 22–23).
13. Krauskopf and Bird (1995, p. 154) and Brady and Weil (1999, ch. 8).
14. The details of the organic chemistry are beyond the scope of this text and the interested reader is referred to works such as Tan (1994) and Brady and Weil (1999, ch. 12).
15. Habib (1982), Merritt and Gardner (1986), Marshall et al. (1996), and Brady and Weil (1999).
16. Some soil scientists divide the soil particles into coarser elements of the soil (>2 mm) called *skelton grains*, and finer materials (< 2 mm) called *plasma* or *fine earth* (e.g., Osmond, 1958 in Tan, 1998, p. 50; Soil Survey Staff, 1975, ch. 18).
17. Soil Survey Staff (1975, p. 474–475), Brady and Weil (1999, ch. 4), Kay and Angers (2000).
18. Soil Survey Staff (1975, ch. 18), Brady and Weil (1999, ch. 4).
19. Soil Survey Staff (1975, p. 474–475), Brady and Weil (1999, ch. 4), Kay and Angers (2000).
20. Brady and Weil (1999, ch. 4).
21. Brady and Weil (1999, ch. 4.), Kay and Angers (2000).
22. Merritt (1986).
23. Habib (1982, p. 4), Brady and Weil (1999, p. 165).
24. Retallack (1990, ch. 3).

Sedimentary Rocks

INTRODUCTION

Although sedimentary rocks make up only 4.8% of the crust of Earth, sediments and sedimentary rocks blanket more than 66% of the surface.[1] Somewhat like the thin plastic case on the outside of a cell phone covering the thicker board and electronics inside, the sedimentary rocks conceal the more abundant igneous and metamorphic rocks below. Yet, sedimentary rocks are formed by more visible and, hence, more understandable processes. They reveal a great deal about Earth history.

In this chapter, we examine the nature, classification, and origins of sedimentary rocks. Each process that contributes to the formation of a sedimentary rock contributes to the character of that rock. The processes of erosion, transportation, and deposition, and the sites and conditions of deposition, vary widely. Sedimentary environments range from high mountain glacial environments, with ice erosion and transportation, to deep ocean environments, where ocean currents transport sediment that then "rains down" from above to be deposited on the seafloor. The characteristics of the resulting sedimentary rocks are diverse.

In chapter 8 we examined the processes of weathering and found that those processes produced a variety of sedimentary products, from rock fragments to clays, iron oxides, and a variety of ions. The products of weathering and the masses of exposed, unweathered rocks are the "raw materials" of sediments. These raw materials are eroded and then transported, deposited, and altered by the processes of diagenesis on their way to becoming components of sedimentary rocks. Sedimentary geologists continue to work on solutions to puzzles presented by particular rock types (e.g., Ruppel and Loucks, 2008), the details of sedimentation in specific environments (e.g., Dashtgard et al., 2012), and the geochemical roles of specific elements in sediment formation and diagenesis (e.g., Taylor and Macquaker, 2011). In addition, in the 21st century, radiometric analyses of detrital zircon grains in sediments has led to a better understanding of sediment ages, sources, and dispersal (DeGraff-Surpless et al., 2002; Dickinson and Gehrels, 2008; Dumitru et al., 2016). The fundamentals presented in this chapter are intended to help readers understand our overall knowledge of sedimentary rocks.

EROSION, TRANSPORTATION, AND DEPOSITION OF SEDIMENTS

Erosion is defined as the fragmentation and removal of solid earth materials from rock or soil exposures at the surface of the Earth. The agents of erosion include moving water (streams and currents), moving air (the wind), moving ice (glaciers), and gravity. Gravity acts alone or in concert with wind, glaciers, and streams. Because gravity—in addition to being an agent of erosion in its own right—drives the downhill movement of water and ice, it is also quite important to other aspects of erosion.

Gravity, or more accurately *the force of gravity*, is the attractive force between two masses—in this case, the Earth as a whole and a mass of material within the sphere of influence of Earth. For our purposes here, masses of material include anything from the masses of rock and soil that constitute landslides on a hillside to a single particle of sediment, no matter how small it might be. Gravity acts on all masses of material at and near the surface of Earth and pulls them down (in the direction of the center of the Earth). Gravity pulls and moves rock, soil, and loose accumulations of sediment downslope by causing unstable materials to fall, slide, or flow. Once the materials begin to move, they have been eroded and are undergoing transportation. When the force of gravity is no longer strong enough to keep the materials moving, transportation ends and deposition occurs. Gravity-induced erosion, transportation, and deposition produce landslide masses and other **mass wasting** products ranging from sheets of clay mud to blocks of unweathered ("fresh") rock (figure 9.1; figure 9.2; color plate 4a). Each of these products of mass wasting represents sediment and a potential component of a future sedimentary rock.

Wind erodes relatively small particles by moving them from the sites at which they reside in soil, sediment, or rock. Once they are eroded, the wind transports particles in **suspension** and by bed-load transport. Suspended particles are those picked up and carried for significant distances in the air as it moves from one place to another. In general, particles transported in suspension by wind are primarily of silt and clay size (i.e., less than 0.0625 mm in diameter). Sand particles, in the range of 1/16 mm to 2 mm in diameter (see the Wentworth Scale in the next section), are transported by bed-load transport, except in very high-velocity winds, which can suspend larger particles for a period of time. **Bed-load transport** is the bouncing, rolling,

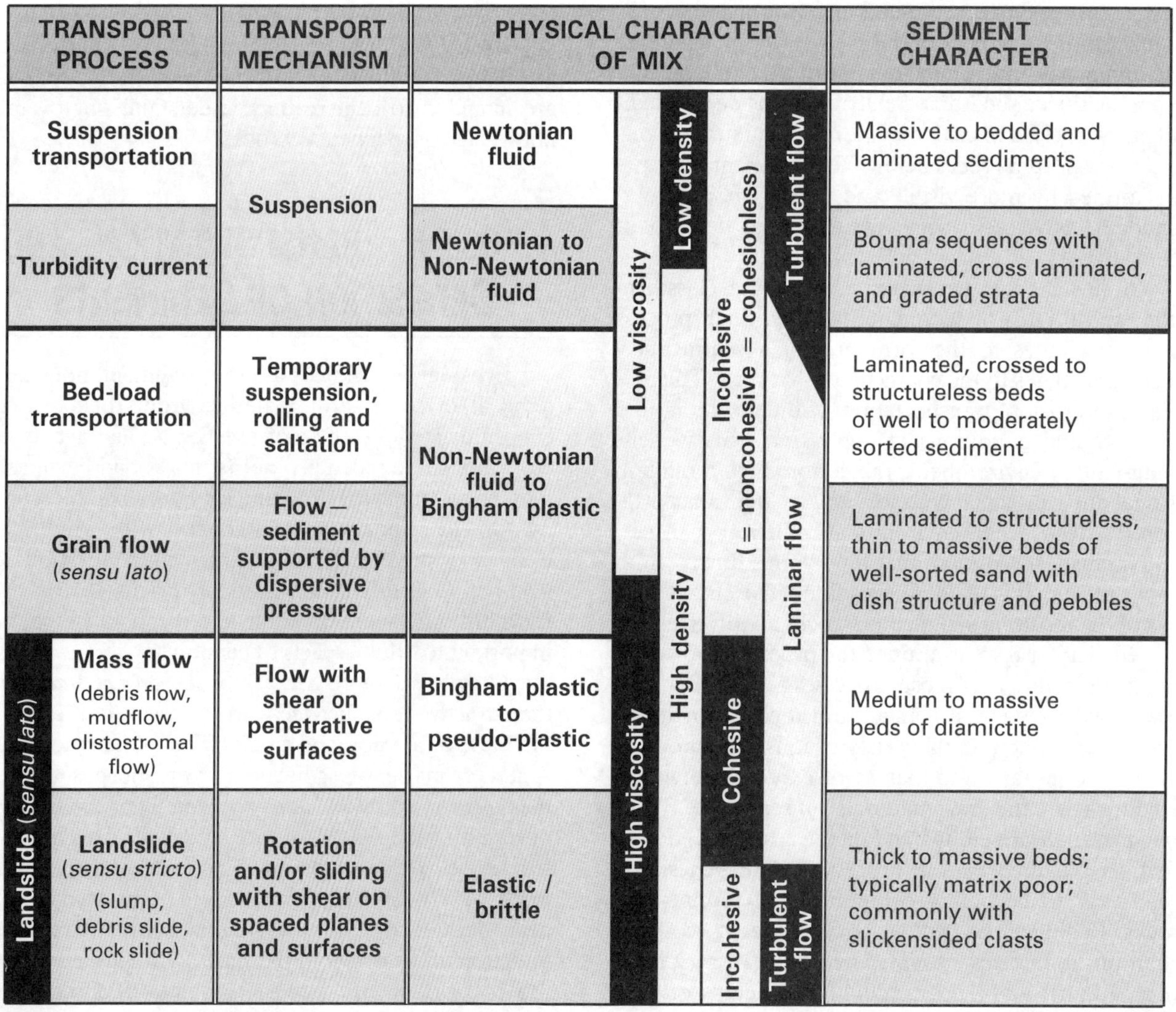

TRANSPORT PROCESS	TRANSPORT MECHANISM	PHYSICAL CHARACTER OF MIX		SEDIMENT CHARACTER
Suspension transportation	Suspension	Newtonian fluid	Low viscosity / Low density / Incohesive (= noncohesive = cohesionless) / Turbulent flow	Massive to bedded and laminated sediments
Turbidity current		Newtonian to Non-Newtonian fluid		Bouma sequences with laminated, cross laminated, and graded strata
Bed-load transportation	Temporary suspension, rolling and saltation	Non-Newtonian fluid to Bingham plastic	High density / Laminar flow	Laminated, crossed to structureless beds of well to moderately sorted sediment
Grain flow (*sensu lato*)	Flow—sediment supported by dispersive pressure			Laminated to structureless, thin to massive beds of well-sorted sand with dish structure and pebbles
Landslide (*sensu lato*) — Mass flow (debris flow, mudflow, olistostromal flow)	Flow with shear on penetrative surfaces	Bingham plastic to pseudo-plastic	High viscosity / Cohesive	Medium to massive beds of diamictite
Landslide (*sensu stricto*) (slump, debris slide, rock slide)	Rotation and/or sliding with shear on spaced planes and surfaces	Elastic / brittle	Incohesive / Turbulent flow	Thick to massive beds; typically matrix poor; commonly with slickensided clasts

Figure 9.1
Summary of transportation processes, mechanisms, and sediment types.
(*Source:* Raymond, 2007, p. 296; modified from Nardin et al., 1979; with additions based on Lowe, 1976 and Postma, 1986.)

Figure 9.2
Landslide consisting of a debris avalanche (above) and a debris flow (lower part) composed of rock fragments and soil, above the village of Matachico, Peru.
(Photo by Loren A. Raymond.)

and sliding of grains along a surface. Evidence of bed-load transport is seen in the erosion of the lower foot or so of posts along beaches, deserts, and other settings where wind-blown sand is present. As the wind velocity decreases, both bed-load transport and suspended-load transport become less sustainable. Bed-load particles are dropped first (figure 9.3). Where minor numbers of grains are being transported, the grains are dropped onto the surface as individual particles, but where large amounts of sediment are deposited they may form sand sheets or dunes of various types (figure 9.4). Suspended clay and silt may be transported for hundreds or thousands of miles before being deposited. For example, wind-blown African sediment reaches the southern United States and the Amazon Basin of South America (Bozlaker et al., 2013).[2] Because it is commonly deposited in small volumes, this fine, suspended sediment simply becomes a component of a sedimentary deposit dominated by other materials (e.g., a clay component of a sand that becomes a sandstone).

Problem 9.1

Using the Internet, (a) find the specific source of much of the dust that crosses the Atlantic Ocean toward the Western Hemisphere. (b) What environmental problems is the dust causing? (c) Will climate change likely decrease the production of dust? (d) What, if anything, can be done to reduce the generation of dust?

Water flowing in submarine currents, in streams (the larger of which are commonly called rivers), and across soil and rock surfaces (where the flow is called *sheet wash*) erodes materials and transports them in suspension, in solution, and by bed-load transport. Erosion by flowing water involves (1) physically moving clay, silt, sand, and larger rock fragments from sites of residence in soil or underlying "bedrock"; and (2) dissolution of materials from the rock to form ions that are removed in **solution**. Thus, larger grains—coarse sand and larger rock fragments—are typically moved by bed-load transport; medium to fine sand, silt, and clay are moved in suspension; and dissolved ions are moved in solution. Deposition of bed-load and suspended particles occurs when stream or current velocities decrease to the point that neither suspension nor bed-load transportation, respectively, is possible (figure 9.3). Deposition from solution occurs when the stream, lake, and ocean water, which are essentially solutions, become saturated. Under the saturated condition, chemical reactions resulting from a change in the chemical environment will cause precipitation. Deposition from solution also occurs where biological agents catalyze or directly cause precipitation—for example, in warm ocean waters in which algae precipitate aragonite.

The Eh-pH fence diagram (Krumbein and Garrels, 1952) provides a semi-quantitative framework for understanding precipitation (i.e., the deposition) of dissolved materials from solution (figure 9.5 on p. 166). The **Eh** or *redox potential* is a number representing the relative ability of solutions to produce oxidation or reduction reactions. Values greater than zero indicate an oxidizing environment in which oxide minerals, such as hematite and the manganese oxide todorokite, are particularly stable and will precipitate. The **pH**, the negative logarithm of the hydrogen ion concentration, is a characteristic of solutions often measured in streams, lakes, oceans, and groundwater. In acid environments, those in which pH values are less than 7.0, organic materials such as peat are stable. At values greater than 7.0, the chemical environment is basic and common minerals such as calcite, silica minerals (e.g., opal), and pyrite are stable and will precipitate. At pH values greater than 7.8, calcite is particularly stable. At a pH greater than 7.8, in environments where the salinity (salt content) is also high, gypsum, anhydrite, and halite are stable phases that will precipitate. Pyrite is especially favored in environments that have low Eh values.[3]

SEDIMENTARY ENVIRONMENTS, TEXTURES, STRUCTURES, AND COMPOSITIONS

Deposited sediments eventually become sedimentary rocks. This is true providing that conditions are right for preservation of the sediment, rather than re-erosion of it. Both precipitated minerals and fragments and grains of materials—including mineral grains, rock fragments, clay particles, fossils or their fragments, and glass fragments from volcanic rocks—accumulate in various settings or environments of deposition. The locale in which sediment accumulates is commonly referred to as a **basin**. Factors influencing the nature of the final sediment include: (1) the source area with its particular rocks, called the **provenance**; (2) the climate in the provenance; (3) the processes of erosion and

Sedimentary Rocks   165

Figure 9.3
Hjulstrom's diagram showing the zones of erosion, transportation, and deposition of various sizes of sedimentary fragments or clasts. Erosion occurs when velocities exceed those shown by the upper curve(s) (curve A). Note that it is more difficult (requires a higher velocity of current) to erode clay than to erode sand. Fine sands are the most easily eroded. In contrast, deposition (curve B) is somewhat proportional to grain size. Cobbles and pebbles are deposited before coarse sand, which is deposited before medium sand, which is deposited before fine sand, silt, and mud. Mud may stay suspended in motionless water for significant periods of time.
(*Source:* Verhoogen et al., 1970, p. 337. Used with permission.)

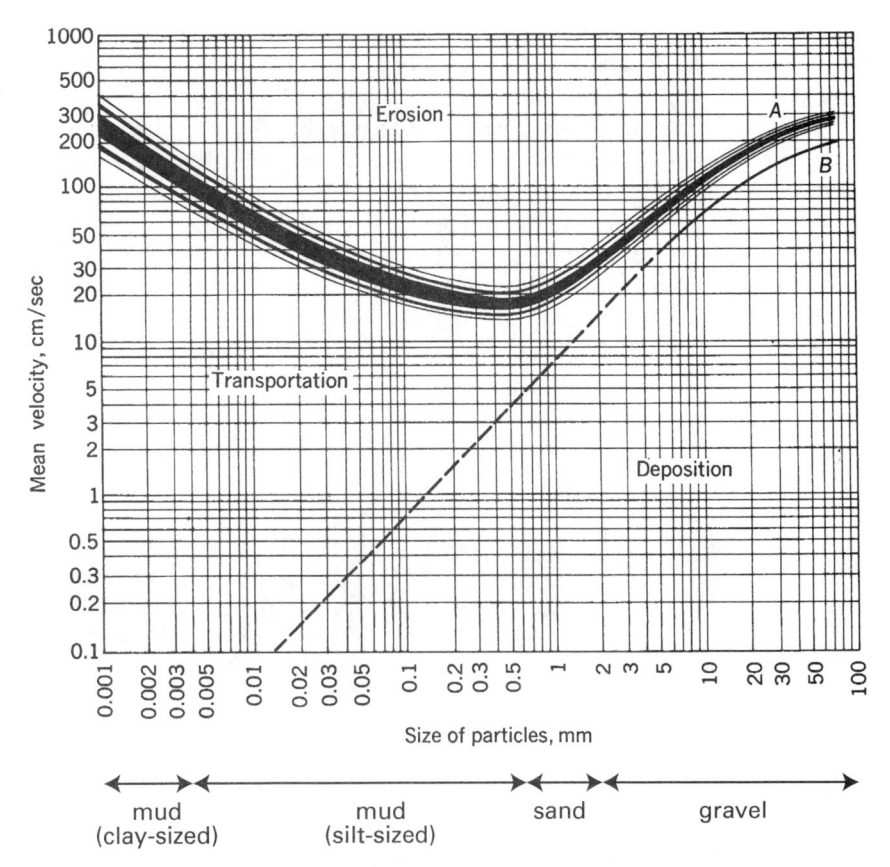

Figure 9.4
Sand dunes near Ancon, Peru, on the coast of the Pacific Ocean.
(Photo by Loren A. Raymond.)

transportation; (4) the distance of transportation; (5) the environmental location of the basin (e.g., ocean floor, riverbed, desert plain); and (6) the conditions under which deposition occurs. The provenance is the principal control of sediment composition, but in each environment, structures or textures may form that are particularly indicative of that environment of deposition. As an example of the compositional control exerted by provenance, quartz-rich sand will not likely be derived from a terrain that contains only basalt, hornblende schist, and ultramafic rocks (all rocks that are poor in quartz). Likewise, lithic sand that is dominated by shale, slate, or volcanic rock fragments is an unlikely product of the erosion of a terrain containing granitoid rocks and quartz-feldspar schists and gneisses.

The influence of the five remaining factors listed above is variable. *Climate*, including the temperatures and amounts of precipitation, is partly controlled by latitude and longitude and local microclimate is certainly controlled by elevation. In cold climates, decomposition rates are lower and disintegration has greater importance in weathering. Warm climates yield the opposite effect. Warm climates that are moist are particularly conducive to decomposition, and they support the

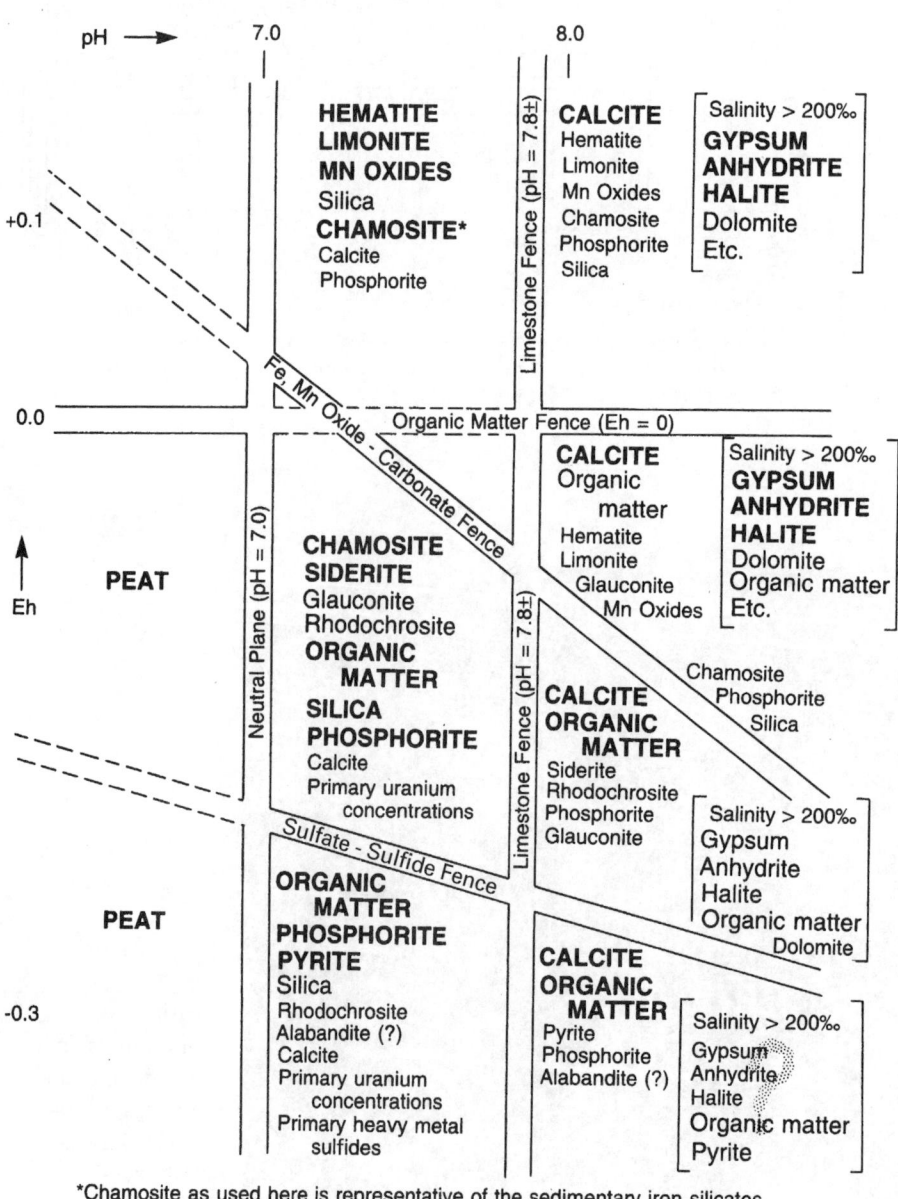

Figure 9.5
Eh-pH fence diagram, showing chemical environments of oxidation (high Eh), reduction (low Eh), acidic (low pH), and basic character (high pH).
(*Source:* Krumbein and Garrels, 1952. Used with permission.)

growth of abundant vegetation that produces organic acids, which also facilitate decomposition. Arid climates favor disintegration. *Erosion* of decomposed (versus disintegrated) rock yields more clay-rich sediments and generally favors finer-grained sediment. Glacial erosion produces both very large boulders and extremely small-sized fragments collectively called *rock flour.* Wind erosion and transportation yield only finer-grained sediment. Aqueous currents generally erode, transport, and deposit sediments of sizes predominantly intermediate between those produced by wind and glaciers.

Once produced and moved, sediment tends to decrease in clast size with distance of transportation. Exceptions exist. For example, sediment-laden, heavy currents called density or **turbidity currents** may carry coarse sediment great distances into deep ocean basins. In addition, in precipitated (crystalline) sediments, grain sizes are generally unrelated to erosion and transportation. As suggested above, the environment and conditions of deposition exert more influence on textures and structures than on composition; but it is also true that precipitation of chemical sediments of particular compositions is controlled by temperature and other physical conditions and the chemical conditions in the basin of deposition. Gypsum evaporites, for example, are deposited under particular chemical and physical conditions, including the condition that the flow of water into the basin must be exceeded by water loss via evaporation during the period of precipitation.

Sedimentary Environments

Sedimentary environments, specific sites in which sediments are deposited, are surface regions of the lithosphere below or above sea level that have a particular set of chemical, physical, and biological characteristics.[4] The rocks that form in a sedimentary environment may be divided into *lithofacies*, characterized by particular textures, structures, and compositions; or *biofacies* characterized by particular groups of fossils. The environments themselves may be grouped into continental environments, transitional environments, and marine environments.[5]

Continental sedimentary environments include **fluvial** (river and stream), desert, glacial, lacustrine (lake), landslide, and paludal (swamp) environments. Many of these can be subdivided into more specific types. For example, the desert environment includes the alluvial fan, erg, and playa environments.[6] *Alluvial fans* are flat, cone-shaped deposits of fluvial sediment deposited at a basin edge where a stream dumps its sediment load as it exits a mountain canyon. *Ergs* are sand-covered desert environments that contain both sand sheets and sand dunes. Playa environments are flat, vegetation-free desert basins that occasionally contain ephemeral lakes called *playa lakes.*

The most common continental environments represented by sedimentary rocks are the fluvial, lacustrine, and glacial environments. Fluvial environments include any environment where rivers or streams are actively or intermittently depositing sediment. Depositional sites include the stream channel as well as associated bars and the surrounding floodplain. Lacustrine environments include (1) the freshwater lake floor and associated delta and shoreline environments, (2) cryolacustrine (glacial lake) environments of the same kinds, and (3) similar playa lake environments. Glacial environments include various environments beneath (subglacial), on top of (supraglacial), within (englacial), and in front of (proglacial) the glacial ice mass. In addition to these are glacifluvial, cryolacustrine, and proglacial aeolian (downwind) environments. The sediments deposited in each of these environments are discussed below.

Transitional sedimentary environments include coastal-deltaic, estuarine-lagoonal, and littoral-beach environments. The coastal-deltaic environments occur where rivers, such as the Mississippi, Nile, Amazon, and Ganges, enter into a sea or ocean. Sediment is deposited as a result of a loss of energy and resultant slowing of the river waters and currents. Estuarine-lagoonal environments occur along coastlines, where drowned river and stream channels (called *estuaries*) or blocked channels (called *lagoons*) serve as mixed fresh-marine water environments of deposition. The **littoral-beach environment**, the most commonly represented transitional environment in the sedimentary record, is the environment along the shore face, where the land meets the sea. Here, sediment transported by waves and longshore (coast parallel) currents is deposited where transporting agents lose energy.

Marine sedimentary environments include shelf-shallow sea, reef, slope and rise, submarine canyon, trench, rift-fracture zone, and deep ocean (pelagic) types. Shelf-shallow sea environments include low-energy open sea environments like that in the Great Bahama Bank east of Florida, low-energy restricted environments like the Black Sea, high-energy environments like those near Vancouver Island in western Canada, and glacimarine environments such as those along the south Alaskan coastline at Glacier and Yakutat bays, where glaciers enter the sea.[7] Reef environments include those of the reef itself, the forereef on the seaward side of the reef, and the reef lagoon in the protected areas behind the reef. Slope and rise environments occur in the deeper parts of the marine-continental margin setting, at depths greater than those bounding the shelves (about 124 meters).[8] Pelagic environments include those of the deep basin and abyssal plains of the oceans, as well as the elevated submarine plateaus within the oceans. **Trench environments** include the trench-floor environment, trench-slope environments, and trench-slope basin environments (figure 9.6a). The trench environments typically host large and thick submarine fans composed of sand and mud that become sandstone and mudrock.

The sediment deposited in each environment is distinguished by particular compositions, textures, and structures (table 9.1). In the rock record, in order to be able to recognize sedimentary deposits representing the various sedimentary environments, it is important to be able to recognize those particular textures and structures, as well as any distinctive compositions that are indicative of particular environments.

For example, consider the contrast between two superficially similar sand-rich environments. The meandering river (fluvial) environment is characterized by sand(stones) lacking large amounts of clay matrix, and by gravels (conglomerates), silt (siltstones), and mud (shales) that together exhibit fining-upward cycles of beds (higher in the set of beds, the grain sizes get smaller). Fossils of land plants, cross-bedded sands and gravels, and lenticular sand and gravel bodies are also characteristic of these environments. Local lenticular paludal (swamp) deposits may be associated with the muddy fluvial deposits formed on flood plains adjoining rivers. In contrast, ergs are distinguished by well-sorted sand (i.e., sand of more or less uniform grain size), with frosted sandgrains in cross-bedded units (see the structures section). There is no regular upward or downward change in grain size in the beds. The sand is typically almost pure quartz but in exceptional circumstances may be composed of another mineral, such as gypsum. Laminated sand-sheet deposits and fine-grained, salt-bearing playa lake deposits may be associated with the cross-bedded

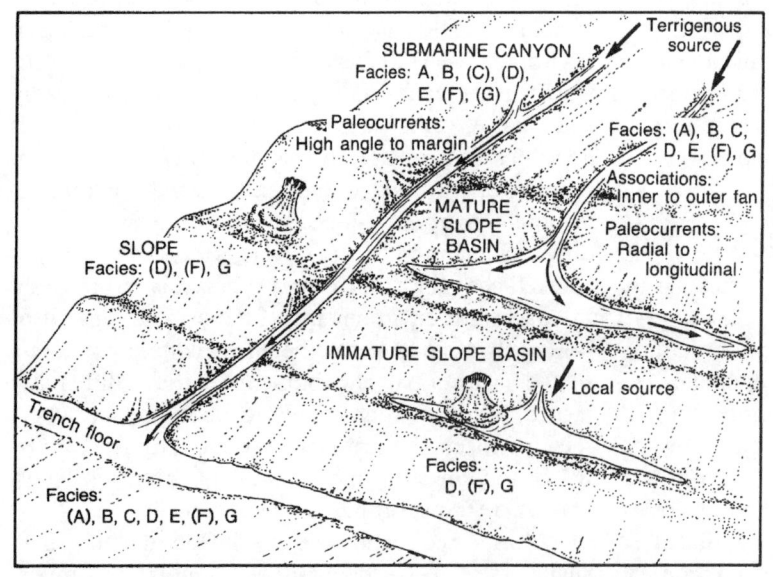

(a) Environments. Various convergent margin sedimentary environments—the trench, slope basin, slope, submarine canyon, and trench floor—showing submarine fan facies for each setting, indicated by letters that correspond to submarine fan lithofacies shown in (b).

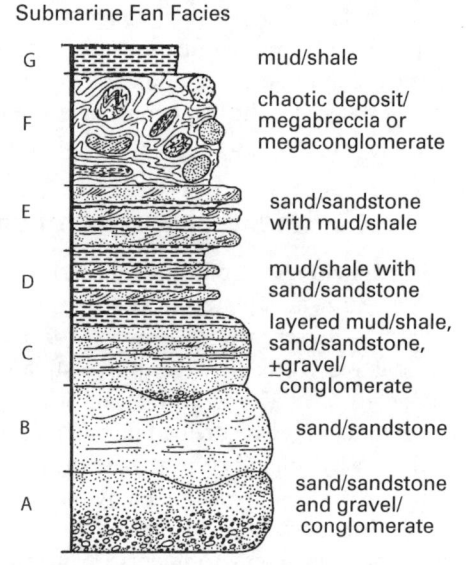

(b) Bedding, some structures, and sediment and rock types (reflecting grain sizes) for submarine fan lithofacies.

Figure 9.6
Sketches of some marine sedimentary environments and associated sedimentary lithofacies.
(*Sources:* (a) From Raymond, 2007, p. 333; after Underwood, Bachman, and Schweller, 1980; and Underwood and Bachman, 1982; (b) From Raymond, 2007, p. 331, after Mutti and Ricchi-Lucchi, 1972.)

Sedimentary Rocks

Table 9.1 Common Rock Types and Characteristics of Various Sedimentary Environments

Environment	Common Rock Types	Notable Characteristics
Continental		Fossils of land plants and animals
Meandering river	cg, ss, sltst, sh	Fining upward cycles, xbdd ss and cg, channels, lenticular-linear units
Braided river	cg, ss, sh, sltst	xbdd cg and ss, channels, lenticular deposits, slight fining upward cycles
Glaciofluvial	cg, ss, sh, sltst	xbdd ss and cg with striated clasts
Glacial	Tillite	Striated clasts; poor sorting
Landslide	Breccia	Lacks sorting; angular clasts
Alluvial fan	fgl, cg, ss, sltst, sh, dm	xbdd cg and ss with dm interbeds; coarsening upward sequence with fining upward cycles
Erg	ss	xbdd ss with tabular laminated ss bds; well-sorted frosted sand grains
Playa/sabkha	Evaporites, sh, sltst, ss, ls, dlst, ch	Thin-laminated, tabular bds; mudcracks
Lacustrine	sh, sltst; local ss, cg, ls, coal	Varves ± turbidites
Cryolacustrine	sh, sltst, ss, cg, dm	Varves, striated dropstones, dm interbeds ± turbidites
Inland basin	fgl, br, cg, ss, sh, sltst ± dm	Interbedded river, lake, and landslide deposits
Spelean (cave)	ls	Dripstones
Deltaic	ss, sltst, sh, coal, and mdst; local cg, and dm	xbdd ss; coal; coarsening upward sequences
Transitional		
Deltaic	ss, sltst, sh, coal, and mdst; local cg, and dm	xbdd ss; coal; coarsening upward sequences
Estuarine	ss, sltst, sh ± coal	xbdd ss; wavy to parallel laminated mudrocks; limited faunal diversity
Lagoonal	sh, mdst, sltst, ss, coal, lmst, pkst, gst	Bioturbated mdst; wavy bddg ± xbdd or laminated ss; limited faunal diversity; stromatolites
Littoral	ss, stlst, gst, pkst, dlst, local cg	Low and high angle xbdd ss ± sltst with marine fossils
Tidal flat	ss, sltst, mdst, sh, lmst, pkst dlst, ch, flat-pebble cg	Laminated to lenticular bddg; bioturbation; limited biological diversity; stromatolites, mudcracks; evaporites; xbds in tidal channels
Marine		Marine fossils
Glacimarine shelf—shallow sea	ss, dm, sltst	Interlayered ss and dm with marine fossils; ± turbidites and laminites; dropstones
Open low- to high-energy (shelf)	sh, lmst, ss, gst, dlst	Tabular-parallel bds; xbdd ss; bioturbation; diverse fauna
Restricted	sh, sltst, ss	Laminated and thin bdd, ± pyrite
Reefs	bst, ls br, with pkst, gst, lmst, sh	Fossils in growth position; locally thick ls; stromatolites
Slope-rise	sh, sltst, ss, ± dm	SFF G, laminated ss-sh beds or SFF B-E with local SFF F
Submarine canyon	dm, ss, sltst, ± sh	SFF A, B; xbdd ss; linear-lenticular units
Trench	ss, sltst, sh, cg, ± dm	SFF A-G, lenticular to tabular units
Rift-fracture zone	Mafic volcanic breccia, ss, sltst, sh, ch, ls, dm	SFF A-G with basalt interbeds
Pelagic (abyssal)	mdst, sh, ch, lmst, ss, sltst	Tabular units; laminated SFF G and other fine-grained rocks, ± SFF B-E

Abbreviations:
bddg = bedding
bds = beds
br = breccia
bst = boundstone
cg = conglomerate
ch = chert
dlst = dolostone
dm = diamictite
evap = evaporites
fgl = fanglomerate
gst = grainstone
lmst = lime mudstone
ls = limestone
mdst = mudstone
pkst = packstone
SFF = Submarine Fan Facies
sh = shale
sltst = siltstone
ss = sandstone
xbdd = cross-bedded
xbds = crossbeds

(*Source:* Raymond 2007, p. 335–336).

sands. Thus, although both environments are characterized by sand (and sandstones), the particular structures and compositions of the rocks formed in the two different environments differ in detail.

Sedimentary Textures

The textures of sedimentary rocks can be divided into clastic and crystalline types. The clastic types are far more abundant. Clastic textures, those dominated by fragments of rock, mineral, glass, or organic remains, derive their name from the Greek root *klastos*, meaning broken. Because they form at the surface (epi-), sedimentary clastic textures—those with rounded to angular grains that are somehow stuck together—are called **epiclastic textures** (figure 9.7a, e; color plates 4b; 25c). **Crystalline textures**, in contrast, consist of masses of interlocking crystals precipitated from solutions or modified to crystalline form by diagenesis (figure 9.7d; color plate 28a).

(a) Epiclastic texture in oligomict quartz pebble conglomerate of the Mississippian Cloyd Member, Price Formation, Broadford Quadrangle, VA.

(b) Epiclastic (?) texture in chert breccia, Cambrian Knox Group, Broadford Quadrangle, VA.

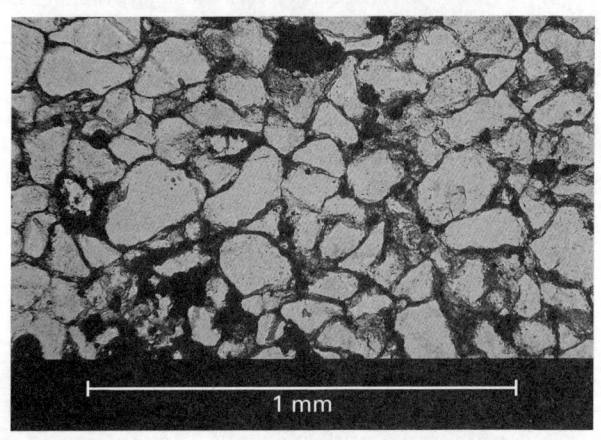

(c) Photomicrograph of epiclastic texture in fine-grained quartz wacke, Silurian Rose Hill Formation, Saltville-Broadford area, VA.

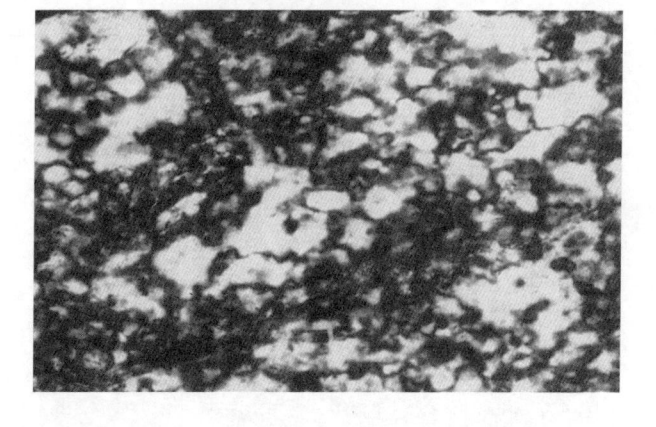

(d) Photomicrograph of crystalline equigranular-sutured texture in chert. Franciscan Complex, Diablo Range, CA.

Figure 9.7
Photographs and photomicrographs of some sedimentary textures in small outcrop, hand-specimen, and thin-section views (microscopic views = photomicrographs).
(*Sources:* Photos (a), (b), and (d) by Loren A. Raymond. Photomicrographs (c) and (e) by Anthony B. Love and Loren A. Raymond.)

In epiclastic textures, the texture can be further described by specifying the grain sizes, grain shapes, grain sorting, and grain roundness versus grain angularity. Grain sizes categories used in North America are those of the Wentworth Scale or the *phi scale* (Wentworth, 1922; Krumbein, 1934; McManus, 1963) (table 9.2). On the Wentworth Scale, gravel sizes range from 2 mm up, sand sizes fall between 2 mm and 1/16 mm, silt sizes fall between 1/16 mm and 1/256 mm, and clay sizes are less than 1/256 mm. Phi (ϕ) values are derived from the formula

$$\phi = -\log_2(\partial/\partial_0)$$

where ∂ is the diameter in millimeters of any grain being considered and ∂_0 is 1mm. **Sorting** is a measure of the distribution of grain sizes. If all the grains are the same size, the sediment is *very well sorted*. If the grains are of radically different sizes, the sediment is *very poorly sorted*. Wentworth grain size categories can be used as a measure of sorting (figure 9.8). Grain shapes are of three types: equant (equi-dimensional), tabular (plate-like), and rod-shaped (figure 9.9 on p. 173). Roundness refers to the degree to which sharp edges and corners have been removed from grains by abrasion, with the categories ranging from very angular to well rounded (figure 9.10 on p. 173). The sphericity differs from the roundness, referring instead to how closely grains approach equant versus elongate or tabular shapes.

In epiclastic carbonate rocks, particles are divided into very small grains referred to as **micrite** (< 0.004 mm) or **lime mud** and larger grains called **spar** (calcite or dolomite crystals). In addition, there are larger rounded clasts called ooids and pellets, plus typically irregularly-shaped skeletal fragments (fossils). Small spherical grains (< 0.2 mm) that represent the feces of mud-eating organisms are called **pellets**. In certain very shallow marine settings, small (0.25–2.0 mm), very round grains called **ooids** may be precipitated in the sediment being sloshed back and forth by waves

(e) Photomicrograph of epiclastic texture with crystalline calcite poikilotopic cement (gray, on right) and hematite cement (black) surrounding rounded to subrounded grains of quartz and filling previous voids in quartz arenite; Silurian Rose Hill Formation, Saltville Quadrangle, VA.

Figure 9.7 *(cont'd.)*

Table 9.2 Wentworth /Sediment Size Classes and Corresponding Rock Names and Textures

φ	Wentworth Scale	Grain Size Names		Group S Rock Names	Texture
−8	256 mm	Boulders			
−6	64 mm	Cobbles	Gravel	Conglomerate, breccia	Epiclastic ruditic
−2	4 mm	Pebbles			
−1	2 mm	Granules			
0	1 mm	Very coarse sand			
1	1/2 mm	Coarse sand			
2	1/4 mm	Medium sand		Sandstone (arenite, wacke)	Epiclastic arenitic
3	1/8 mm	Fine sand			
4	1/16 mm	Very fine sand			
8	1/256 mm	Silt	Mud	Siltstone, shale, mudstone, claystone	Epiclastic lutitic
		Clay			

Sources: Modified from C. K. Wentworth (1922), Krumbein (1934), McManus (1963).

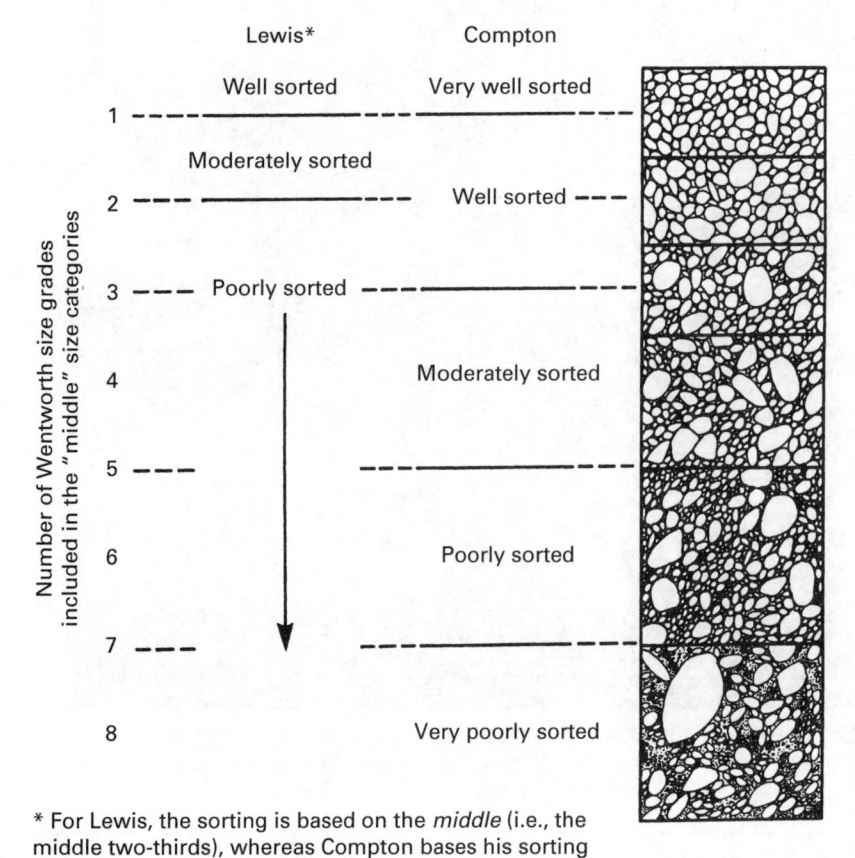

Figure 9.8
Sorting of grains as defined by Lewis (1984) and Compton (1962).
(*Source:* Diagram from Raymond, 1984b, after Lewis, 1984 and Compton, 1962.)

* For Lewis, the sorting is based on the *middle* (i.e., the middle two-thirds), whereas Compton bases his sorting on the *great bulk*, or middle 80%.

(figure 9.11). Rocks characterized by ooids are called **oolitic** (color plate 28b). Rock precipitated by inorganic or biologically controlled processes may be fragmented to make sedimentary clasts called **allochems**. Abundant fragments of fossil animal skeletons in a rock can be designated by the descriptive term **skeletal** (color plate 27c, d). All of these terms can be used as modifiers to detail the textural character of the rock.

Crystalline textures are described using terms relating to grain sizes, grain shapes, and intergrain relationships. The size categories for crystalline textures are the same as those used in igneous rocks (table 9.3). Grain shapes are described as equant, acicular (rod-like), or platy, but these terms may be incorporated into more inclusive categories such as equigranular (equant grains making a granular mass)

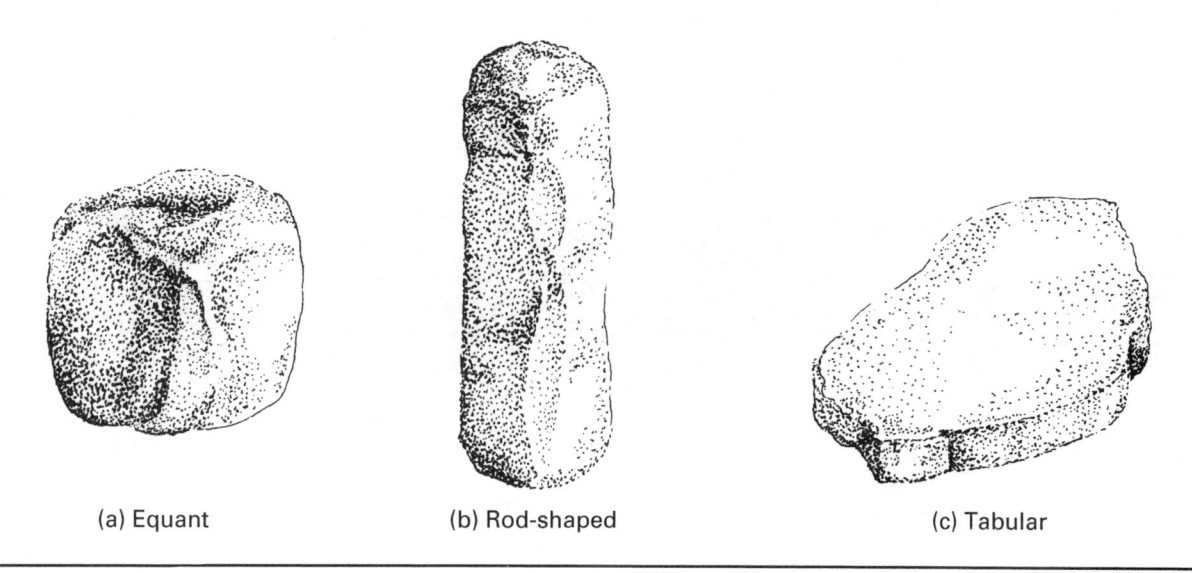

(a) Equant   (b) Rod-shaped   (c) Tabular

Figure 9.9
Grain shapes.
(*Source:* Raymond, 1984b.)

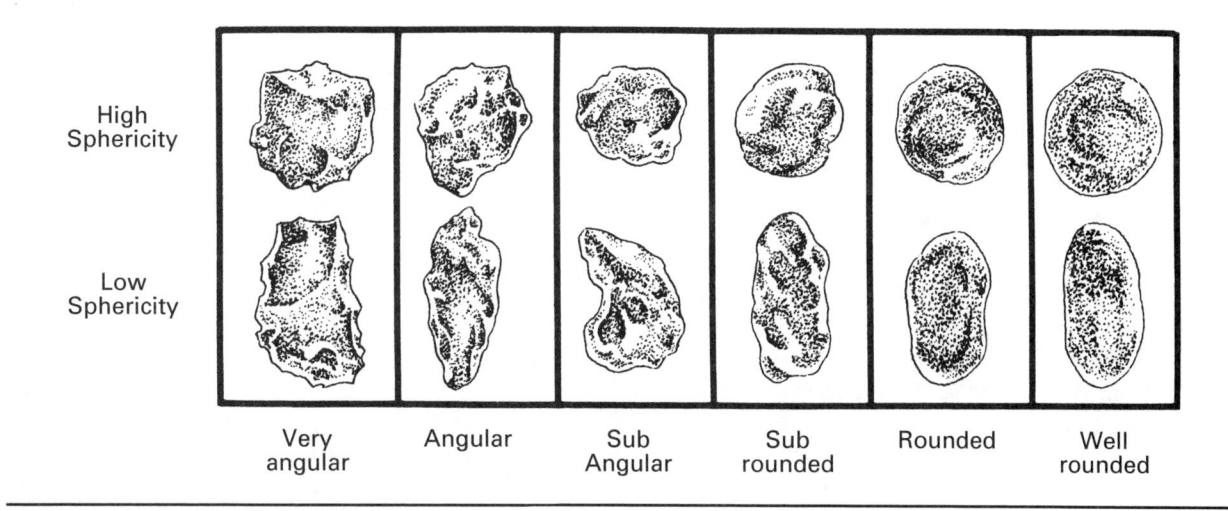

High Sphericity

Low Sphericity

Very angular | Angular | Sub Angular | Sub rounded | Rounded | Well rounded

Figure 9.10
Grain roundness and sphericity.
(*Source:* From Raymond, 1984b based on and modified from Powers, 1953. Used with permission of SEPM, Society for Sedimentary Geology.)

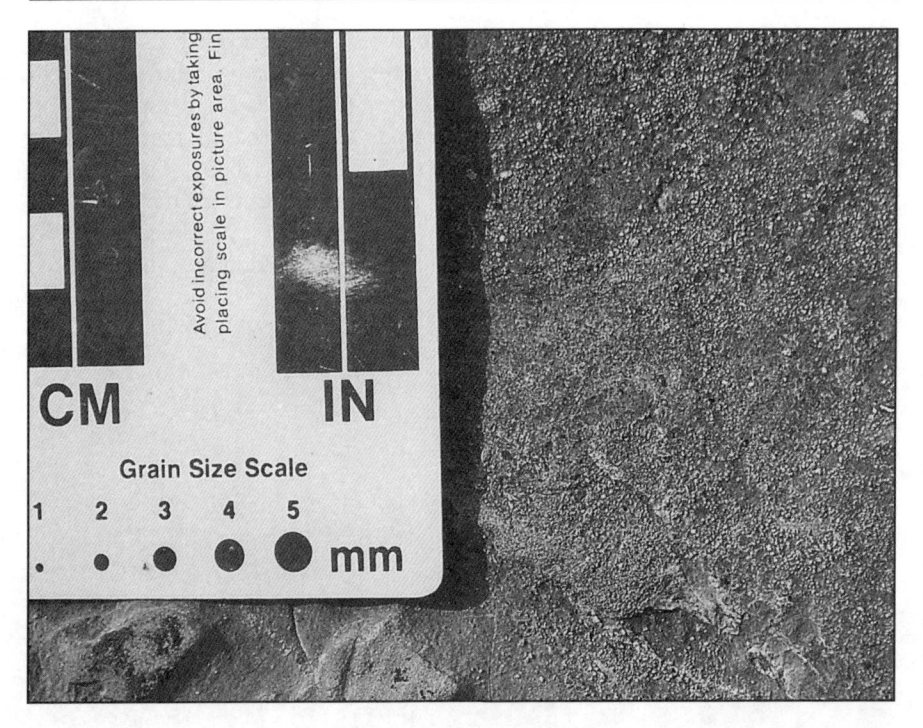

Figure 9.11
Small ooids (tiny white spheres smaller than 1 mm) in dark limestone of the Cambrian Nolichucky Formation, Broadford Quadrangle, Virginia.
(Photo by Loren A. Raymond.)

Table 9.3 Crystalline Grain Size Categories

Phaneritic grains	grains large enough to see
Very coarse-grained	> 3 cm
Coarse-grained	5 mm–3 cm
Medium-grained	1 mm–5 mm
Fine-grained	< 1 mm
Aphanitic grains	grains too small to see with the unaided eye

or foliated (aligned with subparallel tabular surfaces). Terms such as mosaic and sutured (figure 9.7d) specify intergrain relationships.

Sedimentary Structures

Recall that structures are curviplanar surfaces and other features that are generally large enough that only one or a few may occur in a hand specimen or outcrop. Sedimentary structures may be grouped into four main categories: (1) bed and formation shapes, (2) internal structures, (3) surface structures, (4) and other structures (Raymond, 2007, ch. 12) (figures 9.12; 9.13 on p. 179). Of all of the structures, *beds* are by far the most characteristic structures of sedimentary rocks.

Beds are sedimentary layers greater than 1 cm thick that are distinguished from adjoining layers by differences in color, composition, or texture (figure 9.12a; color plates 5a; 26d). Layers thinner than 1 cm are called **laminations** (figure 9.12g; color plate 5b). A group of beds that forms a mass thick enough to map on a scale of 1:24,000 and is characterized by a distinctive rock type or group of rock types is called a **formation**. Subdivision of formations is possible, if the unit contains subgroups of beds that have distinctive rock types. These subgroups, which may or may not be mappable, are called *members*. There is a great variety of bed, member, and formation shapes (figure 9.13) (Krynine, 1948; Peterson and Osmond, 1961; Potter, 1963). In general, however, beds are sheet-like in shape.

Many different kinds of internal structures exist in sedimentary rocks. **Crossbeds** and cross-laminations are sets of inclined layers that are locally truncated by sets of layers inclined in a different direction or at a lower or higher angle (figure 9.12b). Some crossbeds are tabular, whereas others are lenticular, curved, and trough shaped. Ripple marks, which appear as structures on bedding surfaces (figure 9.12c) may appear as internal, small-scale cross-laminations. **Convolute bedding** and laminations are layers that have the form of little basins, domes, and folds (figure 9.12d; below pick point in color plate 5b). Normally, **graded bedding** is layering in which there is a change in grain size from coarse at the bottom to finer at the top of an indi-

(a) Beds of interlayered sandstone (light and thicker) and mudrock (dark and thinner) in the Jurassic Sumerville Formation, Goblin Valley State Park area, UT.

(b) Cross-bedding in sandstone of the Jurassic Navajo Sandstone, Zion National Park, UT.

Figure 9.12
Sedimentary structures.
(Photos by Loren A. Raymond.)

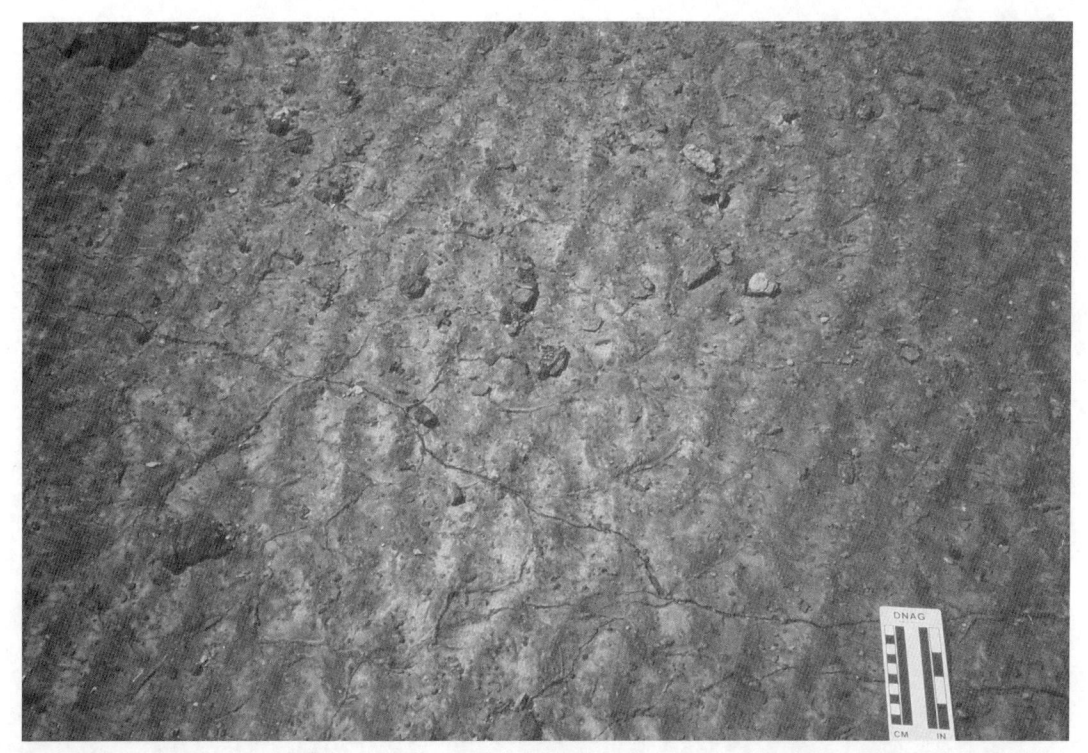

(c) Ripple marks on a bedding surface of Silurian Clinch Sandstone, Powell Mountain, west of Duffield, VA.

(d) A Bouma sequence containing a basal graded bed and a middle convoluted bed, in the Permian Keeler Canyon Formation, Keeler Canyon, CA.

Figure 9.12 *(cont'd.)*

(e) Flame structure (below hammer head) in sandstone-shale turbidite of the Cretaceous Panoche Formation, northeastern Diablo Range, CA.

(f) Tubular burrows (*Scolithus sp.* fossils) cutting across a sandstone bed in the Silurian Tuscarora Sandstone, Clinch Mountain, north of Ward Cove, VA. (cf. *Arthrophycus* in Miller et al., 2009). Scale divisions on left side of scale = 1 cm.

Figure 9.12 *(cont'd.)*

(g) Stromatolites in the Cambrian Nolichucky Formation, Broadford Quadrangle, VA, showing laminated beds. Hammer at lower left, with 32 cm handle, provides scale.

(h) Concretions in massive Cretaceous Panoche Formation sandstone, Del Puerto Creek, northeastern Diablo Range, CA. Hammer handle at lower right, approximately 28 cm long, provides scale.

Figure 9.12 *(cont'd.)*

Sedimentary Rocks 179

(a) Plan shapes of beds.
  1 Linear or shoestring
  2 Circular or disk-shaped
  3 Elliptical
  4 Parabolic
  5 Irregular

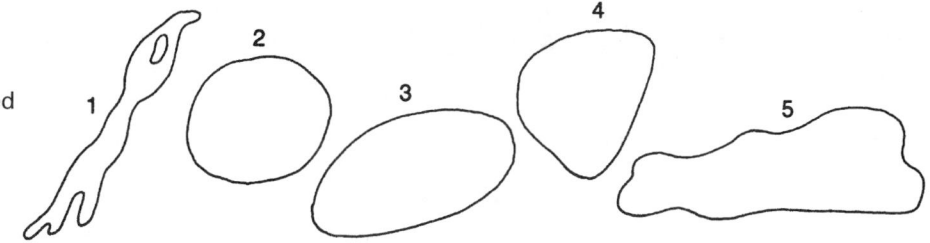

(b) 3D perspectives on formation, member, and bed shapes.

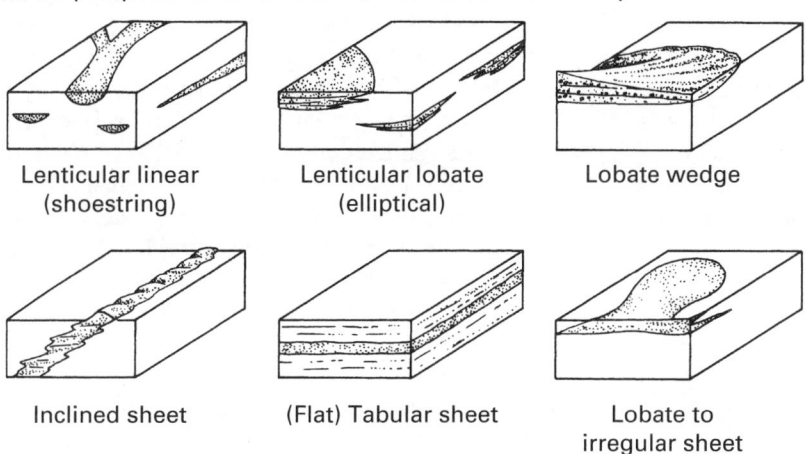

Lenticular linear (shoestring)   Lenticular lobate (elliptical)   Lobate wedge

Inclined sheet   (Flat) Tabular sheet   Lobate to irregular sheet

(c) Internal appearance of bedding types.

Planar bedding   Wavy bedding   Lenticular bedding

Domal bedding   Parabolic trough cross bedding (lenticular)   Irregular (nodular) bedding

Graded bedding   Tabular planar cross bedding   Convolute bedding

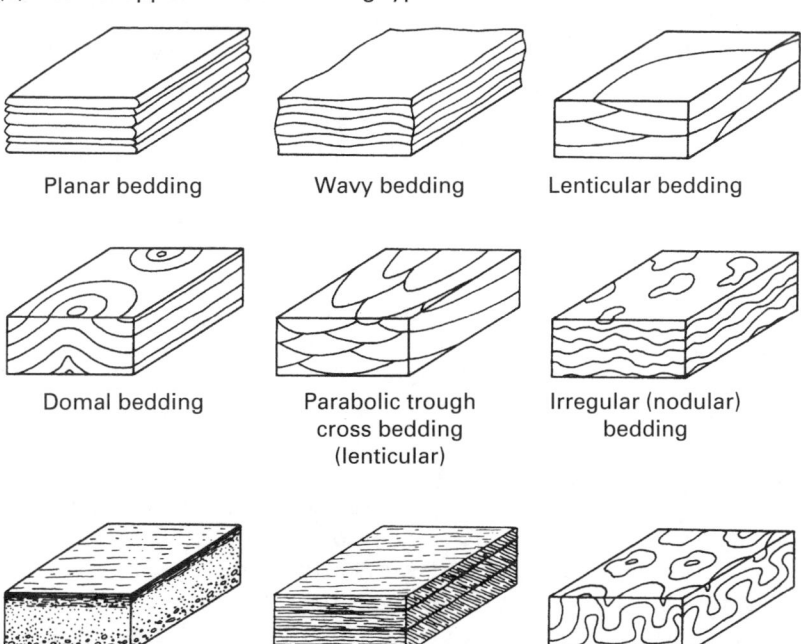

Figure 9.13
Formation and bed shapes.
(*Source:* Raymond, 2007, p. 263.)

vidual bed (figure 9.12d; bottom of bed in color plate 5b). It should not be confused with fining-up sequences in which each successive bed in a group of beds has finer grains than the bed below. Graded beds are particularly characteristic of Bouma sequences, distinctive sets of five beds formed by turbid density currents that sweep down slopes and across basin floors (figure 9.12d; figure 9.14; color plate 5b). **Flame structures** are deformed siltstone or shale layers that extend up from a bed in a curved layer ending with a point, a form reminiscent of the shape of a flame in the wind (figure 9.12e). **Burrows** are irregular to cylindrical filled tubes produced by burrowing organisms (figure 9.12f). Where burrowing has destroyed laminations or bedding, rocks are said to have **bioturbation structure**.

Other structures occurring in sedimentary rocks include some formed during sedimentation and others imposed soon afterwards. Soft sediment folds and faults, bends, and breaks in laminations or bedding may occur within a single bed. Such structures form soon after deposition but before lithification. **Stromatolites** are laminated carbonate masses, commonly domal in form, produced by microbial precipitation (Hofmann, 1973) (figure 9.12g). If the microbial precipitates are spherical to irregularly rounded and concentrically laminated they are called **oncolites**. Like oncolites, **concretions** are typically rounded forms, although they may be spherical or irregular to disk-shaped masses of cemented sedimentary rock that range from small to large sizes (1 cm to >2 m) and occur within a host sedimentary rock (figure 9.12h). Concretions may lack laminations and are not necessarily composed of carbonate rock. Internally, concretions and other weathered rocks may have a series of colored rings, called **liesegang rings,** produced by oxidation or reduction. In contrast to oncolites, concretions develop after sediment has accumulated and result from later cementation of grains by precipitated minerals, such as calcite, quartz, or goethite (Selles-Martinez, 1996). **Stylolites**, surfaces coated or marked by dark mineral accumulations and appearing as jagged lines in rocks, are formed after lithification by dissolution of carbonate or other minerals, leaving the dark organic or oxide residue.

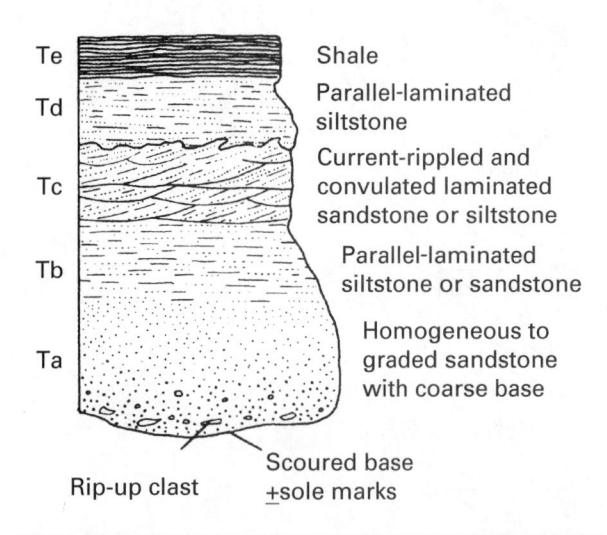

Surface structures in sedimentary rocks occur on bedding or lamination surfaces. Surface structures include ripple marks, mudcracks, sole marks of various types, and tracks and trails of animals. **Ripple marks** are regularly undulating surfaces produced either by oscillating waters (waves) or currents (figure 9.12c). **Mudcracks** are formed by shrinkage, where drying mud masses contract, producing polygonal cracks that later become filled with sediment that preserves the trace of the crack (figure 9.15a; color plate 5c). **Sole marks** are bulbous to linear features that are usually observed on the bottom (sole) of a bed, after erosion has removed the weaker, underlying bed (figure 9.15b). One process of sole mark formation involves the action of currents that may carve a depression into a layer, a depression that later becomes filled with sediment to form a protruding cast called a **flute**. Similarly, if currents drag sticks or stones along a surface, making depressions that later are filled to become casts on the base of a bed, the sole mark is called a *tool mark*. Rounded, bulbous sole marks formed during compaction are called **load casts**.

Among the important structures common in sedimentary rocks are the fossils—evidences of past life. Hard parts of animals such as shells or bones, casts, molds, carbon traces, footprints, and replaced shells are among the forms that fossils can take in a sedimentary

Figure 9.14
Sketch of a Bouma sequence. Layers Ta, Tb, Tc, and Td are deposited by the turbidity current. Ta is characterized by basal coarse sand, with gravel, rip-ups, or armoured mud balls (balls of mud with embedded pebbles), that grades up into coarse to medium or fine sandstone. Tb is a parallel laminated sandstone or siltstone. Tc is a cross-laminated to convolute-bedded layer of medium to fine sandstone or siltstone. Td is a parallel-laminated fine sandstone or siltstone. Te is not deposited by the turbidity current but is mudrock that accumulates as a result of sediment rain, in the period of time between turbidity current deposition.
(*Source:* Raymond, 2007, p. 265; concept from Bouma, 1962.)

Plate 1

(a) Schematic block diagram showing the three types of plate boundary.
   SZ - Subduction Zone
   MOR - Midocean Ridge
       (spreading center)
   TF - Transform Fault

(*Source:* Modified from Raymond, 2007, p. 5, and Isacks et al., 1968.)

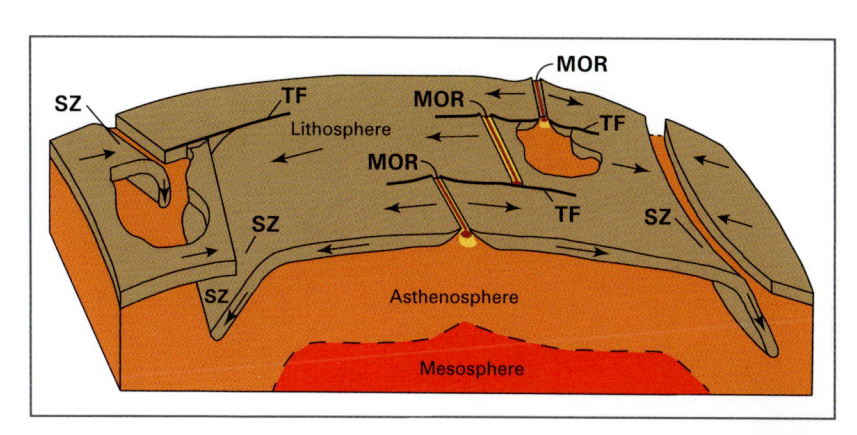

(b) Granite with nearly vertical tensional joints (fractures) and subhorizontal exfoliation joints. Acadia National Park, ME.

(*Source:* Photo by Loren A. Raymond.)

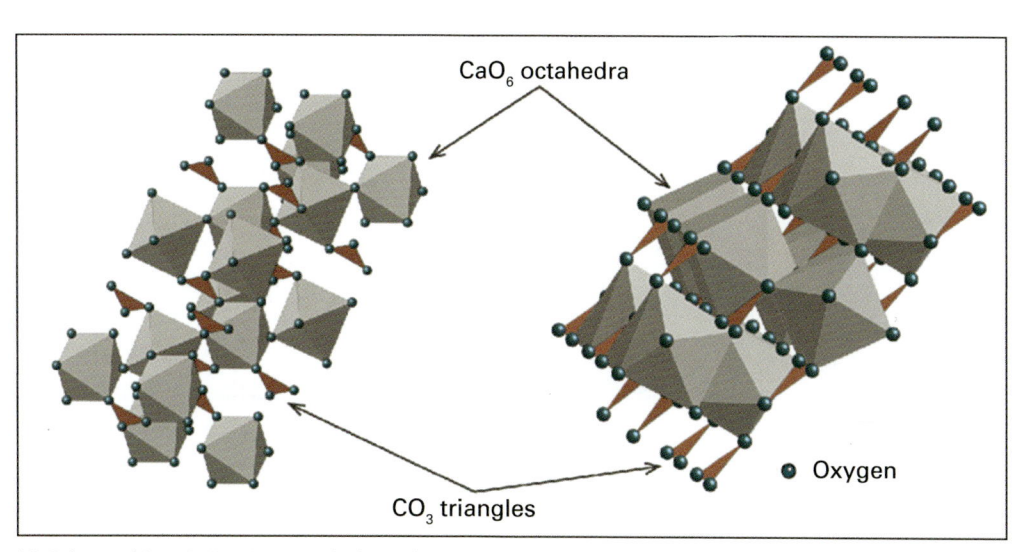

(c) Polymorphic variations in minerals shown by the crystal structures of calcite (left) and aragonite (right). The CaO_6 octahedra and CO_3 triangles in the structures of calcite and aragonite are arranged in very different ways.

Plate 2 Igneous Structures and Textures

(a) Xenoliths of mafic igneous rock (large dark-colored mass of hornblendite in center, plus smaller masses in the surrounding rock) in granodiorite (lighter rock) of the Sierra Nevada batholith, Wilson Reservoir, CA. Granitoid dikes, including white granite aplite, intrude the granodiorite and the xenoliths. Pen (13.5 cm) resting on xenolith-granodiorite contact at lower center provides scale.

(b) Hypidiomorphic-granular texture in slightly weathered biotite-hornblende granodiorite, Carson-Iceberg Wilderness near Mosquito Lake, CA. Notice the zoned, white to gray, nearly euhedral plagioclase crystal in the lower left and the diamond-shaped and rectangular euhedral hornblende crystals. Most plagioclase is subhedral. Anhedral, glassy gray quartz is scattered throughout, as are a few anhedral alkali feldspar grains. Biotite forms thin, sheet-like crystals, but locally appears as very small hexagonal crystals.

(c) Photomicrograph of thin section of hypidiomorphic-granular hornblende-biotite Quartz Diorite, Alpine, CA, in crossed Nichols.
  B – Biotite   H – Hornblende
  P – Plagioclase feldspar   Px – Clinopyroxene (Augite)
  Q – Quartz
Notice that the hornblende surrounds and appears to be replacing the augite. Also notice the subhedral plagioclase and anhedral quartz.

(d) Photomicrograph of gabbro from the Coast Range Ophiolite of CA showing diabasic texture. Photo in crossed Nichols.
  P – Plagioclase feldspar   Px – Clinopyroxene (augite)
The dark materials in and along the twin planes of the plagioclase are clay minerals produced by alteration of the plagioclase. Hornblende in blues and greens replaces pyroxene. The perfect circle at the top is an air bubble in the cement of the thin section.

Weathering   Plate 3

(a) Exfoliation joints in granitoid rocks. Joints dip (tilt) left in foreground and middle ground parallel to the slope and dip right in the background, parallel to the slope. Photograph was taken near Olmstead Point in Yosemite National Park.

(*Source:* Photo by Loren A. Raymond.)

(b) Tree root participating in physical weathering by growing in a joint and forcing the rocks apart, in Linville Gorge, NC.

(*Source:* Photo by Loren A. Raymond.)

(c) Slightly decomposed bedrock (D) and overlying soil, with A, B, and C horizons marked. Notice the roots in the A-horizon near the right and left edges of the photo. The B-horizon is poorly developed and appears to contain some materials that may have been transported to the site prior to the latest soil development period. The bedrock is Mesoproterozoic Gneiss. Cranberry Mines, Newland, NC. The exposure is approximately 3 m high.

(*Source:* Photo by Loren A. Raymond.)

Plate 4   Sedimentary Processes and Rocks

(a) A landslide consisting of a debris avalanche (above) and a debris flow (lower part) of rock fragments and soil, above the village of Matachico, Peru.
(*Source:* Photo by Loren A. Raymond.)

(b) Photomicrograph of quartz arenite sandstone from the Silurian Rose Hill Formation of the Clinch Mountain Wildlife Management Area of southwestern Virginia. The rock has epiclastic texture with crystalline calcite cement, especially abundant surrounding the grains on the right side of the section, giving this area a poikilotopic texture. Black (opaque) cement is hematite.
(*Source:* Photomicrograph by Anthony B. Love and Loren A. Raymond.)

Sedimentary Structures   **Plate 5**

(a) Beds of interlayered sandstone (reddish tan and thicker) and shale (light gray and dark gray and generally thinner) in the Jurassic Summerville Formation, Goblin Valley State Park area, UT. Note: The bedding is distinct because of differences in color, grain size, and composition.

(*Source:* Photo by Loren A. Raymond.)

(b) A Bouma sequence in a limestone turbidite of the Permian Keeler Canyon Formation, Keeler Canyon, CA. The sequence contains a basal graded bed and a middle convoluted bed. Also note the laminations in the bed.

(*Source:* Photo by Loren A. Raymond.)

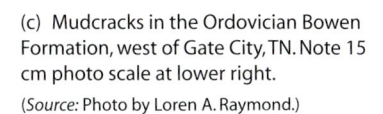

(c) Mudcracks in the Ordovician Bowen Formation, west of Gate City, TN. Note 15 cm photo scale at lower right.

(*Source:* Photo by Loren A. Raymond.)

Plate 6   Metamorphosed Metaultramafic Rocks

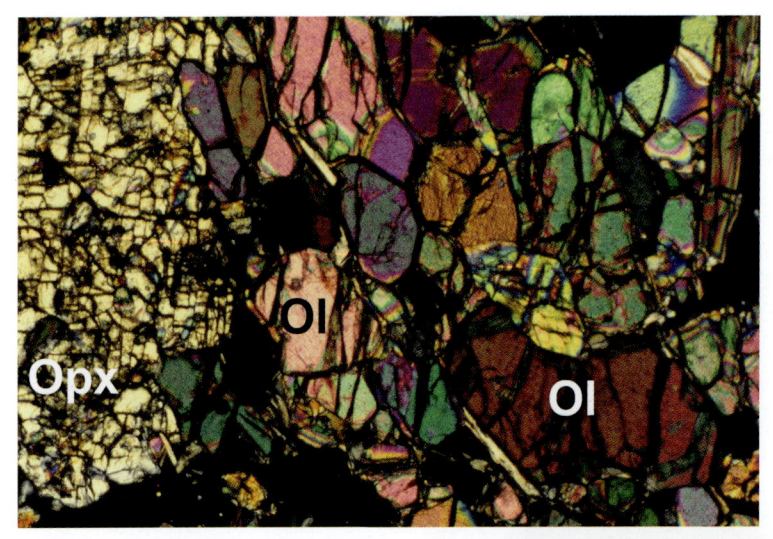

(a) Photomicrograph of metaharzburgite in crossed Nichols, Hoots metaultramafic body, NC. Ol = olivine   Opx = orthopyroxene. The many 120° grain boundaries among olivines indicate equilibrium crystallization. Rock likely crystallized under eclogite facies conditions. Photo width = 2mm. See Raymond et al., 2016 for a discussion of these rocks.

(*Source:* Photo by Loren A. Raymond and Anthony B. Love.)

(b) Photomicrograph of olivine (Ol)-chlorite (Cl)-magnesite (Mgt)-calcium amphibole (CaAm) schist in plain light (PL). Greer Hollow metaultrabasic body, Todd, NC. Rock was metamorphosed under amphibolite facies conditions in the presence of a CO_2-rich fluid. Photo width = 3 mm. See Raymond et al., 2016 for a discussion of these rocks.

(*Source:* Photo by Loren A. Raymond and Anthony B. Love.)

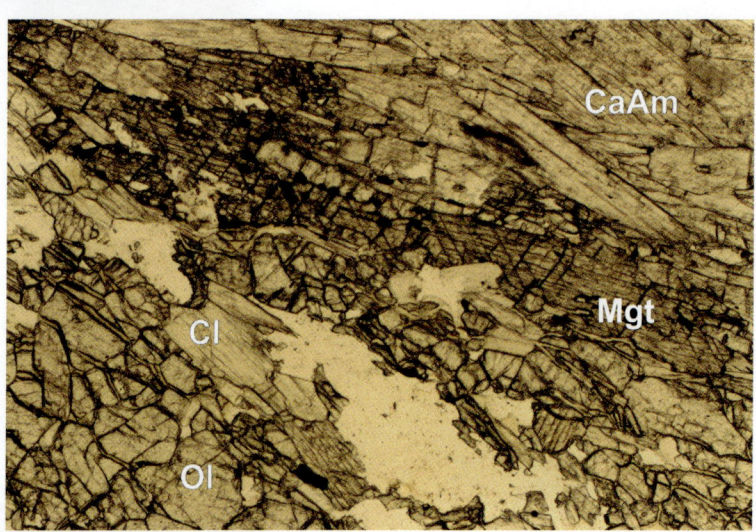

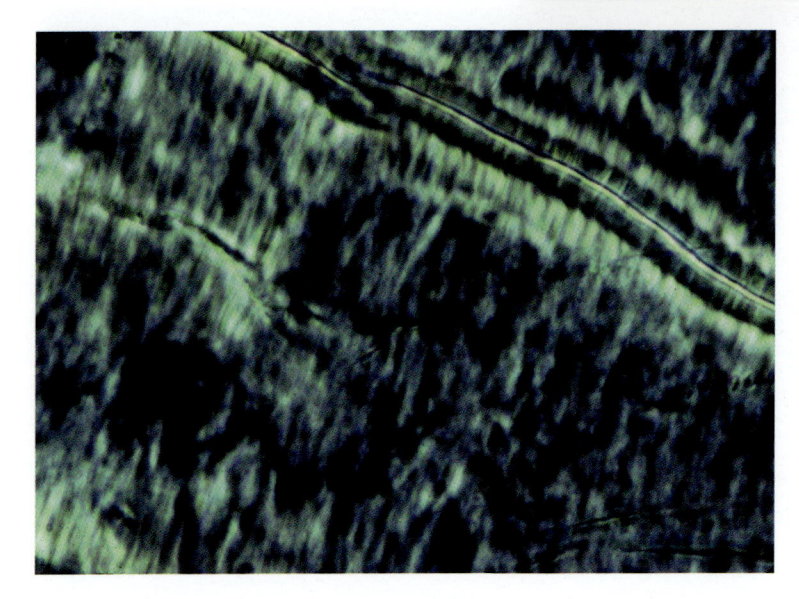

(c) Photomicrograph in crossed Nichols of a magnetite (small black grains) serpentinite (yellow-green to dark greenish gray) Greer Hollow metaultramafic body, Todd, NC. Rock was metamorphosed under greenschist facies conditions. Photo width = 1 mm.

(*Source:* Photo by Loren A. Raymond and Anthony B. Love.)

Tectosilicates — Quartz, Opal, Chabazite, and Nepheline Plate 7

(a) Three varieties of quartz—purple amethyst (Brazil), yellow-orange citrine (locality unknown), and clear quartz (Arkansas). Notice the hexagonal (di)pyramid forms on the crystals and the hexagonal prism on the clear quartz.
(*Source:* Samples courtesy of Maya.)

(b) Opal. Kenwood, CA. Notice the waxy luster.

(c) Chabazite (zeolite group). Baker County, OR.

(d) Nepheline. Dungannon Township, Ontario. Notice the greasy luster.

Plate 8   Tectosilicates — Feldspars

(a) Potassium feldspar. Pink Glaze Mine, Custer, SD. Many of the whitish, thin, irregular lines in the feldspar are exsolution lamellae of albite, giving this crystal perthitic texture.

(b) Potassium feldspar (perthitic). McKinney Mine, near Newdale, NC. The irregular white, mesh-like lines in the feldspar are exsolution lamellae of albite (Na-feldspar) in the dominantly K-rich alkali feldspar (probably microcline).

(c) Albite. A plagioclase feldspar. Locality unknown. Notice the thin, straight, parallel striations delimiting the "albite twins."

(d) Bytownite. Plagioclase feldspar. Crystal Bay, MN. Alternating reflective and non-reflective, straight parallel bands in the lower center and upper right mark the "albite twins."

Phyllosilicates — Micas Plate 9

(a) Muscovite (white mica). McKinney Mine, near Newdale, NC. Notice the flexible, bent cleavage sheet at the top left edge and the single excellent cleavage.

(b) Biotite (black mica). Spruce Pine District, NC. Notice the single excellent cleavage and vitreous luster.

(c) Phlogopite (Mg-rich, brown mica). North Burgess, Ontario.

(d) Stilpnomelane (brown) with white mica (muscovite-paragonite) and quartz. Lone Tree Creek, Northeastern Diablo Range, CA.

Plate 10　Magnesium Phyllosilicates and Prehnite

(a) Chlorite. Frank, NC. Notice the pseudohexagonal form of the cleavage plates in this monoclinic mineral and the excellent {001} cleavage.

(b) Talc. Day Book Mine, NC. Note that the talc here is green, but talc appears in different colors and commonly occurs in a massive aggradational form.

(c) Serpentine, variety chrysotile. New Idria, CA. The fibers are actually sub-microscopically rolled sheets in this phyllosilicate (sheet-structured) mineral.

(d) Serpentine, variety lizardite. Cloverdale, CA. [Note: O'Hanley (1996) discusses the difficulty of correctly distinguishing lizardite and antigorite serpentines, so the identification here is tentative and based on properties and geological setting.]

(e) Prehnite. Paterson, NJ.

Inosilicates (Double Chain)   **Plate 11**

(a) Anthophyllite (orthorhombic "orthoamphibole"). Greer Hollow metaultramafic body, Todd, NC. The acicular (needle-like) crystals here are several centimeters long.

(b) Glaucophane (blue, Na-amphibole). Lone Tree Creek, Northeastern Diablo Range, CA. The glaucophane occurs here in a schist, with brown stilpnomelane, silvery-white white mica, and glassy gray quartz.

(c) Tremolite (monoclinic "clinoamphibole"). Balmat, NY. The tremolite here occurs in short acicular crystals, but may occur in the same long needles as the other amphiboles, such as anthophyllite, Plate 11(a), or hornblende, Plate 11(e).

(*Source:* Sample courtesy of Sonoma State University Geology Department.)

(d) Actinolite. Jenner, CA. Careful examination of the photo will allow recognition of the two cleavages at 120° and 60°, typical of amphiboles.

(e) Hornblende. Locality unknown.

Plate 12 Inosilicates (Single Chain)

(a) Wollastonite (gray to white) with diopside (green) and garnet (dark red). Willsboro, NY.
(*Source:* Sample courtesy of Sonoma State University Geology Department.)

(b) Spodumene. Andover, ME. Note vitreous luster and two cleavage directions.

(c) Augite (monoclinic "clinopyroxene"). Locality unknown. Note typical, slightly dull-vitreous luster and moderately exhibited cleavage. Also see augite in the lower right part of the sample in Plate 16(a).

(d) Enstatite ("orthopyroxene") (brown) with tiny, bright green chromium diopside (clinopyroxene) crystals. Webster, NC.

Cyclosilicates and Sorosilicates Plate 13

(a) Tourmaline, variety schorl. Yunnan, Gejiu, China.

(b) Tourmaline, variety elbaite.

(a) and (b) Note the three somewhat dull pyramidal faces on the lower right end of the crystal in (a), typical of tourmaline. The many faces along the length of both crystals are a combination of a trigonal prism and hexagonal prism faces. Crystals often have a crudely triangular cross-section (b). So-called "watermelon" tourmaline (b) includes pink and green bands of color.

(c) Beryl. Various crystals from Spruce Pine area, NC and an unknown locality. The blue-green and blue specimens here from NC are called aquamarine. Deep green, transparent beryl is called emerald.

(d) Pumpellyite (green) in cavity in glaucophane schist. Marin County, CA. Less common brown varieties of pumpellyite also exist.

(e) Lawsonite (bluish white) with white mica (silvery greenish white) and quartz (glassy gray). From the type locality, now destroyed by development, at Reed Station, Tiburon Peninsula, CA.

(f) Epidote. Samples from various localities. Three smaller prismatic samples are from Garnet Hill, CA. Sample in back shows typical fine-grained, granular to massive character of "pistachio" yellow-green epidote in rocks (here with gray to white quartz).

Plate 14   Aluminous and Other Neosilicates (Orthosilicates)

(a) Andalusite. Locality unknown. The cross-shaped pattern of inclusions characterizes the variety of andalusite called chiastolite.

(b) Kyanite. Samples are from Brazil. Nearly black, green, and almost white crystals are known, but blue is diagnostic and prized by collectors.

(c) Sillimanite. Locality unknown. The habit is fibrous to acicular. The appearance is vitreous silvery-white to vitreous light brown or pale green. Careful examination of the photo will reveal the small acicular crystals in this massive sample.

(d) Staurolite in quartz-white mica schist. Russia. The crossed (cruciform) twins shown here are typical of staurolite and are twinned on the {232} twin plane.

(e) Garnet. Various crystals from various localities. The common varieties pyrope, almandine, and spessartite are commonly red. The yellow-green sample of the variety grossularite here is from Mexico. Black, purple, brown, green, yellow, and white colors occur, as well. Note the typical dodecahedron form of the sample on the upper right, which also has thin, linear trapezohedral faces along the edges.

(f) Olivine. Day Book Mine, NC. Note the equant habit, vitreous luster, and variations in color from yellow-green to green to almost black.

Oxides with Titanium and Other Oxide Minerals   Plate 15

(a) Titanite. Westport, Ontario, Canada. Also see the "honey-brown" titanite crystals typical of titanite occurrences in igneous rocks in the rock shown in Plate 22(d).
(*Source:* Sample courtesy of Sonoma State University Geology Department.)

(b) Ilmenite. Kragero, Norway. Note the submetallic luster of this specimen.

(c) Corundum. Transvaal, South Africa. The prismatic habit with a hint of a triangular character shown here is common. Transparent deep red corundum is the gemstone ruby, whereas similarly transparent blue corundum is sapphire.

(d) Hollandite (?) dendrites on quartz arenite. Mountain City, TN.

Plate 16   Iron Oxides and Hydroxides

(a) Magnetite (black with metallic luster) with augite (dull, dark green) and epidote (pistachio green). Cranberry, NC.

(b) Chromite (black) with quartz (glassy gray) and kammererite (purple chromium chlorite). Day Book Mine, NC.

(c) Hematite. Earthy hematite cement in fossiliferous sedimentary rock. Ontario, NY. Careful examination of the surface of the sample reveals a slightly silvery tone characteristic of "specular" hematite, which crystallizes in a tabular hexagonal habit (not visible here).

(d) Goethite (yellowish-brown to brownish-black). Sample may include some limonite. Tuscaloosa County, AL.

Sulfides   Plate 17

(a) Pyrite. Cube from Spain. Pyritohedron locality unknown.

(b) Galena. Missouri. Note the cubic form and cleavage.

(c) Chalcopyrite. Rouyn District, Quebec, Canada.
(*Source:* Sample courtesy of Sonoma State University Geology Department.)

Plate 18   Sulfates and Carbonates

(a) Gypsum variety selenite (in clear sheets). Carbona Quadrangle, CA. Notice three directions of cleavage (a fourth is not visible).

(b) Barite. Madoc, Ontario.

(c) Calcite. Various samples in various forms and colors from various localities. The "dogtooth" spar scalenohedral form in the yellowish-white crystal at top center is moderately common and rhombohedral cleavage fragments (white and clear) are very common. Less common is the very short prismatic form with rhombohedral terminations (brown on left), shown by this sample from Saltville, VA.

(d) Dolomite. Penfield, NY. Notice the rhombohedral forms and traces of similar cleavage.

Halides and Phosphates Plate 19

(a) Halite. Detroit, MI. Notice the evidence of cubic cleavage.

(b) Sylvite. Carlsbad, NM.

(c) Fluorite. Locality unknown. Crystals are typically cubic (not shown) and cleavage fragments are octahedral (as shown here). Colors range from purple, brown, blue, green, and yellow to white and colorless.

(d) Apatite. Wilberforce, Ontario.

(*Source:* Sample courtesy of Sonoma State University Geology Department.)

(e) Apatite. Locality unknown.

Plate 20   Selected Ore Minerals

(a) Wolframite. Salamanca, Spain.

(b) Scheelite. Timmons, Ontario, Canada.
(*Source:* Sample courtesy of Sonoma State University Geology Department.)

(c) Copper. Michigan.

(d) Sulfur. Louisiana. Note the somewhat resinous luster.
(*Source:* Sample courtesy of Sonoma State University Geology Department.)

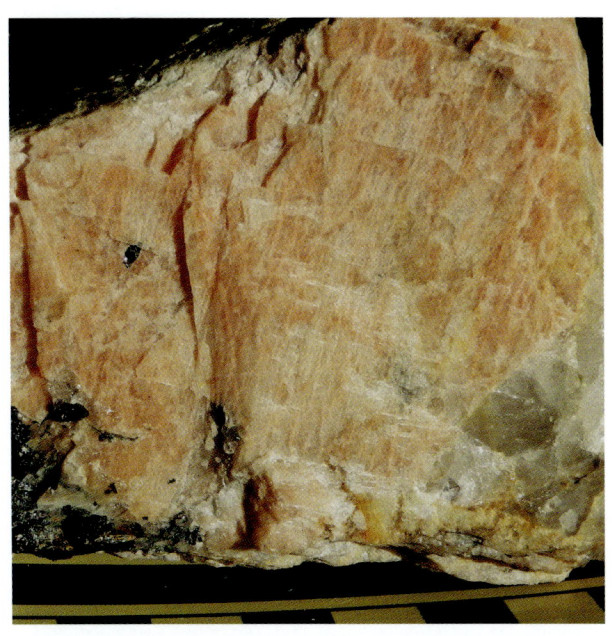

High-silica and Moderate-silica Plutonic Rocks Plate 21

(a) Pegmatitic tourmaline-muscovite granite. Custer, SD. Gray, glassy quartz, black to silvery white muscovite mica (barely visible as thin, several-cm-long crystal on top of sample), and black tourmaline decorate a large, orange-pink crystal of alkali (K-rich) feldspar. Note that the feldspar has perthitic texture.

(b) Biotite granite. Near Woodleaf, NC. Gray quartz, white alkali and plagioclase feldspars, and black biotite form a hypidiomorphic-granular texture.

(c) Biotite quartz monzonite. Salisbury, NC. Gray quartz, pink alkali feldspar, white to pink plagioclase feldspar, black biotite and magnetite, and red to orange hematite comprise the rock.

(d) Hornblende-biotite granodiorite. Sierra Nevada Batholith, US Highway 50 east of Kyburz, CA. The rock consists of gray glassy quartz; small, white plagioclase feldspar grains; large, white, poikilitic alkali feldspar crystals; equant black biotite grains; and rectangular to diamond-shaped, black hornblende crystals. Just below the scale, an amoeboid poikilitic grain of alkali feldspar reflects light from its surface.

Plate 22 Intermediate Composition Plutonic Rocks

(a) Biotite-hornblende granodiorite. Carson-Iceberg Wilderness near Mosquito Lake, CA. Minerals the same as in 21(d), but hornblende is more abundant. Plagioclase generally tends to approach euhedral character more than is the case in the rock in 21(d).

(b) Hornblende-biotite quartz diorite. Locality unknown. Gray, glassy quartz and white to cream- and light gray-colored plagioclase feldspar surround black, equant biotite and rectangular to diamond-shaped hornblende. A bright reflection from a stubby hornblende in the lower center and reflections from twinned plagioclase grains, including that at the left center, reveal the rectangular habits of these crystals. Field of view is 5 cm wide.

(c) Syenite. Locality unknown. White to greenish-brown and grayish brown alkali feldspars with minor, associated black biotite mica, dominate the rock.

(d) Titanite- and quartz-bearing biotite-hornblende diorite. Block in Yosemite Valley, Yosemite National Park, CA. Quartz is glassy, gray, and anhedral, whereas white plagioclase feldspar is anhedral to subhedral. "Honey-brown" euhedral to subhedral titanite grains are scattered through the rock and some, near the center, have a characteristic flattened diamond "wedge" shape. Pseudohexagonal, equant euhedral biotite is visible; and anhedral to euhedral black hornblende is abundant.

Low-silica Plutonic Rocks   Plate 23

(a) Gabbro. Coast Range Ophiolite, Middletown, CA. White, gray, and green anhedral to subhedral plagioclase feldspar is mixed with and surrounds clusters of fine- to medium-grained greenish black augite.

(b) Sodalite-hastingsite nepheline syenite. Red Hill Complex, NH. Gray nepheline and greenish-gray sodalite are associated with abundant white to medium gray alkali feldspar, locally showing pronounced cleavage. The black, rectangular grains are the alkali amphibole, hastingsite, and a few metallic grains of magnetite occur as minor accessory minerals.

(c) Chromitite layer in (meta)dunite. Day Book Mine, NC. Chromite grains are dominantly <1–2 mm and are locally rimmed by the purplish-silver chromium chlorite, kammererite, in the chromitite. In the (meta)dunite and chromitite, the olivine of the (meta)dunite is granular and green of various shades. Kammererite and elongation of chromite pods provides the only evidence of metamorphism in the hand specimen.

(d) Harzburgite. Stillwater Complex, Nye, MT. Minor bluish-gray to greenish-black plagioclase accompanies minor green olivine and abundant brownish-black orthopyroxene in the rock.

(*Source:* Sample courtesy of Sonoma State University Geology Department.)

Plate 24   Clinopyroxinite and Volcanic Rocks 1

(a) Clinopyroxenite. Coast Range Ophiolite, Del Puerto Canyon, CA. This sample is composed almost entirely of light to dark green augite. Traces of olivine are present.

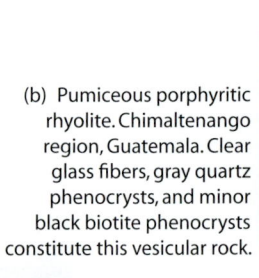

(b) Pumiceous porphyritic rhyolite. Chimaltenango region, Guatemala. Clear glass fibers, gray quartz phenocrysts, and minor black biotite phenocrysts constitute this vesicular rock.

(c) Porphyritic andesite in crossed Nichols thin section view. "Sonoma Volcanics," near St. Helena Road, Sonoma County, CA. Photomicrograph reveals trachytic texture of plagioclase, brightly colored tiny crystals of augite, black magnetite, and phenocrysts of plagioclase feldspar. Width of view is about 1 mm.

(d) Welded lithic lapilli tuff rhyolite. Near Shoshone, CA. Generally brown volcanic lithic clasts (lapilli) and some crystals of quartz are embedded in a glassy, annealed ash matrix of volcanic glass, crystals, and lithic fragments.

(Source: Sample courtesy of David Bero, Sonoma State University.)

(e) Porphyritic hornblende latite. Jamestown, CA. White phenocrysts of feldspars occur in the dark gray aphanitic matrix.

Volcanic Rocks 2 and Coarse Clastic Sedimentary Rocks   Plate 25

(a) Porphyritic hornblende andesite. Merhten Formation, Sierra Nevada, CA. Black, rectangular hornblende phenocrysts, plus cream to white plagioclase phenocrysts, are scattered within the gray aphanitic groundmass.

(b) Vesicular, porphyritic olivine basalt. Locality is near Elko, NV. Large glassy green to black phenocrysts of olivine and a few white phenocrysts of plagioclase (one, on the far right, shows cleavage) are scattered in the gray aphanitic matrix. Numerous vesicles characterize the rock.

(*Source:* Sample courtesy of David Bero, Sonoma State University.)

(c) Boulder-cobble conglomerate. Heavens Beach, Sonoma Coast State Park, CA. This conglomerate shows a wide range of clast sizes from granules to boulders (such as the large, blue, glaucophane schist boulder, part of which is showing at the left). The moderately- to well-rounded clasts include a wide range of rock types, including green mafic volcanic rocks, chert, and gray sandstones. Scale is provided by the hammer handle, approximately 32 cm long.

(d) Chert breccia. Knox Group, Broadford Quadrangle, VA. Angular clasts of brownish-gray to white chert are the clasts in this white chert-matrix breccia.

Plate 26 Sandstones and Mudrocks

(a) Quartz arenite (sandstone). Tumbling Creek, Clinch Mountain Wildlife Management Area, VA. Moderately sorted, medium quartz sand grains, with a few granules of gray chert and buff-colored mudrock at the bottom of the field of view, constitute this rock.

(b) Lithic quartz arenite (sandstone). Dinosaur National Monument, UT. Medium- to coarse-grained, moderately sorted sand grains of variously colored quartz, plus rock fragments of different types, including black, red, and other colors of chert, comprise this sandstone. Careful inspection reveals the grains to be predominantly well-rounded to subangular.

(c) Granule-bearing, feldspathic lithic wacke. Sonoma Coast State Park, north of Jenner, CA. Black lithic fragments of various sizes, including granule to pebble-size shale chips, occur as clasts, with gray to gray-green quartz and white feldspars, in a greenish-gray matrix of chlorite (?) and clays.

(d) A layered sequence of siltstone (medium gray) with mudshale (dark gray) occurs between gray lithic wacke (above) and medium to light gray lithic wacke (sandstone) (below). Heaven's Beach, Sonoma Coast State Park, Sonoma County, CA. Scale is provided by 12-cm hammerhead, upper left.

Varieties of Limestone   Plate 27

(a) Lime mudstone. Notice the conchoidal fracture of the rock and the near absence of clasts of fossils or rock fragments. Locality near Ciegnes, France.
(*Source:* Sample courtesy of Appalachian State University Geology Department.)

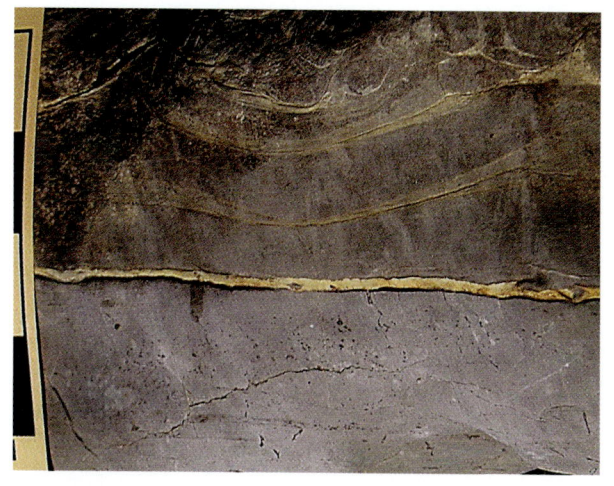

(b) Wackestone (limestone) (below) and lime mudstone (above). Middle Ordovician limestone. Broadford Quadrangle, VA. The wackestone at the bottom of the picture contains little fragments of fossils and limestone in a gray lime mud matrix. The uppermost layers, showing flame structures, are lime mudstone.

(c) Skeletal packstone. Vaughn Gulch Formation, Vaughn Gulch, Inyo Mountains, CA. Brown weathered bryozoans and white crinoid fragments are clasts in the recrystallized, dark gray matrix of this packstone.

(d) Skeletal grainstone. Castle Hayne, NC. A little calcite cement binds the fossil skeleton fragments in this grainstone.
(*Source:* Sample courtesy of Appalachian State University Geology Department.)

Plate 28 Limestone, Dolostone, Evaporite, and Coal

(a) Hematitic, bryozoan, crystalline boundstone. Ordovician. Knoxville, TN. Algal layers, recrystallized and partly stained by red hematite comprise this boundstone.

(b) Ooid dolopackstone and dolowackestone. Cambrian Nolichucky Shale, Broadstone Quadrangle, VA. Cream-colored, weathered ooids and an ooid clump occur in dark gray dolomite. Area of sparse ooids is ooid dolowackestone.

(c) Gypsum evaporite. Mississippian MacCrady Formation, Saltville, VA.

(d) Bituminous coal. Morgantown, WV.

Cherts and Aphanitic Foliated Metamorphic Rocks Plate 29

(a) Chert. Knox Group, Broadford Quadrangle, VA. Laminated white chert. This white chert replaced dolostone, which replaced laminated limestone. Black and yellow dots are lichens growing on the surface of the rock.

(b) Yellow-red radiolarian chert. Franciscan Complex, northeastern Diablo Range, CA. Goethite (?) and hematite color the chert, which has hints of radiolaria appearing as gray spots in the yellow-red chert on the right edge. The black veins are composed of manganese oxide minerals, originally deposited as manganese minerals with the chert and remobilized into veins. Gray quartz veins also cross the chert bed, and the included, faint laminations. Cream-colored to reddish lichens are present on the upper surface.

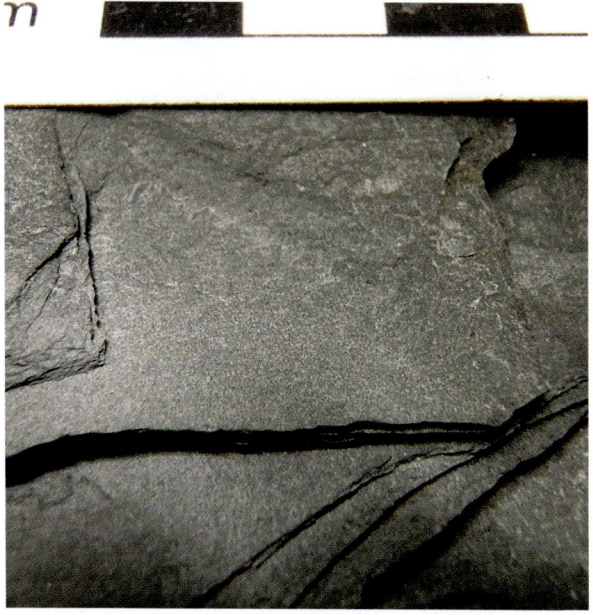

(c) Dark gray slate. Central VA. This aphanitic rock displays typical slaty structure and a slight red coloration, probably due to oxidation of iron-bearing minerals to hematite.

(d) Glaucophane phyllite. Franciscan Complex, Healdsburg Quadrangle, CA. This sample displays classic phyllitic structure with wavy, aphanitic to very fine-grained layers of minerals; in this case, predominantly blue glaucophane and silvery white white mica (muscovite-paragonite). A few tiny red garnets are present.

Plate 30 Schists

(a) Chlorite-actinolite schist. Franciscan Complex, northeastern Diablo Range, CA. Chlorite (black appearing flaky minerals) and actinolite (green acicular minerals) comprise this schist.

(b) Garnet-staurolite-biotite-quartz-white mica schist. Ashe Metamorphic Suite, Highway 221, Ashe County, NC. Red equant garnet, brown rectangular staurolite, black biotite, gray quartz, and silvery-white white mica are the major minerals of this rock.

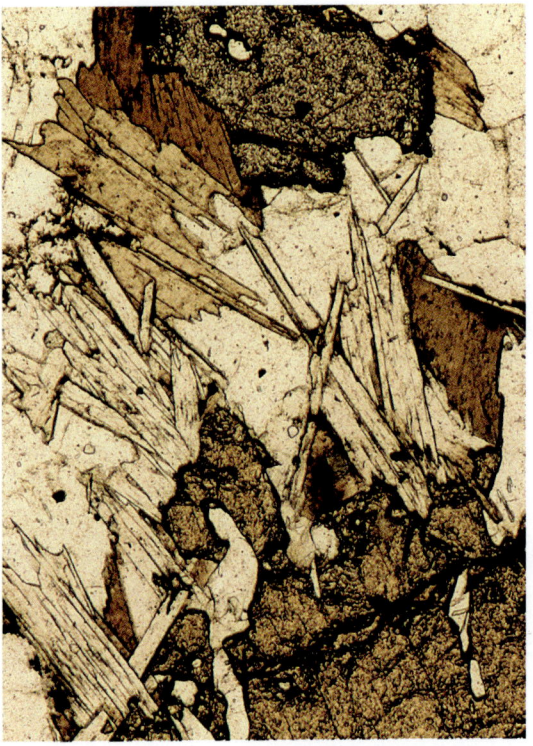

(c) Photomicrograph of a thin section of the schist in (b) in plain light (PL).

(d) Photomicrograph of a thin section of the schist in (b) in crossed Nichols (XN).

(c) and (d) The large grain in upper center is garnet. The large grain in lower right and center is staurolite. Brown grains in PL that appear green, dark green and brown in XN are biotite. Colorless grains in PL (c) that are pink, yellow, blue, and green in XN (d) are white mica. Gray and white grains are predominantly quartz. The long dimension of the photomicrograph is about 7 mm.

Miscellaneous Metamorphic Rocks 1   Plate 31

(a) Semi-schistose jadeite-lawsonite metawacke (or jadeite-lawsonite-chlorite-quartz semi-schist). Franciscan Complex, Hospital Creek area, northeastern Diablo Range, CA. Some gray, glassy quartz grains, a few green chlorite grains, and many white grains that are plagioclase grains partly to entirely replaced by jadeite (pyroxene) are visible. Microscopic and X-ray analyses indicate that the goethite (?)-stained matrix material is a mix of glaucophane, chlorite, white mica, and chlorite-vermiculite minerals.

(b) "Greenstone" (metabasalt). Pillowed. Franciscan Complex, Rock Point, Sonoma Coast State Park, CA. The rock is aphanitc and green, hence the name "greenstone". In this picture, rounded pillows are cut locally by shear planes and white calcite veins. Pillows measure 10–30 cm. For scale, the thin gray layer at the center-left edge is about 1 cm thick.

(d) Dolomite marble. Locality unknown. White dolomite constitutes the entire rock, except for a few tiny patches of brown clay on the surface.

(c) Glaucophane gneiss. Cobble in stream, derived from Franciscan Complex, Austin Creek, Sonoma County, CA. Bluish-black glaucophane dominates the dark bands, whereas white plagioclase and gray quartz dominate the light-colored bands.

(e) Chlorite diablastite. Frank metaultramafic body, Ashe Metamorphic Suite, Frank, NC.

Plate 32 Miscellaneous Metamorphic Rocks 2

(a) Eclogite. Landslide block from Ring Mountain Melange, Reed Station, Tiburon Peninsula, Marin County, CA. Orange and red to black-appearing euhedral garnets occur with granular green omphacites (clinopyroxenes), constituting this rock. A few small, blue, acicular grains of glaucophane are present locally.

(b) Serpentinized harzburgite. Coast Range Ophiolite, Del Puerto Canyon, CA. Serpentine minerals show up as green to black aphanitic materials and locally appear white, where light is reflected from them. The dominant fabric is one of a cross-fracture pattern, with dark serpentine veins filling the fractures that surround small, less serpentinized, cm-scale blocks. Some square grains, distinctly visible because of reflection from relict cleavage (e.g., lower left and lower right), are serpentinzed orthopyroxenes, usually referred to as bastites. Scale division at right is 1 cm.

(c) Epidote hornels. Unnamed unit in the Sierra Nevada east of Alpine Lake, CA along CA Highway 4. Granular yellow-green to dark green epidote is associated here, at a granite-biotite schist contact, with gray quartz, plus associated tiny black grains of biotite and white alkali feldspar.

(d) Quartz-chlorite mylonite. Zone in metaquartzarenite of the Grandfather Mountain Formation, Grandfather Mountain, NC. Zone in upper right is mylonite derived from metamorphosed quartz arenite in lower left. Dark, microscopic, recrystallized grains of quartz and feldspar are mixed with chlorite and white mica in the mylonite and surround syntectonically recrystallized lenses and bands of white to gray quartz and light pinkish-gray alkali feldspar. On the far right, a glassy-gray fragment of a quartz vein is surrounded by mylonitic materials. On the lower left, gray to white grains of quartz and pinkish-gray alkali feldspar with a recrystallized matrix of silvery white mica and green to black chlorite, plus a few iron oxide grains, constitute the meta-sandstone. Scale is provided by the approximately 1.5-cm sharpened end of the pencil.

rock (figure 19.16 on pp. 183–184). Tracks and trails are surface structures that are also fossils. Burrows, tracks, and trails are called trace fossils (figure 9.12f).

Sedimentary Rock Compositions

The compositions of sedimentary rocks can be reported as chemical compositions determined through chemical analysis, or as mineralogical compositions. In terms of chemistry, because sedimentary rocks form in a variety of ways—including (1) as chemical precipitates of many types; (2) as accumulations of various clastic materials, which can have virtually any composition; and (3) as combinations of clastic and precipitated materials—the compositions of sedimentary rocks span a wide range. Table 9.4 on p. 185 presents the compositions of some sedimentary rocks. Note that common rock types such as sandstone and limestone differ radically in chemistry, from rocks that are almost pure silica (SiO_2) to rocks that almost lack silica entirely. In sedimentary rocks, the silica, alumina (Al_2O_3), lime (CaO), and magnesia (MgO) may be very high or very low, depending on the rock type.

The various compositions of sedimentary rocks reflect their diverse mineral composition (table 9.5 on pp. 185–186). Almost any mineral can be present in a sedimentary rock, especially a clastic one. Typical **modes**, the observed mineral compositions of sedimentary rocks, are presented in table 9.6 (on pp. 187–188). Note that quartz varies from zero to more than 90%. Similarly, in carbonate rocks, various components—such as lime mud, spar, fossils, allochems, and the minerals calcite and aragonite—can range from zero to nearly 100%.

(a) Mudcracks in Ordovician Bowen Formation, west of Gate City, TN. Note 15-cm photo scale in lower right. (Photo by Loren A. Raymond.)

Figure 9.15
Some surface features of beds.

(b) Sole marks in Ordovician Knobs Formation sandstone, Lodi, VA, examined in 1973 by the senior author. (Photo by Fred Webb, Jr.)

Figure 9.15 *(cont'd.)*

(a) Acrospirifer brachiopods in silicified limestone, Devonian Huntersville Formation, near Tannersville, VA. (Photo by Loren A. Raymond.)

Figure 9.16
Fossils in sedimentary rocks.

CLASSIFICATIONS OF SEDIMENTARY ROCKS

Using the combined mineral composition, chemistry, textural and structural aspects, and origins of sedimentary rock materials, the sedimentary rocks can be subdivided into three main groups—the **siliciclastic rocks**, the **allochemical** or biogenic **rocks**, and the **precipitates** (called "chemical" sedimentary rocks by some geologists).[9] The common precipitates include the calcite-rich rocks (limestones) and their dolomite-rich derivative rocks (dolomites or dolostones), plus the siliceous rocks (cherts). These types most commonly are biogenic, formed through processes of animal- or plant-mediated precipitation. Other precipitates include the evaporites, such as halite evaporite, gypsum evaporite, and anhydrite evaporite; phosphatic rocks; and ironstones and iron formations. The siliciclastic rocks include rocks composed of quartz, feldspar, clays, and rock fragments, including the sandstones, mudrocks, and conglomerates and related rocks. Allochemical rocks are most commonly limestones composed of clasts and other grains, such as skeletal (fossiliferous) packstones, oolitic grainstones, and allochemical limestone conglomerates (the definitions of which are provided below). Other rock types, such as chert breccias and coals are important in certain environments.

Classification of sedimentary rocks is based, as it was in the phaneritic igneous rocks, on the textures and the mineral compositions of the rocks. Typically, allochemical rocks are divided into carbonate types and non-carbonate types. Limestones and dolostones are the main types of carbonate sedimentary rocks. The carbonate types of both precipitated and clastic

(b) Senior author examines Triceratops scapula and rib in Hell Creek Formation, north of Marmarth, ND. (Photo by F. K. McKinney.)

Figure 9.16 *(cont'd.)*

Table 9.4 Chemical Compositions of Typical Sedimentary Rock Types

	1 (Ss)	2 (Ch)	3 (Ss)	4 (Sh)	5 (Fe-st)	6 (Ls)	7 (Dlst)
SiO_2	96.65[a]	94.7	67.2	61.84	4.21	1.15	0.28
TiO_2	0.17	0.06	0.05	0.83	0.12	nd	nd
Al_2O_3	1.96	1.1	14.6	13.40	4.38	0.45	0.11
Fe_2O_3	0.58	2.7	1.9	3.83	37.72	nd	0.12
FeO	nd	0.22	2.3	1.15	7.27	0.26	nd
MnO	nd	0.05	0.1	0.05	0.18	nd	nd
MgO	0.05	0.14	2.3	2.69	1.68	0.56	21.30
CaO	0.08	0.06	1.8	2.68	22.49	53.80	30.68
Na_2O	0.05	0.01	3.7	0.97	0.01	0.07	0.03
K_2O	0.27	0.37	1.9	2.8	0.00	0.07	0.03
P_2O_5	nd	0.03	0.1	0.44	1.00	nd	0.00
LOI[b]	0.59	0.79	3.4	9.74	20.81	43.61	47.42
Total	100.40	100.23	99.35	100.42	99.87	99.90	99.97

Sources:

1. Quartz arenite (?), Abbott formation (Pennsylvanian), Illinois, sample B19. Analyst: L. D. McVicker (Bradbury et al., 1962).
2. Thin-bedded, red chert, Franciscan Complex (Jurassic-Cretaceous?), California, sample 4, table 9. Analysts: P. L. D. Elmore, I. H. Barlow, S. D. Botts, and G. Chloe (E. H. Bailey, Irwin, and Jones, 1964).
3. "Graywacke" (sandstone), Coastal Belt, Franciscan Complex (Cretaceous?), California, sample ID, table 1. Analysts: P. L. D. Elmore, I. H. Barlow, S. D. Botts, and G. Chloe (E. H. Bailey, Irwin, and Jones, 1964).
4. Shale, Cody Shale (Cretaceous), Wyoming, sample SCo-1. Analyst: S. M. Berthold (L. G. Schultz, Tourtelot, and Flanagan, 1976).
5. Hematitic ironstone, Keefer formation (Silurian), Pennsylvania, sample D. table 9. Analyst: P. M. Buschman (H. L. James, 1966).
6. Limestone, Solenhofen formation (Jurassic), Bavaria. Analyst: G. Steiger (F. W. Clarke, 1924).
7. Dolostone, Royer dolomite (Cambrian), Oklahoma, sample 9294, table 5. Analyst: A. C. Snead (W. E. Ham, 1949).

[a] Values in weight percent.
[b] Loss on ignition and other.
Ch = chert, Dlst = dolostone, Fe-st = ironstone, Ls = limestone, Sh = shale, Ss = sandstone.
nd = not determined or not reported. In the case of iron, total iron may be reported as Fe_2O_3.

Table 9.5 Characteristic Minerals of Sedimentary Rocks

Mineral	Chemical/Structural Group	Distinctive Properties	Commonly Occurs in
Quartz	Tectosilicate	Hexagonal; any color; H=7; vitreous; conchoidal fracture; optically positive	Sandstones, mudrocks, conglomerates, and diamictites; cherts; limestones as a minor accessory mineral
Potassium Feldspars	Tectosilicate	White, pink, or green; H=6; 2 cleavages @ 90°; Carlsbad twinning; low birefringence	Sandstones, conglomerates, diamictites
Plagioclase Feldspars	Tectosilicate	White to colored gray, green, brown, black (by impurities); H=6; polysynthetic twinning (shows striations)	Sandstones, conglomerates, diamictites
Biotite	Phyllosilicate (sheet silicate)	Brown to black colored; one perfect cleavage; H=2–3	Sandstones, conglomerates diamictites
Muscovite	Phyllosilicate (sheet silicate)	Colorless to white and light green, brown, and yellow colored; one perfect cleavage; H=2.5–3	Sandstones, mudrocks, conglomerates, diamictites
Kaolinite (a type of clay)*	Phyllosilicate	Usually aphanitic and massive; dull luster; white; plastic; H=2–2.5	Sandstones, mudrocks, conglomerates, diamictites

(continued)

Table 9.5 *(cont'd.)*

Mineral	Chemical/Structural Group	Distinctive Properties	Commonly Occurs in
Montmorillonite (a smectite clay)*	Phyllosilicate	Usually aphanitic; H=1–2; white to green or brown; musty odor	Sandstones, mudrocks, conglomerates, diamictites
Tourmaline	Cyclosilicate (ring silicate)	Black (variety Schorl) and various colors including red, green, and yellow (variety Elbaite); H=7; habit = trigonal (3-sided) prisms; marked pleochroism optically	Sandstones
Zircon	Nesosilicate	Various colors from red-brown to colorless; typically in elongate crystals; adamantine; high relief and high birefringence optically	Sandstones, conglomerates, diamictites
Calcite	Carbonate (Trigonal)	Any color; H=3; 3-cleavage directions (rhombohedral); effervesces in cold HCl	Various limestones; cement in sandstones and mudrocks; veins in many sedimentary rocks
Aragonite	Carbonate (Orthorhombic)	Usually fibrous to acicular crystals; white, clear, or yellowish colored; H=3.5–4; Effervesces in cold HCl	Some limestones
Dolomite	Carbonate (Trigonal)	Any color; H=3.5–4; 3-cleavage directions (rhombohedral); effervesces in hot HCl or with difficulty, if powdered	Various dolostones
Siderite	Carbonate (Trigonal)	Yellow to red-brown; H=3.5–4; 3-cleavage directions (rhombohedral); effervesces in hot HCl	Some iron-rich sediments; but also occurs uncommonly as cement in some sandstones
Halite	Halide (Cubic, Isometric)	Colorless to white; H=2, cubic cleavage; salty taste	Evaporites
Sylvite	Halide (Cubic, Isometric)	Colorless to white; H=2; cubic cleavage; bitter taste	Evaporites
Gypsum	Sulfate (Monoclinic)	Colorless to white, red, brown; H=2; 4 directions of cleavage	Evaporites
Anhydrite	Sulfate (Orthorhombic)	Colorless to bluish; H=3–3.5; 3 directions of cleavage	Evaporites
Magnetite	Oxide	Black with black streak; magnetic; irregular to octahedral crystals; submetallic to metallic luster	Sandstones, conglomerates, diamictites, mudrocks; ironstones and iron formation
Hematite	Oxide	Silver, black, to maroon and red-brown colored; red-brown streak; metallic to dull earthy luster	Weathered sandstones, conglomerates, mudrocks, diamictites; ironstones and iron formations
Goethite	Oxide	Yellow-brown streak and color; typically aphanitic and earthy in appearance	Weathered sedimentary rocks
Apatite	Phosphate	Brown, green, yellow, blue, and colorless; vitreous luster; H=5; hexagonal prisms common	Cryptocrystalline material in the black phosphate-rich rocks called phosphorite
Pyrite	Sulfide	Yellow-gold colored; metallic luster; in cubes and pyritohedrons; H=6–6.5	Black shales; some carbonate rocks; some iron-rich rocks

* X-ray diffraction and other tests are usually required to distinguish one clay from another.

Sedimentary Rocks 187

Table 9.6 Modes of Typical Sedimentary Rocks

	1 (Ch)	2 (Ss)	3 (Ss)	4 (Ss)	5 (Ss)	6 (Sh)	7 (Ls)	8 (Ls)
Quartz[a]	84.3[b]	65.3	39.7	41.6	22.2	—	—	—
Monocrystalline	(57.6)[c]	(63.0)	(38.7)	(40.6)	(17.8)	—	—	—
Polycrystalline	—	(2.3)	(1.0)	(1.0)	(4.4)	—	—	—
Fossils[d]	(26.7)	—	—	—	—	—	—	—
Alkali feldspar	—	1.0	4.0	0.3	0.4[e]	—	—	—
Plagioclase	—	2.0	6.0	0.3	23.4	—	—	—
White mica	8.0[f]	—	tr	tr	—	—	—	—
Biotite	—	—	5.7	—	—	—	—	—
Chlorite	0.3	—	5.0	—	—	5	—	—
Clays	—[f]	25.3	—	—	—	95	—	—
Kaolinites	—	(25.3)	—	—	—	(15)	—	—
Montmorillonites	—	—	—	—	—	(70)	—	—
Illites	—	—	—	—	—	(10)	—	—
Other minerals	—	1.3	0.7	7.6[g]	1.0	—	—	—
Rock fragments	—	4.7	4.0	0.3	16.6	—	—	—
Shale and Ss	—	(1.0)	—	(0.3)	(1.6)	—	—	—
Chert	—	(2.0)	(2.0)	(tr)	(8.2)	—	—	—
Siliceous volcanic	—	(1.7)[h]	(1.0)	—	(4.2)	—	—	—
Basic volcanic	—	—[h]	(tr)	—	(0.8)	—	—	—
Metamorphic	—	(tr)	(1.0)	—	(1.6)	—	—	—
Other/unknown	—	—	—	—	(0.2)	—	—	—
Matrix	(—)[i]	1.7	0.7	—	35.8	na	(—)[j]	(—)[j]
Cement	7.0[k]	—	33.3[l]	33.3[l, m]	0.6	na	na	na
Carbonate materials								
Mud	—	—	—	—	—	—	58.5	4.5
Spar								
Calcite	—	—	(33.3)	(19.0)	—	—	—	20.0
Dolomite	—	—	—	—	—	—	13.5	—
Allochems[n]								
Oolites	—	—	—	1.3[o]	—	—	—	34.5
Intraclasts	—	—	—	5.0	—	—	—	1.0
Fossils[p]	—	—	—	10.0	—	—	28.0	40.5
Total	100.0	100.0	100.0	100.0	100.0	100	100.0	100.0
Points counted	300	300	300	300	500	X	200	200

Sources:

1. Red radiolarian chert, Franciscan Complex (Jurassic?), The Geysers, California, sample RF-77A. Point count by L. A. Raymond.
2. Anauxite arenite, Testa formation (Eocene), Carbona Quadrangle, California, sample C18a (Raymond, 1969). Point count by L. A. Raymond.
3. Micaceous feldspathic arenite, Panoche formation (Cretaceous), Carbona Quadrangle, California, sample Cl (Raymond, 1969). Point count by L. A. Raymond.
4. Fossiliferous, hematitic quartz arenite. Rose Hill formation (Silurian), Clinch Mountain Wildlife Management Area, Virginia. Point count by L. A. Raymond.
5. Feldspathic wacke, Coastal Belt, Franciscan Complex (Cretaceous), west of Willits, California. Point count by L. A. Raymond.
6. Clay shale, Moreno formation (Cretaceous), Carbona Quadrangle, California. (Raymond, 1969). X-ray analysis by L. A. Raymond.
7. Dolomitic, bioclastic wackestone, Monteagle formation (Mississippian), near Huntsville, Alabama. Point count by H. Gault.
8. Oolitic fossiliferous grainstone, Monteagle formation (Mississippian), near Huntsville, Alabama. Point count by H. Gault.

[a] May include chalcedony and opal.
[b] Values in volume percent.
[c] Values in parentheses are not included in the total because they are included in another number on the table (e.g., quartz types are included in the total quartz).
[d] The fossils listed here are radiolaria. Since they are siliceous, they are included as "quartz"; they are not included as fossils under "carbonate materials."
[e] Untwinned feldspar is included with plagioclase.

(continued)

f In column 1, white mica includes all fine-grained colorless phyllosilicates (micas, clays). Therefore, clays are not listed as separate minerals.
g This value includes 4.3% hematite oolites and 3.3% hematitic fossil fragments.
h All volcanic rock fragments in this sample are included under "siliceous volcanic rocks."
i Technically, the silica occurring between the radiolaria in this rock is matrix rather than monocrystalline quartz.
j Calcite mud forms a matrix but is listed under "calcite."
k The cement here is largely the iron oxide mineral hematite. The hematite in this rock may be overestimated as it coats most other grains.
l Calcite spar forms cement material and is therefore listed under both "cement" and "spar."
m Includes 14.3% hematite cement.
n The allochems listed here are only carbonate allochems. Some rocks include siliceous fossils, hematitic fossils, hematitic oolites, and other materials not included here.
o The percent of oolites listed here includes only carbonate oolites. The rock also includes 4.3% hematite oolites.
p Only carbonate fossils are included. Other fossils are listed in other compositional categories.

Ch = chert, Ls = limestone, Sh = shale, Ss = sandstone.
tr = trace amounts.
na = not applicable.
X= analysis by X-ray diffraction.

Source: Raymond, 2007, p. 273).

character are classified together. Most carbonate rocks composed of calcite (the limestones) are actually clastic rocks composed of lime mud and various allochemical clasts. Most dolostones are actually altered from limestones during lithification (diagenesis) and are thus diagenetically altered rather than primary sedimentary rocks. A few dolostones are primary, but most are not.

There are several classifications of limestones, but the one provided here is based on the suggestions of Dunham (1962, with modifications by Raymond, 2007, p. 286). Dunham divided limestones into five types, four of which are based on the amounts of lime mud versus grains (figure 9.17). The grains can be of any size and include fossils, allochems, ooids, pellets, and some siliciclastic grains. Rocks with less than 10% grains are called *lime mudstone* (color plate 27a). Rocks with more than 10% grains, but in which abundant mud separates the grains (i.e., the rock is "lime mud-supported") are called *wackestone* (color plate 27b). Rocks in which there is a framework of grains touching one another (grain-supported rocks) with mud filling in between grains are called *packstones* (color plate 27c). *Grainstones* are grain supported and contain little mud (color plate 27d). The fifth type of rock is the *boundstone*, a rock composed of fossil elements that are in the positions they occupied (relative to other materials in the rock) at the time the fossil organisms were living (color plate 28a). Limestone that has recrystallized into equigranular-mosaic textured or similarly textured masses of spar during lithification is called *crystalline limestone*. If recrystallization was accompanied by major replacement by dolomite, the rock is a *crystalline dolostone*. Other dolomitic rocks have names corresponding to the Dunham names—such as dolomitic lime mudstone and dolomudstone—depending on whether dolomite constitutes less or more than 50% of the rock, respectively (figure 9.17).

The siliciclastic rocks, which are composed primarily of quartz, feldspars, various rock fragments, and clays, are divided into groups based on the Wentworth size categories (table 9.2; figure 9.18). The rocks with more than 25% of the clasts being larger than 2 mm (the coarse-grained rocks) are conglomerates, breccias, and diamictites (figure 9.19 on p. 190; color plate 25c). *Diamictites* have a muddy matrix and are mud-supported. The **matrix** is the siliciclastic material that is *deposited with the clasts*. Matrix contrasts with **cement**, which is crystalline material precipitated in the voids between the grains, *after* deposition (color plate 4b). **Conglomerate** has small to large rounded clasts, usually a sandy matrix, sometimes a clast-supported texture (larger grains are touching), and clasts that comprise 25% or more of the rock (figure 9.20a, b on p. 191; color plate 25c). *Breccia* has large angular clasts constituting 25% or more of the rock (figure 9.7b; 9.20c; color plate 25d).

The sandstones have grains that are dominantly in the 2 mm to 1/16 mm range. More than 20 classifications of sandstones have been proposed.[10] Here, we provide the Gilbert-Dott classification, in which there are two main types of sandstone: **arenites**, which lack significant matrix but may contain cement; and **wackes**, which contain significant matrix (figure 9.21 on p. 192; color plates 4b; 26a–d). The arenites are divided into three types—quartz arenites, feldspathic arenites, and lithic arenites—based on the abundance of the three components quartz, feldspars, and lithic (rock) fragments. Likewise, wackes are subdivided into quartz wacke, feldspathic wacke, and lithic wacke. The student who ventures into the professional geologic literature will likely encounter the anachronistic term *graywacke*. This name, which had a place in older classifications, is generally used to designate matrix-rich sandstones with significant amounts of feldspar and rock fragments, but it is one that has no place in the Gilbert-Dott classification. Most rocks called graywacke likely have matrix that is, in part, pseudomatrix formed after deposition through diagenetic processes (Whetten, 1966; Dickinson, 1970).

Mud consists of clay- and silt-sized mineral grains. **Mudrocks**, derived from mud, are divided here

Dunham Classification		Extension of Dunham Classification for Dolomitic Rocks			
	Predominantly calcite (Cc >95%)	Dominantly calcite (95% > Cc > 50%)	Dominantly dolomite (Do >50%)	Thoroughly recrystallized rocks with some relict structures	
				Dominantly dolomite	Dominantly calcite
Lime mud, <10% grains	Lime mudstone	Dolomitic lime mudstone	Dolomudstone	Crystalline dolostone	Crystalline limestone
Mud supported, >10% grains	Wackestone	Dolomitic wackestone	Dolowackestone		
Grain supported, contains mud	Packstone	Dolomitic packstone	Dolopackstone		
Grain supported, lacks mud	Grainstone	Dolomitic grainstone	Dolograinstone		
Original components bound together	Boundstone	Dolomitic boundstone	Doloboundstone		

Figure 9.17
Dunham-based classification of limestones and dolostones.
(*Source:* Raymond, 2007, p. 286.)

GRAIN SIZE	GRAIN SIZE NAME	SEDIMENT NAME	SEDIMENTARY ROCK NAMES
> 256 mm 256–64 mm 64–4 mm 4–2 mm	Boulders Cobbles Pebbles Granules	Gravel	Conglomerate and Breccia
2–1 mm 1–1/2 mm 1/2–1/4 mm 1/4–1/8 mm 1/8–1/16 mm	Very coarse sand Coarse sand Medium sand fine sand Very fine sand	Sand	Sandstone (Arenite, Wacke)
1/16–1/256 mm < 1/256 mm	Silt Clay	Mud	Siltstone, Mudrock, Shale, Claystone, Mudstone

Figure 9.18
Wentworth classification of grain sizes with corresponding sediment and sedimentary rock types.

CLASSIFICATION OF CONGLOMERATES, BRECCIAS, AND DIAMICTITES			
MATRIX / SUPPORT	CLAST SHAPE	CLAST COMPOSITION	NAME*

Gravelly or Sandy (Generally clast-supported)	Rounded	• Single composition
		   Quartz ± chert
		   Calcareous
		• Varied composition
			Name of clast types and "conglomerate"
			   Quartzitic conglomerate
			   Calcirudite or limestone conglomerate
			   Polymict (Petromict) conglomerate

	Angular	• Single composition
		   Quartz ± chert
		   Calcareous
		• Varied composition
			Name of clast types and "breccia"
			   Quartzitic breccia
			   Limestone breccia
			   Polymict breccia

Muddy (clay ± silt) and mud-supported	Rounded, angular, or both	• Single composition • Varied composition	Oligomict diamictite** Polymict diamictite**

* Prefix the word conglomerate, breccia or diamictite with the clast size designation (e.g., pebble conglomerate; boulder breccia).

** Where the rock is known to be of glacial origin, the term *tillite* is substituted for diamictite.

Figure 9.19
Classification of coarse-grained sedimentary rocks. Oligomict means clasts are all of the same kind of rock or mineral, whereas polymict means that clasts are of several different types.
(*Source:* Raymond, 2007, p. 280.)

into six types (figure 9.22 on p. 192). The classification parameters are the presence or absence of laminations and the relative amounts of silt and clay. Non-laminated rocks include *siltstone, mudstone,* and *claystone*—names dependent on whether the rocks are more than two-thirds silt, have subequal amounts of clay and silt, or are greater than two-thirds clay, respectively. The laminated mudrocks are *laminated siltstone, mudshale,* and *clayshale,* based on the same three relative amounts of silt and clay (color plate 26d). The general name "shale" is used for nonspecific designation of fine-grained, laminated siliciclastic rocks in place of the general term "mudrock" or any of the specific mudrock names.

The precipitated, non-carbonate rocks are classified on the basis of the dominant chemical component (table 9.7 on p. 193). These rocks include **cherts**, diatomites, and siliceous sinter (dominantly silica minerals—color plates 29a, b); **evaporites** (dominantly evaporite minerals such as halite, sylvite, gypsum, anhydrite, or colemanite—color plate 28c); the *ironstones* and *banded iron formations* (BIFs); the phosphorous-rich rocks (*phosphorite* and phosphatic rock); and the manganese-rich rocks (e.g., *manganolite*).

Coal is a rock in the broad sense of the term (color plate 28d). It is composed mostly of organic fragments of plants (clasts) called *macerals,* the organic equivalent of minerals (Stopes, 1935, in Boggs, 1992).[11] In addition to macerals, coals can include up to 50% siliciclastic grains of quartz, clays, and other minerals. Macerals begin as plant and fungal clasts—such as pieces of wood, bark, resin, leaves, and spores—and are altered by processes of diagenesis to become the maceral types *vitrinite, inertinite,* or *liptinite.* With continued diagenesis, these materials become a fine, dark, insoluble powder called *kerogen.*

Sedimentary Rocks 191

(a) Polymict conglomerate, German Rancho Formation, Stillwater Cove Regional Park, CA. Scale is approximately 15 cm.

(b) Polymict cobble-boulder conglomerate, Jenner Headlands, CA.

(c) Chert breccia, Melange of Blue Rock Springs, Franciscan Complex, Carbona Quadrangle, CA. Pencil of ~15 cm provides scale.

Figure 9.20
Photos of conglomerate and breccia.
(Photos by Loren A. Raymond.)

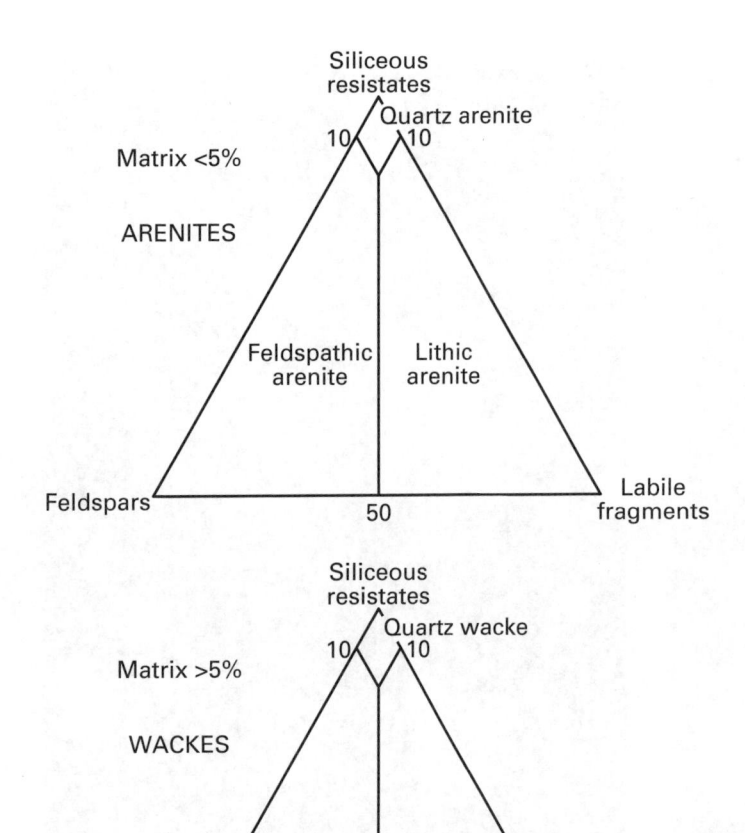

Figure 9.21
Gilbert-Dott classification of sandstones.
(*Source:* Raymond, 2007, p. 282, based on Dott, 1964.)

Figure 9.22
Classification of mudrocks.
(*Source:* Raymond, 2007, p. 284.)

	\\\\\\	MUDROCKS Rocks containing >50% mud			Rocks with <50% mud
		Silt dominant (>67% mud)	Clay and silt	Clay dominant (>67% mud)	Sand-sized or larger grains dominant
	Nonlaminated	Siltstone	Mudstone	Claystone	Conglomerates, breccias, diamictites, and sandstones
	Laminated	Laminated siltstone	Mudshale	Clayshale	

Table 9.7 Key to Common Precipitated Sedimentary Rocks

	Composition	Rock Description	Rock Name
Softer than 5 ½	calcite	aphanitic to phaneritic sedimentary rock composed dominantly of calcite (effervesces readily in HCl).	1. Limestone
	calcite	aphanitic to phaneritic, layered rock, usually light colored and concretionary, deposited by ground or surface waters	2. Travertine
	calcite	cellular travertine (i.e., with holes)	3. Tufa
	calcite	white, chalky travertine-like rock that forms in soil layers	4. Caliche
	dolomite	aphanitic to phaneritic sedimentary rock composed dominantly of dolomite (effervesces slowly or where powdered in HCl)	5. Dolostone
	clay	aphanitic; white, brown, to black; commonly musty smell	6. Mudrocks
	halite	aphanitic to phaneritic, usually light colored rock; salty taste, hardness = 2½	7. Rock Salt (Evaporite)
	gypsum	aphanitic to phaneritic, commonly layered rock with a hardness of 2	8. Gypsum Evaporite
	anhydrite	aphanitic to phaneritic rock with a hardness of 3 – 3½, found in association with other evaporites	9. Anhydrite Evaporite
	apatite or other phosphate minerals	typically aphanitic, variously colored, but typically brown to black rock with more than 50% phosphate minerals; hardness = 5	10. Phosphorites
	organic material	aphanitic, brown to black, typically light weight rock with local fossils of vegetation	11. Coal
	iron oxides	typically aphanitic; yellow, orange, red-brown, or brown; iron oxide-rich rock	12. Ironstone
H > 5½	iron oxides + silica minerals	aphanitic; yellow to red-brown, brown, or green with abundant iron oxides; "cherty"	13. Iron-formation
	silica minerals	aphanitic, dull to waxy, any color or multi-colored rock, locally containing tiny, spherical (radiolarian) fossils	14. Chert

Source: Raymond (2009).

THE STRATIGRAPHIES OF SEDIMENTARY ENVIRONMENTS

Each sedimentary environment is a setting in which several processes operate over time to produce a **lithofacies**, a body of sedimentary rocks characterized by particular rock types, textures, and structures. The study of sedimentary rocks has the advantage that we can witness most sedimentary environments somewhere on the Earth today and can observe the processes that give rise to the lithofacies. The present serves as the key to the past.

In studying the rock record, we are able to apply our observation of present sedimentary environments and their distinctive features to develop an understanding of past sedimentary environments. One of the important structural features evaluated in this kind of analysis is the detailed **stratigraphy**; that is, the detailed description of the rock layering and associated characteristics. For example, sections of rock may consist of alternating beds of packstone and mudshale; fining-up bedding sequences of conglomerate, sandstone, and mudshale; or bed groups of mudrock, coal, and fossiliferous sandstone. Each of these sequences of layers is indicative of a particular environment.

Continental Environments and Rock Associations

Continental environments of deposition are those settings in which sedimentary basins, the general areas of subsidence in which there is a net accumulation of sediment, occur within a continent.[12] Continental environments include various types of river (fluvial) basins; glacial settings; valleys and troughs where landslides are preserved; alluvial fan settings in valleys and along mountain flanks; the desert environments of alluvial fan, erg, and dune field types; lake basins; caves; and deltaic environments within lakes and rivers. Fluvial, lacustrine (lake), and glacial environments yield the largest sedimentary record of the continents.

Along meandering rivers, the stratigraphy that is formed is characterized by fining-up sequences of layers (cycles) composed of conglomerate, arenite and wacke sandstones, siltstone, and other mudrocks, with local areas of coal (figures 9.23a; 9.24 on p. 196; 9.25 on p. 197; table 9.1). Channels are particularly distinctive features and the sandstones filling the channels are cross-bedded (figure 9.25). Any fossils that might exist are those of land animals and plants. The well-known dinosaur fossil locality of Dinosaur National Park represents such a setting, a fluvial sand bar decorated with dinosaur bones.

Lacustrine stratigraphies are characterized by very tabular layers of thin-bedded and laminated mudrocks. Rocks that may be interlayered or associated in interfingering adjacent stratigraphies include sandstones, conglomerates, evaporites, carbonate precipitates, and coal. In many lacustrine stratigraphies, the most distinctive feature is the repetition of dark and light layers called **varves**, formed in response to alternating seasonal changes in sedimentation. The light layers represent spring and summer siliciclastic sedimentation, whereas the dark layers mark the fall and winter accumulation of organic debris produced by cold-weather induced death or organisms.

Glacial stratigraphies are marked by thick beds of diamictite (glacial tillite) (figures 9.23c; 9.26 on p. 198). Associated rocks from nearby streams and lakes include fluvial deposits of conglomerate and sandstone with striated (glacially scratched) clasts, and lacustrine varved mudrocks with dropstones (isolated pieces of gravel dropped from melting ice).

Desert environments have diverse stratigraphies. The rocks range from alluvial fan conglomerates (fanglomerates) and sandstones with coarsening-up sequences, and playa lake mudrocks with interlayered evaporites, to erg (sand-sheet) laminated quartz arenites and sand dune quartz arenites with striking cross-bedding (figures 9.23b; 9.27 on p. 199).

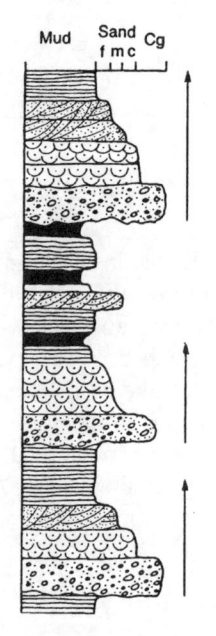

(a) Fluvial. Arrows show fining up sequences.

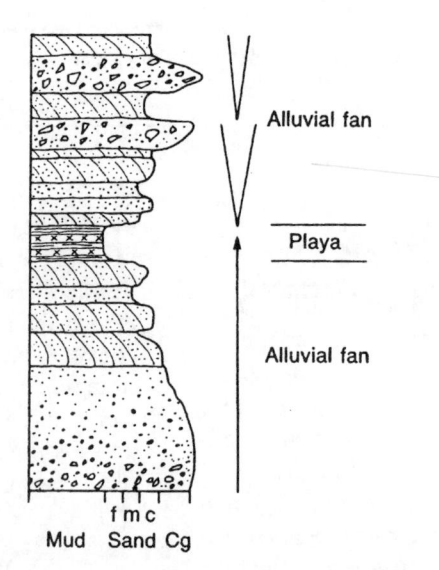

(b) Desert, with alluvial fan, playa lake, and dune. Arrow shows fining up sequence. Diverging lines show coarsening and bed thickening upward sequences.

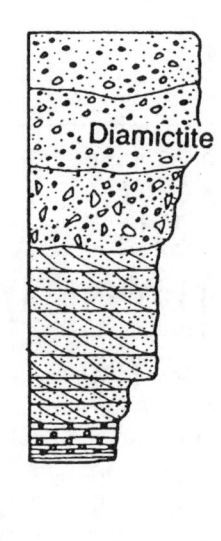

(c) Glacial.

Figure 9.23
Characteristic stratigraphies of some sedimentary environments.

Sedimentary Rocks   195

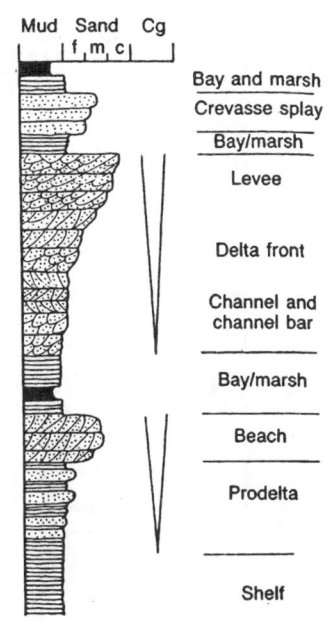

(d) Deltaic. Diverging lines as in (b).

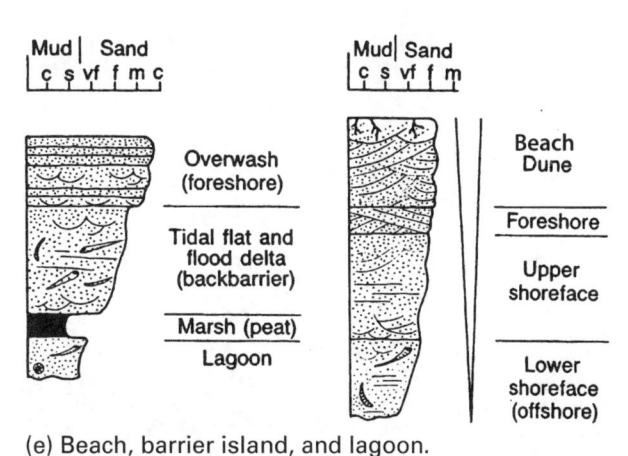

(e) Beach, barrier island, and lagoon.

Sources: Modified from Raymond, 2007, Ch. 15. Sources include: (a) Raymond, 2007, p. 318, based in part on Selley, 1976 and Walker and Cant, 1984; (b) Raymond, 2007, p. 319; (c) Raymond, 2007, p 340; (d) Raymond 2007, p. 322; (e) Raymond, 2007, p. 324, after Heron et al., 1984; (f) Raymond, 2007, p. 328, modified from Galloway and Hobday, 1983; (g) Raymond, 2007, p. 328, modified from McKinney and Gault, 1980; (h) Raymond, 2007, p. 331, after Mutti and Ricci-Lucchi, 1972.

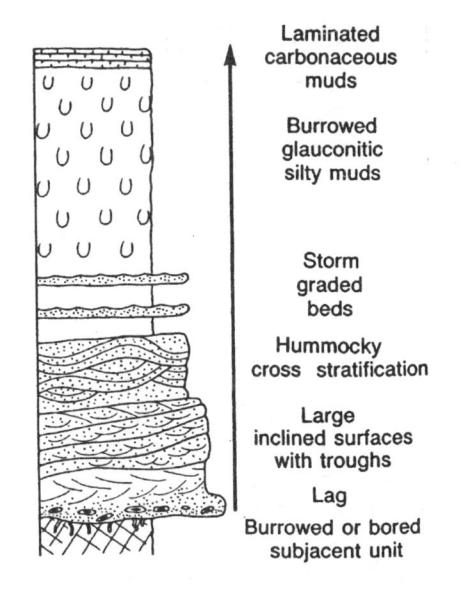

(f) Storm-dominated, clastic shelf with humocky cross-stratification.

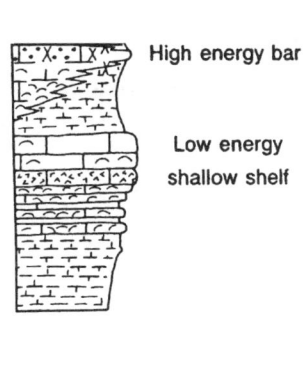

(g) Low-energy carbonate shelf.

(h) Submarine fan.

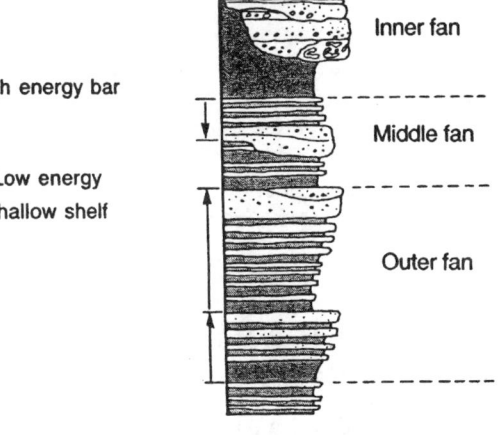

Figure 9.23 *(cont'd.)*

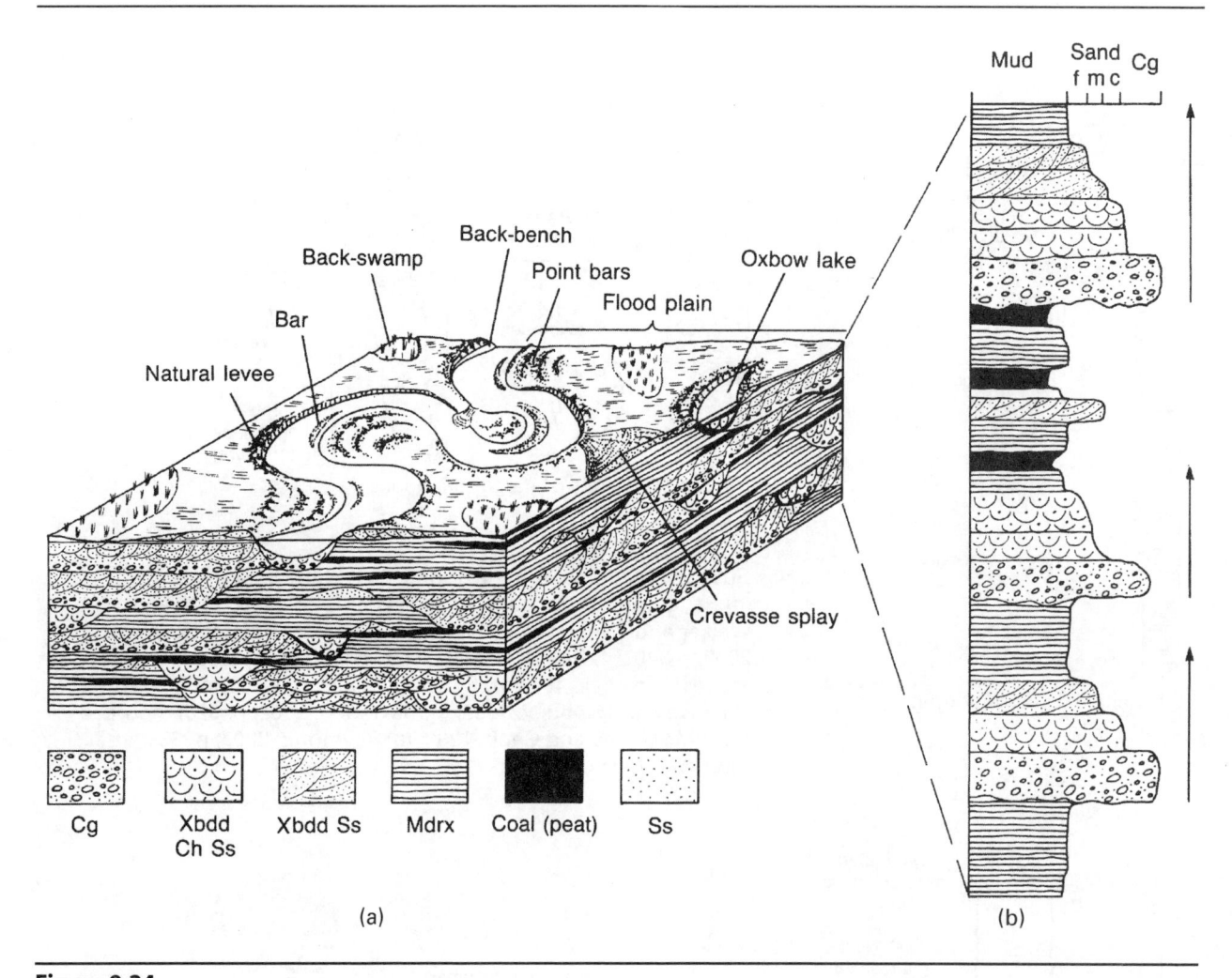

Figure 9.24
A three-dimensional view of a meandering stream environment (a) and the stratigraphy typical of that environment (b). Cg = conglomerate; Xbdd Ch Ss = cross-bedded channel sandstone; Xbdd Ss = cross-bedded sandstone; Mdrx = mudrocks; Ss = sandstone.
(*Source:* Raymond, 2007, p. 318; based in part on Selley, 1976 and Walker and Cant, 1984.)

Transitional Environments and Rock Associations

Rocks formed in transitional environments have been affected by a variety of agents and processes. The rocks of littoral (shoreline) environments are influenced by the actions of waves and wind, as well as by marine organisms. Cross-bedded quartz arenites with low-angle crossbeds and marine fossils, including burrows, distinguish beach stratigraphies (figure 9.23e). In subtropical to tropical climates, carbonate rocks, such as grainstones and whole fossil carbonate conglomerates, may mark the beach and associated storm channels.

The beach stratigraphies differ significantly from lagoonal and estuarine stratigraphies. In the *lagoonal* and *estuarine settings*, quiet waters that are periodically freshened by runoff from precipitation and salinated by encroaching seawater have a limited biotic (fossil) element, but contain local accumulations of coal within stratigraphies dominated by the mudrocks that form in this setting (figure 9.23d, e). The tidal-fluvial setting, however, has complexities of sedimentation dependent on freshwater-saltwater mixing and chang-

Figure 9.25
Photograph of fluvial stratigraphy, showing a 4-meter thick channel with cross-bedding (light lens in lower center) cutting into mudrock overbank deposits (thin, dark and light layers), Mississippian Hinton Formation, near Princeton, West Virginia.
(Photo by Loren A. Raymond.)

ing flows between seasons and, hence, locally contains fine to coarse sand(stone) (Dashtgard et al., 2012). *Tidal flats* adjoining beaches and lagoons are marked by laminated, lenticular, or bioturbated beds of mudrock, packstone, lime mudstone, stromatolitic limestone, and local flat-pebble conglomerate and evaporites. Mudcracks characterize this environment. Estuarine mudrocks are interlayered with local cross-bedded sandstones deposited by flood waters that flow through the estuary. Quartz arenites of the beaches and fluvial deltaic deposits of sandstone may interfinger with mudrocks and carbonaceous mudrocks deposited in lagoons. Where fossils are present, transitional environment rocks may show local accumulations of marine animal, continental animal, freshwater animal, and continental land-plant assemblages that represent the fluctuating marine and continental conditions.

Marine Environments and Rock Associations

Marine environments are responsible for the formation of some very large masses of sedimentary rock. These environments include shallow sea-shelf, including glacimarine, and slope-rise environments along the continental edges; reef environments; submarine canyon, forearc basin, and trench environments; and basin plain environments.

The **shelf environment** extends seaward from the littoral/beach environments out to a point, called the *shelf break*, where the submarine slope steepens and

Figure 9.26
Photograph of Precambrian (Neoproterozoic) tillite of the Konnarock Formation, Highway 58, Konnarock, Virginia.
(Photo by Loren A. Raymond.)

slopes down towards the deep seafloor. Shallow seas, such as the northern Adriatic Sea located between Italy and Croatia, in essence are partially landlocked shelf areas that have similar depth and depositional conditions to the shelves (McKinney, 2007). The wide variations in rock formation on the shelves result from differences in climate, biological activity, wave and current action, sediment input, tectonic setting, and sea-level changes (*highstand* vs. *lowstand* of sea-level).[13] Typical sediment sequences are composed of some combination of shales, siltstones, sandstones, limestones, and dolostones (figure 9.23f, 9.23g; figure 9.28 on p. 200). Bedding ranges from parallel-tabular types to lenticular, hummocky beds with cross-stratification (tempestites) formed by storms (figure 9.23f). In warm climates, skeletal wackestones and packstones locally interbedded with shales are common. Cool climate settings may also yield carbonate rocks, but some clearly have complex histories resulting from changing climate and sea levels (James, 1997; Rivers et al., 2007). In settings in which siliciclastic sedimentation dominates, laminated beds of siltstone and shales are common on the distal shelves, whereas closer to land, storm-deposited tabular interlayers of sandstone are more abundant. Still nearer the shoreline, burrowed, silty mudrocks with lenticular to tabular interlayers of graded to cross-laminated sandstone tempestites and sandbar and sandwave sandstones are characteristic (e.g., Leeder, 1999, p. 269). On glaciated coastlines, glacial sedimentation on the shelf produces tillites, varved mudrocks with dropstones, and turbidites, the distinct bedded sequences with graded sandstones at the base deposited by sediment-loaded turbidity currents (Powell and Molnia, 1989).

Along tropical to subtropical shelves, both reefs and carbonate buildups develop as a result of rapid

Figure 9.27
Large-scale Aeolian cross-bedding in sand-dune deposits of the Jurassic Navajo Sandstone, Checkerboard Mesa, Zion National Park, Utah. The individual at the lower right, providing scale, is about 1.7 meters tall. The vertical lines are joints (fractures) in the rock.

organic sedimentation. *Reef* environments can be subdivided into subenvironments.[14] The *reef front*, at and below wave base, is composed of boundstone. The *reef crest* is marked by skeletal and oolitic grainstone and crystalline limestone, whereas the *reef flat* behind the reef, a quiet water-to-surf zone, is characterized by laminated packstones, grainstones, and carbonate conglomerates. Where reefs form lagoons, the lagoonal environment is marked by laminated stromatolitic lime mudstones and wackestones, with local bioturbation. Seaward of the reef front is the *forereef*, a slope marked by carbonate breccias and conglomerates, packstones, and grainstones.

Seaward of the shelf are both the *slope* environment, representing the steep slope from the shelf into deeper water, and the *rise*, a zone of gentler gradient at the base of the slope, leading into the abyssal plain. Slope basins form where folding and faulting modify the slope. The sedimentary rocks that form in these environments are derived from (1) thinly laminated muds and lime muds that form from sediment rain, (2) cross-bedded sand deposited by contour currents (that flow along the contour of the slope rather than downslope), (3) Bouma sequence sandstones and related rocks deposited by turbidity currents, and (4) submarine landslide deposits (**olistostromes**) formed by slope failures and the resulting debris flows (figure 9.6; figure 9.29). On tectonically active subduction margins, slope basins are a normal feature of the slope. A trench is commonly present at the base of the slope (figure 9.6a).

Submarine canyons cut the slope and rise and serve as conduits for sediment, which flows through the canyons in turbidity currents, grain flows, and

Figure 9.28
Shelf limestone-shale sequence, I-470 and Raytown Road, Kansas City, Missouri.
(*Source:* From Raymond, 2007.)

Figure 9.29
Olistostrome (= submarine landslide deposit) in the Great Valley Group, Cretaceous Venado Formation, Monticello Dam, Lake Berryessa, California. Matt Raymond ~1.3 m tall, provides scale.
(Photo by Loren A. Raymond.)

debris flows. In this canyon setting, the rocks that form are commonly olistrostomal diamictites; massive to thick-bedded, coarse- to medium-grained, locally pebbly wackes and arenites; medium-bedded to laminated, rippled and convolute bedded wackes and siltstones; cross-bedded arenites and wackes; and mudrocks. Rocks with Bouma sequences of sandstone and mudrock—initially deposited as sand and mud by turbidity currents—occur here as well. Yet, complexities of sediment sources and environmental setting mean that a number of variations in composition are possible (Bouma and Stone, 2000; Lomas and Joseph, 2004). Similar rocks occur in submarine fan deposits, formed where the canyons meet the trench, abyssal plain, or the trench slope basin (figure 9.6a).

On the deep ocean floor or *abyssal plain*, various muds, clays, and oozes are deposited. These sediments are lithified into mudrocks, lime mudstones, and cherts. Locally, where turbidity currents carry sediment far out onto the abyssal plains, turbidite sandstones of the Td type will form.

Problem 9.2

Examine the column in figure 9.23h. What might cause the rocks to change up section from outer to inner fan?

DIAGENESIS AND LITHIFICATION OF SEDIMENTS

The sediment transported and deposited in basins is typically changed to rock as it is *lithified* (turned into stone) via processes of diagenesis. As discussed earlier, diagenesis is a collective term for all of the post-depositional processes—physical, chemical, and biological—that result in the transformation of sediment into rock. (Students should be aware that diagenetic processes may also affect igneous and, less commonly, metamorphic rocks by causing changes in textures and mineral and rock compositions.)

There are seven diagenetic processes (e.g., Folk, 1974; Raymond, 2007):

1. compaction
2. recrystallization
3. cementation
4. (dis)solution
5. authigenesis (neocrystallization)
6. replacement
7. bioturbation

Each of these processes may or may not affect a sediment depending on the nature of the sediment, the pressure, the temperature, biological activity, the nature of the fluids present, and the amount of fluid flow through the rock.[15]

Compaction is a process in which pressure on the sediment reduces the pore space volume between the grains of the rock. Typical porosities in sediments are between 25 and 50%, and may range as high as 85%.[16] Pore spaces may be occupied by fluid or gas, and compaction drives these materials out of the pores as the pore volume is reduced. Sedimentary rocks may have as little as 0 to 2% pore space.

Recrystallization occurs where particular pressure, temperature, and fluid phase conditions at locations within a rock or sediment cause reorganization of the crystal lattices of minerals present there. The impetus for change is instability resulting from conditions or chemistry different from those existing in the rock, and the recrystallization represents a reaction within the rock toward a condition of lower free energy. The reorganization will typically result in the development of interlocking grains, such as those characteristic of equigranular-sutured textures (figure 9.7d).

Solution (dissolution), the dissolving of minerals, occurs as a result of instabilities resulting from chemical conditions within the rock versus those in its surroundings. Minerals may dissolve at the mineral-fluid interface, because they are no longer stable in the rock, when external or *extrinsic variables* such as P,T, or fluid composition and character change (e.g., Eh and pH changes). The rates, microscopic locations, and controls of dissolution reactions are not entirely understood but are likely partly controlled by *intrinsic variables*, such as structural irregularities *within* the mineral crystals (Luttge et al., 2013). The ions that break free from the surface will move away in the intergranular fluid. Solution can actually create more porosity (called *secondary porosity*). A particularly important type of solution is called *pressure solution*. High pressures that develop at the points of grain contact create elevated ΔG, and the resulting solution at that point of contact reduces the ΔG. The ions produced by dissolution typically move just a short distance away into adjoining pores, where they re-precipitate as cement, under the lower pressure conditions there.

Cementation is the process by which minerals precipitate within the pores of the rock and bind the grains together (figure 9.7e). Pressure solution may result in re-precipitation of a mineral in pore spaces around that mineral. Commonly, however, minerals unstable at one place in a body of rock may be stable at another, and as the fluids carrying ions of those minerals move through the rock, the minerals that are stable at any given point will precipitate. Typically precipitation occurs in the pores, starting on the grain surfaces surrounding the pore spaces. The most common cements are quartz, calcite, and hematite, but several other minerals such as goethite, gypsum, and dolomite also precipitate as cements.

Authigenesis (also known as neocrystallization) is the process of new mineral formation. A fluid phase containing various ions, passing through sediment, will commonly promote chemical reactions within and among minerals, resulting in the formation of new minerals. For example, fluids of low Eh containing sulfur may extract iron from iron-bearing minerals and precipitate pyrite. The presence of a fluid usually means that water is available, and minerals such as the feldspars may react with that water to produce zeolite minerals. Clearly, the specifics of authigenesis are dependent upon the nature of the fluid phase, the compositions and structures of the minerals that are present, the presence of microbes that may mediate reactions, and the P,T, pH, and Eh condition in the sediments—all the extrinsic variables that influence solution and re-precipitation.

Whereas authigeneic reactions yield new minerals in voids and elsewhere in a sediment, **replacement** is a process by which a mineral is simply changed, in place, to another mineral. As was the case with authigenesis, fluids are important, because they carry reactant ions from place to place. In the case of a sulfur-bearing fluid described in the preceding paragraph, the sulfur-bearing fluid may well react directly with the iron-bearing mineral (i.e., react in place), so that the iron-bearing mineral is changed into (replaced by) the pyrite. In a like manner, during weathering, limonite pseudomorphs after (in the shape of) pyrite form where limonite replaces pyrite.

Finally, *bioturbation* may facilitate diagenesis. Recall that bioturbation refers to the disruption of the original sediment by burrowing organisms. Organisms may mix sediment of different types as they bore into the layers, causing diagenetic reactions to occur, and may create organic compounds that facilitate changes. Bioturbation also can facilitate compaction.

Together, the processes of diagenesis transform sediment into rock. Diagenesis can occur *at the surface* soon after deposition, where the process is referred to as *eogenesis*; it may occur *after significant burial* during *mesogenesis*; or it may transform the sediment (or rock) after later uplift and exposure at the surface in a process referred to as *telogenesis*.

The rock dolomite or dolostone is very common in many stratigraphic sections, yet we see little dolostone forming in modern environments. As a result, the origin of dolostone is vigorously discussed. The origin of most dolostone is attributed to diagenetic reactions. A number of hypotheses for the formation of dolostones via such reactions have been proposed. In general, these hypotheses involve a reaction between a Mg-bearing fluid and the calcite present in limestone. As the fluid passes through sediment, a dolomite forming reaction

$$\text{calcite} + Mg^{++} \rightarrow \text{dolomite} + Ca^{++}$$

takes place, and grain by grain the limestone is replaced by dolomite to become a dolostone. The source of the Mg in the fluid is likely a form of calcite called *high-Mg calcite*, a mineral common in biogenic carbonate sediments. This dolomite-forming reaction may be induced by the interaction of fresh and salt water, by changes in P and T as the sediment is buried, by the mixing of methane with the sediment, or simply by the reaction of Mg ion with Mg-poor calcite in warm, evaporative environments along coastal plains (i.e., in sabkhas).

Ultimately, the sedimentary rocks we see across the surface of Earth have likely resulted from several of the diagenetic processes acting on initially deposited sediments. The resulting rocks preserve a modified record of their earlier history and a newer record of their diagenesis.

Problem 9.3

Examine figure 9.7e. A large rounded surface on the (white) quartz grain near the upper center of the photomicrograph has a triangular patch of calcite (gray with fine, straight lines) attached to its lower left side. Large areas of calcite also occur in the left side of the photo. Examine the various rounded quartz grains, their relationships to the gray calcite cement and the (black) hematite cement, and explain the sequence of events that occurred during diagenesis of this sandstone. Compare your answer to the sequence in figure 11 in Raymond et al. (2014). Are your conclusions consistent with Raymond et al.? Why or why not?

Summary

Sedimentary rocks cover about 66% of the surface of Earth, and they form via processes visible at the surface. Soils, formed both by weathering of preexisting rock and by reactions and organic activity that produce additional materials, are eroded by the aqueous currents of streams, by glaciers, by wind, and by gravity. Eroded, transported, and deposited materials are sediments. Some sediments are moved invisibly as materials in solution, whereas others are suspended or are moved by bed-load transport. The same agents that erode sediments are the agents that transport and deposit them. Characteristics of the deposited sediments reveal the depositional agent and transportation mechanism; and those characteristics are, in part, controlled by provenance of sediment, climate of the provenance, distance of transportation, and the depositional setting.

Sedimentary environments in the lithosphere include marine, transitional, and continental environments. Sediments and the rocks derived from them that are formed in these various environments may be divided into *lithofacies* characterized by various textures, structures, and compositions. In contrast, sediment *biofacies* are characterized by assemblages of fossils. Thus, in the rock record, particular lithofacies and sequences of sedimentary layers in rocks reveal the sedimentary environment of deposition, as do distinctive fossil assemblages. A wide range of structures and textures are used to identify lithofacies. Sedimentary rock compositions also vary widely, from extremely high-silica types to those entirely lacking in silica. Through processes of diagenesis, sediments are lithified (i.e., converted into sedimentary rocks). The processes of diagenesis include compaction, dissolution, cementation, recrystallization, authigenesis (neocrystallization), replacement, and bioturbation. Like all rocks, the resulting sedimentary rocks are classified on the basis of their textures and compositions.

Selected References

Boggs, S., Jr. 2009. *Petrology of Sedimentary Rocks* (2nd ed.). Cambridge University Press, New York. 609 p.

James, N. P. 1997. The cool-water carbonate depositional realm. In N. P. James and J. L. Clarke (eds.), *Cool-water Carbonates*. Society of Economic Mineralogists and Paleontologists Special Publication 56, pp. 1–20.

Lomas, S. A., and Joseph, P. (eds.). 2004. *Confined Turbidite Systems*. The Geological Society of London, Special Publication 222. 328 p.

Pettijohn, F. J., Potter, P. E., and Siever, R. 1987. *Sand and Sandstone* (2nd ed.). Springer-Verlag, New York. 553 p.

Raymond, L. A. 2007. *Petrology: The Study of Igneous, Sedimentary, and Metamorphic Rocks* (2nd ed.). Waveland Press, Long Grove, IL. 720 p.

Scholle, P. A., Bebout, D. G., and Moore, C. H. (eds.). 1983. *Carbonate Depositional Environments*. American Association of Petroleum Geologists (AAPG), Memoir 33. 708 p.

Walker, R. G., and James, N. P. (eds.). 1992. *Facies Models: Response to Sea Level Change*. Geological Association of Canada, St. Johns, Newfoundland. 409 p.

Notes

1. Ehlers and Blatt (1982, p. 6).
2. Also see http://www.miami.edu/index.php/news/releases/impact_of_african_dust_storms/; http://www.worldwatch.org/node/514; and Koren et al. (2006).
3. See Krauskopf and Bird (1995) and Garrels and Christ (1965) for more detailed discussions on mineral stabilities in solutions.
4. Raymond (2007, p. 316).
5. Students often think, on the basis of life experiences, that the word *marine* refers to any body of water (after all, the marina is commonly at the lake). In scientific parlance, however, marine refers only to the large bodies referred to as oceans and seas. Reading (1996) and Raymond (2007, ch. 15) discuss sedimentary environments.
6. These terms are discussed in Raymond (2007, ch. 15) and Reading (1996).
7. For example, Cowan and Powell (1990), Cai et al. (1997), Curran et al. (2004).
8. Bouma et al. (1982).
9. The sedimentary rocks are subdivided in different ways. In this book, we use a threefold subdivision. Some introductory books divide the rocks into just two types, clastic and chemical, whereas others (e.g., Plummer et al., 2007) and some sedimentary geology books (e.g., Prothero and Schwab, 2004) use three or four categories similar to those used here, but designated by names such as Detrital, Organic, Chemical or Siliciclastic, Biogenic, and Chemical (plus Other). Blatt et al. (2006) note that more than 95% of sedimentary rocks are mudrocks, sandstones, and carbonate rocks, and so designate them, with an additional category of Other.

10. Scholle (1979) illustrates many of these classifications and provides references to them. Also see Brewer et al. (1990).
11. See Forsman and Hunt (1958), Degens (1965), and Boggs (1992), the sources for this summary, for more thorough discussions of organic sediments.
12. For more information on stratigraphies of continents, see chapters in Reading (1996).
13. For more information on controls and stratigraphies of shelves, see Bouma et al. (1982), Walker (1984), Johnson and Baldwin (1986), Wright and Burchette (1996), and Leeder (1999, chs. 24 and 25), on which this section is based.
14. For additional information, see the references on which this section is based, including J. L. Wilson (1974), Longman (1981), Enos and Moore (1983), James (1983, 1984), Ausich and Meyer (1990), and Wright and Burchette (1996).
15. Longman (1982), Scoffin (1987), Raymond (2007), and Cuadros et al. (2013).
16. Choquette and Pray (1970), Pryor (1973), Choquette and James (1987).

Metamorphic Rocks

10

INTRODUCTION

Metamorphic rocks are important components of mountain belts and continental cores. They are rocks that have been changed from a preexisting rock, a **protolith** (literally, first stone) by the action of metamorphic processes that produce changes in the textures, mineral assemblages, and chemical compositions of the rocks.

In this chapter, we examine the nature of metamorphic rocks, the processes of metamorphism, and the products of those processes. Like igneous rocks, the metamorphic rocks form in accordance with thermodynamic laws, and their minerals reflect the controls of composition, pressure, and temperature. Metamorphic textures result from various processes of recrystallization.

Of particular interest to metamorphic petrologists in recent years has been the discovery and analysis of ultra-high-pressure metamorphic rocks that in some cases contain diamond or the high-pressure polymorph of SiO_2, coesite. These rocks were formed very deep in the Earth and have been carried back to the surface relatively quickly without being changed to rocks more stable at the surface. Understanding how these and other rocks form, how fast they form, and the role of fluids in metamorphic rock formation are among the subjects challenging those who study these rocks.

METAMORPHISM AND METAMORPHIC PROCESSES

Metamorphic rocks have been changed (meta) in form (morph). **Metamorphism** may be defined as changes in the texture, the chemical composition, the minerals, or some combination of these characteristics of a rock in response to changes in the conditions, such as the pressure and temperature, under which the rock exists. These metamorphic processes include recrystallization, neocrystallization, fluid-driven chemical change or *metasomatism*, cataclasis, and minor localized melting. **Cataclasis**, which is the crushing and breaking of grains in rocks, is largely confined to fault zones. In contrast, **recrystallization**—a process of textural change without the production of new minerals, and **neocrystallization**—the production of new minerals in a rock via geochemical reactions, are both widespread. Neocrystallization may be accompanied by **metasomatism**, a process in which the composition of a rock is altered by fluids that bring in or remove elements from the rock. Some chemical change, notably the gain or loss of modest amounts of water, is common during metamorphism; and such changes in the water content of rocks are not generally designated as metasomatic. Large additions of water, however, or major dehydration of rocks, accompanied by gain or loss of other chemical constituents, do represent the process of metasomatism. *Melting* occurs at high temperatures, where the transition from igneous to metamorphic rock occurs, and melting is common in certain settings, such as in deep mountain roots, where the rocks called migmatites may form. Such rocks are generally assigned to the metamorphic category, although technically they contain both igneous and metamorphic components. Melting also occurs locally along fault zones, but here it is relatively rare.

Metamorphic reactions are brought about by the operation of four agents: *temperature*, *pressure* (uniform stress), directed force per unit area (F/A) or *deviatoric stress*, and *chemically active fluids*. *Temperature* (T) generally increases with depth. In volcanically active regions, the temperature increase may be 50°C per kilometer or more. In subduction zones, along the subducting plate, the temperature increase is rather low and is generally on the order of 5–10°C/km. A somewhat average value for increase in temperature beneath the continents is 20°C/km. *Pressure* (P) (uniform force/unit area) increases with depth in the crust at a rate of about 0.1 GPa (1 kilobar) for every 3.3 km of burial. Thus, for every ten kilometers of burial, the pressure increases by 0.3 GPa (3 kilobars). Such pressures are referred to as P-load. Fluids also generate pressures, which are referred to as P-fluid, P_{H2O}, or P_{CO2}, whichever is appropriate. Metamorphism that generally involves an increase in P and T (i.e., a change in conditions *away* from 0°C and 0 GPa) is called *prograde metamorphism*. *Retrograde metamorphism* involves movement toward zero P,T conditions. These terms have no direct relationship to *deviatoric stress*, the values of which vary widely. Both tensional and compressional deviatoric stresses can produce textural changes in rocks. **Chemically active fluids** are fluids that are activated by moving from regions in which they have thermodynamic *stability* (including chemical stability) into areas where this condition is different. Fluids are dominantly water-based (H_2O-based), but CO_2-rich and S-bearing fluids are also relatively common in metamorphic rocks.

Each of the agents of metamorphism may dominate the metamorphic process in particular places. Metamorphism controlled by a specific agent or agents is given a name based on the dominant agent(s) of metamorphism (table 10.1). Hence, where addition of heat dominates the metamorphism (i.e., where increases in temperature dominate), the metamorphism is called **contact metamorphism** because it occurs at the contacts between igneous intrusions and the surrounding rocks, which are called *country rocks*. Metamorphism in which uniform pressures dominate is called **static metamorphism** or **burial metamorphism**, the latter because it typically occurs in deeply buried rocks. Where deviatoric stresses dominate, the metamorphism is referred to as **dynamic metamorphism**. Chemically active fluids dominate in **metasomatism**. In mountain belts, **dynamothermal metamorphism**—metamorphism in which P,T, deviatoric stress, and fluids are all important—affects entire regions and is sometimes called *regional metamorphism*.

As we study mountain belts and other metamorphic settings, we study metamorphism of the crust. Occasionally, within the crust we also find mantle rocks that have been faulted up and returned to crustal conditions. The conditions that prevail in the crust include pressures in the range of 0.1 GPa to 1.0 GPa (1–10 kilobars) and temperatures that typically fall in the range of 100°C to 750°C (figure 10.1). Higher and lower pressures and temperatures are known to yield metamorphic rocks under certain conditions. It is particularly important to understand that the chemical composition of a rock has a profound affect on the final mineral assemblage that makes up each metamorphic rock.

Problem 10.1

A long-standing controversy in metamorphic petrology involved the question of whether the origin of serpentine (serpentinite)—the State Rock of California—involved metasomatism. Use the Web to discover the alternative to metasomatic formation of serpentinite, "constant volume" serpentinization, and cite at least one line of evidence for the two alternative processes of serpentinization. Summarize a more modern view of serpentinization.

Problem 10.2

If serpentinization is ongoing today, are there any societal implications of modern-day serpentinization?

Table 10.1 Types of Metamorphism

Dominant Agent	Local Metamorphism	Regional Metamorphism
Temperature (T)	Contact metamorphism	—
Pressure (P)	—	Static (burial) metamorphism
Deviatoric stress (DS)	Local dynamic metamorphism	Regional dynamic metamorphism
Chemically active fluids (CAF)	Local metasomatism and local hydrothermal alteration	Regional metasomatism
T, P, ±DS, ±CAF	Medium to high pressure contact metamorphism	Dynamothermal (regional) metamorphism

Source: Modified from Raymond (2007, Table 21.1, p. 470).

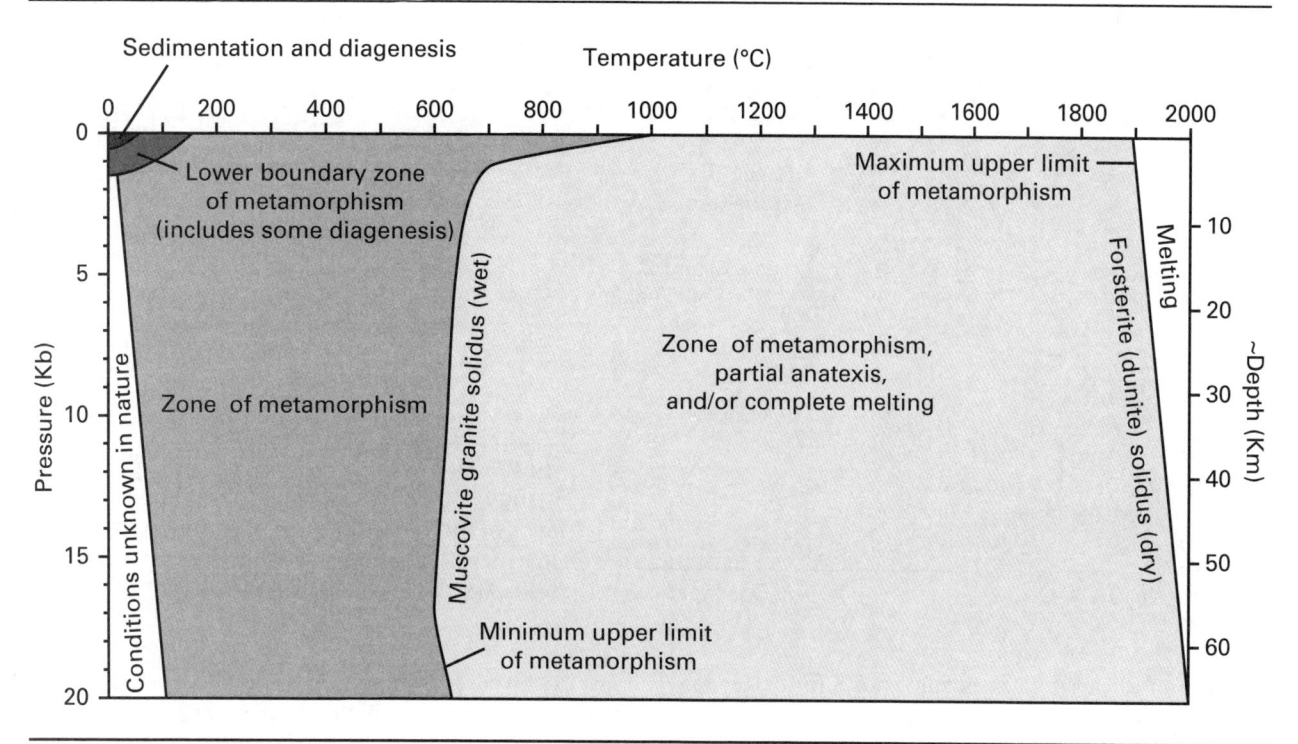

Figure 10.1
P,T grid showing the pressure and temperature limits of metamorphism.
(*Sources:* Raymond, 2007, p. 469, with the forsterite solidus based on Bowen and Anderson, 1914, Davis and England, 1964, and Ohtani and Kumazawa, 1981; and the muscovite granite solidus based on Huang and Wyllie, 1981.)

METAMORPHIC ROCKS AND THEIR CHEMICAL COMPOSITIONS

Over much of the past century, geologists who studied metamorphic rocks commonly focused their studies on particular rocks, such as metamorphosed shales or metamorphosed basalts. Thus, particular rock types or rock compositions were isolated for study. No widely adopted subdivision of all metamorphic rock compositions exists in the metamorphic rock literature and although some attempts have been made to organize compositional categories and names, compositional groups are difficult to define. Naming of rocks is discussed below, but metamorphic rocks are here assigned to six compositional groups (figure 10.2).

The compositional groups of metamorphic rocks encompass the specific rock types and rock compositions discussed in the chapters on igneous and sedimentary rocks and, with some rare exceptions, most others.[1] Each compositional group generally reflects the chemical composition of protoliths of igneous rock, sedimentary rock, or both. *Metabasites* (Group 1) are relatively low-silica, high-iron and magnesia rocks, generally lacking quartz, but containing plagioclase feldspar. Thus, as is the case for all metamorphic rocks, the minerals present reflect the bulk chemistry of the rock (table 10.2 on pp. 209–211 and table 10.3 on p. 212). The metabasitic rocks are derived from the igneous rocks basalt, basaltic andesite, diorite, and gabbro. Rocks quite low in silica but high in iron and magnesia and generally lacking both feldspars and quartz represent the ultrabasic/ultramafic igneous rocks, like dunite or pyroxenite. These are called *metaultrabasic rocks* (Group 2). In contrast, rocks high in alumina and silica, such as mudstones, are typically assigned to the metamorphic compositional group called *metapelites* (Group 3).

Rocks of the three other groups encompass a range of protoliths, from metagranite to metamorphosed limestone and metachert. Metamorphic textbooks generally lack group names for these compositional groups

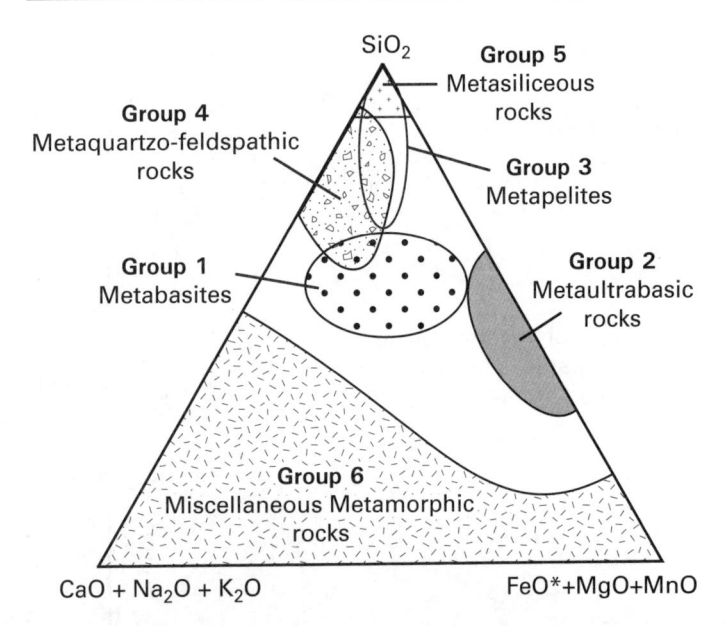

Figure 10.2
Triangular diagram showing the compositional fields of the various groups of common metamorphic rocks. Values are in weight percent.
(*Source:* Modified from Raymond, 2007, p. 492.)

of rocks and instead have specific names for particular types. Thus, metamorphosed limestones and other metamorphosed carbonate rocks are called **marbles**, whereas calcium- and silica-rich rocks at the margins of igneous intrusions are called *scarn* and *hornfels*. The range of rocks containing abundant silica, calcium, and alkalis, such as the volcanic rocks rhyolite and latite, the granitoid igneous rocks (e.g., granite, quartz monzonite, syenite), plus metamorphic rocks of similar composition derived from quartz-feldspar-rich sandstones, are here collectively called the *meta-quartzo-feldspathic rocks* (Group 4). For such rocks, some geologists use the designations feldspar-bearing meta-psammites or calc-silicate rocks, but the latter term has a dual meaning and the former only refers to rocks with sandstone protoliths. The minerals of the meta-quartzo-feldspathic rocks reflect the high-silica, alkali, and calcium content of the rocks and include such types as quartz, alkali feldspars, Na-plagioclase feldspars, and muscovite mica (table 10.2). Group 5 contains very high-silica metamorphic rocks (the *metasiliceous rocks*) that are derived from quartz sandstones, the sedimentary rock chert, and high-silica igneous rocks, such as the rock called quartzolite. The large variety of rocks that are commonly low in silica, typically poor in alumina, but distinctively high in one or two other elements such as calcium, iron, carbon, or phosphorous, are assigned to Group 6, the *miscellaneous metamorphic rocks*. Marbles (color plate 31d) are the most common rocks of this group. Organic rocks such as coal that are metamorphosed to anthracite can be assigned to this group as well. It is important to remember that each of the specific metamorphic rocks has a chemical composition that controls which minerals develop during metamorphism under particular conditions of pressure and temperature, as shown partly in table 10.3.

Metamorphic Structures and Textures

Metamorphic rocks commonly exhibit distinctive structures, several of which are not unique to metamorphic rocks. In the field, structures are particularly helpful in first recognizing rocks as being metamorphic. Like the igneous and sedimentary rocks, however, metamorphic rocks are generally classified into groups on the basis of textures and mineralogy.[2]

Recall again that **structures** are physical features of the rock that are of hand-specimen or larger scale but do not pervade hand-specimen size samples (i.e., they are not repeated over and over in each small part of a hand specimen).[3] Textures, in contrast, are of microscopic to mesoscopic (hand-specimen) scale and generally are pervasive (penetrative)—that is, they are numerous and repeated within small samples. Both metamorphic structures and most textures in metamorphic rocks result from deformation, and together textures and structures form the *fabric* of a metamorphic rock.

Table 10.2 Characteristic Minerals of Metamorphic Rocks

Mineral	Chemical / Structural Group	Distinctive Properties	Commonly Occurs in
Silicate Minerals			
Quartz (SiO_2)	Tectosilicate	Hexagonal; any color; H=7; vitreous; conchoidal fracture; optically positive	Slates, phyllites, schists, gneisses, mylonites, hornfels, granoblastites, skarns, metaquartzites
Potassium Feldspars	Tectosilicate	White, pink, or green; H=6; 2 cleavages @ 90°; Carlsbad twinning; low birefringence	Schists, gneisses, mylonites, granoblastites, metaquartzites skarns, metawackes
Plagioclase Feldspars (Na, Ca feldspar)	Tectosilicate	White to colored gray, green, brown, black (by impurities); H=6; polysynthetic twinning (shows striations)	Phyllites, schists, gneisses, mylonites, hornfels, granoblastites, metabasites, metawackes
Zeolites (e.g. Laumontite, Heulandite)	Tectosilicate	Typically white, grainy masses to tabular crystals	Metawackes, veins in metawackes and metabasites, cavity fillings in metabasites
Muscovite/Aragonite (K-/ Na-White mica)	Phyllosilicate (sheet silicate)	Colorless to white and light green, brown, and yellow colored; one perfect cleavage; H=2.5–3	Slates, phyllites, schists, gneisses, mylonites, metawackes, metaquartzites
Biotite (K, Mg, Fe mica)	Phyllosilicate (sheet silicate)	Brown to black colored; one perfect cleavage; H=2–3	Slates, phyllites, schists, gneisses, mylonites, metawackes
Phlogopite (K, Mg mica)	Phyllosilicate (sheet silicate)	Brown to bronze colored; one perfect cleavage; H=2–3	Marbles, meta-ultrabasic rocks (e.g., meta-peridotite)
Chlorite (Mg-silicate)	Phyllosilicate	Green to brown; one perfect cleavage, H=2–2.5	Metabasites, meta-ultrabasic rocks, slates, phyllites, schists, mylonites, metawackes
Antigorite/Lizardite (Mg-silicates)	Phyllosilicate	White to green (also colored by impurities making it blue, red, orange, or black); H=2.5–4; waxy to dull luster	Meta-ultrabasic rocks, metabasites
Chrysotile (Mg-silicate)	Phyllosilicate	Green; fibrous; silky luster	Meta-ultrabasic rocks
Talc (Mg-silicate)	Phyllosilicate	Greasy feel; H=1; white to gray to green; one perfect cleavage	Meta-ultrabasic rocks, marbles
Tourmalines	Cyclosilicate (ring silicate)	Black (variety Schorl) and various colors including red, green, and yellow (variety Elbaite); H=7; habit = trigonal (3-sided) prisms; marked pleochroism optically	Schists, gneisses
Anthophyllite (Mg-silicate)	Inosilicate (Double chain silicate)	Fibrous to acicular; white to light brown; H= 5.5–6	Meta-ultramafic rocks, schists, gneisses
Tremolite (Ca, Mg-silicate)	Inosilicate (Double chain silicate)	Colorless to white; typically acicular to prismatic; H=6; 2 cleavages @ ~56° & 124°	Marbles; meta-ultramafic rocks, including metadunites; schists
Hornblende/Actinolite (Ca, Fe, Al-amphibole)	Inosilicate (Double chain silicate)	Black (Hbl) to green (Hbl/Act) colored; 2 cleavages @ ~56° & 124°; H=5–6; acicular to rectangular prismatic crystals	Schists, gneisses, metabasites, some meta-ultramafic rocks

(continued)

Table 10.2 Characteristic Minerals of Metamorphic Rocks *(cont'd.)*

Mineral	Chemical / Structural Group	Distinctive Properties	Commonly Occurs in
Riebeckite (Na-rich amphibole)	Inosilicate (Double chain silicate)	Black to dark blue colored; 2 cleavages @ ~56° & 124°; H=6; typically in acicular crystals	Metachert, meta-ironstone
Glaucophane (Na-rich amphibole)	Inosilicate (Double chain silicate)	Light to dark purplish-blue colored; 2 cleavages @ ~56° & 124°; H=5; acicular to rectangular prismatic crystals	Metabasites, schists, metawackes
Enstatite/Hypersthene (Mg-rich pyroxene)	Inosilicate (Single Chain)	Gray, green, brown, or bronze colored; blocky shape with 2 cleavages @ 90°; H=5.5–6	Metagabbros, meta-orthopyroxenite, meta-websterite, meta-peridotite; high-T granoblastites and gneisses
Diopside (Ca, Mg-pyroxene)	Inosilicate (Single Chain)	White to green; blocky crystals; H=~6; 2 cleavages @ 90°	Marbles, meta-ultramafic rocks
Augite	Inosilicate (Single Chain)	Brown, greenish-brown, or black colored; blocky shape with 2 cleavages @ 90°; H=5–6	Metabasites, meta-peridotite
Jadeite (Na, Al pyroxene)	Inosilicate (Single Chain)	White to green; H=6.5–7; typically in fibrous acicular crystals and radiating crystal clusters	Metasandstones, schists, metabasites
Omphacite	Inosilicate (Single Chain)	Dark to bright green; H=5–6; heavy, granular to blocky; 2 cleavages @ 90°	With red garnets in eclogite
Wollastonite (Ca-silicate)	Inosilicate	White to gray; 2 cleavages at ~ 84°; H=5–5.5	Marbles, Ca-rich hornfels
Zoisite (Ca, Al-silicate)	Sorosilicate (Orthorhombic)	Colorless to white and green; bladed crystals; H=6	Glaucophane schists, other metabasites
Epidote/Clinozoisite (Ca, Al-silicate)	Sorosilicate	Light to dark green and greenish black; H=6–7; striated prismatic to acicular crystals	Metabasites, hornfels, schists, phyllites, slates
Lawsonite (Ca, Al-silicate)	Sorosilicate	Typically in tabular crystals; white to light blue; H= 6; two perfect cleavages	Schists, metabasites, metawackes
Pumpellyite	Sorosilicate	Green (rarely brown); needle-like to fibrous crystals; granular masses, and radiating clusters; H=5.5	Metawackes, metabasites, schists
Olivine	Nesosilicate	Dark to light green, yellow; typically lacks cleavage	Meta-ultramafic rocks, especially meta-dunite; marble
Zircon	Nesosilicate	Various colors from red-brown to colorless; typically in elongate crystals; adamantine; high relief and high birefringence optically	A wide range of metamorphic rocks
Titanite (Sphene)	Nesosilicate	Gray, brown, yellow, or black; resinous to adamantine luster; wedge-shaped to equant crystals	Metabasites, meta-peridotites

Table 10.2 Characteristic Minerals of Metamorphic Rocks *(cont'd.)*

Mineral	Chemical / Structural Group	Distinctive Properties	Commonly Occurs in
Garnet	Nesosilicate (Isometric)	H=6.5–7.5; equant crystals; commonly red and dodecahedral, but also green, black, and brown	Phyllites, schists, gneisses, mylonites, granoblastites, skarns, marbles
Andalusite (Al-silicate)	Nesosilicate (Orthorhombic)	Square prisms; red to green color; H=7.5; cross-shaped inclusions	Slates, phyllites, schists, hornfels
Kyanite (Al-silicate)	Nesosilicate (Triclinic)	Long tabular crystals; usually blue, but also white to green and black; H=5–7	Schists, gneisses, quartz-kyanite veins
Sillimanite (Al-silicate)	Nesosilicate (Orthorhombic)	Needle-like acicular crystals; white to gray and green; H=6–7	Schists, gneisses
Staurolite	Nesosilicate (Monoclinic)	Yellow-brown to red- or dark brown; H=7; prismatic crystals often in X- or + -shaped twins	Phyllites, schists
Non-silicate Minerals			
Ilmenite	Oxide	Black; submetallic to metallic luster; slightly magnetic; equant to elongate crystals	A wide range of metamorphic rocks, especially metabasic rocks
Magnetite	Oxide	Black with black streak; magnetic; irregular to octahedral crystals; submetallic to metallic luster	A wide range of metamorphic rocks
Chromite	Oxide	Black to brownish black colored; metallic to submetallic luster; nonmagnetic; octohedral crystals or granular, anhedral grains	Meta-dunite, meta-peridotite, meta-pyroxenites
Spinel	Oxide	Variously colored (e.g., black, brown, pink, green); vitreous luster; octahedral crystals	Some metabasic to meta-ultramafic rocks
Rutile	Oxide	Light yellow to red or black; typically in acicular crystals; red needles with adamantine luster are notable; tetragonal	Metabasites including eclogite; marbles
Pyrite	Sulfide	Yellow-gold colored; metallic luster; in cubes and pyritohedrons; H=6–6.5	A variety of metamorphic rocks
Apatite	Phosphate	Brown, green, yellow, blue, and colorless; vitreous luster; H=5; hexagonal prisms common	A wide variety of metamorphic rocks

Table 10.3 Examples of Mineral Assemblages Formed in Various Rock Types under Selected T, P Conditions

Rock Type	Peridotite	Basalt	Shale	Granite (or Feldspathic wacke)	Limestone (impure)
Low T & P	Serpentine (=Lizardite) + Magnetite	Albite + Chlorite + Actinolite + Calcite + Titanite	White mica + Albite + Quartz + Chlorite + Magnetite	Quartz + K-feldspar + Albite + Muscovite + Biotite + Magnetite	Calcite + Dolomite + Quartz
Low T, High P	Serpentine (=Antigorite) + Magnesite + Dolomite	Jadeite + Lawsonite + Quartz + Chlorite + Glaucophane + Aragonite + Titanite	White mica + Quartz + Lawsonite + Glaucophane + Garnet	Quartz + White mica + Jadeite + K-feldspar + Chlorite + Magnetite	Aragonite + Chlorite + Quartz + Hematite
Med. T, P	Anthophyllite + Chlorite + Tremolite + Talc + Magnetite	Plagioclase + Hornblende + Biotite + Epidote + Magnetite	White mica + Biotite + Quartz + Plagioclase + Garnet + Staurolite + Ilmenite	Quartz + White mica + Plagioclase + K-feldspar + Biotite + Magnetite	Calcite + Dolomite + Tremolite
High T, Low P	Olivine + Orthopyroxene + Diopside + Chromite	Ca-Plagioclase + Orthopyroxene + Augite + Hornblende + Ilmenite	K-feldspar + Quartz + Biotite + Garnet + Cordierite	Quartz + Plagioclase + K-feldspar + Biotite + Cordierite + Garnet	Calcite + Dolomite + Olivine

The common structures of metamorphic rocks include relict structures, cleavage, layering, veins, boudins, mullions, joints, faults, and folds. Relict structures are structures that are retained from the protolith. Common relict structures include sedimentary beds, including cross-beds, and pillows and amygdules of igneous origin. **Rock cleavage** is the tendency of a rock to break in a particular or "preferred" direction. The two main categories of rock cleavage are *continuous cleavage*—cleavage that either has a spacing of 0.01 mm or less *or* is pervasive in the microscopic to small hand-specimen range, and *spaced cleavage*—which lacks a penetrative character at those levels (Powell, 1979) (figures 10.3, 10.4 on p. 214; color plates 29c; 31a). Spaced-cleavage types include stylolitic, anastamosing, rough and smooth types as well as crenulation cleavage, a type of cleavage marked by small-scale folds in hand specimens (figure 10.3b; color plate 29d). Continuous cleavage is both a structure and a texture. *Banding* or *layering* consists of alternating bands of light and dark rock, reflecting alternating layers of different mineral composition (figure 10.5a on p. 214; color plate 31c). Some layers represent relict bedding, whereas others are a result of metamorphic processes. In the field, it is sometimes difficult to distinguish one type from the other. For example, apparent fine-scale cross-laminations observed in the Grandfather Mountain Formation metasandstones of western North Carolina were thought for decades to be relict cross-beds that revealed the directions of current flow. Now they are known to be metamorphic structures, mylonitic laminations (color plate 32d), that mimic the sedimentary features (Raymond and Love, 2006). **Veins** (figure 10.5b, c on p. 215) are generally tabular but locally irregular, blob-like, to curviplanar accumulations of one or more minerals constituting mineralogically simple rocks that have filled a crack in preexisting rock. Quartz and calcite veins are the most common types. **Boudins** are structures that consist of a series of sausage-like lenses of material (figure 10.5c). Layers commonly become boudinaged during metamorphism, as do veins and dikes. *Mullions* are linear, rod-like structures in metamorphic rocks that, in plan, are crudely similar to ripple marks but are generally larger. **Folds** are layers that have been bent into curviplanar shapes (figure 10.5d on p. 215). **Joints** are fractures in rocks, along which the only movement is perpendicular to the fracture surface (the fracture has "opened"). In contrast, **faults** are fractures or zones along which movement has occurred parallel to the surface or zone of failure. Near the surface of Earth, characteristic features along faults include broken rocks (breccias and gouge) and slickensided (smooth and grooved) surfaces. In contrast, faults that form deep within metamorphic belts during metamorphism are *ductile shear zones* composed of fine-grained, recrystallized, foliated rock called *mylonite* that may be characterized by plastically deformed and recrystallized grains (figure 10.6 on p. 216; color plate 32d).

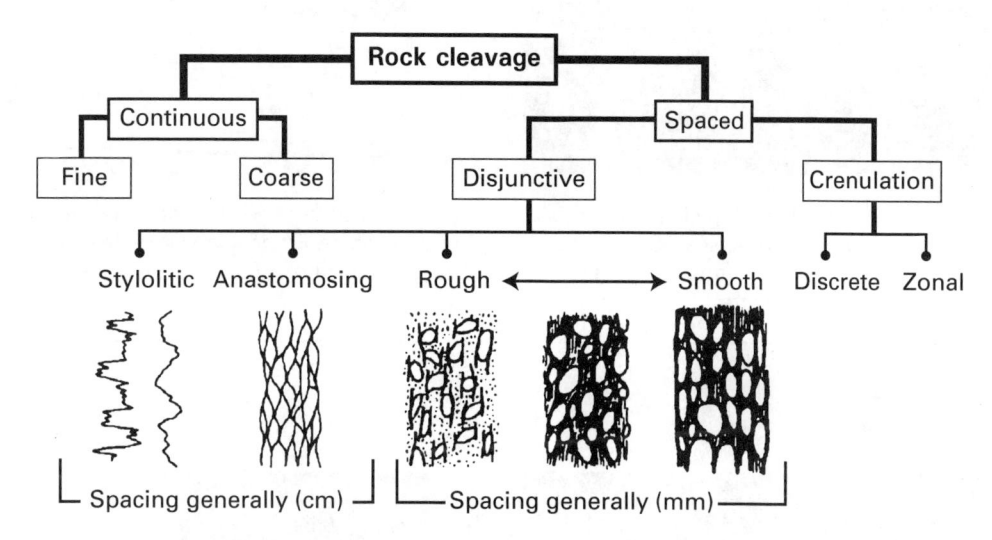

(a) Classification of cleavage types.
(*Source:* After C. M. Powell, 1979, as modified from Raymond, 2007, p. 474. Used with permission.)

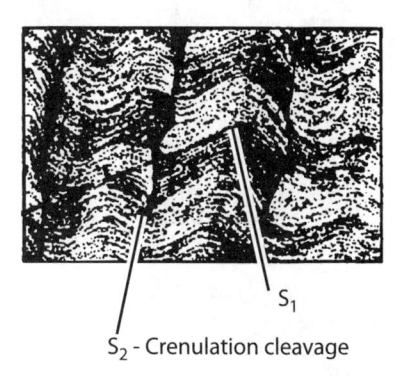

(b) Microscopic view of crenulation cleavage, showing the main micaceous layers (S_1) with crenulation cleavage (S_2) surfaces marked both by concentrations of opaque (black) minerals and fold hinges.
(*Source:* Raymond, 2007, p. 474.)

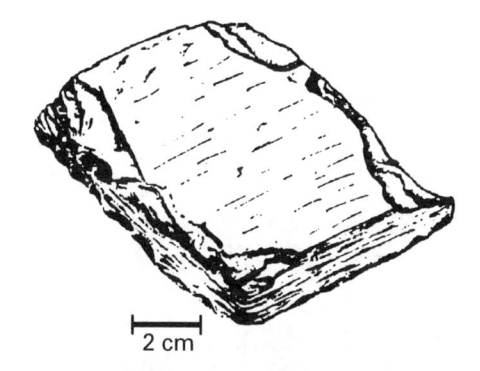

(c) Slate with continuous fine cleavage.
(*Source:* Raymond, 2009, p. 109.)

Figure 10.3
Classification and some diagrams of cleavage types.

(a) Spaced disjunctive anastomosing cleavage in serpentinite-matrix mélange. Franciscan Complex, Coleman Valley Road, Sonoma County, CA. Black and white divisions are centimeters.

(b) Continuous, fine cleavage in fine-grained glaucophane schist, Franciscan Complex, western Healdsburg Quadrangle, CA.

Figure 10.4
Photographs of cleavage.

(a) Layers in gneiss, Madison Range, east of Cameron, MT. Senior author provides scale.

Figure 10.5
Photographs of some metamorphic structures.
(*Sources:* Photo (a) by James Deni; photos (b), (c), and (d) by Loren A. Raymond.)

(b) White quartz vein (diagonal white lines) cutting through folded gneiss in the Jedediah Smith Wilderness, Alaska Basin, ID.

(c) Meta-quartz arenite boudins in metasiltstone with white quartz veins, Grandfather Mountain Formation, NC.

(d) Folds in biotite-feldspar-quartz gneiss from Newdale, NC.

Figure 10.5 *(cont'd.)*

216 Chapter Ten

As noted previously, textures are penetrative physical features defined by the shapes, sizes, and interrelationships of grains. The four main categories of metamorphic texture are foliated, granoblastic, diablastic, and cataclastic.[4] The foliated textures are the most characteristic textures of metamorphic rocks. **Foliated textures** are those that are leaf-like to banded and, due to the alignment of mineral grains, give rocks a tendency to break into somewhat flat pieces. The grains are said to be oriented. Such textures are produced particularly by deviatoric stresses (i.e. stresses oriented in particular directions). Foliated textures are divided into weakly and strongly foliated types, and the strongly foliated types are further subdivided into slaty, phyllitic, schistose, mylonitic, and gneissose textures (table 10.4 on p. 219). Slaty and phyllitic textures are aphanitic to very fine grained. **Slaty texture** is an aphanitic to very fined-grained texture with micaceous minerals strongly aligned, causing the rock to break into flat pieces (figure 10.3c; color plate 29c). **Phyllitic texture** is like slaty texture in grain size, but the layers exhibit a crenulation cleavage so that they appear corrugated on the surface (figure 10.3b; color plate 29d). **Schistose texture** is one in which flaky, bladed, or acicular grains are phaneritic (large enough to see) and are arranged in a sub-parallel arrangement (figure 10.7a; color plate 30a, b). **Gneissose textures** are phaneritic textures in rocks with distinct dark and

Figure 10. 6
Layering in mylonite. Linville Falls Fault Zone, Linville Falls, Blue Ridge Parkway (National Park), North Carolina. Pen (14 cm at right) provides scale.
(*Source:* Photo by Loren A. Raymond.)

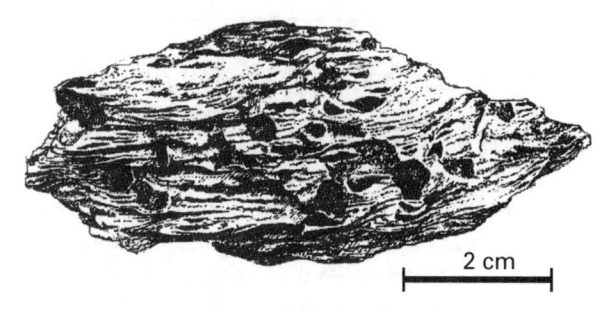

(a) Foliated, schistose porphyroblastic texture in garnet-staurolite-biotite-white mica-quartz schist. Lighter color is schistose white mica, biotite, and quartz; small dark grains are porphyroblasts of garnet and staurolite. From Ashe County, NC.

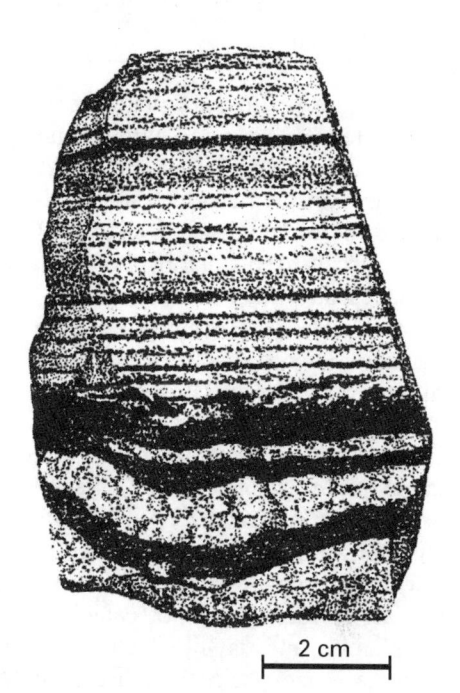

(b) Gneissose texture in granular magnetite-quartz-felspar gneiss from Llano county, TX. A photograph of the lower half of this sample is provided in the online resources.

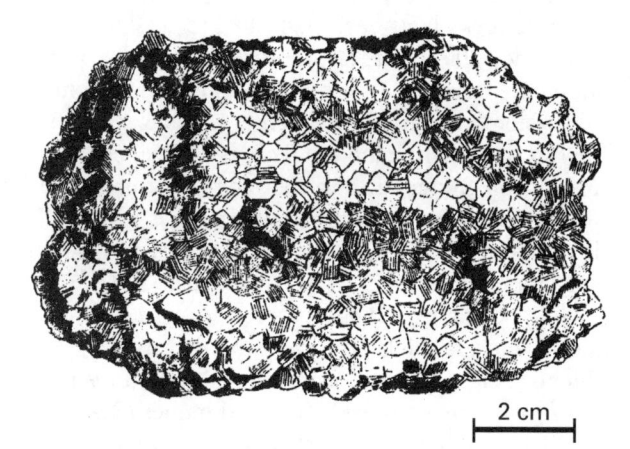

(c) Granoblastic (allotrioblastic-granular) texture in calcite marble from Felton, CA.

(d) Granoblastic (hypidioblastic-granular) texture in quartz-plagioclase-hornblende granofels from Winding Stair Gap, NC.

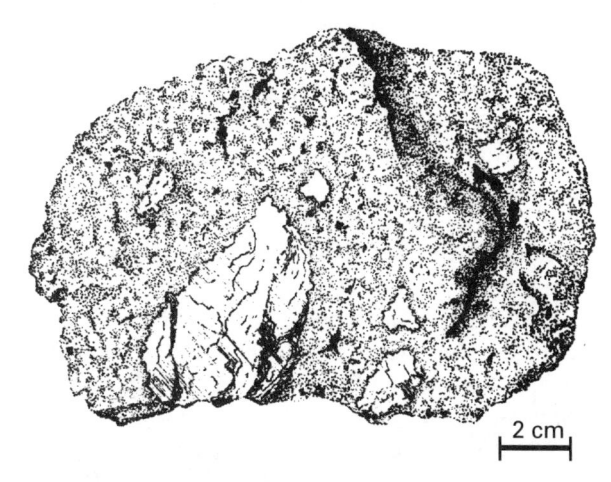

(e) Porphyroclastic texture in metadunite from Day Book Mine, NC.

Figure 10.7
Sketches of typical textures in metamorphic rocks.
(*Sources:* (a) and (b) from Raymond, 2009, figure 43; (c), (d), and (e) from Raymond, 2009, figure 44.)

light bands (figure 10.7b; color plate 31c).[5] Common forms of this texture have alternating bands that are mica rich and quartz + feldspar rich.

Other metamorphic textures have grain arrangements that do not yield foliated textures. **Diablastic textures** consist of needle-like or flaky minerals that are not aligned but rather are arranged in a random fashion (color plate 31e). **Mylonitic texture** is aphanitic to phaneritic and is characterized by deformed, elliptical, and elongated grains, such as very long grains of quartz referred to as ribbon quartz grains. **Granoblastic textures** are those characterized by relatively equant grains (figure 10.7c, d). **Cataclastic textures** are those characterized by broken grains or rock fragments that are simply stuck together and include **breccia textures**, similar in appearance to sedimentary breccias. A second type of cataclastic texture, which is aphanitic and clay-like (aphanitic cataclastic texture), characterizes the rocks called *gouge*. Among the other metamorphic textures is the common **porphyroblastic texture** consisting of two groups (two populations) of grains in the same rock, with one population of a few larger grains surrounded by a generally larger population of substantially smaller grains. Foliated schistose and gneissose textures commonly exhibit porphyroblastic texture, as well as the dominant schistose or gniessose character (figure 10.7a). Porphyroblastic textures also occur in granoblastic rocks. Garnets, staurolite, and potassium feldspar are minerals that commonly form the larger grains or porphyroblasts in metamorphic rocks (color plate 30b). **Porphyroclastic textures** are superficially similar to porphyroblastic textures, but they have larger broken or deformed grains in a finer-grained matrix (figure 10.7e). The porphyroblasts in porphyroblastic rocks "grow" in the rock during metamorphism. In contrast, the porphyroclasts in porphyroclastic rocks represent large grains, formed in pre-existing rocks, that are deformed by breaking, bending, fragmentation, and stress-facilitated recrystallization of grain margins during metamorphism.

Problem 10.3

Compare the textures in figures 10.7a and 10.7e. How are they different?

CLASSIFICATION OF METAMORPHIC ROCKS

Naming of metamorphic rocks has not been a rational process in the past.[6] Historically, the names of metamorphic rocks evolved as an eclectic set of terms that appeared in the literature over time. For certain rocks, historical use was based on inferred origin and that history influences modern classification. Comprehensive and structured classifications of metamorphic rocks, with corresponding names, are few in number and none is widely used. Some classifications of metamorphic rocks have been published (e.g., Bard, 1986; Raymond, 2007), but no one classification is widely adopted.[7] Relatively recently, the Subcommission on the Systematics of Metamorphic Rocks, a branch of the International Union of Geological Sciences' Commission on Systematics in Petrology, has proposed a scheme for naming rocks that incorporates many traditional names, such as gneiss, schist, marble, mylonite, and granofels (Fettes and Desmonds, 2007).[8] The scheme begins by adopting three root names—schist, gneiss, and granofels—and argues that any metamorphic rock can be named using one of these three terms as a root name, prefixed by mineral names (Schmid et al., 2007). It then follows history in allowing many exceptions designated as "specific names," such as slate or amphibolite, which have a past rooted in traditional use rather than rationale. The naming scheme is more of a flow chart than a visual classification, and whether or not it will be widely adopted is unknown.

All classifications are based on texture, mineralogy, or chemistry, or all three. The classification of metamorphic rocks used in this text is presented in table 10.4. Like other classifications, it adopts some names used historically but abandons others, while attempting to place all names in a rational scheme. Metamorphic rocks are divided into two types—crystalline (the dominant type) and clastic. Crystalline rocks are characterized by interlocking grains, whereas clastic rocks consist of broken grains. In the rare rock type *pseudotachylite*, the broken grains may be partially recrystallized and may be associated with small amounts of glass (melt produced by friction along the fault in which the rock has formed). Crystalline rocks include foliated, granoblastic, and diablastic types. In many cases, the basic or "root" names are based on the textures. In other cases, following tradition, names are based on the chemical or mineral composition.

Table 10.4 Abbreviated Classification of Metamorphic Rocks

Texture and Composition	Root Name	Examples of Names
CRYSTALLINE ROCKS		
Strongly Foliated		
Non-banded		
Slaty	*Slate*	Black slate
Phyllitic	*Phyllite*	Quartz-chlorite phyllite
Schistose	*Schist*	Biotite-quartz-white mica schist
Mylonitic	*Mylonite*	Quartz-chlorite mylonite
Serpentine-rich	*Serpentinite*	Talc-bearing serpentinite
Banded		
Gneissose	*Gneiss*	Biotite-quartz-plagioclase gneiss
Mylonitic	*Mylonite*	Quartz-muscovite mylonite
Weakly Foliated		
Semi-schistose	*Semi-schist*	Quartz-white mica-jadeite metawacke
		Actinolite semi-schist
Granoblastic		
Aphanitic (near igneous contacts)	*Hornfels*	Tourmaline hornfels
Aphanitic (with relict texture)	*Meta*-plus protolith name	Metabasalt
Compositional types, generally phaneritic		
Serpentine-rich	*Serpentinite*	Magnetite-bearing serpentinite
Carbonate-rich	*Marble*	Diopside dolomite marble
Calc-silicate-rich	*Skarn, Granofels,* or *Granoblastite*	Garnet epidote skarn
		Epidote-quartz granoblastite
Garnet-omphacite rock	*Eclogite*	Kyanite eclogite
Quartz-feldspar-rich	*Granofels*	Quartz-feldspar granofels
Diablastic		
Aphanitic or *Phaneritic*		
Serpentine-rich	*Serpentinite*	Magnetite-bearing serpentinite
Any composition	*Diablastite*	Actinolite-chlorite diablastite
CATACLASTIC ROCKS		
Brecciated texture (any composition)	*Cataclastic breccia*	Quartz-feldspar cataclastic breccia
Aphanitic, clay-like	*Gouge*	Gray gouge
Aphanitic, typically black, with strained grains ± glass	*Pseudotachylite*	Black, quartz pseudotachylite

(*Source:* Modified from Raymond, 2009, pp. 116–117.)

The most common crystalline rocks are the strongly foliated rocks *slate* and *phyllite* of aphanitic to fine-grain character; the *schists*, which are phaneritic but nonbanded, foliated rocks; and *gneiss*, which is foliated and banded. *Mylonites* are highly deformed crystalline rocks, common in the cores of mountain ranges. They may be banded or nonbanded, range from aphanitic to phaneritic in grain size, and are characterized by deformed grains, including ribbon quartz. Granoblastic and diablastic crystalline rocks are less common but not rare. Granoblastic rocks are dominated by equant grains and do not have a pronounced foliation. Diablastic rocks consist of elongate to tabular grains, and these grains are not aligned in a foliation but instead occur in a somewhat randomly oriented arrangement. Weakly foliated semi-schistose crystalline rocks and the cataclastic rocks—characteristic of fault zones and defined by broken and highly deformed grains and, rarely, glass—are the other major types of metamorphic rock.

METAMORPHIC FACIES, FACIES SERIES, AND DEPICTIONS OF METAMORPHIC MINERAL ASSEMBLAGES

A **metamorphic facies**[9] is a group (or set) of mineral assemblages reflecting a particular and limited range of pressure and temperature conditions under which the mineral assemblages formed. Of course, each naturally occurring mineral assemblage is a rock. In contemporary use, the term *facies* has a dual meaning, referring to (1) "*a set of rocks* representing the full range of possible rock chemistries, with each rock characterized by an equilibrium assemblage of minerals that reflects a specific, but limited, range of metamorphic conditions," and (2) *the limited range of P, T conditions* under which that set of mineral assemblages is stable (Raymond, 2007, p. 497–498). Thus, each facies represents all the mineral assemblages, and therefore all the rocks, that are stable under specific pressure and temperature limits, as well as *an area* on a P,T diagram (figure 10.8). The name of the facies is used for both. The P,T diagram showing areas of stability was called a petrogenetic grid by Bowen (1940) and hence, the pressure-temperature conditions of metamorphism for each facies are depicted on a *petrogenetic grid*.

Several schemes of facies, some with more facies and some with less, have been created (e.g., Turner, 1958; Dobretsov and Sobelev, 1972; Liou et al., 1987).[10] Many have conflicting P,T spaces and some have larger or smaller numbers of facies than are shown in figure 10.8. The most widely used schemes are somewhat modified forms of the petrogenetic grid of Turner (1981).[11] The grid in figure 10.8, which shows 11 facies, is one of these.[12]

During metamorphism, the conditions of pressure and temperature generally will vary from place to place, from lower to higher P and T and drier to more fluid-rich conditions. As a result, the stable minerals or mineral assemblages may progressively vary in response to the variations in P, T, and fluids. Thus, in metapelites, clays (+ some quartz and feldspar) may change to white mica and chlorite. Under incrementally higher P,T conditions, this assemblage changes to white mica-chlorite-biotite assemblages, then to white mica-chlorite-biotite-garnet assemblages, and at even higher grade conditions to a white mica-biotite-garnet-staurolite assemblage. In this example, each new mineral added to the assemblage marks a mineral *zone* (chlorite zone, biotite zone, garnet zone, staurolite zone). Certain of the reactions that give rise to new assemblages or minerals have been chosen to mark the boundaries of the restricted P,T fields in the petrogenetic grid or scheme. For example, in metapelitic rocks the position of the reaction in P,T space for the garnet-bearing assemblage changing to the assemblage containing staurolite is typically used to mark the boundary between the greenschist facies and the amphibolite facies.

The position of selected reaction curves for reactions plotted on a petrogenic grid—indeed, the positions of a very many reactions in P,T space—have been determined in the laboratory by experimentation and thermodynamic calculation.[13] The particular curves (reaction boundaries) used to bound the facies fields are arbitrarily selected by metamorphic petrologists, who choose boundaries they feel are important. Since different rocks have different compositions, reaction curves such as the garnet-to-staurolite curve for metapelites will differ in position from that for the boundary reaction chosen for metabasites. As a result, consideration of all possible compositions results in boundaries on the petrogenetic grid that must be depicted as broad and fuzzy or blurred (figure 10.8). Zones represent smaller sub-areas within each facies.

For example, the amphibolite facies contains staurolite, kyanite, and sillimanite zones present from lower to higher grade conditions. For most facies, such mineral zones within facies can be mapped in the field and plotted on mineral-facies charts (figure 10.9).

Curviplanar lines or trajectories through the petrogenetic grid represent hypothetical geothermal gradients within the Earth and intersect a group or series of individual facies. The line represents changing (generally increasing) conditions of metamorphism. Such a series of facies is called a **facies series** (Miyashiro, 1961). Five facies series—the Contact, Buchan, Barrovian, Sanbagawa, and Franciscan Facies Series—are shown in figure 10.8. Four of the five facies series are named for places or rock units in which the kind of gradient represented by the curve is revealed by a study of the mineral changes in the rocks across the region. The Contact Facies Series takes its name from the concept of contact metamorphism. As an example of a facies series, the Franciscan Facies Series takes its name from the Franciscan Complex of California. Across northern California within Franciscan rocks, from west to east, broad belts of metamorphic rocks were each metamorphosed under successively higher grade conditions, from zeolite facies to prehnite-pumpellyite facies to blueschist facies (Ernst, 1971).[14] The metamorphism of these belts thus defines the curve for the Franciscan Facies Series (FFS, figure 10.8).

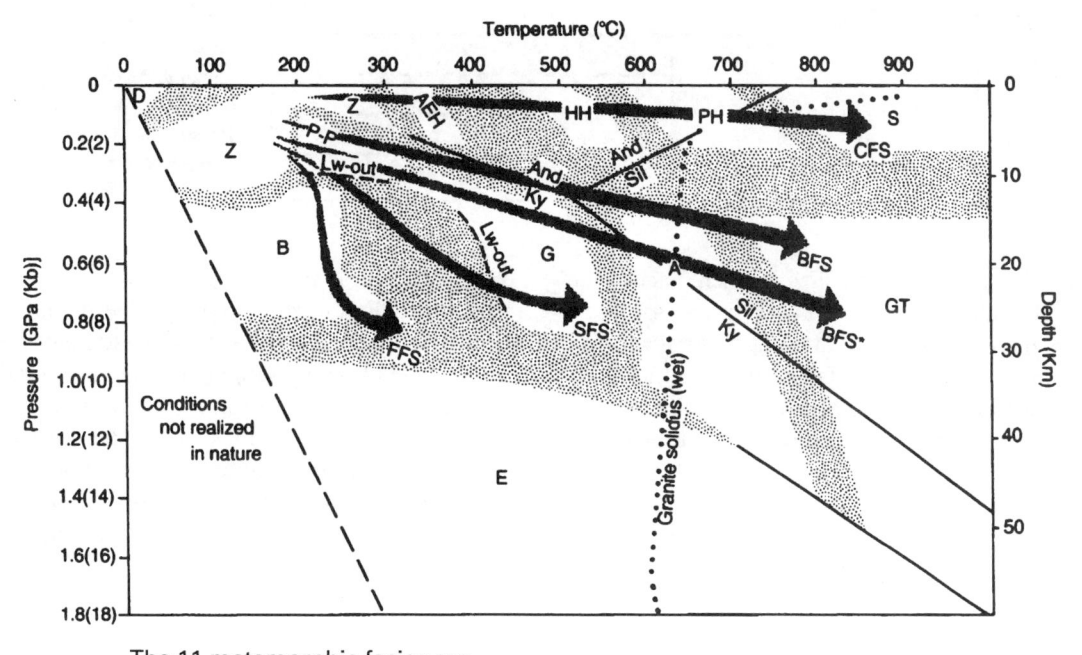

The 11 metamorphic facies are

Z = Zeolite facies
B = Blueschist facies
E = Eclogite facies
P-P = Prehnite-Pumpellyite facies
G = Greenschist facies
A = Amphibolite facies
GT = Granulite facies
AEH = Albite-Epidote Hornfels facies
HH = Hornblende Hornfels facies
PH = Pyroxene Hornfels facies
S = Sanidinite facies

The boundaries between facies are broad and gray, because critical reactions in different compositions of rocks occur at different P and T. The facies series are

FFS = Franciscan Facies Series
SFS = Sanbagawa Facies Series
BFS* = Barrovian Facies Series
BFS = Buchan Facies Series
CFS = Contact Facies Series

Figure 10.8
Petrogenetic grid of Raymond (2007, p. 502) showing facies fields and facies series curves.

Figure 10.9

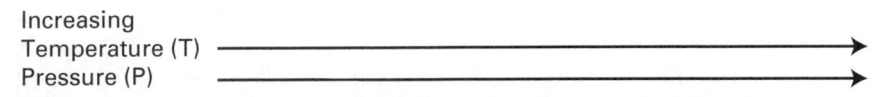

Metamorphic facies	GREENSCHIST			AMPHIBOLITE		
Zone	Chlorite	Biotite	Garnet	Staurolite	Kyanite	Sillimanite
MINERALS						
Quartz	──
White mica	──
Albite	────────────────				
Oligoclase				────────────────────
Chlorite	──────────────── - - -			
Biotite		────────────────────────────────
Almandine			────────────────────────────
Staurolite				────── - - - - - - -
Kyanite					──── - - -
Sillimanite						- - -

Increasing Temperature (T) →
Increasing Pressure (P) →

Mineral facies/zone chart showing mineral assemblage changes with increasing T,P conditions for classical Barrovian Facies Series site in the Highlands of Scotland. A vertical line drawn through the chart at any point shows the stable mineral phases in that particular zone and facies. Dashed mineral lines indicate stability only for certain rock compositions.
(*Sources:* Based on data from Barrow, 1893; Harker, 1932; Chinner, 1960, 1978; Harte and Hudson, 1979; and McLellan, 1985a; 1985b.)

Particular minerals and mineral assemblages that characterize each metamorphic facies on the grid (and facies or zones on the ground) are called *critical minerals* or *critical mineral assemblages* (table 10.5). In the field, the first appearance of a critical mineral or assemblage may appear on the map (and ground) as a line across which the metamorphic grade changes. Hence, in metapelites of the amphibolite facies, one can map out zones of rock marked by staurolite schist, other zones marked by kyanite schist, and still other zones with sillimanite schist, reflecting a change from lower to higher temperatures. The zones can be depicted diagramatically as shown in figure 10.9. On a map, the line marking the place of change from one critical mineral (or assemblage) to another (e. g., the change from the kyanite zone with kyanite schist to the sillimanite zone with sillimanite schist) is called an **isograd** (in this case, the sillimanite-in isograd). The name isograd is used because these lines once were thought to represent lines of equal or the same (=iso) grade (=grad). Each isograd represents one or more metamorphic reactions.

Metamorphic reactions are commonly depicted on triangular diagrams called metamorphic phase diagrams. Unlike the phase diagrams involving melt discussed in chapters 2 and 7, these phase diagrams depict stable mineral phases below the solidus. Several types of phase diagrams exist, each of which is used for a different chemical composition of rock (e.g., metapelites, metabasites, marbles). The principles of use are the same, but some details vary. Here we use triangular phase diagrams; and each triangle is divided into triangular sub-areas, the corners of which show the minerals that are present in a particular rock (bulk) composition for a specific metamorphic zone. Each rock will have a single bulk composition and will plot at a single point. Similarly, many minerals plot as points. Some minerals, such as hornblende, plot as areas on a general diagram because they have wide ranges in composition. Of course, in a specific rock, the hornblende will have a very specific composition and can be plotted as a point. In diagrams with minerals of variable composition, as we show in subsequent sections, a series of closely spaced lines is used to indicate

---

Problem 10.4

Using the Internet and a bibliographic search engine, petrology texts, or other resources, locate an alternative facies scheme not based on Turner (1958). Plot the Turner scheme or figure 10.8 scheme plus the scheme located via the Internet search on the same P,T grid. Discuss how the schemes differ and explain why communication between geologists using different schemes might be confusing.

Metamorphic Rocks 223

Table 10.5 Selected Critical (Index) Minerals and Mineral Assemblages for Various Metamorphic Facies

Facies	Group Bulk Composition	Selected Index Mineral or Mineral Assemblage*
Zeolite	Metapelites	Illite + Chlorite + Quartz + Albite + Laumontite
	Quartzo-feldspathics	Laumontite
	Metabasites	Heulandite + Analcite + Chlorite + Quartz + Titanite
Prehnite-Pumpellyite	Metapelites	White Mica + Chlorite + Stilpnomelane + Quartz + Albite
	Quartzo-feldspathics	Quartz + Albite + Prehnite + Chlorite + White Mica
	Metabasites	Prehnite + Pumpellyite
	Metaultrabasics	Antigorite (serpentine) + Brucite + Magnetite
Blueschist	Metapelites	White Mica + Stilpnomelane + Chlorite + Glaucophane
	Quartzo-feldspathics	Jadeite + Lawsonite
	Metabasites	Glaucophane + Lawsonite + Jadeite + Aragonite
Greenschist	Metapelites	White Mica + Chlorite + Biotite + Quartz + Albite
	Quartzo-feldspathics	Quartz + Chlorite + Biotite + Epidote + Alkali Feldspar + Albite
	Metabasites	Chlorite + Actinolite + Epidote + Albite
Amphibolite	Metapelites	White Mica + Biotite + Garnet + Staurolite + Kyanite
	Quartzo-feldspathics	Quartz + Biotite + White Mica + Garnet + K-Feldspar + Plagioclase
	Metabasites	Hornblende + Plagioclase + Garnet + Epidote
	Metaultrabasics	Anthophyllite + Chlorite + Tremolite + Magnetite
Granulite	Metapelites	Quartz + Orthoclase + Plagioclase + Biotite + Sillimanite
	Quartzo-feldspathics	Quartz + Orthoclase + Plagioclase + Orthopyroxene + Ilmenite
	Metabasites	Hornblende + Augite + Orthopyroxene
	Metaultrabasics	Olivine + Orthopyroxene + Spinel
Hornblende hornfels	Metapelites	Andalusite + Cordierite ± Biotite
	Quartzo-feldspathics	Hornblende ± Garnet
	Metabasites	Hornblende + Augite + Andesine
	Impure Marble	Tremolite + Diopside + Calcite

* Note: The minerals listed, in most cases, do not represent complete mineral assemblages.

a two-mineral (2-phase) field. Triangular areas represent three-mineral (3-phase) fields. Only triangular and lined areas are allowed to represent a set of minerals in equilibrium—a stable assemblage—on these diagrams.

As an illustration of phase diagrams and reactions depicted on phase diagrams, we here use the CMS (CaO-MgO-SiO$_2$ [+H$_2$O+CO$_2$]) diagram shown in figure 10.10. The CMS diagram may be used to depict reactions in both carbonate (marble) and metaultrabasic rocks. In the following discussion, we will assume that CO$_2$ saturates the system and may come and go from the rocks as needed. Two types of reactions are depicted on this diagram: *terminal* and *nonterminal* reactions. In terminal reactions, a new mineral appears or a mineral that has been present disappears from the diagram. In nonterminal reactions, the specific minerals depicted on the diagram do not change, but the lines connecting minerals, lines that depict equilibrium between minerals, do change.

A terminal reaction is depicted in figure 10.10 for a rock with the bulk composition of an impure calcite marble (marked by the star) composed mostly of calcite and a little diopside and quartz (the three corners of the subtriangle in which the bulk composition plots). If a terminal reaction takes place, such as

calcite + quartz = wollastonite + carbon dioxide
$$CaCO_3 + SiO_2 = CaSiO_3 + CO_2$$

(reaction 10.1)

the new phase diagram containing the new mineral wollastonite is represented by the triangle at the right. One of the "rules" of triangular phase diagrams is that only a maximum of three phases can be depicted to be in equilibrium. Hence, the triangle (system) on the right has a tie line connecting Di and Wo, separating the area that otherwise would have had four minerals after wollastonite was added. The subtriangle containing the bulk composition now has corners (minerals in equilibrium) that are calcite (Cc), diopside (Di), and wollastonite (Wo). In mapping in the field we would see, as we walk into higher-grade rocks, that quartz no longer appears in the marble, but instead there is wollastonite. Note that the two minerals (the two phases) that reacted are connected in the phase diagram on the left, by a line (called a tie line). On the right, wollastonite appears between these phases and they are no longer directly connected, indicating that they are no longer in equilibrium and do not occur together in the rock. Note also that diopside did not participate in the reaction. It remains in equilibrium with calcite, as indicated by the tie line still connecting calcite and diopside. Diopside is also in equilibrium with the new phase wollastonite (and is connected to it with a tie line). The final rock is a wollastonite-diopside-calcite marble.

In nonterminal reactions, tie lines connecting minerals in equilibrium "flip." Figure 10.11 depicts a nonterminal reaction. The bulk composition of the rock is indicated by the star. The initial rock, depicted for the system (the star in the triangle on the left), is a marble composed of calcite, dolomite, and quartz (the three corners of the subtriangle in which the star [bulk composition] is plotted). With increasing T, the reaction

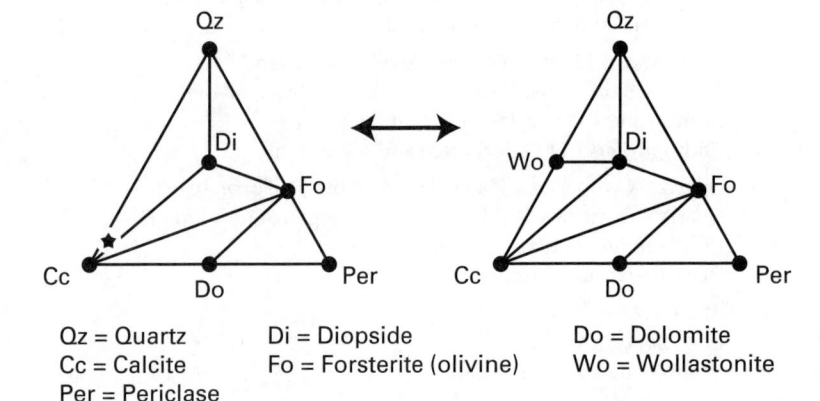

$$3 \text{ dolomite} + 4 \text{ quartz} + H_2O = \text{talc} + 3 \text{ calcite} + 3 CO_2 \quad \text{(reaction 10.2)}$$

Qz = Quartz   Di = Diopside   Do = Dolomite
Cc = Calcite Fo = Forsterite (olivine)   Wo = Wollastonite
Per = Periclase

Figure 10.10
A pair of CMS phase diagrams showing a terminal reaction.
(*Source:* Raymond, 2007, p. 511, after Bowen, 1940. Used with permission.)

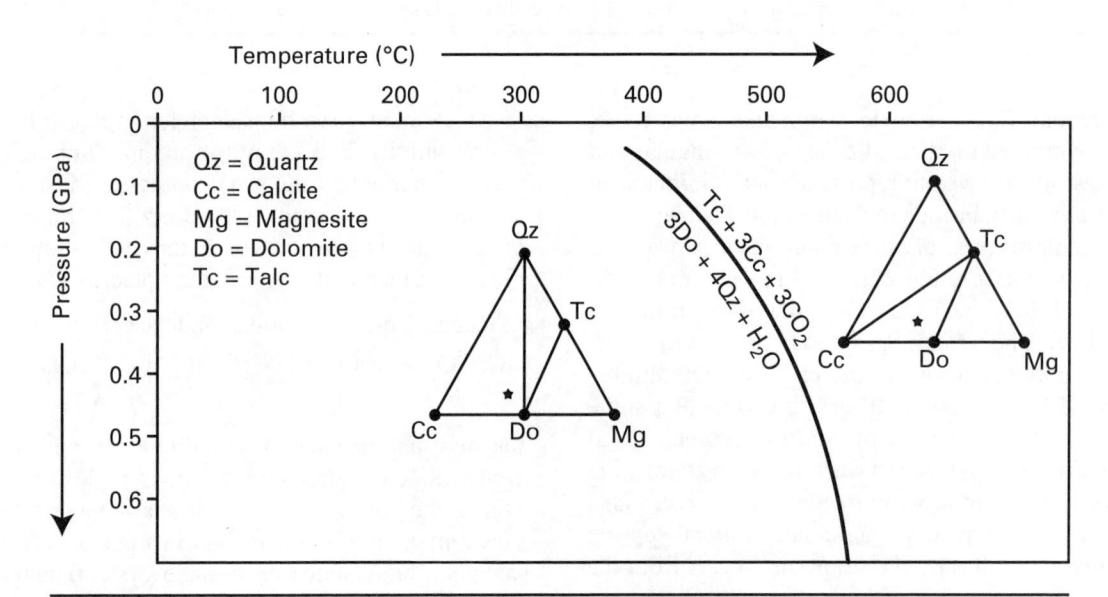

Figure 10.11
A set of CMS phase diagrams on a P,T grid, showing a nonterminal reaction in the CMS system. Star represents the bulk composition of a rock undergoing metamorphism.
(*Source:* Modified from Raymond, 2007, p. 510 and based in part on Metz and Puhan, 1971.)

takes place. As a result, the tie line connecting minerals formerly in equilibrium (dolomite and quartz, connected by a tie line on the left triangle) changes to the new minerals in equilibrium (calcite and talc, connected by a tie line on the right triangle). The final rock is now a talc-dolomite-calcite marble. Note that the minerals on the left side of reaction 10.2 are those connected by the tie line in the left diagram, and that the minerals on the right side of reaction 10.2 are those connected by a tie line on the right triangle. Note also that on the triangle as a whole, no new minerals have appeared and no minerals have disappeared. The tie line connecting Do and Qtz has "flipped" and is now connecting Cc and Tc. On the right triangle, if the bulk composition had been richer in quartz and had therefore plotted closer to quartz in the triangle, quartz would still remain in the rock. In that case, the final rock would have been a talc-quartz-calcite marble. In either case, walking in the field from lower to higher grade rocks, we would note the change from talc-absent to talc-bearing rocks.

METAMORPHIC SITES, ASSOCIATIONS, AND SPECIAL TYPES

Metamorphism of specific types occurs in particular tectonic settings or sites. For example, static (burial) metamorphism may occur in such locations as fore-arc basins at the active margins of continents between subduction zones and volcanic arcs (figure 1.7b) or in the trench seaward of the arc, both of which often receive large volumes of sediment that bury early-formed layers. In contrast, contact metamorphism occurs beneath the arc, where magma contributes high temperatures to the rocks. The relationship between sites of metamorphism and the tectonic setting aids in reconstruction of eroded mountain belts, a major objective for those studying rocks of the continents.

Contact and Hydrothermal Metamorphism

Intrusion of magma causes contact and hydrothermal metamorphism. Contact metamorphic rocks, those formed where intrusions of magma metamorphose the intruded country rocks, are typically fine-grained, nonfoliated rocks called **hornfels** (color plate 32c). At the boundaries (contacts) of the intrusion, metamorphism results from direct input of heat into the rocks.

At significant distances from the contacts, metamorphism may be caused by hot fluids, derived either from the intrusion itself or from fluids in the country rock that are heated by the intrusion. These fluids are mobilized by the heat and pass through larger regions of surrounding country rock. In the case of fluids, the reactions are hydrothermal (metasomatic) reactions. Hydrothermal metamorphism is common at mid-ocean ridges and in active geothermal fields. In both contact and hydrothermal metamorphism, magmas are the ultimate source of heat for the metamorphism.

The zone of contact metamorphism around an intrusion is called a contact aureole (figure 10.12). At the contact with a basic (silica-poor) intrusion, temperatures may reach 1000°C or more. Siliceous intrusions are cooler, whereas silica-poor intrusions are hotter. Generally, the temperatures in the aureole decrease markedly within short distances from the intrusion; hence, contact aureoles are relatively thin. Pressures of contact metamorphism generally fall below 0.4 GPa (4 kb) and such metamorphism typically occurs where deviatoric stress does not make a significant contribution to the metamorphism. These P,T conditions may produce zeolite, prehnite-pumpellyite, albite-epidote hornfels, hornblende hornfels, pyroxene hornfels, and sanidinite facies of metamorphism. Not all of these facies will be present in each aureole, but depending on fluids and other factors, a select group of facies will appear.

Three AFM phase diagrams, a type of phase diagram different from those described above, are depicted here in figure 10.13 (on p. 227). These represent the metamorphism of metapelites of the contact aureole at Onawa, Maine (figure 10.12). They depict the mineral assemblages for the greenschist facies country rock (top diagram, with a rock composition = star) composed of chlorite + quartz + white mica + ilmenite. Chlorite is a mineral that has a wide range of composition due to solid solution, so it is depicted as an area rather than a point. Note that since no more than three minerals may be depicted for each triangle on the diagram (and in this case only one mineral, chlorite, is represented by the bulk composition because it plots directly on the composition of chlorite), other minerals in the equilibrium assemblage are listed in a box adjacent to the phase diagram.

The middle phase diagram represents a mapped zone of porphyroblastic schist of the hornblende hornfels facies. This schist has the assemblage quartz + white mica + biotite + cordierite + andalusite + ilmenite. The lower diagram represents a pyroxene hornfels facies *Inner Hornfels* zone with rocks containing the assemblage quartz + alkali feldspar + biotite + cordierite + sillimanite + ilmenite. Note that for the bulk composition of this country rock (the star), some terminal reactions occur between the two hornfels rocks (e.g., andalusite changes to sillimanite and white mica changes to alkali feldspar). Also note that for the porphyroblastic schist, another reaction has occurred in which chlorite has changed to biotite, so that the rock composition that was plotted in a single-phase field (a chlorite solid solution), now plots in a three-phase field (biotite [solid solution]-cordierite-andalusite). In addition, a terminal reaction in which cordierite appears has taken place. Theoretically, each individual mineral change can be drawn using a single diagram, but the rocks on the ground contain mineral assemblages representing an array of changes. (Interested students might attempt to draw a series of triangles representing the series of individual mineral changes.)

Because fluids can pass relatively easily through rocks, hydrothermal metamorphism may affect a much larger area than contact metamorphism. Areas of tens of thousands of km^2 may be affected by hydrothermal metamorphism. Another contrast with contact metamorphism is that temperatures in hydrothermal metamorphism are generally low and facies above the albite-epidote hornfels facies are relatively uncommon and localized.

Static (Burial) Metamorphism and Metamorphism in Subduction Zones

Belts of metamorphic rocks around the world can each be assigned to a metamorphic facies series. In several places, notably in Japan and California, mountain belts occur in pairs, as do the facies series representing the metamorphic rocks of those belts. The landward (inner) belt is a high temperature (HT) belt with a Buchan or Barrovian Facies Series (± Contact Facies Series) and the seaward (outer) belt is a low temperature, low- to high-pressure belt (HP belt) with Sanbagawa or Franciscan Facies Series. These "paired" metamorphic belts represent a subduction couplet with the HP belt representing the relatively shallower (and colder) part of the subduction zone and the HT belt representing the arc and its magma-rich roots above the deep subduction zone (figure 10.14 on p. 228). In rare situations, the HP belt may contain rocks that have gone to depths of more than 100 km, yielding ultra-high pressure rocks (UHP rocks). Each belt of a pair has formed at depth, has been uplifted later during mountain building and associated uplift, and has been exposed by erosion. Erosion makes observation of the belts possible at the surface.

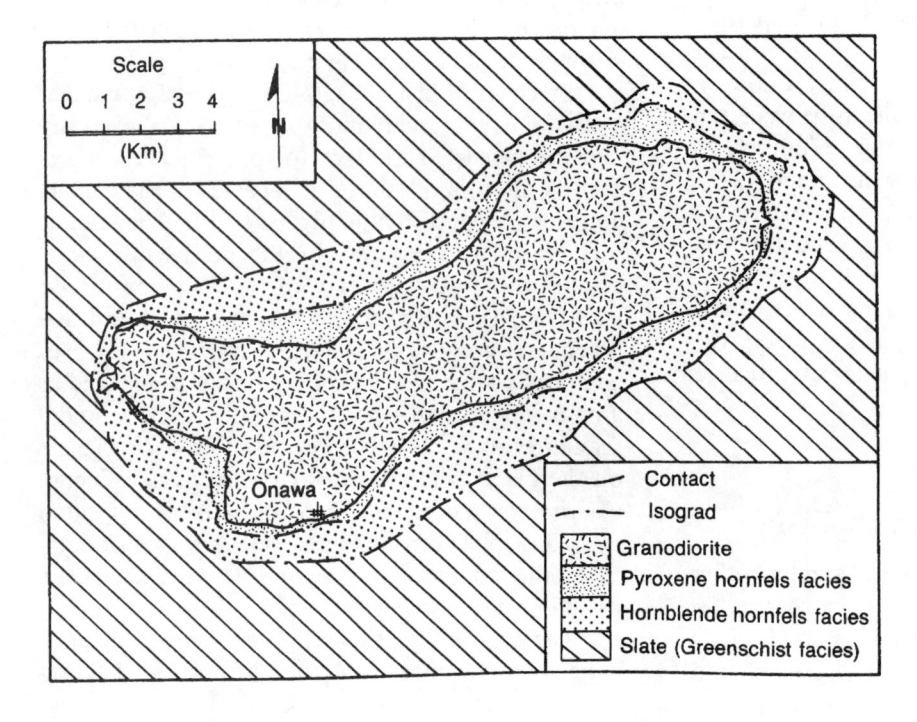

Figure 10.12
Metamorphic map of the Onawa pluton contact aureole, Maine. Note the somewhat concentric metamorphic zones.
(*Source:* from Raymond, 2007, p. 521; modified from Philbrick, 1936; Moore, 1960.)

Static metamorphism occurs in the outer regions of the subduction zones.[15] In large parts of the shallower subduction zone and in the sedimentary basins above those parts (the fore-arc basin), most of the deposited sediments (and the associated rocks) are lithified after burial. Here they are subjected to increasing pressure with increasing burial, without simultaneously being subjected to significant deviatoric stresses. At these sites, temperatures are generally low because the down-going plate carries heat with it and, in essence, refrigerates the deeply buried materials. Pressures become increasingly higher at increasing depth because of the load of overlying rock, but eventually, at considerable depths, deviatoric stresses may become important in the metamorphism.

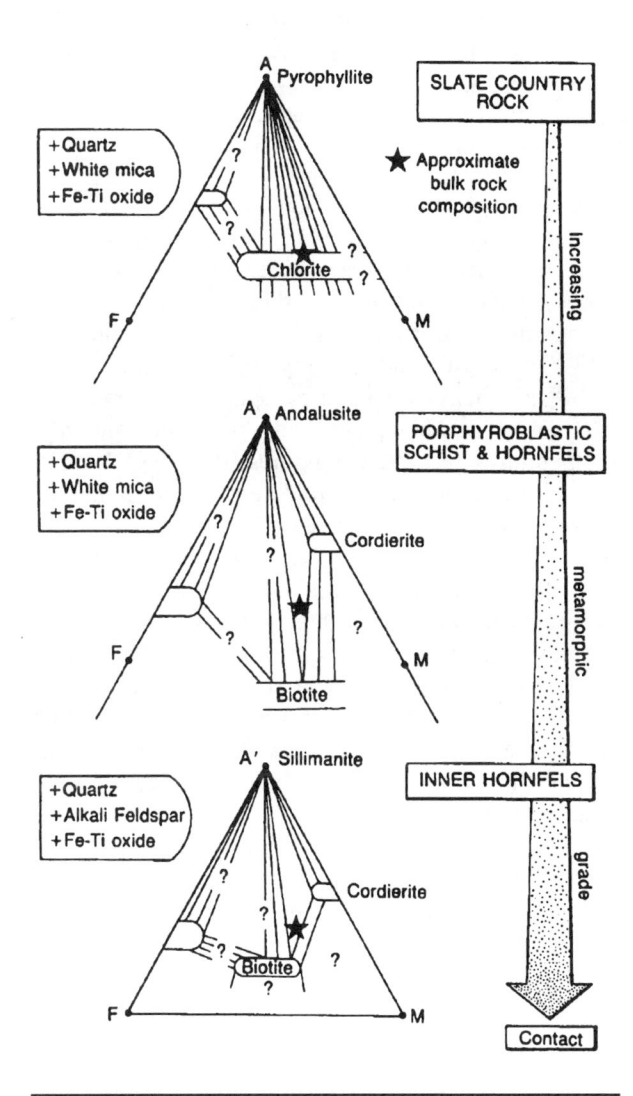

The Franciscan Facies Series consists of the zeolite, prehnite-pumpellyite, blueschist, and eclogite facies. The Sanbagawa Facies Series contains the zeolite, prehnite-pumpellyite, blueschist, greenschist, amphibolite, and locally, the eclogite facies. Both the Sanbagawa and Franciscan Facies Series contain the blueschist facies—a high-P, low-T facies of metamorphism characteristic of the HP facies series and HP metamorphic belts. Critical minerals of this facies are aragonite, lawsonite, and jadeitic pyroxene. Ironically, the blue mineral glaucophane that gave the facies its name also occurs in parts of the greenschist facies and is therefore not a critical phase in the blueschist facies.[16]

The rocks of the HP metamorphic belts commonly consist of (1) jadeite- and lawsonite-bearing metawackes, (2) Na-amphibole metacherts, metashales, and glaucophane phyllites (color plate 29d), and (3) glaucophane metabasites. Where the three types occur together as a sequence of layers, this group of rocks represents subducted and metamorphosed oceanic crust and overlying sediment and is called an oceanic plate stratigraphy (Chipping, 1971; Wakita, 2012). Melanges—bodies of mixed rocks containing rocks of various origins in a matrix of metashale or sheared serpentinite—are also common locally. In the sedimentary basins above the subduction zones, the basal rocks of which may undergo static metamorphism, arenites and conglomerates are more common, whereas metabasalts are rare.

Rocks buried deeply in the Earth tend to heat up. Thus, over time, blueschist facies rocks have a tendency to become warmer and convert to rocks of the greenschist facies or amphibolite facies (depending on the T and P). How blueschist facies rocks are preserved for long periods of time and during uplift, so that we are able to observe them at the surface, is a debated question. A number of processes of rapid uplift, preceded by a period of "refrigeration" by the subducting plate, provide possible solutions to the dilemma (e.g., Ernst, 1971; Cloos, 1982; Wakabayashi and Unruh, 1995).[17]

The discovery of the UHP minerals *coesite* in eclogite (see eclogite in color plate 32a), *diamond* in

Figure 10.13
AFM phase diagrams for contact metamorphic zones surrounding the Onawa pluton, Maine. The star shows the approximate bulk composition of the country rocks. Note the differences in the mineral assemblage triangles in which that bulk composition occurs and the associated different sub-assemblages (to the upper left) for the three metamorphic zones. Sodic plagioclase occurs in some low-grade assemblages.
(*Source:* from Raymond, 2007, p. 522, based on Philbrick, 1936; Moore, 1960; Symmes and Ferry, 1995.)

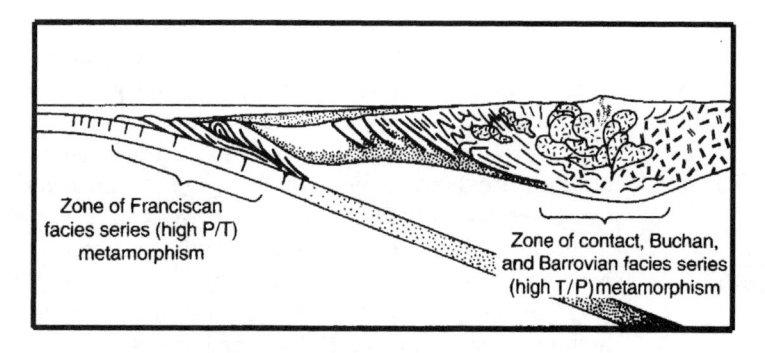

Figure 10.14
Diagram showing positions of paired, low-T and high-T metamorphic belts in a subduction zone.
(*Source:* After Raymond, 2007, p. 503.)

mafic rocks, and the critical assemblage *magnesite + diopside* in carbonate rocks within several mountain belts during the 1990s led to the conclusion that some rocks had been carried by subduction all the way to mantle depths (Chopin and Schertl, 1999; Liou, 2000).[18] Conditions of metamorphism are T > 400°C and P = 1.2–2.8 GPa. The UHP rocks, like the blueschist facies rocks, had to be refrigerated and rapidly uplifted in order to be preserved. Clearly that has taken place, and the same kinds of processes used to explain blueschist facies preservation have likewise been used to explain the existence of UHP rocks at the surface.

Metamorphism in Arcs and Moderate-to-High-Temperature Metamorphic Belts

Barrovian and Buchan Facies Series represent the moderate-to-high-temperature metamorphic parts of paired metamorphic belts. These facies series develop in and below volcanic arcs that commonly become major components of orogenic mountain systems. Here, heat flow is increased by rising magmas and metamorphic fluids. Buchan Facies Series also develop in continental rifts, where heat flow is high due to the presence of shallow mantle beneath a thinned crust. In these settings, rising magmas and metamorphic fluids may also contribute to high heat flow.

Low pressures, characteristic of Buchan Facies Series, lie between those of Contact and Barrovian Facies Series, and these low pressures are commonly accompanied by high temperatures. Andalusite and cordierite, in the middle grades of metamorphism, are critical minerals in aluminous rocks (especially pelitic schists). The high-pressure mineral kyanite is absent. The minerals are similar to those of the Contact Facies Series, but Buchan rocks are predominantly foliated whereas Contact rocks are not. Thus, deviatoric stress is a factor in metamorphism of the Buchan type.

The Barrovian Facies Series, in a sense, is the average facies series because the geothermal gradient represented by this facies series is average (about 20°–30°C/km) (figure 10.8). The typical aluminous minerals of this facies series are pyrophyllite, kyanite, and sillimanite, which represent low, medium, and high temperatures, respectively (figures 10.8; 10.9). Kyanite is a critical mineral at the middle grades of metamorphism in rocks of pelitic type (fomer shales). In metabasites, at low grades, chlorite-actinolite schists are typical; whereas in the middle grades of metamorphism, biotite-garnet-hornblende schists (traditionally called *amphibolites*) are characteristic (e.g., table 10.2). At the highest grades of metamorphism, in the granulite facies, the rocks may appear similar to gabbros, with assemblages such as quartz-alkali feldspar-plagioclase-orthopyroxene-ilmenite or hornblende-augite-plagioclase (figure 10.7d).

At the highest grades of metamorphism in both Buchan and Barrovian Facies Series, migmatites may develop. Migmatites are complex, partially melted, crystallized, and recrystallized rocks characterized by the residue of an older dark part (the melanosome) and a lighter crystallized part (the leucosome) (figure 10.15). The leucosomes are typically granitic and may have compositions equivalent to eutectic melts, but they may consist of early crystallized or cumulate minerals (e.g., Brown, 2013).[19] Layering is common in migmatites, but other structural patterns such as breccia structure and ptygmatic (highly contorted) structure occur in some migmatites. Metasomatism may contribute to the formation of migmatites locally.

Metamorphism of Ultrabasic Rocks

Metamorphosed ultrabasic rocks incorporated in mountain belts are typically called *alpine ultramafic (or*

Figure 10.15
Photograph of a migmatitic rock near Hazelwood, North Carolina along Highway 26, west of Balsam Ridge. Thin white dikes are granite and dark gray rock is predominantly hornblende schist (amphibolite). Field of view is approximately 15 meters wide.
(*Source:* Photo by Loren A. Raymond.)

ultrabasic) rocks. These rocks are metamorphosed dunites, pyroxenites, and peridotites—rocks typical of both the mantle and fractionated mafic magmas in the crust. The rocks are characterized by a range of minerals, from anhydrous minerals such as olivine, orthopyroxene, and clinopyroxene to hydrated minerals, such as amphiboles and Mg-phyllosilicates, including chlorite, talc, and serpentine minerals (Evans, 1977; Raymond et al., 2016). The rocks typically have foliated fabrics and commonly are folded. At the highest grades of metamorphism, in the eclogite and granulite facies, foliated rocks are less common and the rocks tend to be "dry," with olivine, orthopyroxene, and accessory chromite as common minerals (figure 10.16a; color plate 6a). At the middle grades of metamorphism (e.g., amphibolite facies), assemblages such as magnetite-anthophyllite-chlorite and magnesite-chlorite-Ca-amphibole are typical (figure 10.16b; color plate 6b). In the greenschist facies, various serpentinites and talc-chlorite schists characterize metaultrabasic rocks (figure 10.16c; color plate 6c).

Serpentinization is the process of metamorphism by which mafic and ultramafic rocks, particularly those that are olivine-bearing, are converted to serpentinites (O. Hanley, 1996; B.W. Evans et al., 2013). These reactions occur under greenschist, prehnite-pumpellyite, or zeolite facies conditions; and the resulting combination of minerals is a function of the pressure, temperature, and character of the serpentinizing fluid, including its Eh. The processes of serpentinization have been long debated, initially with regard to whether serpentinization was a constant vol-

Figure 10.16
Photomicrographs of metaultramafic rocks.
(*Source:* Photos by Loren A. Raymond with technical assistance by Anthony B. Love. See Raymond et al., 2016 for a discussion of these rocks.)

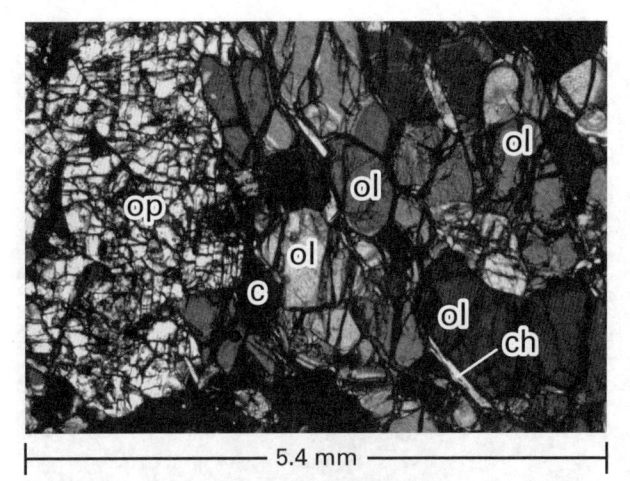

(a) Metaharzburgite, Hoots metaultramafic body, NC.
ol = olivine   op = orthopyroxene
c = chromite   ch = chlorite

(b) Chlorite (Chl)-Magnesite (Mgt)-Calcium Amphibole (CaAm) schist, Greer Hollow metaultrabasic body, Todd, NC.

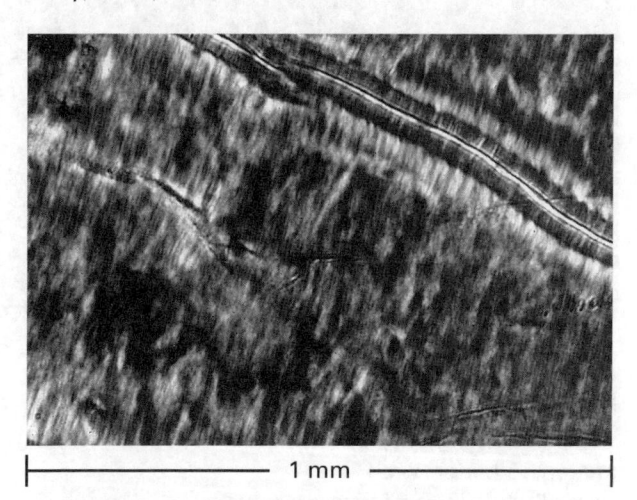

(c) Serpentinite, Greer Hollow metaultramafic body, Todd, NC. Small black grains are magnetite.

ume or constant composition process (see problems 1 and 2 earlier in the chapter). The complex textures and structures of serpentinites make determination of alternative processes of formation difficult.[20]

The most common types of alpine ultramafic rock bodies are the ophiolites (Moores et al., 2000; Dilek and Furnes, 2014). These layered sequences of rock are thought to represent the oceanic or back-arc crust-composed basalt, gabbro plus related granitoid rocks, layered gabbro and ultramafic rocks, and mantle metamorphic rock. The ultramafic parts of these rock masses are distinctive elements of mountain-belt rock assemblages. Serpentinites, in particular, are derived from the basal ultramafic parts of ophiolites, may be detached from the ophiolite during tectonism, and may, as a result, form isolated masses of serpentinite and serpentinized peridotite (color plate 32b).

Metamorphism Along Faults

Rocks formed in fault zones are metamorphic rocks in which deviatoric stress is the main agent of metamorphism. These rocks are called *dynamoblastic* rocks (Raymond, 2007) or *fault rocks* (Brodie et al., 2007; Kirkpatrick and Rowe, 2013).[21] Dynamoblastic rocks form both in narrow fault zones (during earthquakes and creep episodes) and in wide zones of penetrative (or pervasive) shear in the cores of mountain belts. All dynamoblastic rocks have been changed in texture, mineralogy, or both. For these reasons, it is appropriate to discuss the rocks of the faults and other shear zones as metamorphic rocks.

If we examine the rocks that occur along exposed fault zones, we find that they contain one or more of four types of materials: (1) deformed and/or aligned grains, (2) broken grains and rock fragments, (3) glass, and (4) undeformed grains and rock materials. The ratios of these various materials determine the rock name (figure 10.17). Fault rocks with glass in the matrix are called **pseudotachylites**. Rocks dominated by broken grains belong to the general category of rocks called cataclasites, containing the two specific rock types gouge and cataclastic breccia. The term *breccia* has its usual meaning in that the rock to which it is applied consists of angular fragments of material greater than 1/16th mm in diameter or length in a fine-grained matrix. The prefix *cataclastic* is used for fault zone breccias (hence, *cataclastic breccia*) to distinguish them from rocks formed in sedimentary or igneous settings. **Gouge** is a clay-like rock consisting of broken fragments smaller than 1/16th mm in diameter or length. **Mylonites** are weakly to strongly foliated rocks (figure 10.6) consisting of highly deformed grains, but in some cases these rocks also contain undeformed to moderately deformed porphyroclasts.

Aligned micas, lens- (or football-) shaped grains of feldspar and quartz forming porphyroclasts, and highly elongated "ribbon" quartz grains are typical.[22] Micas and elongate, thin, deformed "ribbon" quartz grains create the foliation. In many mylonites, the materials between the aligned and deformed grains are less deformed and consist of porphyroclasts and tiny lenses of rock called *microlithons*. Rocks with very thin, highly sheared, typically slickensided, weakly foliated materials between abundant microlithons are called *quasimylonites*. In these rocks, shear failure and cataclasis dominate over recrystallization and deformation. A subset of the quasimylonites is represented by pseudotachylites, which may form as small zones of glass with embayed grains or recrystallized material marked by mats of crystallites within larger zones of quasimylonites.

The rocks of fault zones contain minerals indicative of pressures and temperatures of metamorphism, just as do other metamorphic rocks. In metamorphic terranes, the metamorphic grade of the fault zone is commonly somewhat lower than that of the surrounding rocks, both because faulting is post-peak metamorphism and because abundant fluids flow through and cool the fault zone during faulting. Thus, in an amphibolite facies terrane, a mylonite zone may be of greenschist facies mineralogy.

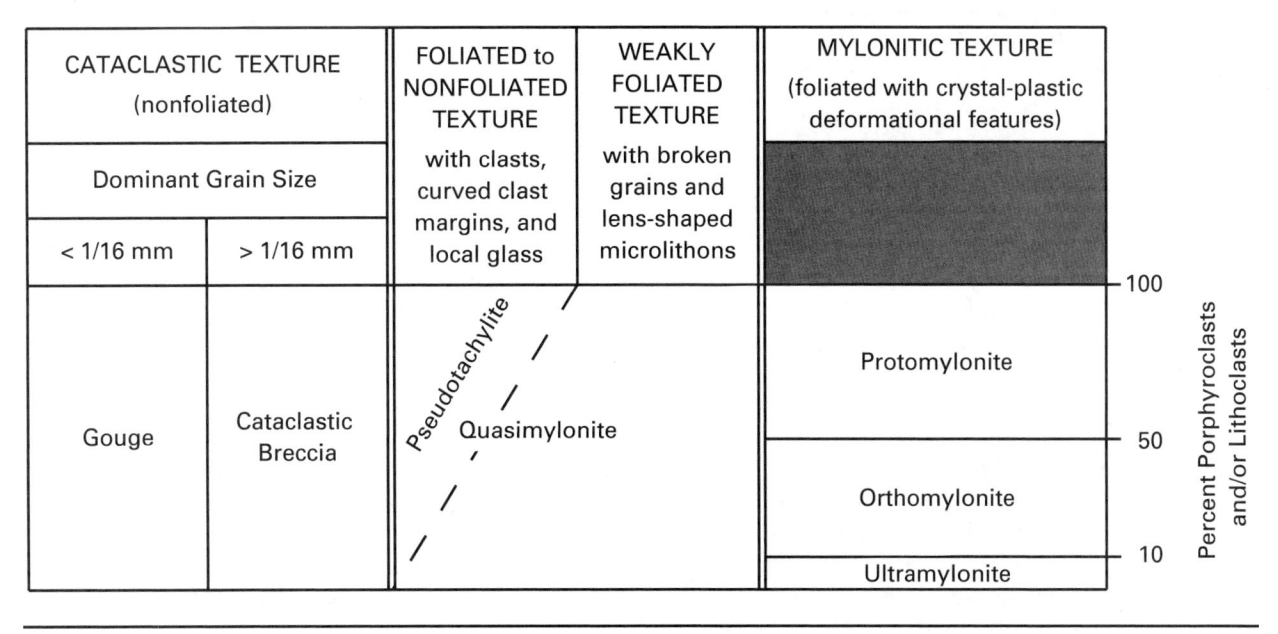

Figure 10.17
Classification of fault zone rocks.
(*Source:* from Raymond, 2009, p. 117.)

232   Chapter Ten

Analysis of both the minerals and the deformation of the grains that gives dynamoblastic rocks their textures helps to reveal the depth of faulting (figure 10.18). Rocks formed near the surface are cataclasites. Those formed at intermediate depths may be protomylonites or quasimylonites, depending on the temperatures and deformation (strain) rates. Those at depth are typically mylonites of various types (figures 10.6; 10.17). The pseudotachylites form at shallower depths under high strain rates.

Several processes yield the constituents of dynamoblastic rocks. These include (1) brittle deformation of grains (breaking or cataclasis), (2) syntectonic recrystallization (processes of coincident mineral growth via neocrystallization and recrystallization, and plastic deformation, for example by stretching and twinning), (3) rotation of grains and microlithons, and rarely, (4) melting. Which of these processes dominates is a function of the pressure, temperature, presence of fluids, and strain rates.

Figure 10.18
Dynamoblastic rocks in relation to fault depth. Circles show schematic thin section microscope-scale views of the rocks. Upper sections show broken clasts in gouge. The 4–6 km depth thin section shows weakly foliated quasimylonite with disjunctive anastomosing cleavage and clasts. Thin section from greatest depth shows well-developed mylonite with stretched porphyroclasts.
(*Source:* Raymond, 2007, p. 633, based on the ideas of Sibson, 1977, and Anderson et al., 1983.)

Problem 10.5

Over the past several years, a project involving drilling into the San Andreas Fault Zone in California has been carried out and the rocks brought up from that zone have been studied in detail. Using the Internet, locate articles on this project. (a) What grades of metamorphic rock have been discovered? (b) What practical value do scientists involved in this project hope to provide to society?

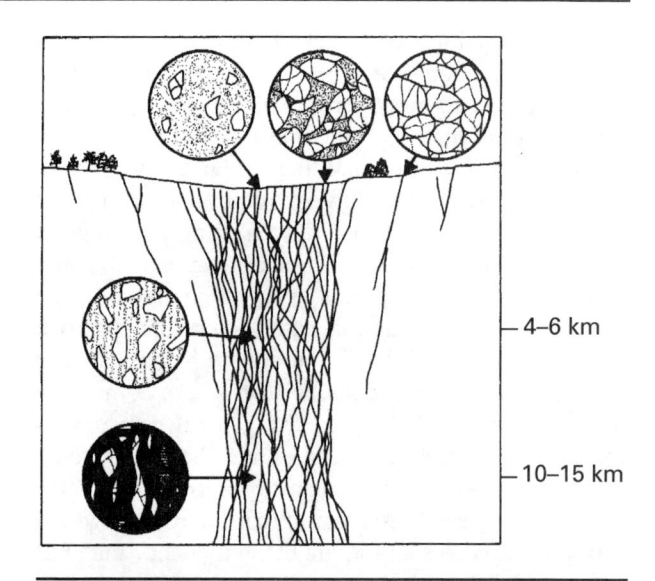

Summary

A metamorphic rock is a rock changed from any preexisting igneous, sedimentary, or metamorphic rock into a new rock of different texture, mineral composition, or both. The preexisting rock is the protolith of the metamorphic rock. The processes of metamorphism are recrystallization—a change in texture; neocrystallization—the formation of new minerals; and metasomatism—a fluid-driven change in chemical composition. These processes are caused by four agents: pressure, deviatoric stress, temperature, and chemically active fluids. Contact metamorphism is caused primarily by increases in temperature near intrusions, whereas static metamorphism is induced by increases in pressure. Deviatoric stresses cause dynamic metamorphism. Fluids, pressure, temperature, and deviatoric stress typically play a role in dynamothermal (or regional) metamorphism.

Metamorphic rocks may be assigned to one of six groups that differ from one another in their chemical compositions, with each group generally reflecting the chemical composition of igneous or sedimentary rock protoliths. Metamorphic rocks derived from basalts and gabbros are called metabasites; whereas those rocks very low in silica, high in iron and magnesia, but generally lacking both feldspars and quartz are metaultrabasic rocks. The metamorphic rocks that are high in silica and alumina derived from muds and mudstones are called metapelites. Rocks of the three other groups are commonly named on the basis of their protoliths, where it is possible to do so (e.g., metagranite or metachert), but some with similar chemical compositions can also be lumped together in a group. Thus, high-silica rhyolites, quartz arenites, and high-silica granitoid rocks are assigned to the metasiliceous rocks. Rocks with abundant silica, calcium, and alkalis derived from rocks such as granites, rhyolites, latites, and arenites can collectively be called meta-quartzo-feldspathic rocks. Metamorphic studies typically lack broad group names for many rocks, especially those that do not fit in any of the above-mentioned groups. Miscellaneous rocks that are commonly low in silica and typically poor in alumina but high in one or two other elements have been given specific names (e.g., marble, skarn, and anthracite), and these are collectively assigned to the miscellaneous metamorphic rock group.

Like other classes of rock, metamorphic rocks have a range of characteristic minerals and textures, in some cases used as a basis for classification. Foliated textures are the dominant textures of the metamorphic rocks. Structures include folds, boudins, and rock cleavage. Both newly formed common minerals (e.g., quartz) and uniquely metamorphic minerals (e.g., glaucophane) are produced by terminal, nonterminal, and continuous reactions.

The set of all rocks with distinctive mineral assemblages formed under particular pressure and temperature conditions (and the region of those P,T conditions in P-T space) is called a metamorphic facies. Eleven facies are recognized here, and these can be grouped into facies series representing hypothetical geothermal gradients in Earth. The Franciscan and Sanbagawa Facies Series represent low to high pressure, with low-temperature conditions being characteristic of the "outer" parts of subduction zones. The Barrovian and Buchan Facies Series form in mountain belts above "inner" subduction zones, where intermediate pressure and temperature gradients exist. The Contact Facies Series forms in those same belts, where zones of contact metamorphism develop around shallow to intermediate depth plutons. Dynamic metamorphism is associated with transform and other faults.

Selected References

Anderson, J. L., Osborne, R. H., and Palmer, D. F. 1983. Cataclastic rocks of the San Gabriel Fault: An expression of deformation at deeper crustal levels in the San Andreas Fault Zone. *Tectonophysics*, v. 98, pp. 209–251.

Fettes, D., and Desmons, J. (eds.). 2007. *Metamorphic Rocks: Classification and Glossary of Terms*. Cambridge University Press, Cambridge. 244 p.

Powell, C. M. 1979. A morphological classification of rock cleavage. *Tectonophysics*, v. 58, pp. 21–34.

Raymond, L. A. 2007. *Petrology: The Study of Igneous, Sedimentary, and Metamorphic Rocks* (2nd ed.). Waveland Press, Long Grove, IL. 720 p.

Raymond, L. A. 2009. *Petrography Laboratory Manual: Hand-specimen and Thin Section Petrography* (3rd ed.). Waveland Press, Long Grove, IL. 170 p.

Spear, F. S. 1993. *Metamorphic Phase Equilibria and Pressure-Temperature-Time Paths*. Mineralogical Society of America, Washington, DC. 799 p.

Spry, A. 1969. *Metamorphic Textures*. Pergamon Press, Oxford. 350 p.

Thompson, A. B. 1976. Mineral reactions in pelitic rocks: Part 1, Prediction of P-T-X (Fe-Mg) phase relations. *American Journal of Science*, v. 276, pp. 401–424.

Turner, F. J. 1958. Concept of metamorphic facies. In Fyfe, W. S., Turner, F. J., and Verhoogen, J. (eds.), *Metamorphic Reactions and Metamorphic Facies*. Geological Society of America Memoir no. 73, pp. 3–20.

Winkler, H. G. F. 1979. *Petrogenesis of Metamorphic Rocks* (5th ed.). Springer-Verlag, New York. 348 p.

Notes

1. The International Union of Geological Sciences Subcommission on the Systematics of Metamorphic Rocks (IUGS-SCMR) (Schmid et al., 2007, in Fettes and Desmons, 2007) recognizes terms such as the established and widely used *metabasite* (but not metamafite) and *metapelite*, which are used here. SCMR does not recommend any chemical category names for rocks of highly siliceous, quartzo-feldspathic (moderately high SiO_2-Na_2O-K_2O-CaO), or other bulk rock compositions. Turner (1968) long ago recognized the quartzo-feldspathic rocks as a major group but included the siliceous rock chert with an eclectic group of "ferruginous and manganiferous rocks," which are here included in the miscellaneous metamorphic rocks.

2. The Eurocentric IUGS-SCMR recognizes the basic validity of using mineralogy and texture to classify metamorphic rocks but defines the word *structure* to encompass what is typically called *texture* in the United States, partially defining structure as "arrangements of parts of a rock mass irrespective of scale, including spatial relationships between the parts . . ." Compare

this to the definition of texture in the text below and to the definition of texture in the American Geological Institute Glossary of Geology (e.g., Jackson, 1997).
3. See note 3 above regarding different uses of the term *structure*.
4. The IUGS-SCMR recognizes three basic structural categories (corresponding somewhat with schistose, gneissose, and granoblastic textures in this book) to which they assign the rock root names schist, gneiss, and granofels (Schmid et al., 2007). However, they accept an array of additional "specific" or special textures in various categories of rock (e.g., spaced cleavage in fault rocks). Special textures, in addition to the common textures in metamorphic rocks described here, are depicted in various books, such as Spry (1969) and Yardley et al. (1990).
5. Gneissose texture (and structure) are similarly defined in Jackson (1997), Raymond (2007), and the IUGS-SCMR, but SCMR defines the term as a structure only. The IUGS-SCMR defines gneissose structure as poorly developed to broadly spaced schistosity. Poorly developed schistosity has been called *semi-schistose* texture and broadly spaced schistosity has been called gneissose texture (e.g., Raymond, 1984b; 2009). Alternative terms for gneissose texture are gneissic texture (e.g., Blatt et al., 2006, p. 368) and gneissic layering (Hyndman, 1985, p. 428).
6. Actually, it is almost impossible to construct a rational and structured classification of metamorphic rocks for many reasons. Most importantly, geologists do not want to change from the various long-used terms with which they have become familiar, regardless of the facts that (1) the traditional names do not fit into a structured classification and (2) many traditional names are poorly defined. Evidence of this is provided by every classification proposed, including the one in this text and the new "classification" recommended by the IUGS Subcommission on the Systematics of Metamorphic rocks (IUGS-SCMR) (Fetters and Desmons, 2007), both of which accept the reality that geologists will not abandon long used terms like "marble." Like other proposed classifications, the IUGS classification retains several general, anachronistic, or ambiguous older names such as amphibolite, calc-silicate rock, and skarn. Actually, the IUGS classification is more of a flow chart, rule book, and explanatory text than a classification and it basically accepts many of the old names for common rocks (e.g., marble, amphibolite, ultramylonite) and offers special classifications for particular groups of rocks (e.g., fault rocks, migmatites, impactites).
7. Also see Spock (1962), Best (1982), and Raymond (2007, 2009).
8. As noted, the IUGS-SCMR scheme is not a structured single classification, but rather is a flow chart with a number of rules and subclassifications.
9. See the original definition of Eskola (1915, 1920) as reported in Turner (1958).
10. Also see Miyashiro (1994), Bucher and Frey (1994), and Schmid et al. (2007).
11. For example, Hyndman (1985) and Philpotts (1989).
12. Also see Turner (1958, 1968) and Raymond (2007, p. 498).
13. See Holdaway and Mukhopadhyay (1993) and summaries in Spear (1993) and Raymond (2007, Appendix C).
14. Also see Bailey et al. (1964), Berkland et al. (1972), Blake et al., 1988, and Ernst and McLaughlin (2012).
15. Blueschist facies metamorphism in particular has been attributed to burial (structural or stratigraphic), but also to high pressures resulting from fluid overpressures and to metasomatism and metastable recrystallization. In specific locales, arguments against these latter three hypotheses have been offered, but they are viable as concepts (see Raymond, 2007, p. 589–595 for a review). Thus, naming high-pressure metamorphism *burial metamorphism* implies, before study, that the origin of the metamorphism is known. Stating a conclusion (via a name or otherwise) prior to study is not good science. In this text, therefore, the name burial metamorphism is not used, except in a few places where it is used to remind readers that "static" has a meaning similar to burial, without the implications for origin.
16. Confusion has existed in the past with regard to the distinction between blueschists (sensu stricto; e.g., glaucophane-rich schists) and blueschist facies rocks, most of which are *not* blue in color. Thus, some geologists still refer to gray-brown metasandstones containing lawsonite as "blueschists." Regrettably, the IUGS-SCMR has chosen to incorporate and newly define the ambiguous term "blueschist" into its recommended list of rock names. In this case, SCMR uses blueschist for a rock of blue color, with the color due to the presence of blue amphibole.
17. Also see Page (1966), Krueger and Jones (1989), and Karig (1980); see also Raymond (2007, ch. 26) for a review.
18. The literature on UHP rocks is voluminous, as a Web search will reveal. See Chopin (1986), Carswell and Zhang (2000), Ogasawara (2005), and Liou et al. (2006) for older references and Burov et al. (2014) for more recent references.
19. Also see Mehnert (1968), Brown (2001), and Solar and Brown (2001).
20. For more on serpentinites, see O'Hanley (1996), Ernst (2004), and Evans et al. (2013).
21. The IUGS-SCMR proposes a much more extensive list of rock names for fault rocks than that proposed here. See Brodie et al. (2007). Also see Anderson et al. (1983) for a discussion of fault rocks associated with the San Andreas Fault system and see Kirkpatrick and Rowe (2013) for a discussion of pseudotachylites.
22. See figure 10.7e for an image of a porphyroclast in a nonmylonite rock.

Resource Geology

INTRODUCTION

Studies of resource geology have often been left to specialized courses to which most geoscience students are not exposed. Yet the processes that create and concentrate resources are the same as those that produce other Earth materials. Since development of most of the metallic, nonmetallic, and energy resources that are extracted for modern civilization depends on the work of geoscientists, an introduction to the materials of Earth would be incomplete without some discussion of resources.

There is an old saying, "If you can't grow it, you have to mine it." Although this might seem like overt mining propaganda, a simple look around illustrates the literal truth of the statement. All of the metallic and nonmetallic materials that we use, as well as many of the energy sources, are extracted from places where they have been concentrated in the crust of Earth. Geoscientists play a critical role in every aspect of resource exploitation, from understanding how concentrated deposits form to locating deposits and planning the process of extraction with a minimal impact on the surrounding environment.

Traditionally, research and teaching on or about resources has been divided up between various subfields of the geosciences, from sedimentary geology (for petroleum, coal, sand, and gravel) to traditional "economic geology" focused on metal and precious stone deposits. Given the breadth of the topics and the division of studies among disciplines, no single textbook chapter can begin to approach full coverage of the depth or the breadth of the geology of resources. This brief survey will provide an introduction and, we hope, the inspiration to further investigate this important applied field.

RESOURCES AND DEPOSITS

What is it that makes an Earth material a resource? It may seem obvious enough: Extract some material from the crust that someone will pay for and it becomes a resource. Yet, such a simple statement hides enormous complexity. Obviously, the material must be sold at a profit, since most people would not go to the effort and expense of mining solely for charitable purposes. The profitability limitation means that a variety of factors play a role in determining if resource extraction can proceed.

Most Earth materials of interest to the entrepreneurs who extract them from the Earth's crust occur naturally in such low abundances that it would not be profitable to mine them and sell the mined material "as is." For any given material to be considered for profitable extraction, it must be enriched in the Earth well beyond its average abundance. Such an enriched material, particularly one containing a valuable metal, is commonly referred to as an **ore**. The level at which the enrichment is sufficient for profitable mining is the *ore grade*. What constitutes an ore-grade material depends on the particular resource and the concentration factor (table 11.1). The **concentration factor** is a number multiplier by which an elemental abundance must be multiplied to increase its normal crustal abundance to one of ore grade, so that it is commercially extractable. The concentration factor is not only a function of value or crustal rarity, it also depends on differences in extraction technologies. Clearly, ore-grade deposits of resources that do not require high degrees of concentration or processing will be more common than those deposits that must be mined, concentrated, and processed to an exceptional degree. On average, mercury is 20 times more abundant than gold in rocks, yet gold can be profit-

Table 11.1 Crustal Abundance, Typical Ore Cutoff Percentages, and Ore Concentration for Some Elements

Metal	Crustal Abundance (%)	Typical Ore Cutoff Grade (%)	Concentration Factor
Aluminum	8.2	30	4.9
Iron	5.6	25	4.5
Titanium	0.57	1.5	2.6
Copper	0.006	0.5	83
Nickel	0.0084	0.13	15
Uranium	0.00027	0.02	74
Chromium	0.01	1.5	150
Tungsten	0.000125	0.1	800
Gold	0.0000004	0.0008	500
Lead	0.0014	3	2143
Molybdenum	0.00012	0.2	1667
Mercury	0.0000085	0.2	23529

ably mined at 1000 times lower concentrations because of the well-developed technology for gold extraction and its high price.

The degree of concentration required is not the only factor considered in determining whether a resource can be extracted from a given location. The amount of resource available; the market price of the resource; the costs of locating, mining, processing, extracting, and shipping the product; and the costs of decommissioning the mine and monitoring the site for years or even decades after mining has ceased all contribute to this decision. The path to an operating project is therefore a complicated one, involving not only aspects of the geology but also those of mining engineering, mineral economics, public relations, law, and politics. As initially described by McKelvey (1972), these factors can be divided into three broad categories: (1) the value of the resource, (2) the state of knowledge about the resource, and (3) the accessibility of the resource. These characteristics are summarized in figure 11.1, originally developed by the U.S. Geological Survey.[1]

Resource Value

As any retailer knows, there are two ways to make a profit: sell a few expensive items or sell a lot of inexpensive ones. The resource field is no different. Some materials, like precious metals or gems, have a very high value per gram, and these resources can be extracted either from very small deposits that have a high grade or from larger deposits of very low grade. Other resources, such as crushed stone, sand, or gravel, have very low values per gram. In these cases, the deposit must be both large enough and concentrated enough to make the operations profitable.

An obvious corollary to the high value/low quantity versus low value/high quantity principle is that the value per gram of a given resource must be based on what people will pay for it. Because the market value may change, sometimes on a minute-to-minute basis, deposits of any type can be classified according to their profitability. If a resource can be extracted and sold for a profit, it is considered to be **economic**, but a resource that cannot be produced profitably is termed **sub-economic**. Between these two extremes are those deposits that are unprofitable currently but would become profitable, if the market value increased. Such deposits are referred to as *marginal*.

Resource Knowledge and Accessibility

Because few resources occur entirely on the ground surface where they can be precisely measured, the amount of knowledge about any given deposit is obviously variable. That knowledge depends on how thoroughly the deposit (or the region hosting the deposit) has been studied. An **identified resource** is a resource for which the location and grade, as well as the quality and quantity of the valuable material, are known or are carefully estimated based on well-specified geologic evidence. This broad category is subdivided into measured and indicated resources (figure 11.1).

Measured resources are those of calculated quantity, in which dimensions are determined by direct observations in outcrops, drill holes, trenches, or existing mine workings. For a resource to be *mea-*

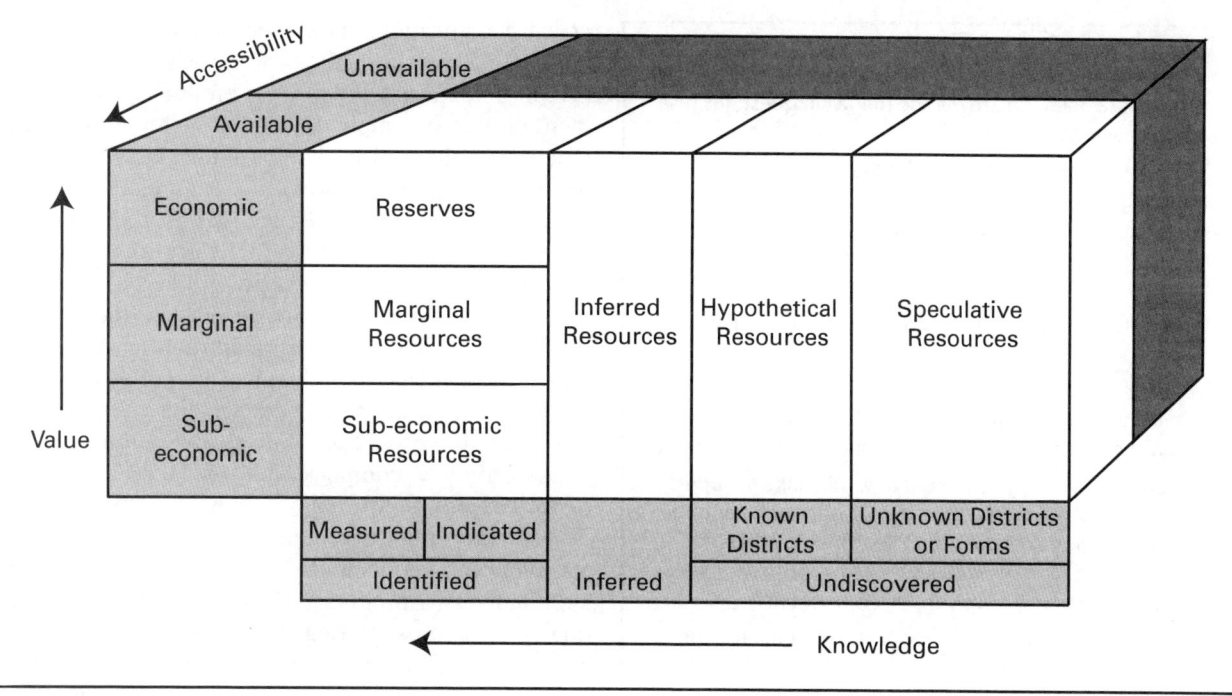

Figure 11.1
McKelvey diagram illustrating how the relationship between the value of, knowledge about, and accessibility of a resource all combine to determine what is a reserve for a given resource, which then can be used to determine the quantity of the reserve of that resource.
(*Source:* Modified from U.S. Geological Survey Circular 831.)

sured, the spacing of observations must be sufficiently close that the underlying geology is well understood. For a resource to qualify as an *indicated* resource, the quantity and grade (or quality) of the resource are determined in the same ways as for a measured resource, but the spacing between observations is greater. This lowers the level of confidence in any calculation or interpolation, but generally the confidence is still considered to be high enough for continuity of the deposit beneath the surface to be assumed. **Inferred resources** are those for which the level of confidence in the quantities is still lower, due to the fact that continuity between or beyond measured or indicated resources may not be fully supported by existing measurements.

Undiscovered resources contrast with identified resources, because undiscovered resources are just that—undiscovered. Rather, their presence is postulated. This fundamental uncertainty notwithstanding, these resources can also be subdivided into two categories: hypothetical and speculative resources. **Hypothetical resources** are those that share similarities with known deposits. Because of these similarities, it is reasonable to anticipate the existence of a deposit in the same districts or regions containing known deposits. *Speculative* resources fall into two separate categories: deposits that are similar to existing deposits, but are hosted in less-explored areas that have geologic settings similar to those hosting the known deposits; or new types of deposits, the existence and potential of which have yet to be recognized.

Even if a deposit can be extracted profitably and the location and quantity of the resource are known precisely, it still may not be possible for mining operations to commence. The deposit may occur in a region that is subject to instability due to wars or terrorism. The deposit may occur in a country that prefers not to permit extraction for political reasons, or in a region of sensitive environments that could be negatively impacted by mining. In addition, there may be questions of the exact ownership of the land on which the deposit is located. Finally, the deposit may simply be physically inaccessible—covered by thick ice or deep water, located at the top of a mountain range, or located in areas where the climate is so extreme that major industrial operations are impractical.

Problem 11.1

Use the data below to answer the following questions (use of a spreadsheet is helpful):

Capital (startup) costs: $475,000,000
Operating costs (per ton): $55
Richest ore: 4,000,000 tons at 0.12 ounces per ton
Mid-grade ore: 8,000,000 tons at 0.08 ounces per ton
Poorest ore: 16,000,000 tons at 0.04 ounces per ton

(a) Calculate the amount of gold present in each grade of ore (assume 100% recovery).

(b) Calculate the costs of mining each grade of ore. You must include the capital costs and note that mining a lower grade of ore *requires* the mining of richer ore.

(c) Determine what gold price, if any, will make the operation sufficiently profitable (>10% profit) for mining the different grades of ore. Again keep in mind that mining a lower grade of ore *requires* the mining of richer ore as a counter-balance. Use U.S. dollars per ounce.

(d) What range of gold prices would suggest mining the first two grades but abandoning the lower-grade ore?

(e) Find online how the average annual price of gold has varied between January 2000 and today's date, and determine what percentage of the time this mine would have been profitable.

(f) Gold is rarely if ever found pure; most commonly it is alloyed with silver (which adds value) and mercury (which adds costs). Discuss how the calculations would have to be modified to account for these factors.

Problem 11.2

Discussions of resources in the popular media often describe (in apocalyptic terms) the decrease in the reserves of a given commodity. Describe the arguments implicit in these discussions and address their validity.

CLASSIFICATION, TYPES, AND ORIGINS OF RESOURCE DEPOSITS

Resource deposits form and are altered by the same geologic processes that form and alter other materials of Earth. Nevertheless, resource studies have been considered to be unique, and these studies have been complicated by the fact that there is no widely accepted classification of deposits or processes. Proposed classifications have been based on the type of rock acting as a host, the tectonic setting or geologic age of the deposits, the process producing the enrichment of ores, or simply the kind of material extracted. One of the alternatives is to base the classification on the relative timing of formation of minerals of value, with deposits that form along with host rocks referred to as **syngenetic** and those that form after the host rock designated as **epigenetic**. Although a simple enough distinction in theory, in practice the application of this classification is much more complicated. Some deposits that might seem to be purely of one or the other type can show characteristics of both, leading to controversy.[2] Some alternative approaches to classification are complicated by the fact that different deposit types can form by the same fundamental processes. For example, the igneous processes that form a stratovolcano and its underlying magma chamber can simultaneously produce (1) precious metal-bearing quartz veins cross-cutting the surrounding rocks, (2) stockworks of fractures containing disseminated base metals in the upper reaches of the crystallized plutons, and (3) massive metasomatic replacements of adjacent limestones by magnetite (figure 11.2).

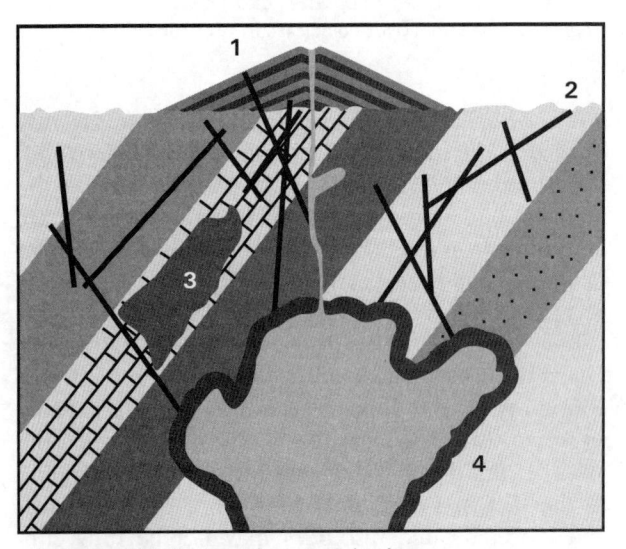

1 High-sulfidation precious metal veins
2 Low-sulfidation precious metal veins
3 Metasomatic replacement of limestone by magnetite
4 Small, cross-cutting stockwork veins of base-metal minerals surrounding the intrusion.

Figure 11.2
Schematic diagram showing different types of mineral deposits formed from an intrusion of magma and the development of a stratovolcano.

The first criterion for subdivision in the classification scheme used throughout this chapter is a genetic one. Deposits are divided into those resulting from igneous (magmatic or volcanic) processes, those that form via sedimentary processes, and those that form by various subsurface and surface-fluid processes.

Magmatic Process Deposits

Magmatic process deposits include volcanic and plutonic types. The plutonic types involve separation of fluids, crystals, or both. The volcanic types result from volcanic eruptions of magma at or near the surface.

Crystal and Fluid Separation

Some ore deposits form when magma separates into multiple components, be they solids or fluids. This separation is referred to as *magmatic segregation*. Fractional crystallization, described in chapter 7, is a magmatic segregation process in which crystals and liquids are separated. Another form of segregation that involves separation of physically incompatible liquids within a magma is called liquid immiscibility magmatic segregation. This process can be visualized through an analogy with oil-and-vinegar salad dressing. Just as droplets of oil are dispersed throughout the vinegar in the dressing but cannot mix with it, dispersed droplets of incompatible melt, such as sulfide melt, do not mix with the silicate melt or the crystallized silicate minerals of typical magmas and separate out as melt droplets or larger melt pools.

One important type of magmatic segregation deposit is the **PGM** magmatic segregation deposit (PGM = platinum group metal; PGMs are also called platinum group elements or **PGE**s). These deposits contain such platinum group elements as platinum, osmium, and palladium, the latter of which is used in the catalytic converters of automobiles. Typically, PGM deposits form where "fertile" basaltic/gabbroic magmas—those containing PGEs and significant amounts of sulfur—intrude the crust, form a magma chamber, and undergo fractional crystallization (Wager and Brown, 1968; Mungall and Naldrett, 2008). Repeated infusions of additional magma facilitate the formation of cumulate-layered rocks composed of various proportions of olivine, pyroxenes, and plagioclase, along with layers rich in magnetite, chromite, and PGM-bearing sulfide minerals (color plate 23c, d). All of these minerals, except the PGM-bearing sulfide minerals, crystallize and fall to the lower parts of the magma chamber (figure 11.3). The PGM-bearing minerals are formed later than the other minerals and crystallize from immiscible sulfide melt (discussed below) that sinks as a fluid to the lower levels of the magma chambers, where the melt solidifies. In addition, some PGM-bearing minerals are also crystallized late in upper levels of the pluton from rising, PGE-enriched hydrothermal fluids.

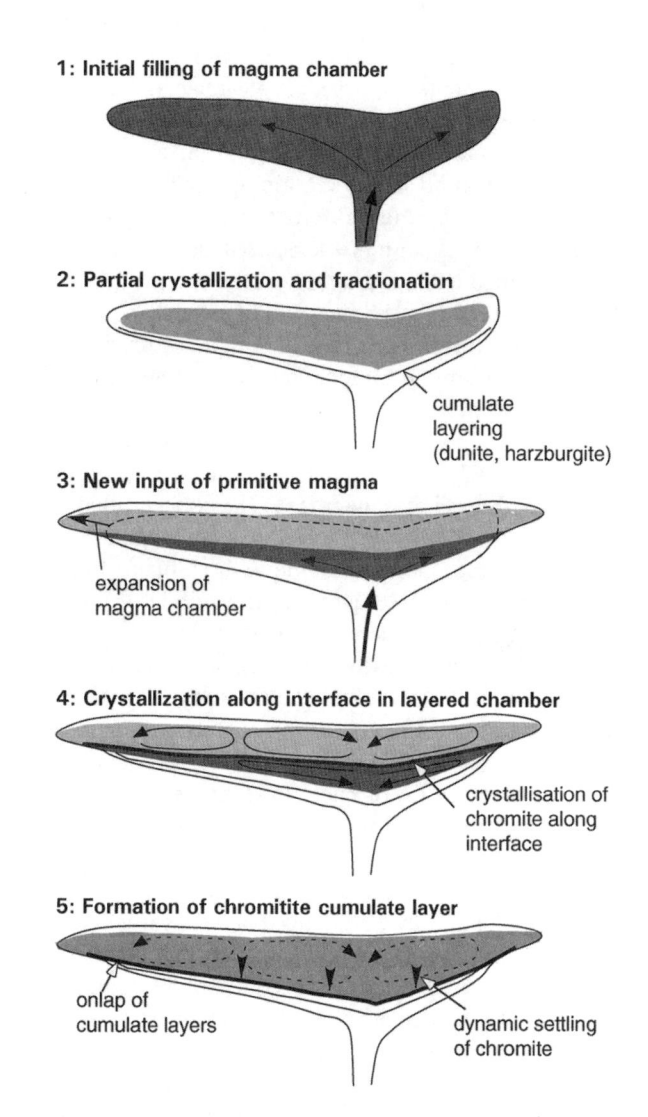

Figure 11.3
Schematic sequence of events showing formation of a layered intrusion consisting of various layered igneous rocks, including a thin cumulate chromitite layer. Curved arrows in sequence 4 and 5 show schematic circulation lines within the layers of the magma chamber.
(*Source:* Ridley, 2013. Used with the permission of Cambridge University Press.)

The melts that crystallize to yield many rocks with economically exploitable PGM contents are simply silicate melts with elevated sulfide contents. The sulfide content of these melts is such that both fractionation processes and assimilation of country rock further increase sulfide concentration to significant levels of saturation (Mungall and Naldrett, 2008). PGMs are incompatible in normal silicate mineral phases such as pyroxene and plagioclase that crystallize from the silicate magma, because sulfide ions are not normal atomic components of these silicate phases. Because they are not normal components of silicate minerals, during fractional crystallization, the PGEs and sulfide ions are left behind in the remaining melt. Sulfide is also driven toward saturation in the melt during assimilation of siliceous country rocks, because assimilation raises the silica content and lowers the temperature of the magma, reducing sulfide solubility.

As crystallization of a melt continues, saturation of the melt in the sulfide component and concentration of PGMs are both increased. When the sulfide reaches full saturation, a sulfide-rich liquid begins to segregate into droplets, which are immiscible (will not mix) with the surrounding silicate magma. PGEs concentrated in the melt surrounding the droplets are chalcophile elements that are prone to combine with sulfur to make ores and therefore are drawn into the sulfide droplets, where their concentration is increased. If the magma remains relatively quiescent, the droplets increase in size; and because of their high density they fall through the magma and concentrate in lower levels, along with crystals of magnetite, chromite, and silicate minerals (figure 11.4). Here, the droplets and larger accumulated masses of sulfide liquid soon crystallize. Because other more abundant elements—including iron, copper, and nickel—also become saturated in the sulfide melt, the PGEs are commonly crystallized as constituents of minerals such as pyrrhotite, chalcopyrite, and pentlandite in which they can substitute for the more common elements. The particular minerals and mineral assemblages that form depend on the crystallization history and crystallization conditions. In some cases, the PGM minerals are disseminated through layers of chromite, magnetite, and other mineral-rich layers in sufficient quantities to make these layers in the intrusive plutons a major source of PGMs. Clearly, the process of PGM-mineral formation is one of liquid-immiscibility magmatic segregation, but the associated silicate minerals with ores of chromite and magnetite form via fractional-crystallization magmatic segregation. Hence, both processes operate to form valuable deposits.

Some layered intrusive bodies contain abundant chromite and are principle sources of chromium. Locally, these deposits contain PGMs within the chromites as inclusions of minerals such as isoferroplatinum (Pt_3Fe). Major layered plutonic bodies having a series of layers that superficially resemble the layering of sedimentary strata are referred to in resource literature as **stratiform** deposits. Major examples of PGM deposits of stratiform type include deposits of the Bushveld (Lopolith) Complex in South Africa, the Great Dyke in Zimbabwe, and the Stillwater Complex in Montana. In some stratiform intrusions, pooling of sulfide creates a massive layer composed of minerals like pyrite, pyrrhotite, pentlandite, and chalcopyrite underneath a layered mass of mafic and ultramafic rock. Deposits of this type include those at Kambalda, Australia; Noril'sk, Russia; and Sudbury in Canada. All are major producers of PGMs and nickel.

In contrast to the major stratiform plutons, layered intrusives and ophiolites do not generally host economic levels of chromite or PGM minerals. Yet, PGM-bearing chromites do occur within dunites, chromitites, and peridotites of both ophiolites and ophiolite-derived placers, but these are less important (and seldom economic) sources of PGMs (Brenan, 2008; Mungall and Naldrett, 2008). Of broadly similar origin but also of lesser importance are podiform deposits of chromite, which are irregularly-shaped bodies that occur within ophiolite complexes. In all of these cases in which chromite is relatively abundant, the chromite accumulations initially form as magmatic segregations via fractional crystallization. These rock bodies can be valuable in and of themselves, however, because the fractionally crystallized olivine (an industrial mineral) that they contain is useful as a refractory material in furnace linings and casting sands.

In *siliceous* silicate magmas, fluid segregation and elemental incompatibility also may play roles in the formation of economic deposits (just as they do in PGM deposit formation). Two types of deposits formed from fluids derived from siliceous magmas are, in particular, major resources—pegmatite-textured granitoid deposits and metal-bearing ore bodies. In silicious magmas, incompatible elements include a variety of metals (Be, Li, Sn, W, Rb, Cs, Nb, Ta, U), rare-earth elements (REE), and volatile elements like phosphorus, boron, and fluorine (Cerny, 1982; London and Kontak, 2012).[3] In some cases water also tends to act as an incompatible element, so that after a period of fractional crystallization, the remaining residual fluid resembles a magmatic or hydrothermal fluid, rather

Resource Geology   241

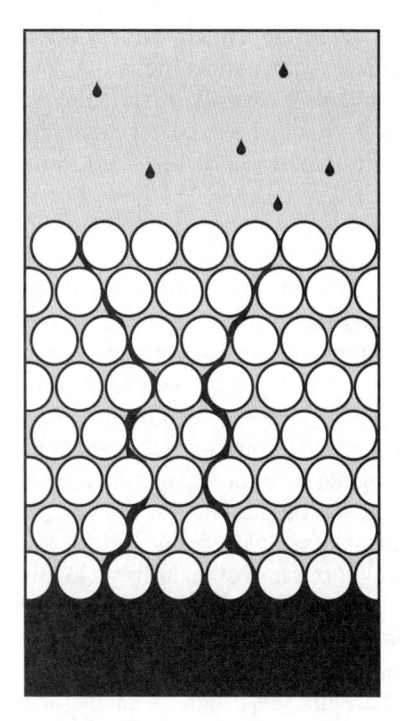

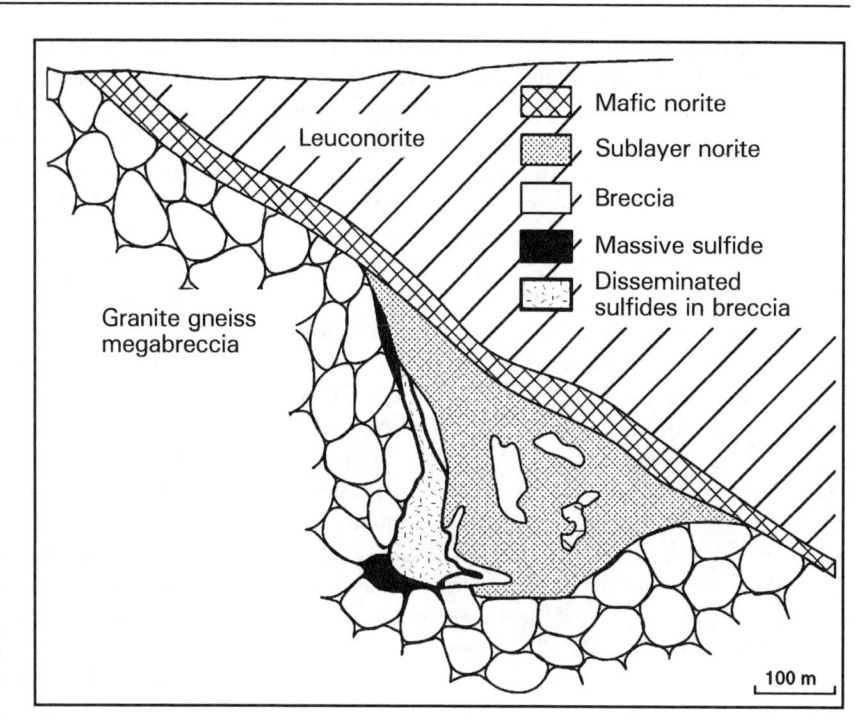

(a) (Left) Sketch showing droplets of dense, immiscible sulfide liquids (black) sinking through a magma chamber (upper part of diagram) and then flowing around previously crystallized silicate mineral grains (represented by spheres) to pool between crystals or in large pools near the bottom of the crystallizing magma chamber. (*Source:* Modified from Naldrett, 1973.)

(b) (Right) Cross-section through the Ni-Cu sulfide ore body #4 at the Levack Mine, Sudbury, Canada, showing massive sulfide pools and a zone of breccia with disseminated sulfides. The configuration of the ore bodies suggests that the rocks shown may have been rotated 30° or more, with the left side of the cross-section rotated upward relative to the right side, after initial formation of the ore bodies. (*Source:* After Pohl, 2011. Used with permission.)

Figure 11.4
Accumulation of sulfide liquids to form sulfide deposits.

than a typical melt. Pegmatitic granites seem to form from melts with compositions that are typically like those of the minimum melt in what is called the hydrous granite system (represented by phase diagrams of the system $NaAlSi_3O_8$ - $KAlSi_3O_8$ - SiO_2 - H_2O), and these melts are clearly residual fluid-like melts. Some data suggest that crystallization of these incompatible element-rich systems may be delayed until temperatures fall more than 150°C below the minimum melt temperatures of that system, a condition caused by element-induced freezing point depression.

Pegmatites, by definition, are dominated by crystals larger than three centimeters in length (figures 7.12b, 11.5b; color plate 21a). Formation of the pegmatitic granitoid deposits with economically valuable minerals begins with fractional crystallization of a siliceous melt. Crystallization of early-formed minerals concentrates volatile elements—such as phosphorus, boron, and fluorine—in the melt. At late stages of crystallization, these elements act as fluxes that disrupt the ability of the SiO_4 tetrahedra to join together in the residual fluid, producing two notable effects. First, the nucleation rate is reduced (thereby lowering the solidus temperature and nucleation density), and second, the fluid viscosity is lowered. In combination, these changes allow large crystals to grow quickly and cause crystallization to proceed sequentially. The resultant rock bodies are frequently zoned and are made up of unusually large crystals (figure 11.5).

The formation of large, and in some cases exceptionally pure, crystals in pegmatitic granitoid bodies results in industrially important deposits of minerals, some of which are gem bearing. Industrially important pegmatite bodies range from sources of the incompatible metals and gems to bodies that consist almost entirely of the industrial minerals feldspar, quartz, and muscovite. Examples of the latter are the granitoid pegmatite bodies exposed near Spruce Pine, North Carolina, which contain pure quartz crystals used to produce the high-purity silica for the production of computer chips. Other significant pegmatitic granitoid deposits in the United States occur in southern California, New Mexico, Colorado, the New England states, and the Black Hills of South Dakota. Pegmatitic granitoid rocks are known in many other countries, including Brazil, Italy, and Vietnam. In many of these localities, some pegmatites contain rare, very high- quality gem and mineral specimens. These minerals formed at the final stages of fractional crystallization, where a rare element-enriched gas or fluid filled a cavity and provided the setting for the minerals to crystallize in a "pocket." The supercritical, fluid-rich state of the residue of fractional crystallization facilitates the growth of near-euhedral crystals of unusual size and extraordinary quality. The most common of these gemstones are crystals of quartz (often smoky quartz) and alkali feldspar (notably green amazonite). In addition, gem crystals of topaz; varieties of tourmaline; aquamarine, heliodor, and beryl (e.g., variety emerald); crystals of spodumene; and other relatively rare minerals characterize the gem-bearing pegmatites (color plates 13c; 21a) (Simmons et al., 2012).

The second major class of economic deposits formed from fractionating siliceous magmas is the metal-bearing ore bodies, formed both from fluids migrating out of plutons and into country rocks and by evolving fluids interacting with nearby rocks and magmas. The stockwork vein deposits associated with granitoid porphyry plutons represent a major type of important ore deposit formed from fluids that emanate from fractionating magmas, and these types of deposits are discussed in the section on fluid deposits below.

A third and unusual example of magmatic segregation is one involving formation of melt that is highly enriched in carbon dioxide. Such melts and associated fluids are important in the formation of the rocks called **carbonatites**, igneous rocks that are composed essentially of carbonate minerals such as calcite and ankerite. Carbonatite parent magmas originate as mantle melts of alkali-olivine basalt character that fractionally crystallize, and ultimately yield CO_2-rich

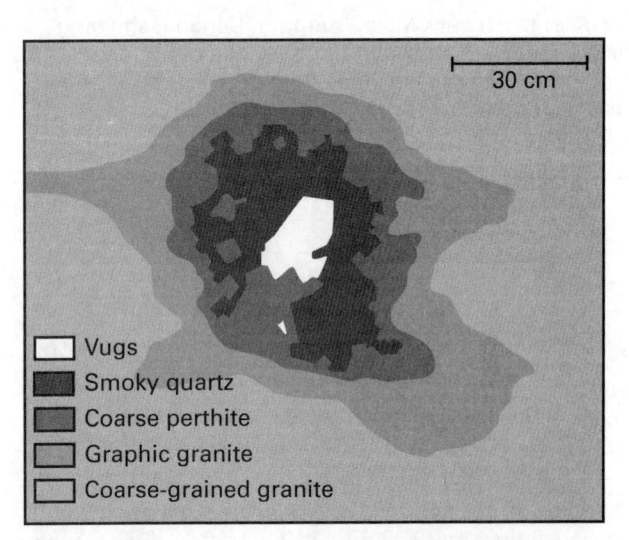

(a) Map of a small pegmatitic granite body from the Whipple Mountains, CA, showing zonation of mineralization and open spaces, called vugs, formed by residual gas.
(*Source:* Modified from Jahns, 1955.)

(b) Pegmatitic granite showing texture of large, intergrown crystals of alkali feldspar (white to light gray) and quartz (medium to dark gray) at the Minpro Mine, Spruce Pine, NC.
(*Source:* Modified from Raymond, 2007, Figure 10.5. Original photo courtesy of J. Callahan.)

Figure 11.5
Pegmatites.

fluids and melts. Because the carbonatites are unusual, disagreement exists about various aspects of their origin, such as whether the carbonate-enriched fluid forms via carbonate-silicate fluid immiscibility or standard fractional crystallization of silicate minerals.[4]

Regardless of the source and history of the fluids and despite their rarity, carbonatite-bearing plutons are an important type of economic deposit that yields such industrial minerals as rare-earth-element minerals, fluorite, and apatite. Trace levels of REEs occur in silicate minerals, but they are much more compatible with carbonate and phosphate minerals common in the carbonatites. Nearly all of the mined REE metals are extracted from carbonatite rocks. REEs are critically important for a variety of high-technology devices including cell phones, fiber optics, and high-field strength magnets used in wind turbines, hard drives, and audio speakers. Examples of REE deposits include those at Bayan Obo, China, and Mountain Pass, California.

Problem 11.3

As discussed above, rare-earth elements tend to concentrate in the water-based fluids that produce pegmatites, as well as in the carbonate-based fluids that produce carbonatites. What implication does this provide about the early crystallization of their parent magmas and the chemistry of these early-formed minerals?

Volcanic Process Deposits

Magmas that intrude near to or erupt at the surface yield volcanic rocks. Various materials produced during volcanic eruptions have economically significant uses. Some volcanic rocks like pumice or scoriaceous (highly vesicular) basalt are mined for their value "as is." Scoria has been used in outdoor grills and landscaping. Pumice, on the other hand, is used for abrasives and is also used in landscaping. Other materials derived from volcanic rocks include *perlite* (used for insulation and as a soil amendment) and, notably, diamonds—the fewer larger specimens valued as gemstones and the smaller, abundant specimens used as industrial grinding materials. Diamonds occur in the volcanic rocks *kimberlite* and *lamproite*, very rare potassium-rich ultramafic rocks. Like carbonatite, kimberlite and lamproite crystallize from magmas that originate from the partial melting of hydrous and carbonate-rich regions within the mantle at temperatures just above the solidus, a process that produces a melt that is enriched in water and carbon dioxide. The volatile components make the melts low in viscosity and density, which allows them to collect diamonds as *xenocrysts* and enables them to rise rapidly through the crust to erupt at the surface.

Magma movement rates for kimberlite and lamproite magmas are as high as 70 km per hour.[5] The rapid magma movement results in explosive eruptions, partly attested to by the presence of structures called shatter cones around the narrow, conical intrusive kimberlite structures called pipes. The pipes lie below sites of volcanic kimberlite eruption (Hawthorne, 1975; Kjarsgaard, 2007) (figure 11.6). These magmas also produce haloes of brecciated country rock with little evidence of contact metamorphism. The most compelling evidence of the rapid rise of the magma lies in the preservation of high-pressure minerals like diamond and *coesite* (the high-pressure polymorph of quartz) contained within the rocks. Such minerals would convert to lower-pressure phases if the magmas ascended at a more typical and appreciably slower rate.

Only about 10% of kimberlites and lamproites contain diamonds, indicating that unique conditions exist in the formation of the diamondiferous rocks.[6] The diamondiferous pipes occur only in old cratons (continental cores) (figure 11.7 on p. 245). The pipes display deep keels of brittle, fractured rock that was relatively cool, presumably due to the depletion of radioactive elements that would have provided heat. The relatively low temperatures allowed the diamond/graphite phase boundary to rise into the lithosphere, making it easier for the high-pressure minerals to survive the ascent to the surface. The classic example of diamond-bearing kimberlites is that at Kimberley, South Africa, but diamondiferous deposits exist worldwide, including in the Northwest Territories of Canada and at Crater of Diamonds State Park in Arkansas.

Hydrous Fluid Deposits

Recall from chapter 3 that the term *hydrous fluids* designates all of the water-based fluids. Several different types of hydrous fluids play an essential role in the origins of particular types of ore deposits.

Hydrothermal Fluids

Veins, the branching and *anastomosing tabular* zones of minerals deposited by fluids, commonly are (1) dominated by quartz, (2) deposited in fractures,

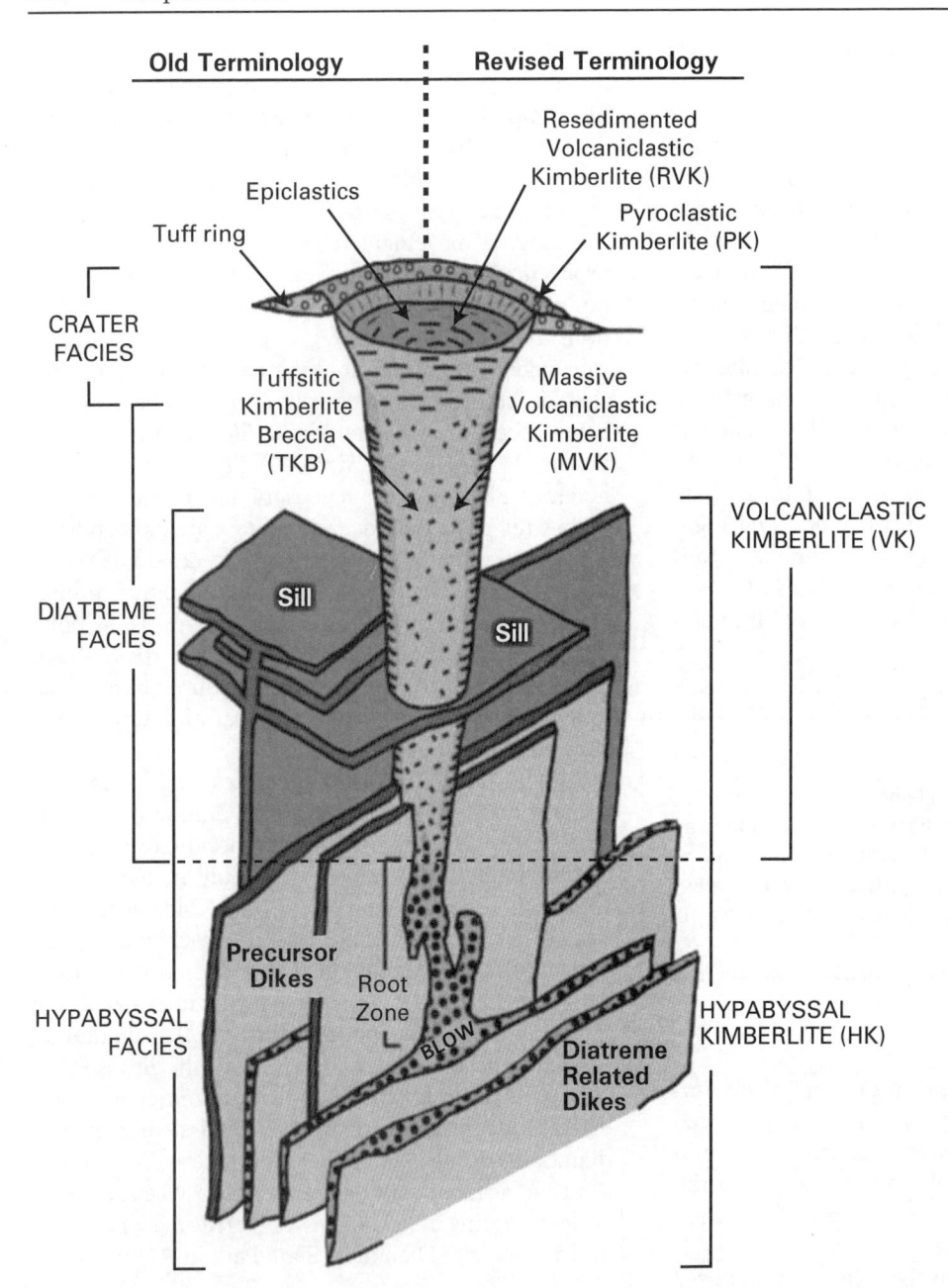

Figure 11.6
Idealized 3-D diagram of a kimberlite diamond pipe and associated surface crater (maar), showing traditional and new terminology.
(*Source:* Kjarsgaard, 2007.)

and (3) predominantly cross-cutting of existing bedrock. The importance of veins in the history of geology is enormous, as many of the materials used by pre-industrial societies were mined by hand from veins. The classic interpretation of vein origin is that metal-rich hydrothermal fluids, ascending from subsurface plutonic bodies, deposit minerals from fluids as the fluids flow through fractures in the rock above the plutonic bodies. This concept dominated resource geology for so long that even today, people may refer to narrow zones of valuable materials as veins, even when they are not such features (e.g., a "vein" of coal). Vein deposits are now a much more narrowly defined category of feature, but the understanding of their formation is essentially the same.

Veins typically form as hot fluid moves upwards along faults and other fractures (figure 11.8). Specifically, the veins develop where hot, water-based fluids containing metal and other element ions, plus dissolved silica—usually as silicic acid molecule (H_4SiO_4)—ema-

Resource Geology 245

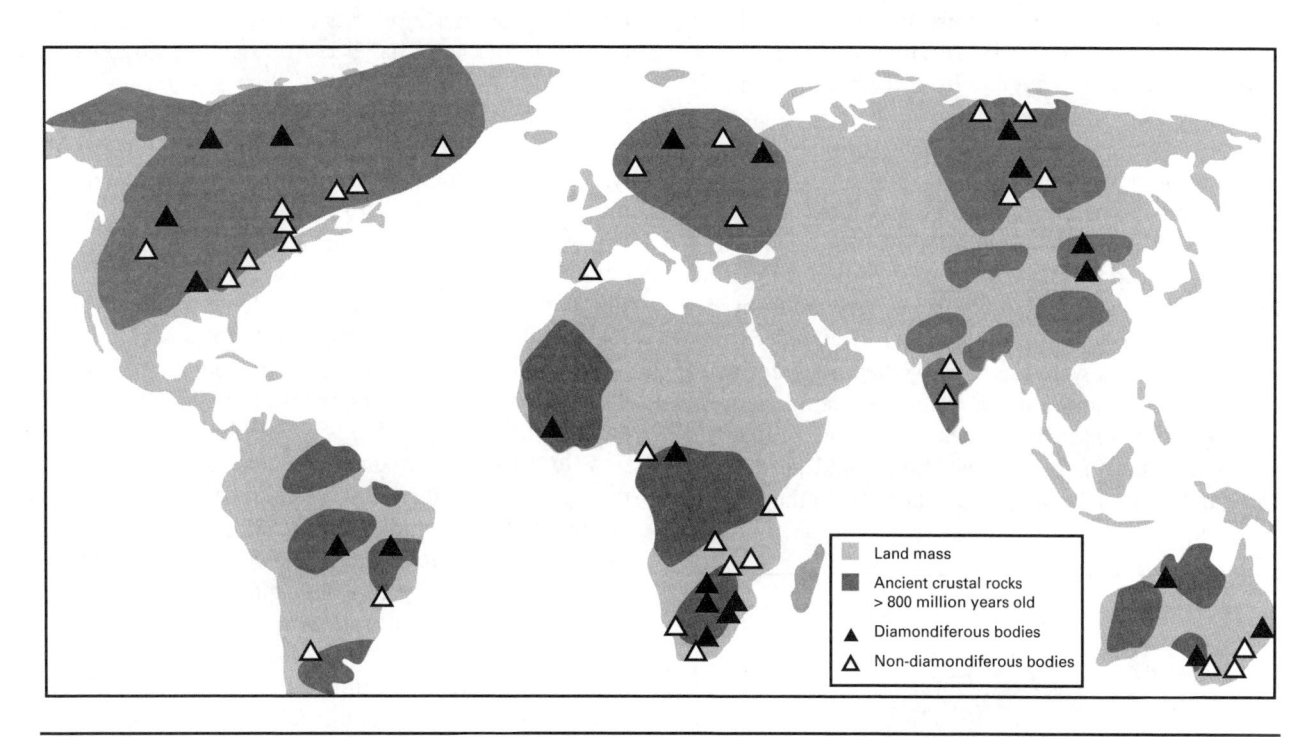

Figure 11.7
Map of the world showing correlation of ancient crustal rocks with the occurrence of diamondiferous and non-diamondiferous kimberlite/lamproite bodies.
(*Source:* Modified from Harlow and Davies, 2005, and Gurney et al., 2005.)

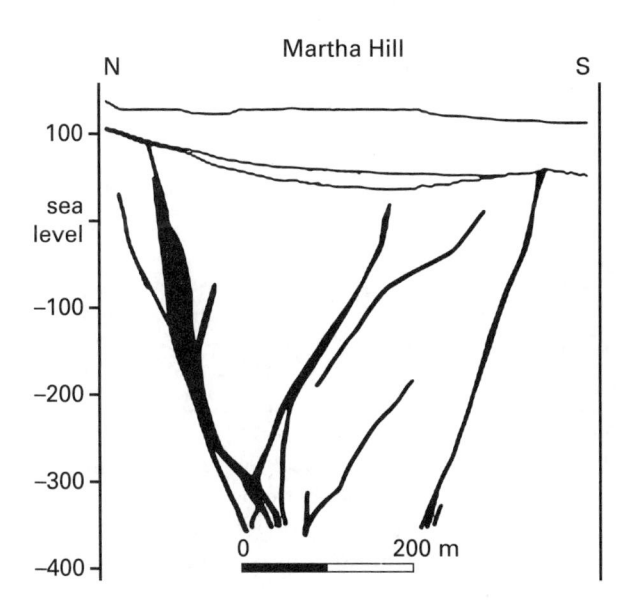

Figure 11.8
Sketch showing precious metal veins in steeply dipping fracture zones in deep bedrock. Mineralization is restricted to the veins.
(*Source:* Simmons et al., 2005.)

nate from igneous bodies or metamorphic regions and then penetrate the fractures in surrounding rocks. Here, they precipitate various minerals. In some cases, fluids also penetrate porous rocks. In either case, the fluids transfer heat as they migrate from their source and, if abundant, may produce a pervasive halo of alteration in the surrounding rock while simultaneously creating an array of veins of precipitated minerals. Over time, the progressive nature of fluid escape, especially from magmatic bodies, and penetration of rock, commonly results in a crudely ringed array of alteration zones and mineral deposits in the country rocks outward from the original source of the fluids (figure 11.9). In older studies, such deposits were classified by their apparent temperature of formation: *hypothermal* (high temperature); *mesothermal* (moderate temperature); and *epithermal* (low temperature). In contrast, they could have been recognized simply as proximal and distal to the source of heat and fluid. Now, the pattern of alteration and zoned mineralization is more clearly understood to reflect a progressive process of fluid escape and penetration of rock, and has become useful as an exploration guide.

The most important type of vein deposit, in terms of metal produced, is the **stockwork vein complex** type deposit, notably the porphyry deposits of copper, molybdenum, tin, and tungsten. Recall that the term "porphyry" refers to a rock texture in which a few large grains are surrounded by a larger number of small grains. Many volcanic magma systems yield rocks with this texture. The term "porphyry deposit" is something of a misnomer, as the name clearly describes the host rock rather than the actual mineral deposit. In porphyry deposits, exsolution of a water-rich fluid from a remaining magma leads to elevated fluid pressures in the early-crystallized outer rock shell of the magma chamber, pressures that fracture this shell and the surrounding rock. The fractures allow the fluid to escape, pervasively altering the surrounding rock as the fractures fill with quartz and mineralized vein rocks.[7] Once the fractures have healed (filled), the magma chamber begins to re-pressurize and the cycle repeats. The result in the rocks is a latticework of small to very small veins that cross-cut and intermesh in the outer, solidified shell of the pluton (figure 11.10). In the larger context, latticeworks are

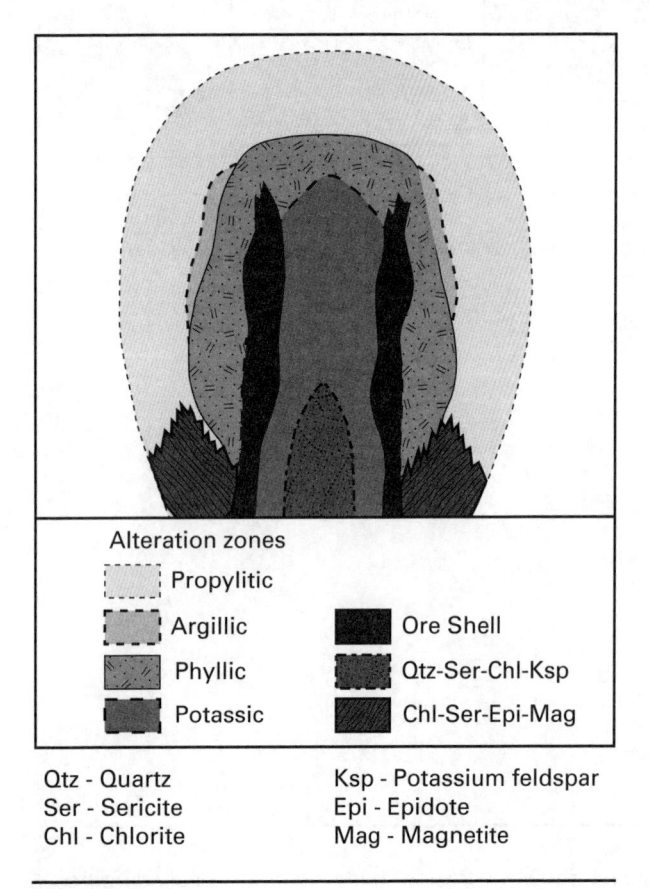

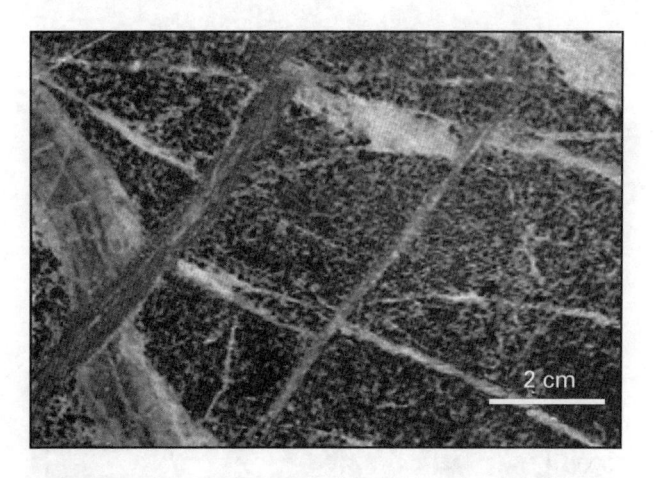

Figure 11.9
Idealized model of ore mineralization and alteration zoning associated with typical porphyry deposits.
(*Source:* Modified from Lowell and Guilbert, 1970; Beane and Titley, 1981.)

Alteration zones
- Propylitic
- Argillic
- Phyllic
- Potassic
- Ore Shell
- Qtz-Ser-Chl-Ksp
- Chl-Ser-Epi-Mag

Qtz - Quartz
Ser - Sericite
Chl - Chlorite
Ksp - Potassium feldspar
Epi - Epidote
Mag - Magnetite

Figure 11.10
Photograph of typical stockwork veining associated with porphyry mineralization, from the Bingham Canyon area, Utah. There are two generations of quartz-molybdenite veins cutting pre-existing, non-parallel, barren veins.
(*Source:* Modified from Porter et al., 2012.)

considered to develop in and along the cylindrical connection between an overlying stratovolcano above and the deep-seated pluton that supplies the volcano with melt. Studies of stratovolcanic activity and stockwork vein deposits reveal that such deposits are closely associated with porphyritic, intermediate to felsic-intrusive plutons formed above active subduction zones.

A second category of vein deposit is the so-called "classic vein deposit." Such deposits may occur in rocks covering large regions and are of low ore tonnage compared to other types of deposits, so that most currently mined deposits of this type are those bearing gold and/or silver—deposits of higher-value ores. The most common classic vein deposits are divided into two types based on vein characterization using the sulfur/chalcophile element ratios of the mineralization (a proxy for the sulfur saturation of the mineralizing fluids). The veins (and vein types) are called *low-sulfidation epithermal veins* and *high-sulfidation epithermal veins*.[8] Differences in the mineralizing fluids result in distinct characteristics in deposits formed from the two types, as listed in table 11.2. The environment of formation of the "classic" veins is at a much shallower depth than that of the stockwork complexes of porphyry deposits, generally forming at no more than about 2000 m (about 6500 feet) below the surface.

In contrast to deposits consisting of classic veins and stockwork veins produced by magmatic-hydrothermal fluids from suprasubduction-zone magmatic activity, there are a variety of ore deposits that form on the ocean floor as a result of the interaction between fluids of either volcanic-hydrothermal or diagenetic type and either the seafloor sediments or deep ocean waters. These include the **SEDEX (sedimentary exhalative) deposits** and the Cyprus-, Besshi-, and Kuroko-type deposits, also referred to as either **VMS (volcanogenic massive sulfide) deposits** or VHMS (*volcanic-hosted massive sulfide*) deposits. In all of these kinds of deposits, the mineralization is generally (1) **stratabound** (confined to a very limited number of sedimentary layers), (2) **stratiform** (layered), and (3) associated with regions of oceanic plate rifting. Mineralizing fluids are hot (200–400°C), rich in metals leached from the surrounding rocks, relatively poor in sulfur, reducing, and acidic in character.[9] Where the fluids convect upwards and encounter seawater and sediments that can be characterized as cold (≈4°C), metal-poor, sulfur-rich, oxidizing, and alkaline, new sulfide minerals like pyrite, sphalerite, galena, and chalcopyrite precipitate from the fluids to form deposits that consist of a series of lens-shaped ore bodies (figure 11.11).

Table 11.2 Features of High- and Low-Sulfidation Vein Gold Deposits

Feature	High sulfidation	Low sulfidation
Typical setting	Active volcanic centers	Geothermal fields
Host rock	Intermediate to felsic volcanic rocks & shallow intrusive rocks	Intermediate to felsic lavas & pyroclastic rocks
Relationship to volcanic center	Proximal (<2 km)	Medial to distal (2–10 km)
Alteration	Core of vuggy silica surrounded by wide argillic (clay-rich) zone	Core of crustiform & colliform silica, narrow argillic zone
Fluid sources	Magmatic fluids dominant	Groundwater dominant
Oxidation state	Oxidizing	Reducing
pH	Acidic to highly acidic	Neutral to weakly alkaline
Fluid salinity	Variable	Low
Mineralized rock characteristics	Blanket of highly altered rock	Interconnected network of veins up to 10 m in thickness
Precipitation process	Direct sublimation or from cooling brines	Fluids undersaturated in metals until confining pressure low enough to permit boiling
Ore location	Mostly disseminated in altered host rocks	Mostly disseminated in veins
Ore mineralization	Au, Au-Cu, Ag by-product	Au-Ag, Ag-Au, Pb, Zn, Cu common
Typical minerals	Pyrite, bornite	Pyrrhotite, chalcopyrite

(*Sources*: Modified from Panteleyev, 1989; White and Hedenquist, 1995; Evans, 1997; and Simmons et al., 2005.)

The two types of deposits develop somewhat differently. VMS deposits form where seawater has descended through fractures in the ocean floor associated with seafloor spreading centers. The water is heated by the underlying volcanic rocks and becomes mineralized as it is driven back through overlying rocks toward the surface. Where the hot waters encounter the colder waters at or near the surface, minerals precipitate to form mineralized zones. As the heat diminishes, movement of the fractured oceanic plate away from the spreading center cuts off fluid flow, stopping deposition in the existing lenses. Simultaneously, a new set of lenses begins to form closer to the spreading center. *Cyprus-type* deposits, in particular, develop at mid-ocean ridges, where circulating fluids rise to the surface and precipitation produces hollow columns emitting clouds of finely powdered sulfide minerals (figure 11.12). The resemblance of the columns to the black-smoke-belching smokestacks of the Industrial Era inspired their nickname, *black smokers*. The columns eventually topple, creating a mound of rubble mixed in with the settled sulfide "smoke" ore. *Besshi-type* and *Kuroko-type* deposits form in association with back-arc basin spreading centers. Such basins develop where subduction beneath and tension in the overlying arc crust cause spreading in the suprasubduction zone arc to back-arc region (e.g., Gibson et al., 2007). Besshi-type deposits occur in a mix of mafic volcanic rocks and terrestrial siliciclastic sediment, whereas the Kuroko-type deposits develop in intermediate to felsic volcanic rocks.

SEDEX deposits differ from previously discussed types of ore deposit in that they develop in continental rift or passive margin basins (figure 11.13). Here, heated fluids leach metals from continental rocks and metal concentrations may be enhanced by contributions from diagenetic fluids. The fluids are exhaled into ocean waters or sediment sequences on the ocean floor in the bottoms of grabens, producing both massive and banded or laminated mineralization. The most likely modern analogue for these deposits occurs

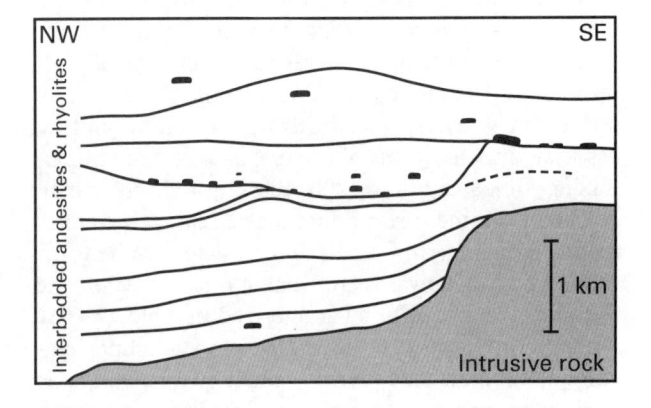

Figure 11.11
Simplified cross-sectional sketch of the composite stratigraphic section of the Noranda volcanic massive sulfide district, Canada, showing VMS lenses (black bodies) throughout the district at the base of felsic to intermediate volcanic sequences.
(*Source:* Modified from Franklin et al., 1981.)

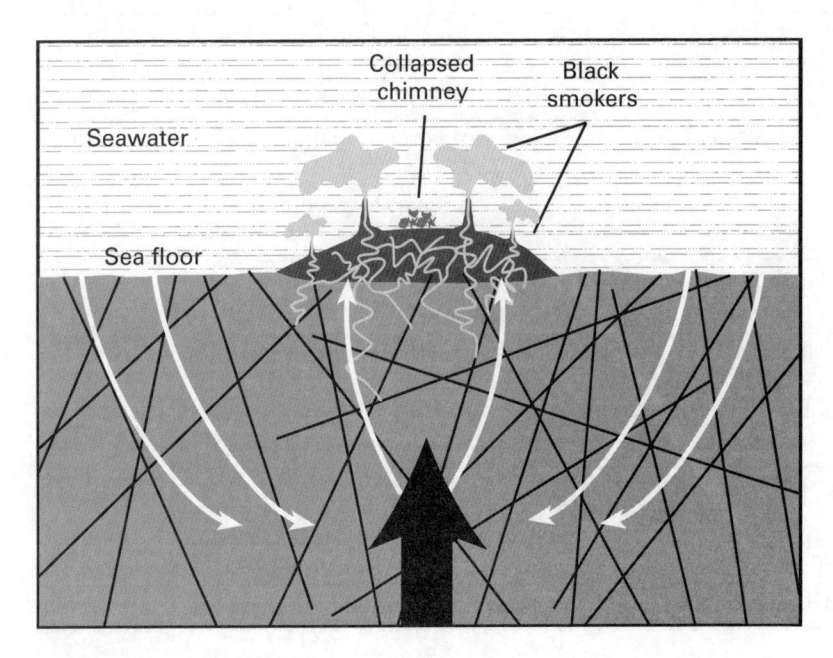

Figure 11.12
Idealized model of a "black smoker" showing formation and the development of a seafloor volcanogenic massive sulfide (VMS) lens. See text for discussion.

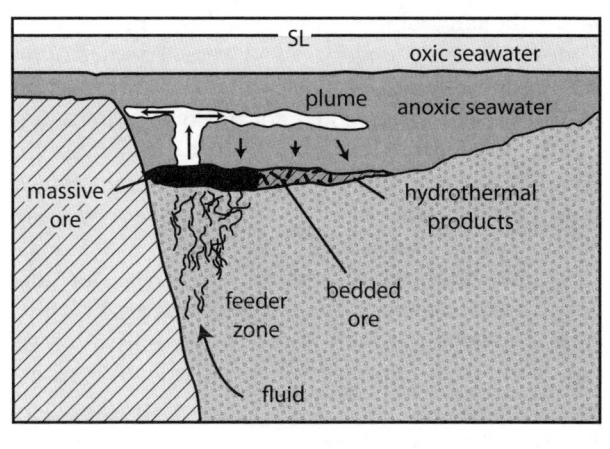

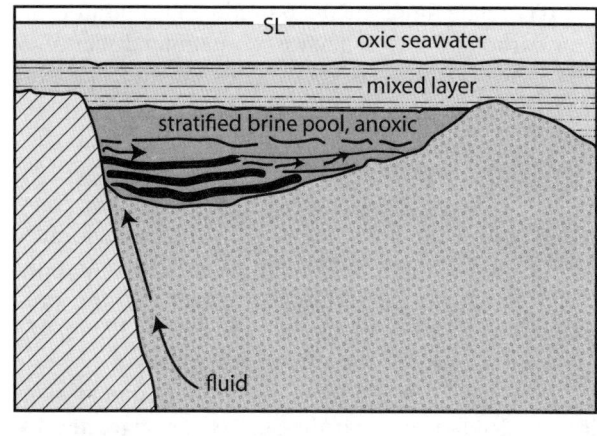

Figure 11.13
Models of vent-proximal and vent-distal formations of sedimentary exhalative (SEDEX) deposits.
(*Source:* Leach et al., 2005.)

in the Red Sea rift between northeast Africa and the Arabian subcontinent.

Connate Fluids and Stratabound Deposits

Connate fluids, also called formational fluids, initially collect as fresh or oceanic water that is buried with sediments as they are deposited. The fluids derive either from those waters trapped within the intergranular pores of the sediments or from those that exist as structural water—water bound as OH^- or H_2O within the chemical makeup of clays or other minerals of the sediments. While they are undergoing diagenesis, such hydrous minerals may change (alter) to new minerals that contain less OH component as a result of changing conditions within their environment. Thus, during diagenesis, water is released into the surroundings. The fluids are typically oxidized, mildly acidic (pH 4–6), low temperature (<150°C), and of high salinity (>100 ppm), and they contain very small quantities (ppm) of dissolved base metals.[10] Connate fluids have heat derived from geothermal (general underground) warmth of rocks rather than an igneous or metamorphic heat source. Thus, they are similar to so-called "oil-field brines" and basinal brines, but they differ from those in that they lack the significant amounts of accompanying hydrocarbons that characterize the oil-field brines.

Connate fluids are driven out of the deeper reaches of sedimentary basins by lithostatic pressure (the load of the overlying rocks) or by compressional tectonic activity. They tend to flow up the dip following particular strata. Where these fluids deposit minerals within the confining strata, the resulting deposits are stratabound. Deposits derived from connate fluids vary in character, and their attributes depend on differences in original fluids and their interactions with the particular host rocks (figure 11.14).

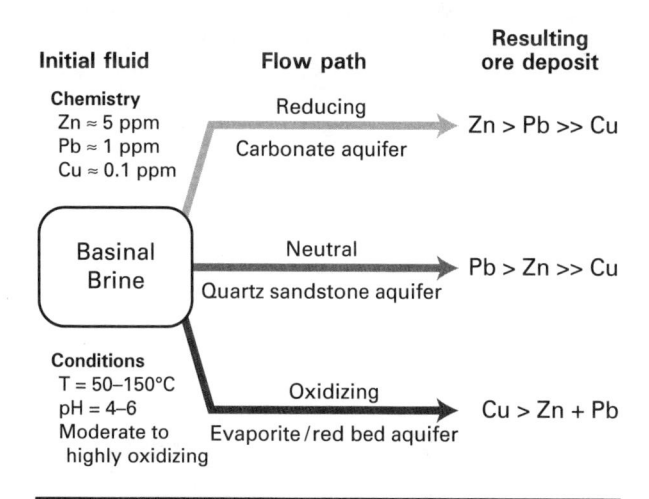

Figure 11.14
Illustration showing the relationship between a single basinal brine, the conditions under which the brine may interact with different sediments, and the resulting stratiform, sediment-hosted ore deposits. Sandstone-hosted copper deposits and carbonate-hosted lead-zinc deposits result from the modified fluids.
(*Sources:* Modified from Metcalfe et al., 1994, with data from Sverjensky, 1989.)

If connate fluids encounter reducing conditions in thick carbonate rock sequences containing discontinuities—such as at reefs rock facies changes, general facies changes, or in zones of paleokarst collapse breccias—ore minerals can precipitate from the fluid (figure 11.15). Typically, such ores are dominated by sphalerite, with lesser amounts of galena and very minor copper minerals, along with calcite, dolomite, fluorite and barite. The latter minerals commonly form cement between breccia fragments in the ores (figure 11.16; see color plates 15d; 16a, b; 17a–c; 18; 19a–c for pictures of some of these minerals).

Minerals formed from connate fluids appear to be deposited relatively near the surface, as suggested by low fluid pressures calculated for the times of deposition. Some of the minerals, particularly sphalerite, show distinct sequences of laminations that can be traced for considerable distances across the deposits. The ore-producing fluids also yield extensive dolomitic alteration of the host carbonate rocks. Ore-bearing carbonate rocks of this type are principally of Ordovician and younger age and were deposited along the marine margins of stable cratons or along failed rift zones. The formal name for this type of deposit is a **carbonate-hosted lead-zinc deposit**, but because the deposits are common in the central United States, they are often referred to as *Mississippi Valley Type* or **MVT** deposits (figure 11.17 on p. 252).

Somewhat infrequently, the host rocks for connate-water lead-zinc deposits are clastic rocks rather than carbonate rocks (as is the case at Laisvall, Sweden, and in some deposits in the Missouri Viburnum trend). Notably, in these cases galena is the dominant ore mineral of the deposit.

The most productive, currently mined uranium deposits also result from metal transport and deposition by connate fluids. Uranium deposits, in contrast to lead-zinc deposits, have formed on unconformable surfaces between Archean/Early Proterozoic crystalline basement rocks containing carbonaceous material

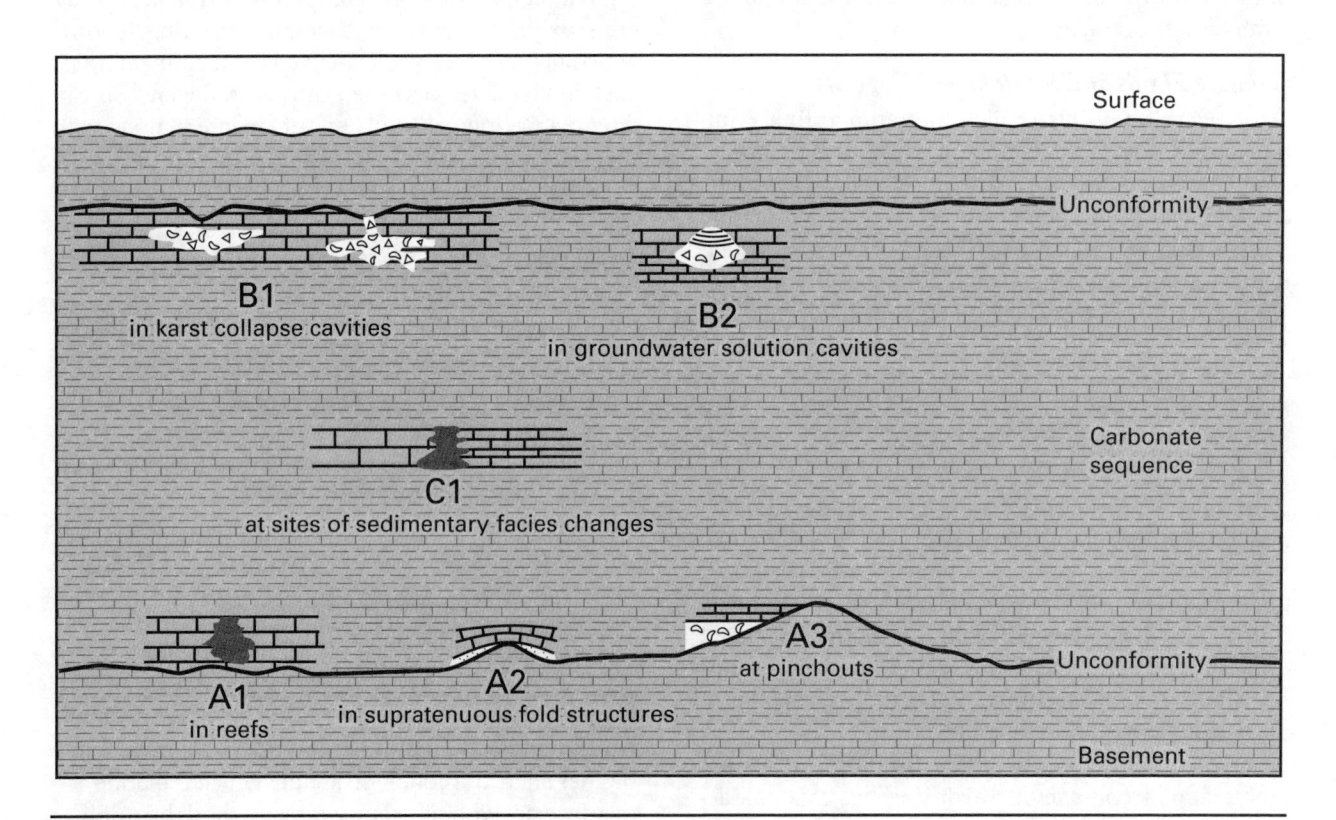

Figure 11.15
Idealized cross-section illustrating the formation of carbonate-hosted lead-zinc deposits at stratigraphic discontinuities: A – above unconformities; B – below unconformities; C – between unconformities. Deposits may also form in veins (not shown).
(*Source:* Modified from Callahan, 1967.)

and the clastic sedimentary rocks that overlie them. These **unconformity-related uranium deposits** demonstrate an unusual property of chemistry critical in the development of all uranium deposits. Oxidized uranium (U^{6+}) forms numerous stable and soluble complexes with a wide variety of anions at all pH values. In contrast, reduced uranium (U^{4+}), which forms uraninite, is highly insoluble under many of the same conditions (figure 11.18 on p. 253). Oxidizing fluids can therefore scavenge uranium from the relatively minor amounts included in various uranium-bearing minerals spread over a wide area and can easily transport the metal until reducing conditions are encountered. Zones of basement permeability, such as faults and fractures, may act as conduits for convecting fluids. Where the fluid encounters carbonaceous material that lowers the oxidation potential (the Eh) of the fluids, the metal is deposited to form the ore (figure 11.19 on p. 253).

A related type of mineral deposit forms in so-called "failed" intracontinental rift zones. Copper is a common ore that develops here, in sections of strongly oxidized, coarse-grained sediments containing scattered organic matter. These sediments and the resulting sedimentary rocks are called *red beds* because of oxidation of iron. Red beds, capped by marine black shales containing considerable organic matter, may encounter connate fluids that enter the sediment- or rock-filled rift system. The organic matter and fluids react to yield distinct types of mineralization. In one type, fluids are restricted to the red-bed sediments and the higher porosity of these sediments allows for rapid fluid flow. Combined with the less-abundant organic material, high flow rates minimize mineralization.[11]

Figure 11.16
Typical breccia ore from a carbonate-hosted lead-zinc (Mississippi Valley type) deposit. Angular fragments of carbonate host rock are cemented by calcite (white) surrounding sphalerite (gray). U.S. Steel zinc mine, Mascot-Jefferson district, East Tennessee.
(*Source:* Photo by Neil Johnson.)

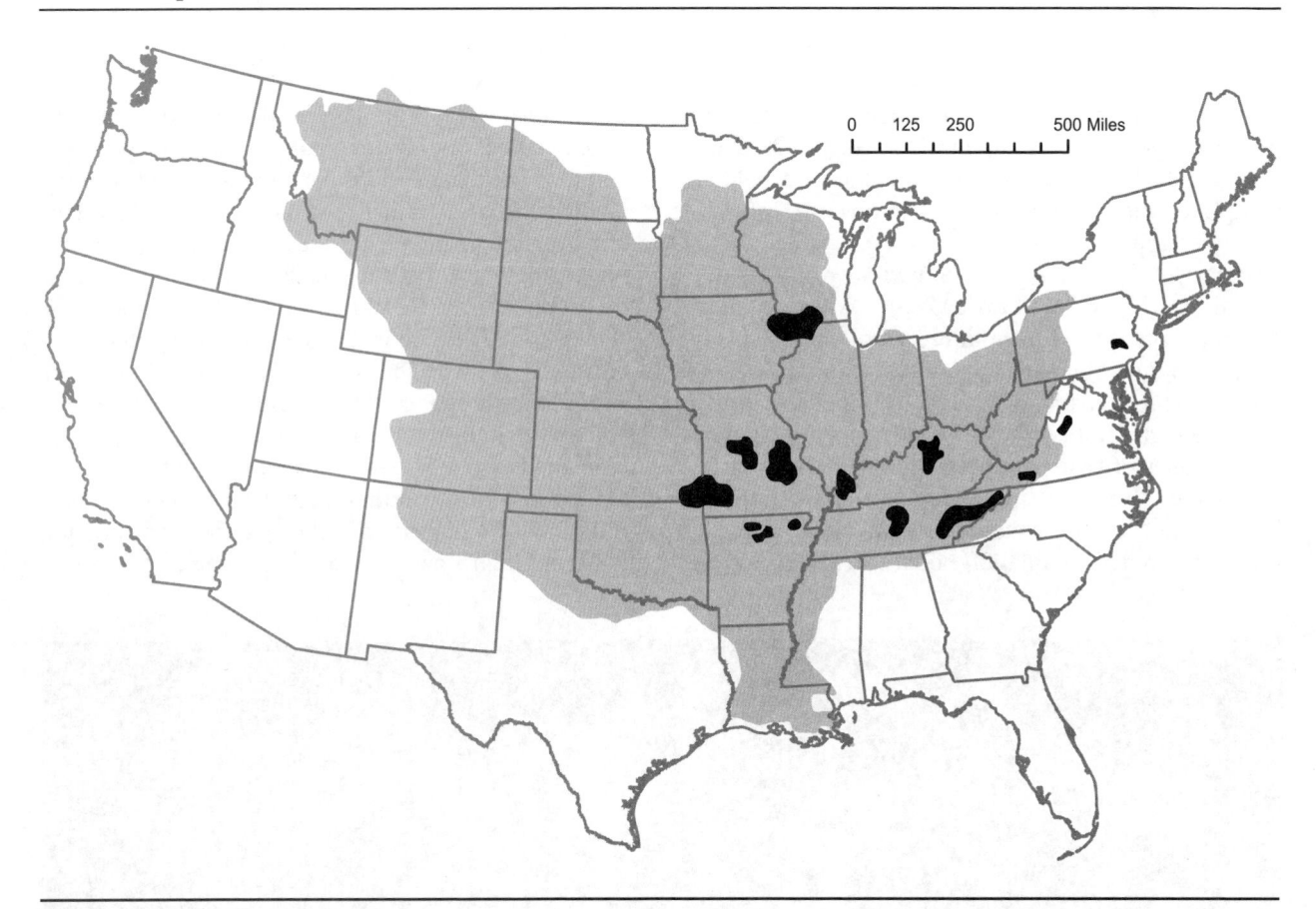

Figure 11.17
Map of the contiguous United States showing locations of major carbonate-hosted lead-zinc districts.
(*Source:* Modified from Craig, Vaughan, and Skinner, 2011. Used with permission of Pearson Education.)

In such cases, precipitation may produce disseminated copper or copper-iron sulfides in relatively small quantities, with ore minerals filling pores or replacing cements. In some deposits, there can be appreciable silver as well as copper. In a second type, connate fluids penetrate the overlying shales and the combination of lower flow rates (in shale) and abundant organic matter results in very large deposits of *copper-rich shale* consisting of laminated copper sulfides with significant quantities of cobalt, silver, nickel, and uranium. Examples of large deposits of this kind include those of the Zambian copper-cobalt belt and the deposits referred to as the European Kupferschiefer.[12]

Groundwater/Meteoric Fluids and Associated Deposits

Groundwater plays a key role in two closely related types of uranium deposit, the **tabular** and the **roll-front** deposits. Tabular deposits form in medium- to-coarse clastic sediments or sedimentary rocks deposited at the edges of arid continental interior basins that hosted playa lakes.[13] The host sediments or sedimentary rocks include significant quantities of amorphous, partly decomposed organic material (humus), which is why they have been called uraniferous *humate* deposits. Just as with the unconformity-related deposits, the keys to deposition in tabular deposits are flow (or convection) of the metal-bearing fluid and changes from oxidizing to reducing conditions. In tabular deposit formation, the oxidizing groundwater from stream valleys mixes with reduced groundwater circulating from playas (figure 11.20 on p. 254). Precipitation (deposition) of uranium occurs geographically in a band roughly parallel to lake shorelines. Mineralization is in flat zones that parallel or nearly parallel the bedding.

Roll-front deposits also form in medium- to coarse-grained clastic sediments, but in the case of

Resource Geology 253

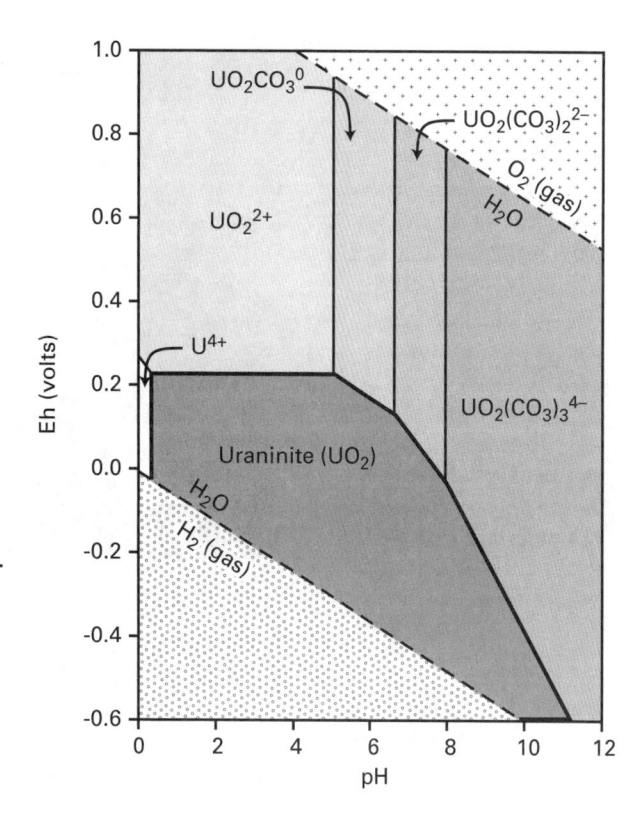

Figure 11.18
Eh–pH diagram that shows uranium species stable under various conditions. Under oxidizing conditions, a variety of uranium complexes are present as soluble species. Under reducing and low positive Eh conditions, solid uraninite forms.
(*Source:* Langmuir, 1978.)

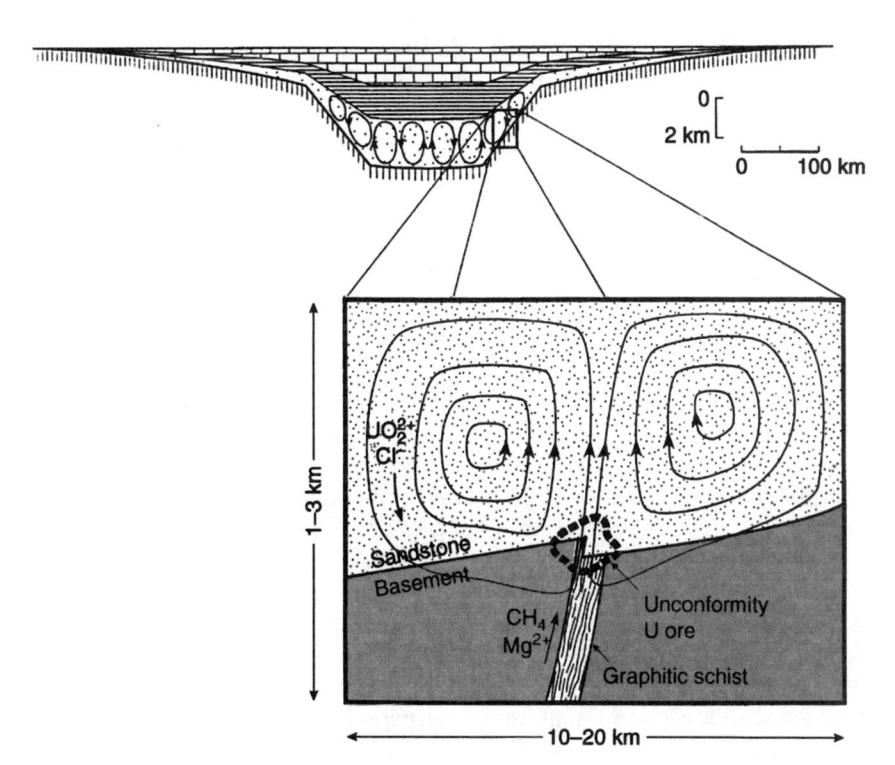

Figure 11.19
Typical setting of a Proterozoic unconformity-related uranium deposit. In old basement rocks, methane and magnesium-bearing fluids flowing up through a layer of graphitic schist (of course, containing carbon that reduces the oxidation potential of associated fluids) encounters the unconformity between the underlying rocks and the overlying sandstone sediment. Here, connate fluids circulating in the sandstones encounter the rising reducing fluids, reducing the oxidation state of the contained uranium and causing precipitation of uraninite ore.
(*Source:* Ridley, 2013. Used with the permission of Cambridge University Press.)

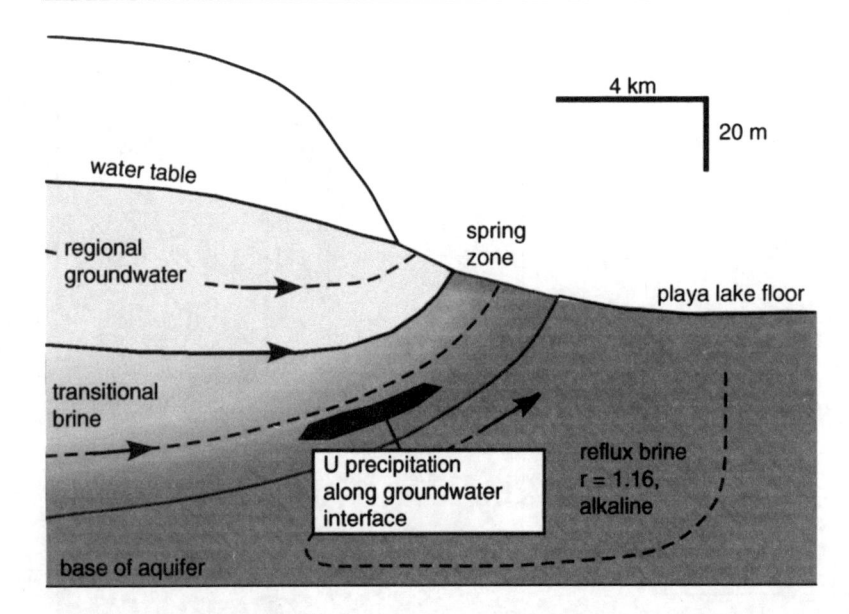

Figure 11.20
Possible groundwater flow paths (arrows) during the genesis of a tabular uranium deposit.
(*Source:* Ridley, 2013. Used with the permission of Cambridge University Press.)

roll-front deposits the clastic sediments or sedimentary rocks act as aquifers (water-transporting layers) between less permeable layers (aquicludes) located above and below the permeable layer. As uranium-bearing groundwater penetrates the aquifer, it oxidizes remnant organic material in the unaltered sediment. Thus, an oxidized-reduced boundary, a *reaction front,* develops between the oxidized zone and the unaltered reduced zone. The front has a vertical, broadly arcuate shape, and in this front the uranium minerals precipitate (figure 11.21). As the oxidizing waters move through the aquifer, the mineralized reaction front "rolls" along with it, moving the mineralizing zone forward and increasing the size of the deposit.[14] Because the chemistry of vanadium is similar to that of uranium in terms of the changes in solubility associated with the redox state, appreciable amounts of vanadium are also deposited with uranium in roll-front deposits. The uranium-vanadium minerals are frequently brightly colored, which greatly aided exploration for uranium in the Colorado Plateau region in the early 1950s. As is widely known, uranium is used both in energy generation and in specialized uses such as in the production of nuclear weapons. Vanadium is used primarily for strengthening steel.

The term **supergene** is used to refer to mineralization resulting from remobilized, *descending* fluids, principally rainwater. Where mineralized rocks occur at the surface, they are subject to weathering. Nearly ubiquitous pyrite in these mineralized rocks weathers to produce iron oxyhydroxide minerals like goethite (FeOOH) that form a gossan—a leached mass of iron oxide—at the surface. As rainwater interacts with pyrite (and other minerals) to yield oxidizing and acidic fluids, the fluids dissolve other economic minerals and carry their metals down toward the water table. There, the minerals are deposited. Supergene enrichment normally occurs at or near the water table, as the descending fluid is reduced upon mixing with groundwater (figure 11.22 on p. 256). Supergene deposits are relatively tiny compared to other deposit types discussed, but they can be rich in ore. A silver enrichment encountered in the Atacama Desert of Chile contained native silver fragments up to 90 kg in mass and a body of silver and embolite (Ag(Cl,Br)) in excess of 20,000 kg.[15] In contrast to supergene deposit formation, fluids *ascending* from deeper levels of the crust yield **hypogene** deposits.

The **residual** deposit, another type of deposit formed near the surface, develops in place (*in situ*) from the chemical weathering of original rock that is close to the surface of Earth. Although at first glance the definition of residual deposit might seem to apply to certain placers (described below), the designation is most commonly restricted to large accumulations of insoluble minerals that either form at or near the surface or are left behind as more water-soluble compounds are transported away. As an example, consider the weathering, in a tropical environment, of an aluminum-rich rock like the igneous rock *syenite*. Under the conditions of warm temperatures, voluminous rainfall, and long periods of tectonic stability, K and

Figure 11.21
Typical setting, shape, and petrographic characteristics of a roll-front uranium deposit.
(*Source:* Ridley, 2013. Used with the permission of Cambridge University Press.)

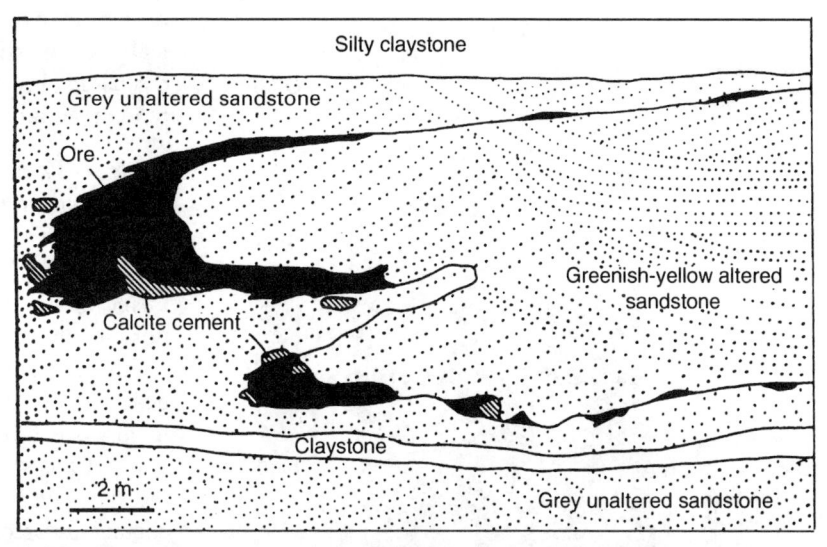

(a) Representative shape and ore distribution, with calcite cement zones, of a roll-front uranium deposit.

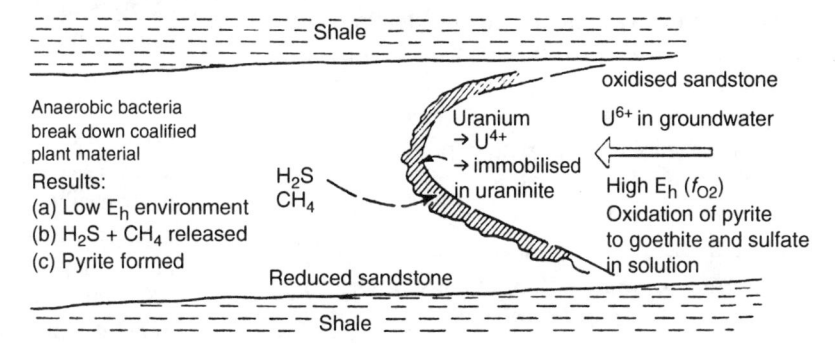

(b) Roll-front deposit formation, with high E_h groundwater fluid interacting with low E_h sandstone.

Na ions are readily dissolved out of the rock and even the silica is removed, albeit much more slowly. Eventually, what remains are highly insoluble iron and aluminum oxides and hydroxides (table 11.3). This remaining material is called **laterite**. Via such a leaching process, an unweathered rock containing 23 weight percent Al_2O_3 can be converted into a laterite residual deposit of 69 weight percent Al_2O_3.

Aluminum-rich laterites are referred to as *bauxites* and consist of varying amounts of the aluminum minerals gibbsite ($Al(OH)_3$), boehmite (g-$AlO(OH)$) and, to a lesser extent, diaspore (α-$AlO(OH)$). The parent rocks for these laterites are typically aluminum-rich plutonic and metamorphic rocks with more than 12 weight percent Al_2O_3 or carbonate rocks containing thick layers of aluminous volcanic ash.[16] Laterite deposits represent the only current economic source of aluminum and typical deposits occur in Jamaica and southwestern Australia. In areas where rainfall is not sufficient to completely remove the silica, residual material consists of large accumulations of clays particularly rich in kaolinite. These clay-rich deposits are also valuable, as the clays are used to make ceramics and to add gloss to paper for printing. Similarly, weathering of ultramafic rocks yields nickel laterites, which supply 30–40% of the world's nickel. Olivine is the principal mineral in the source rocks that yield the nickel, and small amounts of nickel and cobalt that substitute for the magnesium in the olivine structure are concentrated by weathering to make the enriched ore. The largest example of a nickel laterite deposit occurs in New Caledonia.

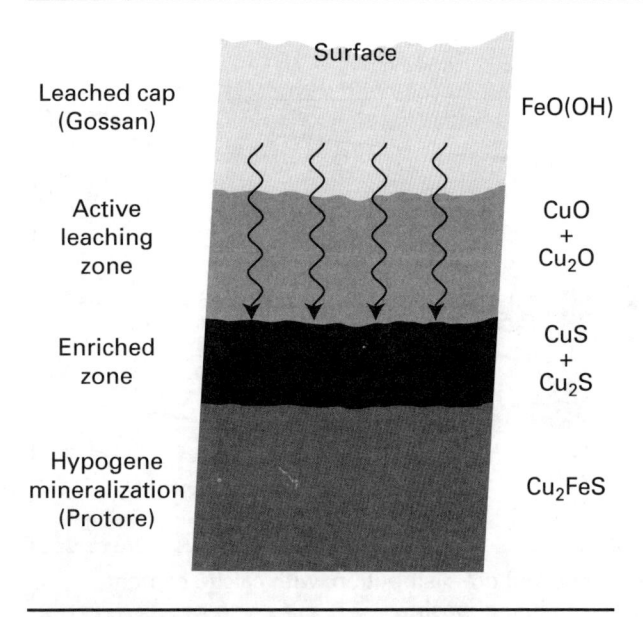

Figure 11.22
Development of a supergene profile. Idealized depiction of the weathering of an exposed chalcopyrite (CuFeS$_2$)-bearing ore deposit, leading to the formation of a gossan and a copper mineralized zone. Note the changes in the mineralogy with depth.
(*Source:* Modified from Titley, 1978.)

Table 11.3 Comparison of Syenite Rock and Laterite Chemistry

Oxide	Original Rock Analysis (%)	Weathered Rock Analysis (%)
SiO$_2$	52.5	—
Al$_2$O$_3$	22.7	69
Fe$_2$O$_3$	2.0	12
FeO	1.6	—
CaO	1.2	–
Na$_2$O	8.2	—
K$_2$O	6.6	—
H$_2$O	4.8	15
Total	99.6	100.0

(*Sources*: Syenite analysis from Nash and Wilkinson, 1970. Typical bulk chemical analysis of a syenite, compared with residual values calculated from this analysis, once soluble components have been removed.)

Sedimentary Deposits

Sedimentary ore deposits form via both physical processes of concentration and chemical reactions. The physical processes of concentration depend on properties such as the mass of the ore minerals. Chemical concentration, as in other types of ore formation discussed above, occurs as a result of reactions such as redox reactions.

Surface Water Precipitation and Associated Ore Formation

Under unusual circumstances, metals that are relatively insoluble under normal surface conditions, including iron and manganese, may be taken into solution in abundance. Unusual conditions, particularly highly acidic conditions in the waters of surface to near-surface continental to deep-ocean basin environments, are favorable for iron dissolution and subsequent transport. When the conditions change, the metals may be deposited from the waters to form chemically precipitated iron- or manganese-rich minerals. Some "bog iron" ores, ironstones, and sedimentary manganese deposits may form in this way, where changes in chemical environments yield deposits associated with shallow continental sea, estuary, swamp, or shallow-passive margin and oceanic sedimentary rocks.[17] Iron deposits of this type are largely sub-economic due to their small size or hard-to-process mineralogy. In contrast, sedimentary manganese deposits are more easily exploited and are a major source of the element manganese. South African Kalahari deposits contain nearly 80% of the world's manganese reserves.[18]

Major sources for iron ore are Precambrian **banded iron formations** or **BIF**s, examples of which occur in Brazil, adjacent to Lake Superior in the United States and Canada, in Labrador, near Kursk in Russia, and in Western Australia. These deposits are individually and collectively enormous, with approximately 10^{15} tons of iron, enough for a million years of consumption at current rates of production. The term BIF derives from the distinctive nature of the rocks, which is characterized by a series of alternating bands of specular hematite and red hematite-rich chert (called jasper) (figure 11.23a).

There are three distinct types and ages of BIFs.[19]

- *Algoman* deposits are of Archean age and are the oldest. Their association with greenstone (volcanic) belts implies that the iron originated from submarine volcanism.
- *Superior* deposits are by far the largest and these occur in marine basin sediments of passive margins of Paleoproterozoic age, suggesting that at that time they were deposited, dissolved iron from continental sources was important in their formation.

(a) Polished slab of banded iron formation (BIF), Hammersley Range, Western Australia. (*Source:* Photo by Neil Johnson.)

(b) Hematite-cemented, oolitic sandstone iron ore (ironstone), Clinton, NY. Scale intervals are in cm. (*Source:* Photo by Loren A. Raymond.)

Figure 11.23
Photos of two major types of iron ore rock.

- *Rapitan* deposits are the youngest BIFs and these are associated with glaciomarine sediments of Neoproterozoic age, during which the so-called "Snowball Earth" glaciation occurred.

Interpretation of the origin of BIF deposits is controversial and initial analyses are based on four key characteristics.[20] (1) BIFs are basically rocks of sedimentary character, considering their laminated nature, location in depositional basin settings, and association with other sedimentary rocks and fossils. (2) The vast majority of BIF deposits date from a period of time, short by Precambrian standards, with deposition between about 2.6 and 1.8 billion years ago. (3) BIF deposits are exceptionally low in aluminum and phosphorus (< 1%). (4) The current atmosphere is too rich in O_2 for large quantities of iron to be carried in solution, so that conditions at the time of deposition must have been more oxygen depleted. One hypothesis of BIF origin is based on evidence that the atmosphere 2.6 billion years ago was poor in O_2 and richer in CO_2. As a result, dissolved iron could develop and form bicarbonate compounds that could be transported to ancient basins by rivers and streams. Here, the iron would be precipitated as iron oxides, silicates, or carbonates—depending on the conditions. The end of much BIF deposition occurred after (1) atmospheric oxygen increased between 2.6 to 1.8 billion years ago, (2) iron solution and transport gradually decreased, and (3) oxygen levels reached prohibiting levels. The Neoproterozoic Rapitan-type deposits that formed later are attributed to sea ice cutting the oceans off from the oxygenated atmosphere, allowing Fe^{2+} to accumulate in local settings.

The textures and chemistry of BIFs are more complex than simple models of sedimentation predict. Sources of ferrous iron in solution may be both hydrothermal and terrestrial. Furthermore, abundant and complex replacement textures indicate post-depositional changes in the ores caused by migrating fluids. These textures also suggest ore enrichment may have been caused by migrating fluids of meteoric, diagenetic, and basinal derivation (via both supergene enrichment and hypogene enrichment processes) (Simonson, 2011; K.A. Evans et al., 2013). These complexities mean that the origins of each deposit must be assessed individually and that development of ore bodies was multifaceted and more complex than one of simple sedimentation.

Manganese, which is abundant in Earth's crust, was largely concentrated during limited and specific periods of geologic time.[21] Notable deposits are associated with BIFs of Paleoproterozic age, with Mesozoic cherts (color plate 29b), and with Cenozoic marine shales and carbonate sedimentary rocks. Hence, sedimentary manganese deposits can be classified into mud/shale types, carbonate rock types, and chert-related types. MnO_2 precipitation and seafloor sediment alteration in anoxic (oxygen-poor) marine basins formed many initial ores, and bacterially mediated diagenetic reactions probably influenced formation of the ores (Du et al., 2013). Like the BIFs, alteration and replacement commonly modified and enriched the ore bodies, which contain minerals such as rhodochrosite and pyrolusite.[22] Chert-hosted manganese deposits have been important in the past in the United States and chert-hosted ore bodies are known elsewhere, where they form deposits that formed either in restricted basins or on the open ocean floor. The latter typically formed as manganese nodules and crusts.

Although the principle current sources of manganese are sedimentary deposits mined on the land surface, by far the greatest potential manganese resource is found in the deep ocean. Concentrically banded nodules, varying in size from 2 to 30 cm, litter areas of the oceanic abyssal plains (figure 11.24). Consisting of roughly 20 weight percent Mn, 15% Fe, and up to 2% Cu+Co+Ni, these nodules are estimated to contain in excess of 1,000 billion tons of extractable material.[23] The nodules occur only on the very deep ocean floor, well away from any terrestrial sediment sources, and have exceedingly slow growth rates—on the order of a few millimeters per million years. The source of the metals is poorly understood, with proposed hypotheses suggesting that they were (1) leached from oceanic basalts, (2) dissolved from the continents, or (3) remobilized from surrounding sediments. How the metals became concentrated is also contentious, with microbial activity, localized seafloor volcanism, or diagenetic reactions among the suggested mechanisms.

A significant reason why the Mn-nodule-forming processes are somewhat poorly understood has to do with the physical and legal difficulties of nodule mining. Unlike the terrestrial or near-shore regions, where ownership issues are well understood, the legal status of mining in international waters has been in dispute for a number of years. In addition, environmental questions about mining large areas of the seafloor have raised serious concerns. Until such disputes are resolved, efforts to more fully understand and perhaps extract this resource will likely proceed at a slow pace.

Phosphorites are defined as rocks containing more than 15–20 weight percent P_2O_5. The majority are layered, marine sedimentary rocks with shallow water fossils. These rocks were deposited at low latitudes during Phanerozoic time. The mineralization occurs as **cryptocrystalline** apatite (referred to as *collophane*), more than 90% of which is used in the production of fertilizer.

Phosphorus is poorly soluble in ocean waters, but this solubility differs between deep ocean waters, which are colder and more acidic (and contain 0.3 ppm), and the warmer, more alkaline near-surface waters (containing less than 0.1 ppm).[24] The differences in temperature and pH mean that zones of oceanic upwelling, such as occur off the west coast of South America, transfer phosphorus-richer waters to warmer, shallow zones, where it precipitates. The upwelling also produces extensive biotic activity, since phosphorus is an essential nutrient, and various organisms enhance phosphate deposition. Although one of the largest known examples of a phosphorite deposit is the Phosphoria Formation that extends from Montana to Utah in the United States, the most currently productive U.S. deposits are found in central Florida and at Aurora, North Carolina (figure 11.25 on p. 260). These eastern U.S. deposits formed in response to upwelling gyres in the Gulf Stream during late Oligocene and early- to mid-Miocene time.

Evaporites are precipitated rocks composed of mineral salts. Typical evaporite minerals include halite (NaCl), sylvite (KCl), gypsum ($CaSO_4 \cdot 2H_2O$), carbonate and bicarbonate compounds including calcite ($CaCO_3$), dolomite ($CaMg(CO_3)_2$), and nahcolite ($NaHCO_3$), plus magnesium and lithium salts and boron minerals (color plates 18; 19a–c; 28c). Natural waters, both fresh and oceanic, will precipitate a variety of salts when they evaporate. The main requirement for precipitation is that water loss via evaporation exceeds inflow from any outside source. Abundant precipitation yields an economic resource.

There are two major types of settings that produce evaporite resources. The first is the fresh-water lake basin with interior drainage. Such basins have no outlets and evaporation concentrates salts in the residual waters. Death Valley, California and the Great Salt Lake, Utah are modern basins of this type. Settings with large amounts of evaporates, in some cases, are the hypersaline remnants of much larger lakes, as is the case with the Great Salt Lake of Utah. Alternatively, they may be ephemeral lakes that evaporate completely, as is

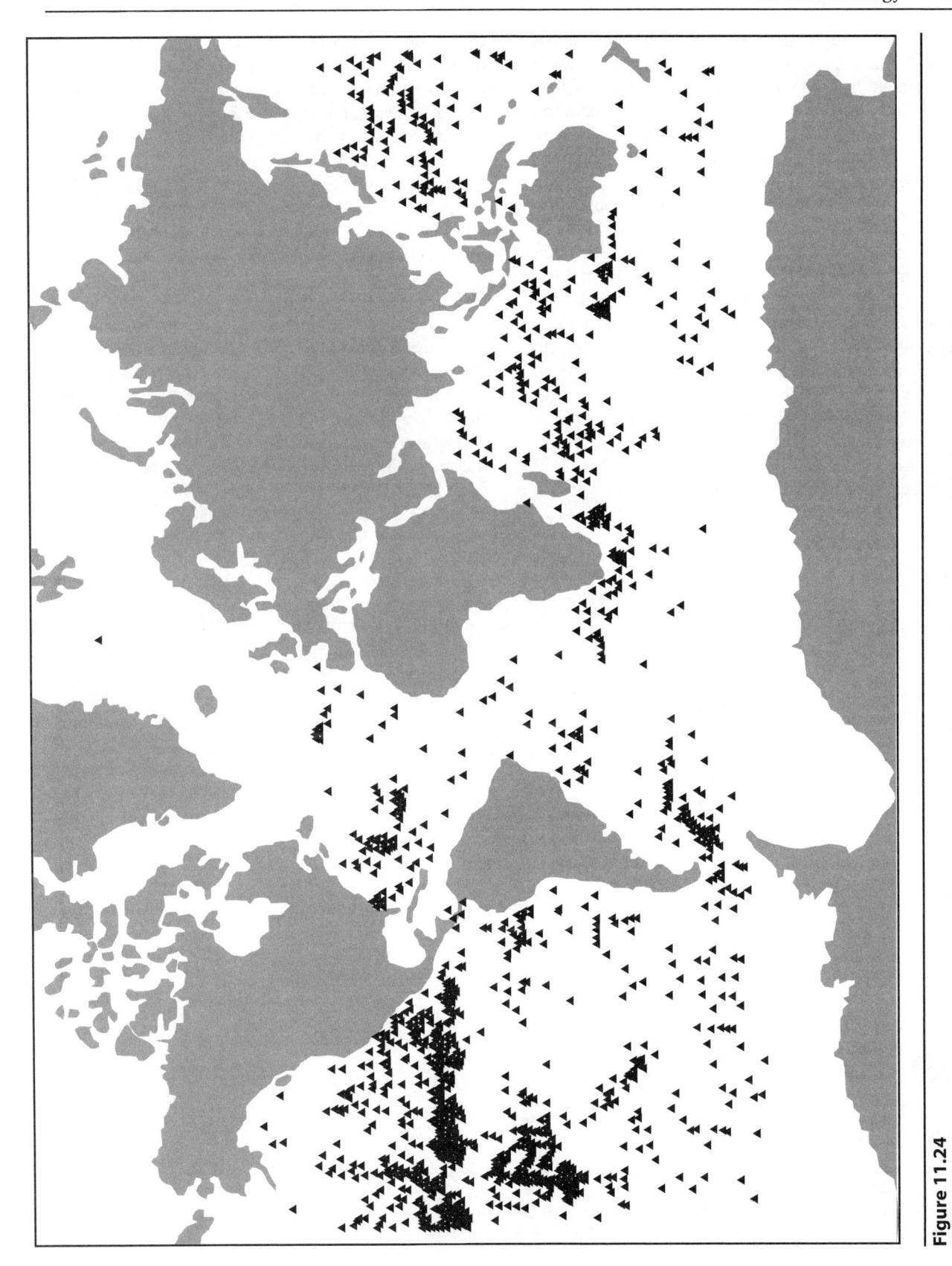

Figure 11.24
Map showing manganese nodule stations listed in the Scripps Institution of Oceanography's Sediment Data Bank.
(*Source:* McKelvey et al., 1983.)

the case in Death Valley, California and many valleys across Nevada. Since hypersaline lakes generally do not evaporate to dryness, the minerals produced are mostly halite, but under appropriate conditions evaporation will yield a mineral suite characterized as $MgSO_4$-poor, containing minerals such as halite, sylvite, aragonite, and trona $(Na_3H(CO_3)_2 \cdot 2H_2O)$.[25] Locally, in addition to halite, ephemeral lakes may deposit the potassium and magnesium salts collectively called *bitterns*. If evaporite-forming lakes develop in continental rifts or other sites of magmatic activity, circulating hydrothermal fluids may transfer magmatic boron to the lakes; and precipitation may yield more exotic minerals such as borax $(Na_2B_4O_7 \cdot 10H_2O)$ and ulexite $(NaCaB_5O_9 \cdot 8H_2O)$. Such continental rift sites are the largest sources of boron currently extracted.

The second class of evaporite setting, which is much larger than the first, is the restricted marine basin. Such "basins" include the *sabkhas* (salt flats) along arid coastlines.[26] As evaporation proceeds in these settings, the basin waters become more concentrated and decrease in volume, causing additional seawater to enter the basin. In other words, evaporation drives the system, causing inflow to continue. In general, the process is continuous and occurs over time, and precipitation is continuous as well. Large deposits of sulfates like gypsum, sylvite, and kainite $(KMg(SO_4)Cl \cdot 3H_2O)$ are produced in this way. Lateral zonation of the salt types is typical in these settings, and the mineral suites may be characterized as $MgSO_4$-rich.

Detrital Deposits

The processes of weathering, transportation, and deposition may combine to concentrate minerals that form economic deposits. In chapter 8 it was noted that minerals weather at different rates, and therefore slower-weathering minerals, especially if they are above average in specific gravity and are physically durable, can be concentrated. **Placer** deposits are created in moving waters by gravity-driven separation

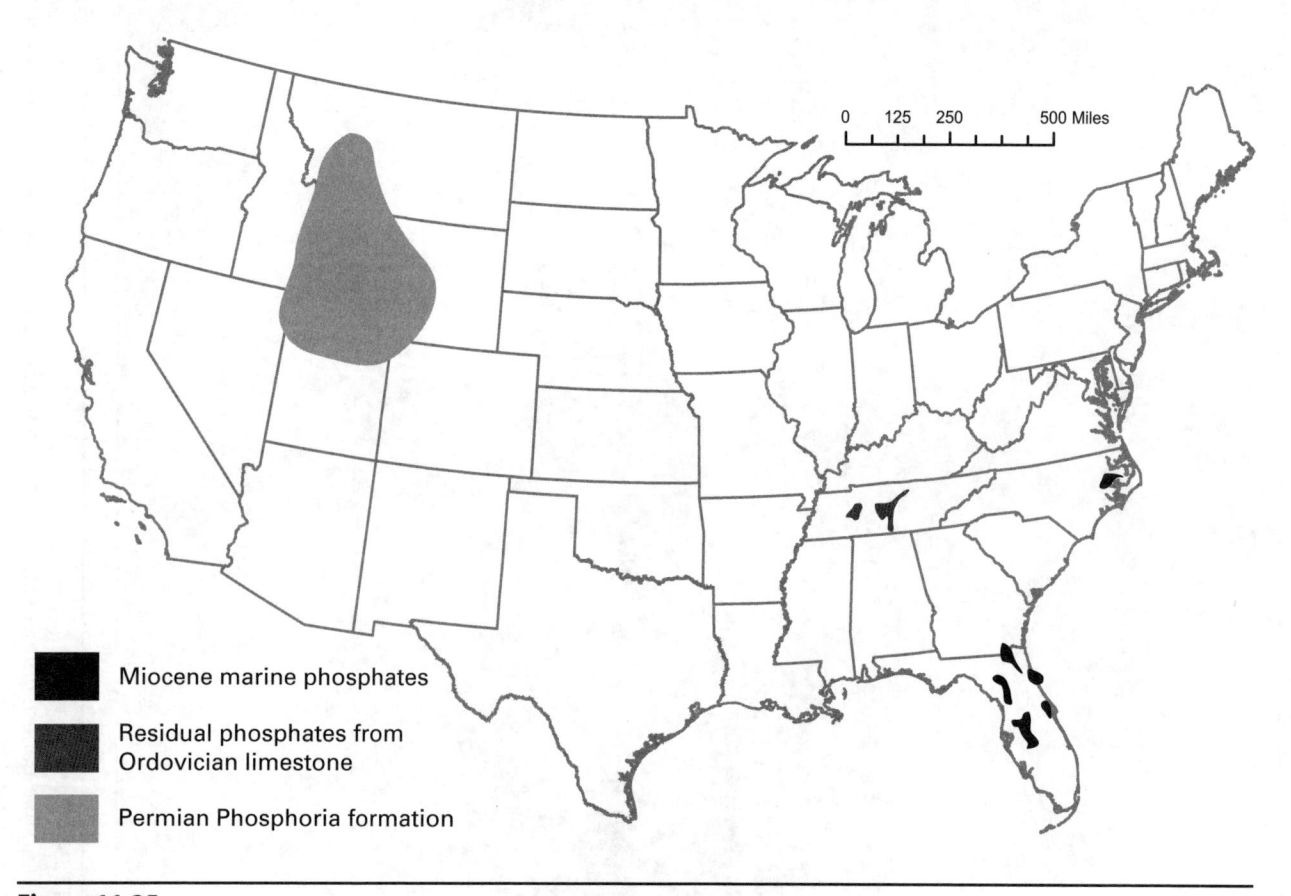

Figure 11.25
Map of major phosphorite deposits in the conterminous United States.
(*Source:* Modified from Craig, Vaughan, and Skinner, 2011. Used with permission of Pearson Education.)

(settling) of heavy, chemically and physically resistant minerals such as gold, platinum, diamond, zircon, rutile, cassiterite, or monazite. These minerals are eroded upstream, transported by flowing waters, and deposited, as flow velocity decreases and gravity pulls the heavier minerals to the bottom of the stream. In the classic picture of a grizzled forty-niner panning gold nuggets from a stream, the miner is mining gold from the easiest deposit to mine, a stream placer.

Other deposits, related to placer deposits, are formed by some or all of the same processes that yield the placers. Resistant mineral deposits that may exist in place above the source are called *residual* deposits (figure 11.26a). Downslope from the source, deposits that form where sediments have moved down are designated as *eluvial*, but if the deposit represents sediments stranded by stream migration or downcutting, it is called a *terrace* deposit.

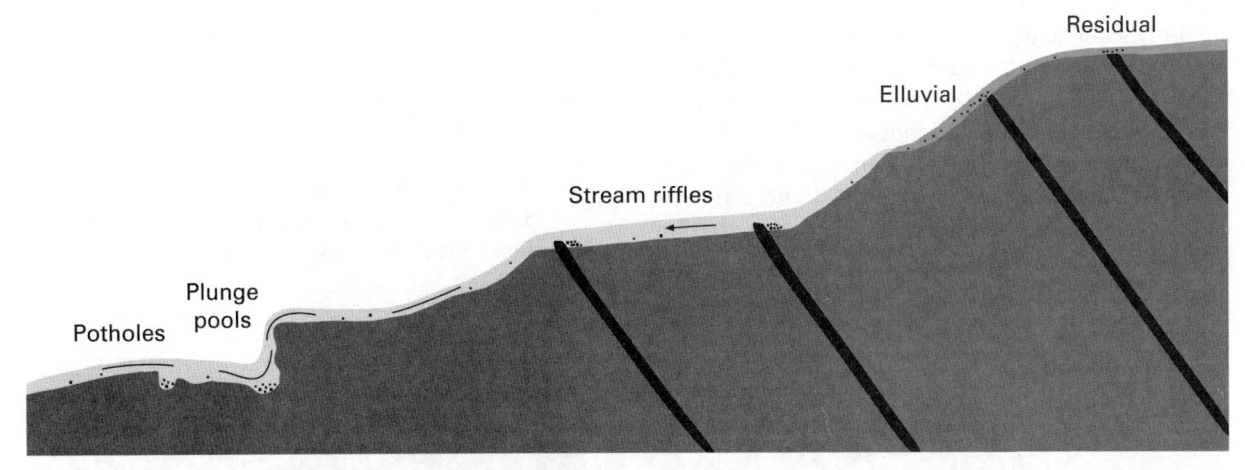

(a) Cross-sectional schematic showing setting of placer formation (dots) in residual sites above veins, in elluvial sites on hillsides, in the riffles and potholes of stream beds, and in the plunge pools of waterfalls.

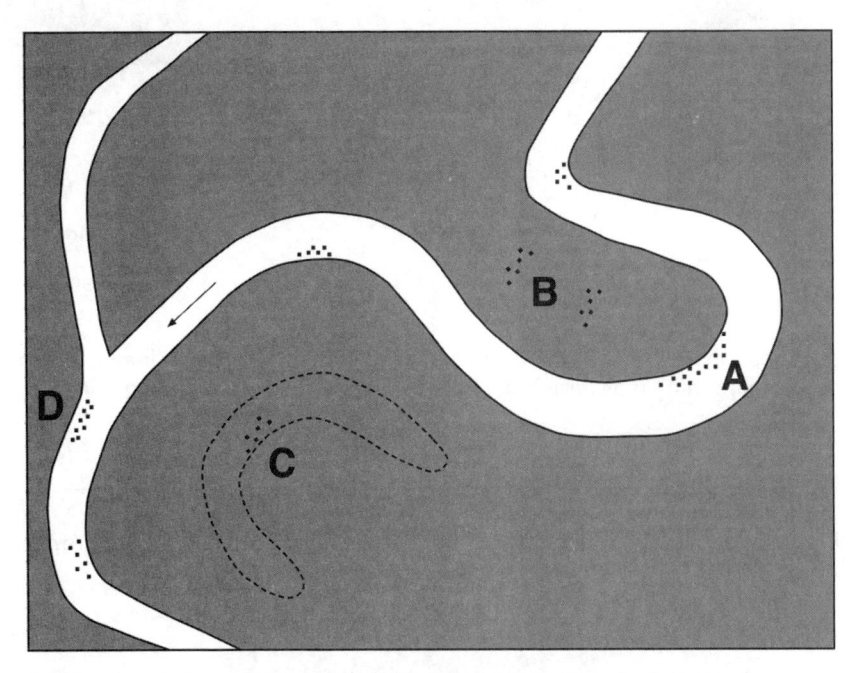

Figure 11.26
Placer and related deposits.
(*Sources:* Modified from Lindgren, 1913 and Craig, Vaughan, and Skinner, 2011. Used with permission of Pearson Education.)

(b) Overhead view of placers (dots) formed at locations in rivers and streams, such as point bars (A), former point bars (B and C), and downstream from the confluence of two streams (D).

The placers *sensu stricto* form in several settings, including behind stream riffles and in potholes that occur in plunge pools of waterfalls (figure 11.26a). Placer deposits can also form in point bar deposits or where sandbars develop at the intersection of fast-moving tributaries and slower-moving main streams (figure 11.26b). Beach placers develop in coastal areas because incoming turbulent waves transport all sediments, but the laminar flow of the retreating waves strands the higher-density minerals at the maximum reach of the wave. While most such deposits are small and only suitable for artisanal mining, the majority of the world's titanium is produced from rutile/ilmenite beach placers in Australia and the East Coast of the United States. Significant amounts of REEs are recovered from monazite placers.[27] The largest deposit of gold, the Witwatersrand deposit in South Africa, is almost uniformly considered by geologists to be a *paleoplacer* deposit (figure 11.27).

OTHER MINERALS, ROCKS, AND ENERGY RESOURCES

The (nonmetallic) **industrial rocks and minerals** include a large variety of materials. Nonmetallic economic deposits range from the gravel and sand deposits that are used in construction and mined from stream beds and river beds, to deposits of nonmetallic minerals and rocks with properties useful in industry and modern life. The latter, called industrial minerals and rocks, are used extensively for construction and the manufacture of such things as paper, paint, and porcelain. While some of these materials seem neither glamorous nor of little value per unit volume, in fact they are major resources used in society.

Some industrial minerals and rocks were described previously, such as pumice, olivine, high-

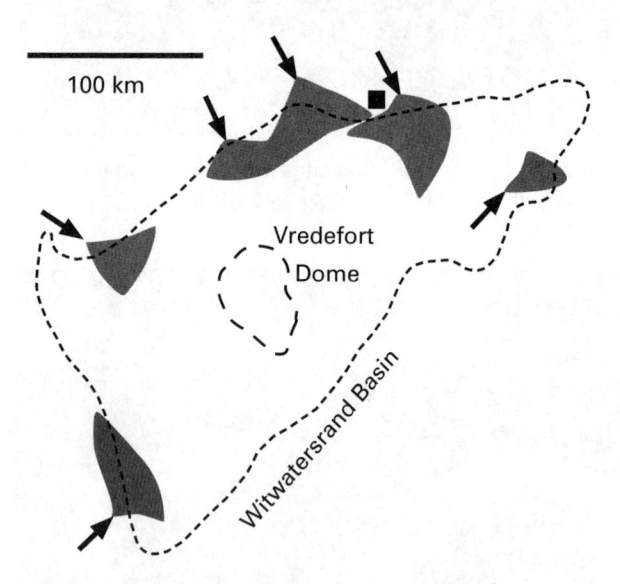

(a) Locations, relative to the basin as a whole, of the fluvial, shoreline fans (gray) that host the major gold deposits.

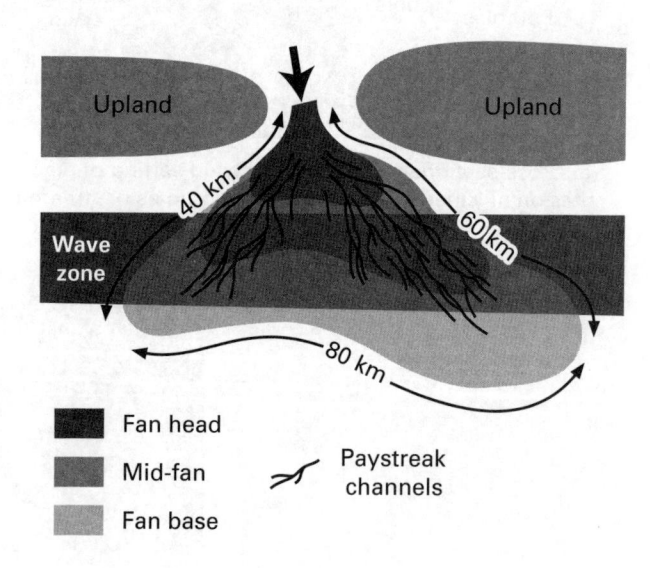

(b) Idealized reconstructed map of the fluvial fans giving rise to Witwatersrand gold deposits, illustrating the relationship of the gold-rich (paystreak) channels to the zone of wave action.

Figure 11.27
Paleoplacer nature of the gold fields in the Witwatersrand Basin, South Africa.
(*Source:* Modified from Pretorius, 1981.)

purity quartz, and diamond. In addition, a variety of igneous, metamorphic, and sedimentary rocks are mined simply for crushed stone used in construction and landscaping and, as noted, sand and gravel sediments are dredged and mined as a major construction resource. Pure quartz sand is used in glass-making and in the processes involved in hydraulic fracturing ("fracking") of bedrock for development of natural gas resources. Many minerals have properties of use, including the zeolites used for water purification; the barite used in oil-drilling muds; the gypsum used for making wallboard employed in home construction; and the alkali feldspar used in production of a variety of commercial products such as paint and porcelain. These are just examples of a diverse array of nonmetallic, industrial economic resources, all of which cannot be covered here.

A final major Earth-material resource is the energy resource. Coal, petroleum, natural gas, and uranium are major energy resources that have powered either personal automobiles and appliances or industrial enterprises for more than 150 years. The first three of these are nonmetallic resources. Coal forms by compression, and in some cases metamorphism, of buried vegetation deposited in continental swamp environments. In contrast, petroleum and related fuels are likely derived from microscopic animal matter deposited in and with marine sediment. Hence, the dominant view is that petroleum has an organic origin. The presence of elements and compounds typical of animals (e.g., nitrogen) within petroleum, plus the extensive association of petroleum with marine sedimentary rocks, favors such an origin. Nevertheless, some individuals argue for an abionic origin in the mantle.

> **Problem 11.4**
>
> Use the Internet to expand your knowledge of global warming and petroleum resources, then answer the following questions.
>
> (a) What percentage of annual worldwide energy consumption is based on hydrocarbon resources (coal, oil and related fuels, and natural gas)?
>
> (b) What viable alternatives to continued fossil-fuel consumption will allow for progressive growth in the developing countries and continuation of a relatively high standard of living in the developed countries?

Petroleum has many important economic uses. It is used as a fuel and for the manufacture of chemicals and plastics, notably synthetic rubber and fibers. In addition, petroleum is used in making some medicines, paint, and pesticides (Textor, 2010). Natural gas is similarly used to generate electricity and power and to manufacture plastics and other materials.

We have finally learned that use of hydrocarbon fuels, especially coal, is having major negative environmental consequences, most notably by enhancing global warming. Combustion of petroleum yields CO_2 as a by-product, and CO_2 traps heat in our atmosphere. Thus, as a world society, we are caught between continuing to power world economic growth with this crustal Earth material or radically altering the socioeconomic structure of society in order to avoid the consequences of rising sea levels, erratic weather, changes in climate, and consequential impacts on world agriculture, ocean acidity, and food resources.

Summary

Most of the metallic and nonmetallic resources, plus the majority of energy resources, that are extracted for modern civilization come from the Earth. What makes an Earth material a resource is entirely based on economics. Factors involved include the market value of the resource at any given time, knowledge of the geology and quantity of the resource in the deposit, and the accessibility of the resource in the Earth.

The study of resources is complicated by the lack of a generally accepted classification scheme. The classification scheme used here is genetic, with deposits divided into those that form due to igneous processes, those that form via sedimentary processes, and those that form as a result of hydrous fluid processes.

Magmatic deposits form where magma separates or segregates into multiple components, where magmas simply crystallize to form valuable rock bodies, or where magmas containing a valuable resource erupt at the surface. Fractional crystallization of magma yields intrusive bodies with layers of magnetite, chromite, olivine, and locally, platinum-group metals. Segregated fluids can produce massive sulfide bodies of pyrite and pyrrhotite (if the segregating fluid is sulfur-rich), pegmatitic rocks containing rare-earth minerals and gemstones (if the segregating fluid is water-rich), or carbonatites with rare-earth carbonate minerals (if the fluid is carbonate-rich). Volcanic processes produce a variety of important resources, the most notable of which is diamond, brought to the surface as xenocrysts in low-density melts that rise rapidly from the mantle to form lamproites and kimberlites.

Hydrothermal fluids produce a number of resource deposits, notably in veins precipitated from fluids that traverse underground fractures. Veins typically include both valuable ore minerals and disposable gangue minerals. Veins composed of ore and disposable gangue minerals include high- and low-sulfidation epithermal deposits of silver and gold in cross-cutting vein "stockworks" of porphyritic plutons. These stockworks may also contain copper, molybdenum, tin, and tungsten. Where hydrothermal fluids with metal ions are released in ocean floor environments, they react with seawater or diagenetic fluids in sediments to produce volcanogenic massive sulfide (VMS) or sedimentary exhalative (SEDEX) deposits of pyrite, pyrrhotite, sphalerite, galena and chalcopyrite.

Connate or formational fluids also yield resource deposits in sedimentary basins. Fluids coming from deep basinal levels commonly carry dissolved metals that are deposited in thick carbonate rock sequences to form carbonate-hosted lead-zinc deposits containing sphalerite, galena, fluorite, and barite. Oxidizing fluids can extract uranium from Archean basement rocks and later deposit it where reducing conditions occur in sediments that unconformably overlie these crystalline rocks. Fluids in failed rifts can deposit some copper minerals in oxidized red-bed sediments containing organic material but can also deposit large volumes of copper cobalt, silver, nickel, and uranium in overlying marine shales.

Resources produced via precipitation or deposition from low-temperature surface or near-surface waters include a variety of metals and nonmetals. Banded iron formations (BIFs), composed of alternating layers of precipitated hematite and chert, were formed early in Earth's history, largely in marine environments and before the atmosphere became enriched in the O_2 that inhibits solubility of iron in large quantities. Nodules of manganese precipitate on the deep ocean floor, but these have not been exploited because political, economic, and environmental factors restrict both investigation and exploitation. Some on-land sedimentary manganese deposits are mined. Phosphate rocks that precipitated from the upwelling, cold, deep ocean water are also a valuable resource. Evaporite salt deposits form by continuous evaporation in restricted marine basins and tidal flats or by evaporation of lakes in interior continental basins. The former process is responsible for the major deposits of salt. Residual or lateritic deposits develop in places where weathering processes remove soluble rock components (including most cations and silica) and leave a residue enriched in materials such as alumina, clays, or nickel oxides. Descending meteoric waters can also dissolve primary minerals as part of weathering, then precipitate new ore minerals at or near the water table to produce supergene deposits of copper or silver. Similarly, meteoric fluids or groundwater form uranium deposits where migrating, oxidizing water with dissolved uranium encounters reducing conditions that cause the uranium to precipitate. Such uranium deposits form either in clastic sediments as tabular deposits at the edges of arid basins or in clastic aquifers, where mineralization occurs at a redox boundary to create roll-front deposits. Finally, detrital or placer deposits, which consist of minerals of greater density and durability than most rock-forming minerals, form where moving water slows enough to deposit concentrations of heavy mineral grains. This occurs in rivers or on beaches, forming deposits of gold, platinum, diamond, zircon, rutile, cassiterite, or monazite.

Nonmetallic industrial minerals and rocks include a huge range of materials, from gravel and sand that are mined from stream beds and used in construction, to specific nonmetallic minerals that have a value based on particular properties. The low monetary value of such materials per unit volume is deceptive, as they are of major value to modern society. A variety of igneous, metamorphic, and sedimentary rocks are mined simply for crushed stone used in construction, and sand and gravel deposits are a major geological and construction resource. The volcanic rock pumice is used in a variety of commercial products. Nonmetallic minerals are used for such things as the manufacture of paper, paint, and computers. Other representative materials and their uses include industrial minerals such as barite, used in oil-drilling muds; and gypsum, used for making wallboard important to home construction. These are a few examples of a diverse array of nonmetallic, industrial economic resources. Finally, the energy resources coal, petroleum, and natural gas have been highly valuable economic resources for more than 150 years. Yet, their use is now known to have major, negative environmental consequences.

Selected References

Beane, R. E., and Titley, S. R. 1981. Porphyry Copper Deposits. Part II. Hydrothermal Alteration and Mineralization. In Skinner, B. J. (ed.), *Economic Geology 75th Anniversary Volume*, pp. 214–269.

Cerny, P. 1982. Anatomy and classification of granitic pegmatites. In Cerny, P. (ed.), *Granitic Pegmatites in Science and Industry*. Mineralogical Association of Canada, Winnipeg, pp. 1–39.

Evans, A. M. 1997. *An Introduction to Economic Geology and its Environmental Impact*. Blackwell Scientific, Malden, MA. 364 p.

Evans, K. A., McCuaig, T. C., Leach, D., Angerer, T., and Hagemann, S. G. 2013. Banded iron formation to iron ore: A record of the evolution of Earth environments? *Geology*, v. 41, pp. 99–102. doi:10.1130/G33244.1

Franklin, J. M., Lydon, J. W., and Sangster, D. F. 1981. Volcanic-associated massive sulfide deposits. In Skinner, B. J. (ed.), *Economic Geology 75th Anniversary Volume*, pp. 485–627.

Leach, D. L., Sangster, D. F., Kelley, K. D., Large, R. R., Garven, G., Allen, C.R., Gutzmer, J., and Walters, S. 2005. Sediment-hosted lead-zinc deposits: A global perspective. In Hedenquist, J. W., Thompson, J. F. H, Goldfarb, R. J., and Richards, J. P. (eds.), *Economic Geology 100th Anniversary Volume*, pp. 561–607.

London, D., and Kontak, D. J. 2012. Granitic pegmatites: Scientific wonders and economic bonanzas. *Episodes*, v. 8, pp. 257–261.

Lowell, J. D., and Guilbert, J. M. 1970. Lateral and vertical alteration-mineralization zoning in porphyry ore deposits. *Economic Geology*, v. 65, pp. 373–408.

McKelvey, V. E. 1972. Mineral resource estimates and public policy. *American Scientist*, v. 60, pp. 32–40.

Ridley, J. 2013. *Ore Deposit Models*. Cambridge University Press, Cambridge. 398 p.

Robb, L. 2005. *Introduction to Ore-Forming Processes*. Blackwell, Oxford. 373 p.

Simmons, S. F., White, N. C., and John, D. A. 2005. Geological characteristics of epithermal precious and base-metal deposits. In Hedenquist, J. W., Thompson, J. F. H., Goldfarb, R. J., and Richards, J. P. (eds.), *Economic Geology 100th Anniversary Volume*, pp. 485–522.

Notes

1. U.S. Geological Survey Circular 831, 1980.
2. A good example occurs in the Witwatersrand gold deposits in South Africa. Although most geologists consider them as syngenetic, some vocal proponents hypothesize an epigenetic origin. See England et al. (2001) and Law and Phillips (2005).
3. Also see Simmons (2007), Raymond (2007), and London and Morgan (2012) for reviews of crystallization histories and pegmatite ore-body formation, as briefly summarized in this section. Swanson (1977) and Fenn (1986) discuss undercooling in granitic magmas, and Fenn (1986) relates graphic textures of pegmatitic quartz-alkali feldspar granitic rocks to undercooling and supersaturation of granitic melts noted in this paragraph. Cameron et al. (1949) is a classic work that includes descriptions of pegmatites.
4. Robb (2005), Pohl (2011). See Raymond (2007, ch. 11) for a review of theories.
5. A. M. Evans (1997).
6. Ridley (2013).
7. McMillan and Panteleyev (1988).
8. White and Hedenquist (1995).
9. Lydon (1988).
10. Robb (2005).
11. Robb (2005).
12. Borg et al. (2012).
13. Cuney (2009).
14. Cuney (2009).
15. Guilbert and Park (1986).
16. Guilbert and Park (1986).
17. Jensen and Bateman (1979).
18. For example, see Gutzmer and Beukes (1995) and Kuleshov (2012). Gutmer and Beukes (1995) describe hydrothermal metasomatism of the sedimentary ores.
19. Button et al. (1982).
20. Refer to reports, reviews, and volumes such as Morris (1987), Beukes and Klein (1990), Morris and Horwitz (1993), Fralick and Barrett (1995), Raymond (2007, ch. 20), Hagemann et al. (2008), Poulton and Canfield (2011), Simonson (2011), and K. A. Evans et al. (2013) for more information and additional references relative to the complex debates on origins of iron-rich rocks.
21. See Guilbert and Park (1986) for a review of manganese ores. Also see works such as Trask et al. (1950), Cox and Singer (1986), Huebner and Flohr (1990), Okita and Shanks (1992), Fan et al. (2006), Munteanu et al. (2004), and Du et al. (2013) for additional information, perspectives on the diversity of deposits and processes, and references.
22. See Cox and Singer (1986), Alexandri et al. (1985), Okita and Shanks (1992), Fan et al. (2006), and Du et al. (2013).
23. For more information refer to Manheim et al. (1975), Heckel (1977), Manheim and Gulbrandsen (1979), Guilbert and Park (1986), and Sheldon (1987).
24. Guilbert and Park (1986).
25. See Hardie (1990), Li et al. (1997), and Schreiber and El Tabakh (2000). Also see Raymond (2007, ch. 20) for additional reviews and discussions.
26. See Jensen and Bateman (1979); Hardie (1990); Schreiber and El Tabakh (2000); and see Raymond (2007, Ch. 20) for reviews and discussions.
27. See Pohl (2011) for information on titanium.

An Overview and Summary

12

INTRODUCTION

This book has introduced the reader to the nature and origins of minerals, rocks, soils, and fluids. Although the chapters simply scratch the surface of extensive bodies of knowledge, they provide a basic overview of materials and processes important in Earth.

The characteristics of rock types that constitute each of the major classes reveal the origins of those rocks. Below, the rocks are assembled into petrotectonic assemblages—groups of rocks formed at and representative of particular tectonic settings. In particular, those settings include spreading centers, outer and inner parts of subduction zones, transform faults, hot-spot (plume) sites, and nontectonic settings within continents and along continental margins.

Each of the rock types is composed of minerals and, in some cases, glass, organic debris, and fragments of preexisting rock. Each of the various minerals forms in response to the conditions present. Nuclei form and crystals grow as atoms are added to the nuclei. As the crystals increase in size, they form distinct symmetrical-compositional units representing each of the minerals that is stable under the particular conditions of formation. The symmetrical-compositional units are grouped into six (or seven) different crystal systems that depend on the various conditions of equality and inequality for crystallographic axes and the permissible angles between them. The unique combination of chemistry and crystal structure defined by particular types of atoms and their atomic arrangements distinguishes each of the more than 4700 known minerals. The most abundant minerals, the rock-forming minerals, are relatively few in number, with silicate minerals being the most abundant minerals in crustal rocks. The atomic arrangements in the minerals give rise to various physical properties—such as luster, hardness, streak, and crystal habit—that are useful for hand-specimen identification. Optical, X-ray, and other more sophisticated techniques of study reveal physical and chemical properties that are more definitive in identification of the minerals. Where the minerals or rocks have monetary value, they become the subject of study of economic geologists.

Igneous rocks crystallize from magmas formed in the mantle or at the base of the crust. Decompression, flux by fluids, and localized heating all result in local melting at these sites. Basaltic magmas are the predominant primary melts of the mantle, whereas rhyolitic melts form primarily at the base of the crust. Magmas rise toward the surface, losing their heat along the way. Plutonic rocks crystallize at depth below the surface, where undercooling is modest. Phaneritic crystalline textures result. At or near the surface, undercoolings are large and volcanic rocks that form are characterized by glassy, aphanitic to aphanitic-porphyritic crystalline textures, or by pyroclastic (fragmental) textures. The latter form as a result of explosive eruptions typically associated with degassing of magmas.

Sedimentary rocks develop at the surface via various processes of sedimentation. Clastic rocks of silicate and carbonate composition form after weathering and erosion of a source terrane or provenance produces sediment that is deposited in basins by sediment rain, currents, or sediment flows and slides (including landslides). Chemical sediments form via crystallization of minerals from aqueous solutions of various Eh and pH, either on the continents or in marine environments. Microbes commonly mediate precipitation. Sediment characteristics such as bedding and composition reflect the source, the transportation processes, and the nature of the depositional environment. On the continents, fluvial environments are dominant, but lacustrine (lake) and glacial environments are locally important. Deltas and shorelines are sites of significant sedimentation in transitional, continent edge-marine settings. Both shallow marine and submarine fan environments are represented by large volumes of rock in the rock record. The sediments become rock through various diagenetic processes, including compaction, cementation, and various processes of mineral growth and recrystallization.

Preexisting igneous, sedimentary, and metamorphic rocks become new metamorphic rocks where deep burial, heating by intrusions, and faulting or other tectonic processes subject them to changed conditions of pressure, temperature, and deviatoric stress. Metamorphic fluids play a major role in transforming preexisting rocks into new metamorphic forms. During these transforming metamorphic processes, new stable minerals crystallize in response to the existing conditions. The principal site of metamorphic rock formation is in the mountain belts that are created where plates collide. Metamorphic rocks also form and reform at depths in the crust and mantle.

Petrotectonic Assemblages and Plate Settings

All rocks formed as crustal Earth materials form at some site within or at the margins of a tectonic plate. Each site has particular conditions of fluid concentration, pressure, stress, and temperature that control the rock-forming processes. The mineral and rock compositions are controlled by those conditions, and the resulting rocks are characterized by the minerals, rock-chemical compositions, textures, and structures that form in the various environments. Thus, there is a link between tectonic setting and the assemblage or group of rock types that forms at each setting—the petrotectonic assemblage (Dickinson, 1971).

The petrotectonic assemblages are shown in figure 12.1. Each site of rock formation (the petrogenetic site) has typical igneous, sedimentary, or metamorphic rocks—or combinations of these—that form at and represent that site. Beginning with the spreading centers, both in the ocean basins and in back-arc basins between arcs and continents, new crust is formed (figure 12.2 on p. 270). The petrotectonic assemblage formed here contains oceanic tholeiitic basalts of the mid-ocean ridges (MORBs) formed at the surface, plus corresponding gabbros and derivative rocks such as diorite and peridotite that develop in the layered crust above residual metamorphic mantle (mantle tectonite) (Dilek et al., 1998). These layered sequences of rock—the ophiolites—are locally exposed as thrust-up masses within mountain belts. The basaltic magmas, the parents of these rocks, are produced by decompression melting of mantle beneath the ridges. Faulting at the spreading center creates a basin where sediments of a range of grain sizes accumulate to yield rocks ranging from basaltic breccias and volcanic sandstones to fine-grained limestone, chert, and shale. Hydrothermal fluids, created by ocean water circulating through the hot basaltic-gabbroic crust, alter and metamorphose the ocean crust and underlying mantle rocks. In addition, repeated intrusions of magma metamorphose the newly formed rocks, creating rocks such as hornblende schist and gneiss (*amphibolite*). Zeolite, albite-epidote hornfels, greenschist, and amphibolite facies rocks result from the various metamorphic conditions. Along the faults bounding the basins, rocks of similar metamorphic grade form as dynamic recrystallization and cataclasis change the ophiolitic rocks into mylonites and breccias.

Transform faults cut the oceanic crust and, locally, the continental crust. In their purest form, these are essentially zones of shear, where gouge and breccias of cataclastic nature form near the surface, and mylonites—formed by recrystallization and neocrystallization under high shear stresses and strain rates at depth—transform mid-ocean ridge rocks into metamorphic derivatives. Some transform faults may be trans-tensional zones that experience both dynamic metamorphism of crustal rocks and minor spreading, resulting in magmatism and the formation of additional crust.

Rocks of the mid-ocean ridge petrotectonic assemblage are the typical and parental petrotectonic assemblage of transform faults. These assemblages are changed by dynamic cataclastic and recrystallization processes, dynamothermal metamorphism, or metasomatism (figure 12.2). Most transform faults have escarpments that serve as sediment sources, as well as rift valleys and base-of-slope sites that serve as basins for sediment derived by erosion and mass wasting of the escarpment. Like sediments of the mid-ocean ridge petrotectonic assemblage, those of the transform-fault basins are dominantly derived from the adjacent oceanic crust. Sediment rain and precipitation may contribute to the sediment formed at these sites.

The subduction zones are the most complex of the plate boundary sites of rock formation. The outer belts of these convergent boundaries, where plates of

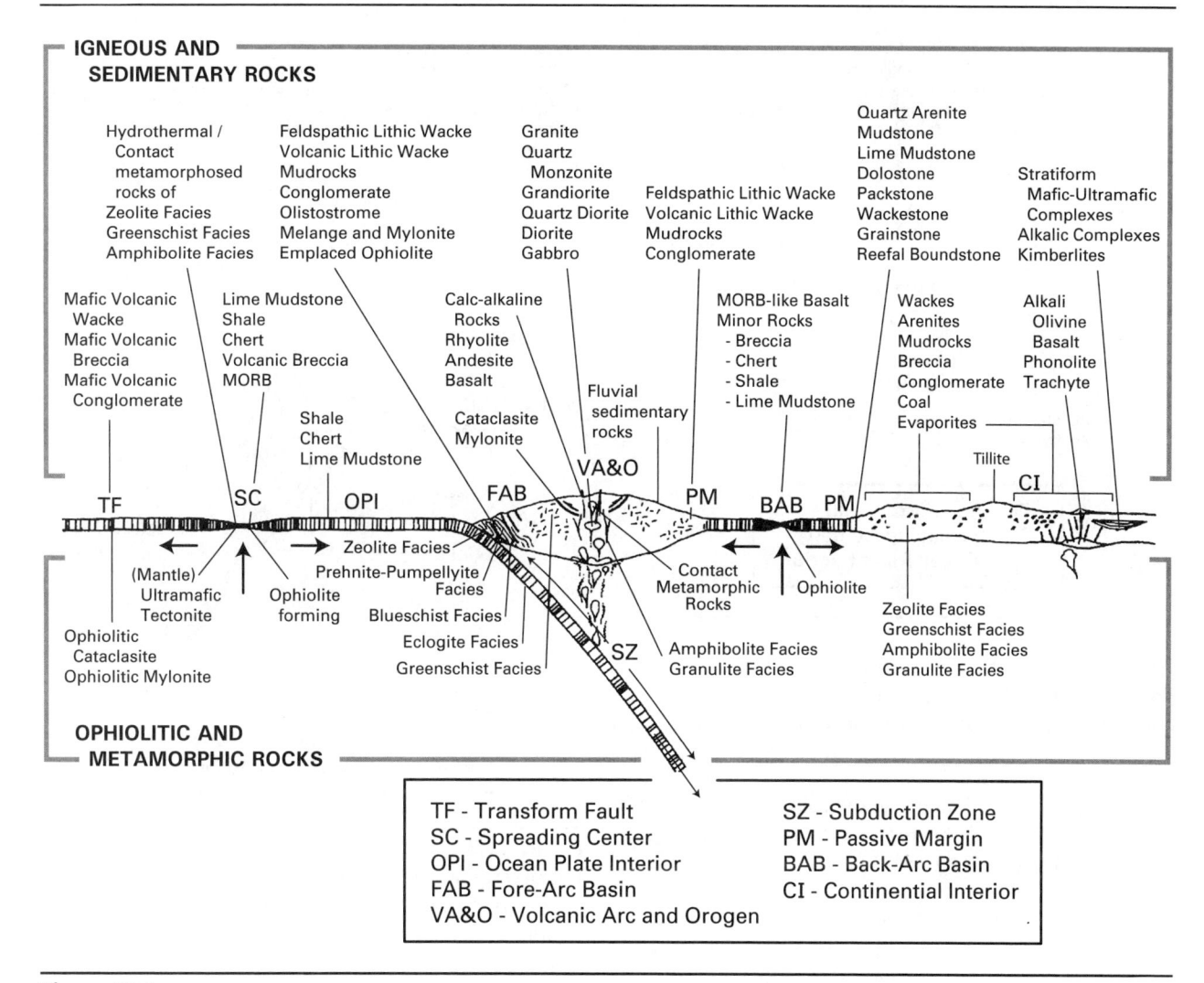

Figure 12.1
Petrotectonic assemblages formed at an oceanic plate/continent-bearing plate collision zone and on the adjoining continental and oceanic plates.
(*Source:* Raymond, 2007, figure 30.1.)

ophiolitic lithosphere with or without attached continental crust are subducted, have both trench basins and upslope fore-arc basins between the trench and the volcanic arc in early stages of subduction (figure 12.3). These basins are sites of accumulation of turbidites plus mudrocks formed by sediment rain. Locally, submarine landsliding creates olistostromal mélanges made up of diamictites composed of a matrix of mudrock with blocks of sedimentary and other rocks derived from the basins, uplifted basin margins, and adjacent continental masses. Diapirs resulting in mud volcanoes may contribute subducted sediment to these olistostromes. The olistostromal mélanges are charac-

teristic of active tectonic margins. On the slope, in the fore-arc basins, sediments are buried, lithified through diagenesis, and, if buried deeply enough, metamorphosed to low grades of metamorphism (zeolite and prehnite-pumpellyite facies). The sediments in the trench are commonly subducted with the underlying oceanic crust to depths where the pressures are high yet temperatures remain low. Prehnite-pumpellyite, blueschist, and in some cases eclogite and amphibolite facies metamorphic rocks result from these conditions. Rapid uplift of blocks along these zones later brings low- to medium-temperature metamorphic rocks to the surface, before they can be heated and

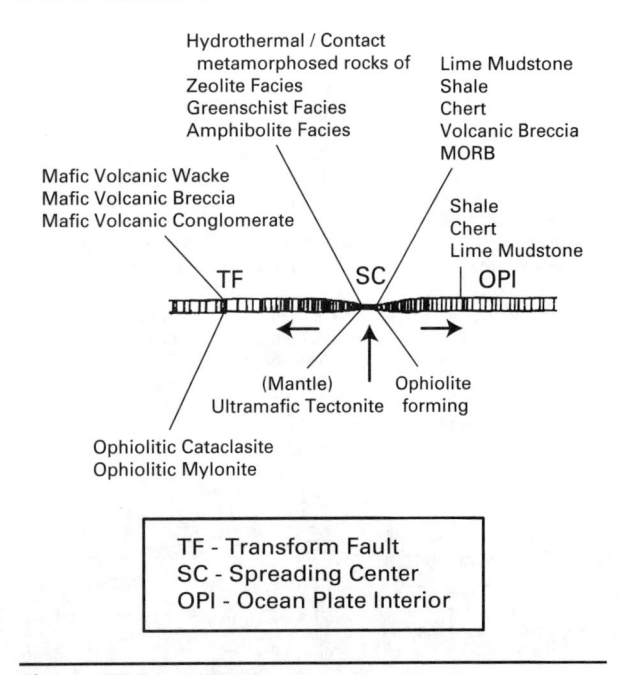

Figure 12.2
Petrotectonic assemblages formed at a mid-ocean ridge spreading center and a transform fault.
(*Source:* modified from Raymond, 2007, figure 30.1.)

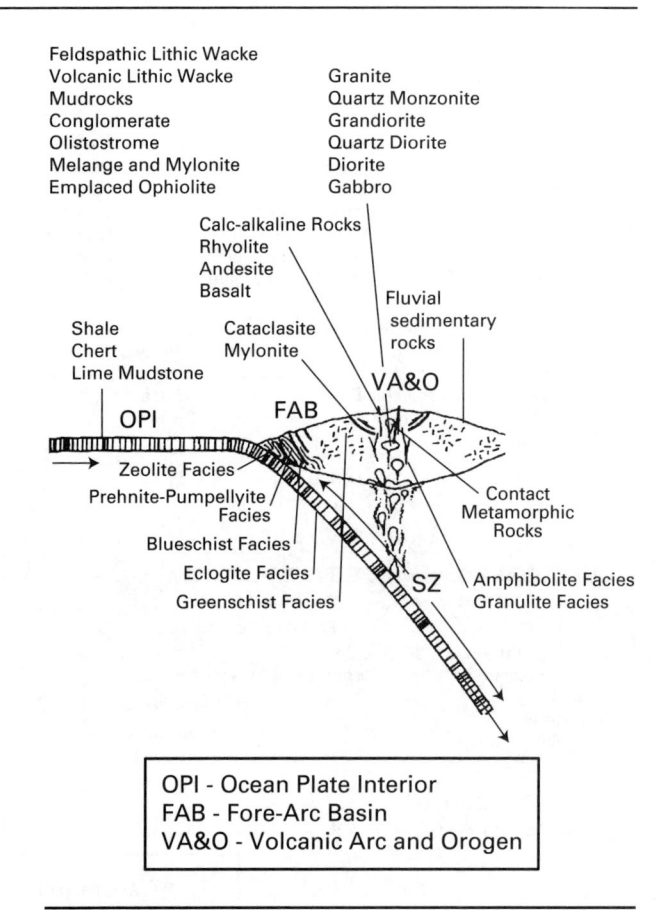

Figure 12.3
Petrotectonic assemblages formed at an active-margin convergent zone with subduction zone and overlying continent.
(*Source:* modified from Raymond, 2007, figure 30.1.)

converted to facies of higher-grade rocks. Where these rocks are mixed and sheared in the subduction zone, tectonic mélanges form (Hsu, 1968; Raymond, 1984a; Festa et al., 2012).

Landward of the subduction zone, an arc develops from rising basaltic magmas formed by flux melting, perhaps locally aided by decompression melting. These magmas may differentiate by fractional crystallization or assimilate crustal rocks on the way to the surface to yield andesitic or other magmas. Where subduction occurs beneath a continental margin, the basaltic magmas may pool at the base of the crust, where they melt the crust by thermal melting to yield primary rhyolitic melts. Basal crust also melts where continents collide. The magmas rise toward the surface and either crystallize at depth or erupt at the surface to form siliceous igneous rocks, or they may mix with basaltic magmas that are assimilating crustal rocks (figure 12.3). From such derivative magmas, rocks such as granodiorite, quartz monzonite, and diorite crystallize at depth to form plutons, or, if the magmas rise to and erupt at the surface, they may yield volcanic rocks such as rhyodacite, quartz latite, or andesite. Thus, the arc volcanoes contain a wide range of volcanic rocks from rhyolite to basalt. The arc-root plutons similarly feature a wide range of rock types from granite to gabbro and peridotites. Via contact metamorphism, the magmas that yield these igneous rocks transform the rocks with which they come in contact; and they heat the surrounding terrain to produce dynamothermal metamorphic rocks of the greenschist, amphibolite, and granulite facies. Fluids derived from and mobilized by the heat of these intrusions facilitate concentration of ores in stockworks. Sediments derived from the arc are rich in quartz, feldspar, rock fragments, and clays. These sediments are deposited in adjacent continental basins, as well as in the arc-trench gap and trench basins offshore, ultimately yielding sedimentary rocks such as mudrocks, siltstones, wackes, and conglomerates. Placer ore may form where the eroded rocks contain valuable and heavy minerals that are concentrated in stream channels by flowing water.

Where two plates, each bearing a continental fragment or well-formed arc collide, processes similar to those just described will occur; but as the lighter continental rock masses impact one another, with one mass underplating another, typical arc development is arrested. Siliceous magmas, created as a result of heating of underthrust rocks, rise to cause metamorphism surrounding the plutons in the overlying plate and may locally erupt to form volcanic rocks. Uplift of the buoyant plates of continental crust and complex scenarios of faulting and folding in these continent-microcontinent and continent-continent collision zones result in mountain ranges that yield large masses of sediment shed to the flanks (e.g., Merschat et al., 2005, 2010; Jain et al., 2009; Henderson et al., 2010). In some suture zones between continents, rocks subducted to extreme depths—ultra-high pressure (UHP) rocks containing coesite, diamond, and other "exotic" minerals—are brought back to the surface through rapid uplift processes (Liou et al., 2006).

A variety of sedimentary rocks forms on the continent. At various continental sites, sediments derived from the arcs and from eroding mountain systems, as well as from the old Precambrian cores of the continents, yield sediments to various basins and depositional sites along rivers, in lakes, around and downstream from glaciers, in swamps, and in transitional environments, such as deltas and estuaries (figure 12.1) (Boggs, 2009). These are either moved to new depositional sites later or are lithified to form the rocks characteristic of these various environments.

Igneous and metamorphic rocks may also form on and within the continents (figure 12.1). Alkaline igneous magmas yielding alkali olivine basalts and alkali gabbros and their derivatives (e.g., syenites and trachytes, plus unusual feldspathoidal rocks such as nepheline syenites), form where mantle-derived magmas of alkaline character intrude the crust or erupt at the surface. Rare carbonate and fluid-rich magmas that rarely contain diamonds erupt to form the rocks called kimberlites that serve as economically valuable sources of those diamonds. Metamorphism, like that in the arcs, is created by these various crust-heating magmas. The lower crust is metamorphosed, in addition, by heat associated with moderate to high pressure at depth. Thus, greenschist, amphibolite, and granulite facies metamorphic rocks form in moderate to deep crustal zones. Fluids mobilized by heat in the crust can move from region to region and may assimilate ore elements, such as lead and zinc, which are deposited as Mississippi Valley-type ores, where the fluids encounter appropriate host rocks.

Clearly, the Earth is dynamic. Dynamic processes create the crustal Earth materials used by geologists to study the planet's history, processes, economic deposits, and geologic hazards.

Summary

Minerals—including rock-forming and economic minerals—and rocks, soils, and fluids all form in particular settings under specific conditions. The features of each of these Earth materials, including the chemical composition, textures, structures, and behaviors (in cases of fluids and rock-forming agents) reveal those conditions of formation. In addition, petrotectonic assemblages give clues to the plate tectonic settings of rock origins.

Igneous rocks form both at the surface and at depth from magmas. These and other rocks are weathered to form soils, usually under the influence of fluids. Soils and rocks are eroded to form sediments, and sedimentary rocks form at the surface through precipitation and sediment accumulation. Diagenesis follows, converting sediment into rock. Metamorphic rocks form, generally at depth, as preexisting igneous, sedimentary, and metamorphic rocks are subjected to new conditions of pressure, temperature, chemical-bearing fluid activity, and deviatoric stress.

Minerals constitute the bulk of the materials that make up the rocks. Each of the more than 4700 minerals known has a unique combination of chemistry and crystal structure defined by atomic arrangements in various symmetries. The abundant rock-forming minerals, however, are relatively few in number. Silicate minerals are the most abundant minerals in the rocks of the crust, but carbonate minerals are locally important. Some relatively rare minerals as well as some common ones have monetary value and form economic deposits of various types, as do certain rocks. In these various economic deposits, sedimentary, igneous, or metamorphic processes or the actions of fluids serve to concentrate the economically valuable materials.

Selected References

Dickinson, W. R. 1971. Plate tectonics in geologic history. *Science,* v. 174, pp. 107–113.

Dilek, Y., Moores, E. M., and Furnes, H. 1998. Structure of modern oceanic crust and ophiolites and implications for faulting and magmatism at oceanic spreading centers. In Buck, W. R., Delaney, P. T., Karson, J. A., and Lagabrielle, Y. (eds.), *Faulting and Magmatism at Mid-Ocean Ridges.* American Geophysical Union, Geophysical Monograph 106, pp. 219–266.

Henderson, A. L., Najman, Y., Parrish, R., BouDagher-Fadel, M., Barford, D., Garzanti, E., and Ando, S. 2010. Geology of the Cenozoic Indus Basin sedimentary rocks: Paleoenvironmental interpretation of sedimentation from the western Himalaya during the early phases of India-Eurasia collision. *Tectonics,* v. 29, TC6015, doi:10.1029/2009TC002651.

Liou, J. G., Tsugimori, T., Zhang, R. Y., Katayama, I., and Maruyama, S. 2006. Global UHP metamorphism and continental subduction/collision: The Himalayan model. In Liou, J. G., and Cloos, M. (eds.), *Phase Relations, High-Pressure Terranes, P-T-ometry, and Plate Pushing: A Tribute to W.G. Ernst.* The Geological Society of America, International Book Series, v. 9, pp. 53–79.

Pearce, J. A., and Peate, D. W. 1995. Tectonic implications of the composition of volcanic arc magmas. *Annual Reviews of Earth and Planetary Sciences,* v. 23, p. 251–285.

Raymond, L. A. 2007. *Petrology: The Study of Igneous, Sedimentary, and Metamorphic Rocks* (2nd ed.). Waveland Press, Long Grove, IL. 720 p.

Appendix A
Determinative Tables for Mineral Identification

On the following pages are a series of tables for use as a guide in mineral identification. The tables are organized into three broad groups based on mineral luster and shade of color. Each of these groups is then subdivided by hardness and cleavage.

To use the tables:

1. Decide whether the sample has a metallic luster, is light-colored with a nonmetallic luster, or is dark-colored with a nonmetallic luster.
2. Test the hardness of the sample with a piece of window glass. Determine whether the sample is clearly softer than the glass, is about as hard as the glass, or is clearly harder than the glass.
3. Decide whether the sample shows any obvious cleavages, may or may not have cleavages, or appears to have no cleavages at all.

Together, these three determinations will allow you to focus on one of the individual tables. Within each table the minerals are arranged in order of increasing Mohs hardness, so to narrow your options, you will need to test this property further using items like a fingernail (Mohs hardness of 2.5), a copper penny (3.5), a steel nail (5), or a streak plate (6). Once you are down to a few possibilities, read the mineral descriptions to assist your final determination. Properties in ***bold italic*** text in the Other/Special categories column are those which are particularly helpful for that mineral.

Because some minerals can show a variety of properties, there are duplicate entries in many of the tables. Names in quotes are not official names of mineral species, but are commonly used in lieu of species names, particularly when precise determinations require more advanced techniques.

Minerals With a Metallic Luster

Softer than glass; Cleavage present & usually visible

Formula	Mohs Hardness	Color	Specific Gravity	Streak	Habit	Other/Special
MoS_2	1 - 1.5	Metallic silvery-gray	4.6 - 4.7	Metallic gray with greenish or bluish tints	Tabular, hexagonal crystals; massive	1 perfect cleavage; sectile; *greasy feel, may write on paper*
CuS	1.5 - 2	Metallic blue to indigo, commonly with brassy and red iridescence	4.6 - 4.8	Metallic gray to black	Massive; hexagonal plates	1 perfect cleavage into thin sheets that can be flexible; sectile; brittle fracture
Sb_2S_3	2	Metallic silvery-gray	4.6	Metallic gray	Elongated prismatic crystals; massive	1 perfect, 2 imperfect cleavages; conchoidal fracture
PbS	2.5 - 3	Lead gray	7.6	Lead gray	Cubic or octahedral crystals; massive	**3 perfect cleavages at 90° to each other**; conchoidal fracture; may show octahedral parting; contact and penetration twins common
ZnS	3.5 - 4	Metallic black to translucent brown, yellow, green	3.9 - 4.1	Brown to pale yellow-white	Tetrahedral or dodecahedral crystals; massive	6 perfect cleavages; conchoidal fracture; contact twinning common
$(Fe,Mn)WO_4$	4 - 5.5	Yellowish-brown to reddish-brown to black	7 - 7.5	Yellowish-brown to reddish-brown to black	Prismatic or tabular crystals with striations on one face	1 perfect cleavage; parting sometimes seen; uneven fracture; contact twinning common; ferberite (Fe) and huebnerite (Mn) make up a solid-solution series

Softer than glass; Cleavage present but often not visible

Formula	Mohs Hardness	Color	Specific Gravity	Streak	Habit	Other/Special
C	1 - 1.5	Metallic gray to black	2.2	Black	Flat, hexagonal sheets; massive	1 perfect cleavage; sectile; *greasy feel, writes on paper*
MoS_2	1 - 1.5	Metallic silvery-gray	4.6 - 4.7	Metallic gray with greenish or bluish tints	Tabular, hexagonal crystals; massive	1 perfect cleavage; sectile; *greasy feel, may write on paper*
Cu_2S	2.5 - 3	Lead grey to black	5.5 - 5.8	Black	Massive	1 poor cleavage; conchoidal fracture; sectile masses

Softer than glass; Cleavage present but often not visible (continued)

Mineral	Formula	Mohs Hardness	Color	Specific Gravity	Streak	Habit	Other/Special
Bornite	Cu₅FeS₄	3	Bronze to black	5.1	Black	Massive	4 poor cleavages; conchoidal fracture; bronze color on fresh surfaces, tarnishes to coating of covellite
Chalcopyrite	CuFeS₂	3.5 - 4	Golden, brassy yellow	4.1 - 4.3	Greenish black	Massive; tetrahedral crystals	6 poor cleavages; uneven fracture; tarnishes to coating of covellite
"Wolframite"	(Fe,Mn)WO₄	4 - 5.5	Yellowish-brown to reddish-brown to black	7 - 7.5	Yellowish-brown to reddish-brown to black	Prismatic or tabular crystals with striations on one face	1 perfect cleavage; parting sometimes seen; uneven fracture; contact twinning common; ferberite (Fe) and huebnerite (Mn) make up a solid-solution series

Softer than glass; Cleavage absent or very poorly visible

Mineral	Formula	Mohs Hardness	Color	Specific Gravity	Streak	Habit	Other/Special
Hematite	Fe₂O₃	2 - 6	Steel grey to black	5.3	Red to reddish brown	Massive; tabular crystals	Conchoidal to uneven fracture; parting due to twinning common
"Psilomelane"	Mixed Mn oxides and hydroxides	2 - 6.5	Black	3.7 - 4.8	Brownish black	Massive; botryoidal or stalactitic; dendritic masses	Often in hard or powdery masses; a mixture of pyrolusite, romanechite, cryptomelane and other difficult to characterize Mn minerals
Copper	Cu	2.5 - 3	Copper red to brown	8.9 - 9	Copper red	Massive; octahedral or cubic crystals; dendritic masses	Hackly fracture; contact twinning common; *very high ductility and malleability*
Gold	Au	2.5 - 3	Golden yellow	19.3	Golden yellow	Massive; octahedral or cubic crystals; dendritic masses	Hackly fracture; contact twinning common; *very high ductility and malleability*
Silver	Ag	2.5 - 3	Silvery white	10.1 - 11.1	Silvery white	Massive; octahedral or cubic crystals; dendritic masses	Hackly fracture; contact twinning common; *very high ductility and malleability*

Softer than glass; Cleavage absent or very poorly visible (continued)

Mineral	Formula	Mohs Hardness	Color	Specific Gravity	Streak	Habit	Other/Special
Tetrahedrite/ Tennantite	$Cu_{10}(Fe,Zn)_2(Sb,As)_4S_{13}$	3 - 4.5	Lead grey to black	4.6 - 5.1	Black to brown	Massive; tetrahedral crystals	Uneven fracture; tetrahedrite (Sb) and tennantite (As) make up a solid-solution series
Pentlandite	$(Fe,Ni)_9S_8$	3.5 - 4	Bronze to pale copper-red	4.6 - 5	Brownish-bronze to black	Massive	Conchoidal fracture; octahedral parting
Pyrrhotite	$Fe_{1-x}S$	4	Brownish bronze	4.6 - 4.7	Black	Massive; hexagonal crystals	Conchoidal to uneven fracture; occasional parting due to twinning; often weakly magnetic

About as hard as glass; Cleavage present & usually visible

Mineral	Formula	Mohs Hardness	Color	Specific Gravity	Streak	Habit	Other/Special
"Wolframite"	$(Fe,Mn)WO_4$	4 - 5.5	Yellowish-brown to reddish-brown to black	7 - 7.5	Yellowish-brown to reddish-brown to black	Prismatic or tabular crystals with striations on one face	1 perfect cleavage; parting sometimes seen; uneven fracture; contact twinning common; ferberite (Fe) and huebnerite (Mn) make up a solid-solution series

About as hard as glass; Cleavage present but often not visible

Mineral	Formula	Mohs Hardness	Color	Specific Gravity	Streak	Habit	Other/Special
Goethite	$FeO(OH)$	2 - 6.5	Yellow to reddish-brown, blackish-brown	3.3 - 4.3	Brownish yellow	Massive; prismatic crystals; botryoidal or stalacticic masses of radiating crystals	1 perfect cleavage; uneven to splintery fracture; referred to as 'limonite' when in yellowish masses; masses frequently powdery or earthy
"Wolframite"	$(Fe,Mn)WO_4$	4 - 5.5	Yellowish-brown to reddish-brown to black	7 - 7.5	Yellowish-brown to reddish-brown to black	Prismatic or tabular crystals with striations on one face	1 perfect cleavage; parting sometimes seen; uneven fracture; contact twinning common; ferberite (Fe) and huebnerite (Mn) make up a solid-solution series

About as hard as glass; Cleavage absent or very poorly visible

Mineral	Formula	Mohs Hardness	Color	Specific Gravity	Streak	Habit	Other/Special
Hematite	Fe_2O_3	2 - 6	Steel grey to black	5.3	Red to reddish brown	Massive; tabular crystals	Conchoidal to uneven fracture; parting due to twinning common
"Psilomelane"	Mixed Mn oxides and hydroxides	2 - 6.5	Black	3.7 - 4.8	Brownish black	Massive; botryoidal or stalactitic; dendritic masses	Often in hard or powdery masses; a mixture of pyrolusite, romanechite, cryptomelane and other difficult to characterize Mn minerals
Chromite	$FeCr_2O_4$	5.5	Black	4.5 - 4.8	Blackish brown	Massive; octahedral crystals	Uneven fracture; resembles magnetite, but not magnetic

Harder than glass; Cleavage present & usually visible

Mineral	Formula	Mohs Hardness	Color	Specific Gravity	Streak	Habit	Other/Special
Arsenopyrite	FeAsS	5.5 - 6	Silvery-white to steel-gray	6.1	Grayish black	Masses of prismatic crystals with striations on faces; massive	1 distinct cleavage; uneven fracture; contact twinning common
Columbite/ Tantalite	$(Fe,Mn)(Nb,Ta)_2O_6$	6 - 6.5	Black to brownish black with reddish internal reflections	5.2 - 8	Black to dark red	Prismatic, tabular, equant or pyramidal crystals; massive	1 distinct cleavage; contact twinning common; Forms two series: ferrocolumbite (Nb) to ferrotantalite (Ta) and manganocolumbite (Nb) to manganotantalite (Ta)
Marcasite	FeS_2	6 - 6.5	Pale yellowish-bronze	4.9	Brownish black	Tabular to prismatic crystals, frequently with curved faces	1 distinct cleavage; uneven fracture; **twinning common producing cockscomb form**; surfaces commonly weather to dull appearance
Rutile	TiO_2	6 - 6.5	Reddish-brown, commonly with silvery tint	4.2 - 4.4	Pale brown to black	Prismatic to acicular crystals; massive	4 distinct cleavages; conchoidal to uneven fracture; adamantine to metallic luster; contact twinning common

Harder than glass; Cleavage present but often not visible

Mineral	Formula	Mohs Hardness	Color	Specific Gravity	Streak	Habit	Other/Special
Columbite/Tantalite	(Fe,Mn)(Nb,Ta)$_2$O$_6$	6 - 6.5	Black to brownish black with reddish internal reflections	5.2 - 8	Black to dark red	Prismatic, tabular, equant or pyramidal crystals; massive	1 distinct cleavage; conchoidal to uneven fracture; contact twinning common; Forms two series; ferrocolumbite (Nb) to ferrotantalite (Ta) and manganocolumbite (Nb) to manganotantalite (Ta)
Marcasite	FeS$_2$	6 - 6.5	Pale yellowish-bronze	4.9	Brownish black	Tabular to prismatic crystals, frequently with curved faces	1 distinct cleavage; uneven fracture; ***twinning common producing cockscomb form***; surfaces commonly weather to dull appearance
Rutile	TiO$_2$	6 - 6.5	Reddish-brown, commonly with silvery tint	4.2 - 4.4	Pale brown to black	Prismatic to acicular crystals; massive	4 distinct cleavages; conchoidal to uneven fracture; adamantine to metallic luster; contact twinning common
Cassiterite	SnO$_2$	6 - 7	Yellowish-brown to reddish-brown to brownish black	7.02	White to brown or gray	Massive, botryoidal, concretions; prismatic or pyramidal crystals	1 imperfect cleavage; distinct parting; conchoidal to uneven fracture; adamantine to metallic luster; contact and penetration twins common

Harder than glass; Cleavage absent or very poorly visible

Mineral	Formula	Mohs Hardness	Color	Specific Gravity	Streak	Habit	Other/Special
Hematite	Fe$_2$O$_3$	2 - 6	Steel grey to black	5.3	Red to reddish brown	Massive; tabular crystals	Conchoidal to uneven fracture; parting due to twinning common
Ilmenite	FeTiO$_3$	5 - 6	Black	4.72	Black	Tabular crystals; massive	Conchoidal fracture; may be weakly magnetic
Magnetite	Fe$_3$O$_4$	5.5 - 6.5	Black to brownish black	5.2	Black	Octahedral crystals; massive	Conchoidal fracture; octahedral parting common; ***magnetic***
Pyrite	FeS$_2$	6 - 6.5	Pale brassy-yellow	5.02	Black with greenish to brownish tints	Cubic, pyritohedral or octahedral crystals; massive	3 indistinct cleavages; conchoidal to uneven fracture; penetration twins common

Light Colored Minerals With a Non-Metallic Luster
Softer than glass; Cleavage present & usually visible

Mineral	Formula	Mohs Hardness	Color	Specific Gravity	Streak	Habit	Other/Special
Talc	$Mg_3Si_4O_{10}(OH)_2$	1	White to greenish or grayish	2.7 - 2.8	White	Massive	1 perfect cleavage; pearly to greasy luster; *soapy feel*
Pyrophyllite	$Al_2Si_4O_{10}(OH)_2$	1 - 2	White to tan, greenish, silvery, brown	2.8 - 2.9	White	Massive; thin radiating crystals	1 perfect cleavage; pearly to greasy luster
Orpiment	As_2S_3	1.5 - 2	Yellow, with brown to reddish tints	3.5	Pale yellow	Massive; prismatic crystals	1 perfect cleavage; pearly to resinous luster; often easily powdered
Gypsum	$CaSO_4 \cdot 2H_2O$	2	Colorless to white, gray, yellow, brown or gray	2.3	White	Massive; tabular or prismatic crystals	1 perfect, 1 distinct cleavage into flexible fragments; conchoidal fracture; vitreous luster; contact twinning common
Halite	NaCl	2	Colorless, white, blue, pink, purple, yellow, gray	2.16	White	Cubic crystals; massive	*3 perfect cleavages at 90°*; conchoidal fracture; vitreous luster; salty taste
"Chlorite"	$A_{4-6}Z_4O_{10}(OH,O)_8$	2 - 2.5	Green to pale green, pink, yellow, white	2.6 - 3.0	White	Massive; tabular crystals	*1 perfect cleavage into flexible fragments*; vitreous to pearly luster; A = Al, Fe, Mn, Mg and Z= Al, B, Fe, Si for more than ten named varieties
Cinnabar	HgS	2 - 2.5	Red, with gray or metallic tints	8.1	Red	Massive; tabular, prismatic or rhombohedral crystals	1 perfect cleavage; conchoidal fracture; adamantine luster; may have coatings of liquid mercury
Brucite	$Mg(OH)_2$	2.5	White to pale green, blue or gray	2.4	White	Tabular crystals, commonly in masses; massive	1 perfect cleavage into flexible fragments; pearly to waxy luster; sectile
Ulexite	$NaCaB_5O_9 \cdot 8H_2O$	2.5	Colorless to white or gray	2	White	Masses of acicular or parallel crystals; botryoidal	1 perfect, 1 good cleavage; vitreous luster surfaces perpendicular to crystal length transmit images via fiber optics ("TV stone")

Softer than glass; Cleavage present & usually visible (continued)

Mineral	Formula	Mohs Hardness	Color	Specific Gravity	Streak	Habit	Other/Special
					Mica Group		
Muscovite	$KAl_2AlSi_3O_{10}(OH)_2$	2 - 2.5	Colorless to white, green	2.8 - 3	White, with greenish tints	Pseudo-hexagonal or parallelogram-shaped crystals; massive	*1 perfect cleavage into elastic fragments*; vitreous to pearly luster
Phlogopite	$KMg_3AlSi_3O_{10}(OH)_2$	2.5 - 3	Light, coppery-brown to dark brown	2.7	White, with brownish tints	Pseudo-hexagonal or parallelogram-shaped crystals; massive	*1 perfect cleavage into elastic fragments*; vitreous to pearly luster
Lepidolite	$KLi_2AlSi_3O_{10}(OH)_2$	2.5 - 4	Purplish-pink, gray-green, yellow	2.8 - 3	White, with pinkish tints	Pseudo-hexagonal or parallelogram-shaped crystals; massive	*1 perfect cleavage into elastic fragments*; vitreous to pearly luster
Calcite	$CaCO_3$	3	Colorless to white, green, yellow, purple, blue, red, gray, black	2.71	White	Scalenohedral, prismatic, rhombohedral, tabular crystals	*3 perfect cleavages, not at 90° (rhombohedral); reacts strongly to acid*; significant double refraction in colorless crystals; commonly twinned
Barite	$BaSO_4$	3 - 3.5	Colorless to white, yellow, brown	4.5	White	Tabular, prismatic or equant crystals; massive	1 perfect cleavage; others less perfect; uneven fracture; vitreous to resinous luster
Celestite	$SrSO_4$	3 - 3.5	Colorless to pale blue or white	3.97	White	Massive; tabular or elongated crystals	1 perfect, 1 good cleavage; uneven fracture; vitreous luster
Aragonite	$CaCO_3$	3.5 - 4	Colorless to white, gray, green, yellow, blue, purple	2.95	White	Prismatic crystals, nearly always in 3-fold twins; massive	1 distinct cleavage; conchoidal fracture; vitreous to resinous luster; *twinned crystals have hexagonal shape; reacts strongly to acid*

Softer than glass; Cleavage present & usually visible (continued)

Mineral	Formula	Mohs Hardness	Color	Specific Gravity	Streak	Habit	Other/Special
Dolomite	$CaMg(CO_3)_2$	3.5 - 4	Colorless to white, pink, tan, brown, gray, black	2.85	White	Rhombohedral or prismatic crystals, commonly curved into saddle shapes	*3 perfect cleavages not at 90° (rhombohedral)*; conchoidal fracture; vitreous to pearly luster; *reacts weakly to acid unless powdered*
Rhodochrosite	$MnCO_3$	3.5 - 4	Pink to red, brown	3.70	White	Rhombohedral, tabular, prismatic or scalenohedral crystals; massive	*3 perfect cleavages not at 90° (rhombohedral)*; conchoidal fracture; vitreous to pearly luster; reacts weakly to acid
Sphalerite	ZnS	3.5 - 4	Metallic black to translucent brown, yellow, green	3.9 - 4.1	Brown to pale yellow-white	Tetrahedral or dodecahedral crystals; massive	6 perfect cleavages; conchoidal fracture; resinous to adamantine luster
Stilbite	$(Ca_{0.5},Na,K)_9[Al_9Si_{27}O_{72}] \cdot 28H_2O$	3.5 - 4	White to yellow, brown	2.1 - 2.2	White	Tabular crystals, intergrown to resemble wheat sheaves; massive	1 perfect cleavage; pearly to vitreous luster
Siderite	$FeCO_3$	3.5 - 4.2	Brown to yellowish or reddish brown, gray, white	3.96	White, often with brown tints	Rhombohedral, tabular, prismatic or scaleneohedral crystals; massive	*3 perfect cleavages, not at 90° (rhombohedral)*; reacts to hot acid
Magnesite	$MgCO_3$	3.5 - 4.5	Colorless to white, gray	3.00	White	Massive; rhombohedral, prismatic or tabular crystals	3 perfect cleavages, not at 90° (rhombohedral); conchoidal fracture; vitreous luster; reacts to hot acid
Wavellite	$Al_3(OH)_3(PO_4)_2 \cdot 5H_2O$	3.5 - 4	Greenish-white to green, yellow, brown, black	2.36	White	Globular or botryoidal masses of radiating fibrous crystals; crusts or coatings	2 perfect, 2 good cleavages; uneven fracture; vitreous to resinous luster

Softer than glass; Cleavage present & usually visible (continued)

Mineral	Formula	Mohs Hardness	Color	Specific Gravity	Streak	Habit	Other/Special
Fluorite	CaF_2	4	Colorless, white, blue, pink, purple, yellow or mixtures	3.18	White	Cubic or octahedral crystals, sometimes combined; massive	**4 perfect cleavages**; conchoidal to uneven fracture; vitreous luster; penetration twinning common
Colemanite	$Ca_2B_6O_{11} \cdot 5H_2O$	4.5	Colorless to white	2.42	White	Massive; equant crystals	1 perfect, 1 distinct cleavage; uneven to conchoidal fracture; vitreous to adamantine luster
Apophyllite	$KCa_4(Si_4O_{10})_2(F,OH) \cdot 8H_2O$	4.5 - 5	Colorless to white, pink, green	2.3 - 2.4	White	Cubic to prismatic crystals	1 perfect cleavage; uneven fracture; pearly luster on base faces, vitreous luster on others
Scheelite	$CaWO_4$	4.5 - 5	Colorless to white, greenish, reddish, orange-yellow	6.1	White	Octahedral crystals; massive	1 distinct cleavage; adamantine luster; strong bluish-white fluorescence
Wollastonite	$CaSiO_3$	4.5 - 5	Colorless to white, pinkish or grayish	2.8 - 2.9	White	Masses of fibrous crystals; equant crystals; massive	2 perfect cleavages at ≈90°; splintery fracture; vitreous luster

Softer than glass; Cleavage present but often not visible

Mineral	Formula	Mohs Hardness	Color	Specific Gravity	Streak	Habit	Other/Special
Talc	Mg$_3$Si$_4$O$_{10}$(OH)$_2$	1	White to greenish or grayish	2.7 - 2.8	White	Massive	1 perfect cleavage; pearly to greasy luster; *soapy feel*
Pyrophyllite	Al$_2$Si$_4$O$_{10}$(OH)$_2$	1 - 2	White to tan, greenish, silvery, brown	2.8 - 2.9	White	Massive; thin radiating crystals	1 perfect cleavage; pearly to greasy luster
"Bauxite"	Mixed Al hydroxides	1 - 3	Tan to reddish-brown	2 - 2.5	Tan to light reddish	Massive, very commonly in pisolitic masses	A mixture of gibbsite, boehmite and diaspore
Realgar	AsS	1.5 - 2	Bright red to orange	3.6	Red to orange-red	Prismatic crystals; massive	1 good cleavage; breaks down with exposure to light into a reddish-yellow powder
Sulfur	S	1.5 - 2.5	Yellow	2.1	White with yellow tints	Tabular crystals; massive	3 imperfect cleavages; 1 parting; resinous luster; brittle
Gypsum	CaSO$_4$•2H$_2$O	2	Colorless to white, gray, yellow, brown or gray	2.3	White	Massive; tabular or prismatic crystals	1 perfect, 1 distinct cleavage into flexible fragments; conchoidal fracture; vitreous luster; contact twinning common
Borax	Na$_2$B$_4$O$_7$•10H$_2$O	2 - 2.5	Colorless to white, gray, green, blue	1.72	White	Massive; prismatic or tabular crystals	1 perfect, 1 nearly perfect cleavage; conchoidal fracture; vitreous to earthy luster
"Chlorite"	A$_{4-6}$Z$_4$O$_{10}$(OH,O)$_8$	2 - 2.5	Green to pale green, pink, yellow, white	2.6 - 3.0	White	Massive; tabular crystals	*1 perfect cleavage into flexible fragments*; vitreous to pearly luster; A = Al, Fe, Mn, Mg and Z= Al, B, Fe, Si for more than ten named varieties
Cinnabar	HgS	2 - 2.5	Red, with gray or metallic tints	8.1	Red	Massive; tabular, prismatic or rhombohedral crystals	1 perfect cleavage; conchoidal fracture; adamantine luster; may have coatings of liquid mercury

Softer than glass; Cleavage present but often not visible (continued)

Mineral	Formula	Mohs Hardness	Color	Specific Gravity	Streak	Habit	Other/Special
Kaolinite	$Al_2Si_2O_5(OH)_4$	2 - 2.5	White, various tints of yellow, red or brown	2.6	White	Massive	1 perfect cleavage, only seen with electron microscopes; powders rather than fractures; dull to earthy luster
"Serpentine"	$Mg_3Si_2O_5(OH)_4$	2 - 5	White to green, yellow, blue, black mottling	2.2 - 2.6	White	Massive; fibrous; botryoidal	Chrysotile is fibrous form with no observed cleavage; antigorite is platy form with 1 perfect and 1 good cleavage; lizardite is massive form with 1 perfect cleavage and waxy, mottled surface; uneven fracture; silky to waxy luster
Brucite	$Mg(OH)_2$	2.5	White to pale green, blue or gray	2.4	White	Tabular crystals, commonly in masses; massive	1 perfect cleavage into flexible fragments; pearly to waxy luster; sectile
Anhydrite	$CaSO_4$	3.5	Colorless to white, gray, yellow, brown or gray	3	White	Massive; equant or tabular crystals	1 perfect, 1 nearly perfect, 1 good cleavage; uneven to splintery fracture; pearly to vitreous or greasy luster
Alunite	$KAl_3(SO_4)_2(OH)_6$	3.5 - 4	White to reddish-brown, yellowish or grayish	2.6 - 2.9	White	Massive; tabular or rhombohedral crystals	1 distinct cleavage; uneven to conchoidal fracture; vitreous to pearly or earthy luster
Azurite	$Cu_3(CO_3)_2(OH)_2$	3.5 - 4	Blue to dark blue	3.77	Blue	Tabular, prismatic or equant crystals; massive	1 perfect, 1 fair cleavage; conchoidal fracture; adamantine luster; reacts strongly to acid
Malachite	$Cu_2CO_3(OH)_2$	3.5 - 4	Green	4.05	Green	Massive, especially botryoidal or fibrous	1 perfect, 1 fair cleavage; uneven to conchoidal fracture; vitreous to adamantine luster; reacts strongly to acid
Pyromorphite	$Pb_5(PO_4)_3Cl$	3.5 - 4	Yellowish-green to dark green, gray, brown	6.5 - 7.1	White, with greenish tints	Prismatic or tabular hexagonal crystals; massive	1 indistinct cleavage; conchoidal to uneven fracture; resinous to adamantine luster; member of apatite group
Magnesite	$MgCO_3$	3.5 - 4.5	Colorless to white, gray	3.00	White	Massive; rhombohedral, prismatic or tabular crystals	3 perfect cleavages, not at 90° (rhombohedral); conchoidal fracture; vitreous luster; reacts to hot acid

Softer than glass; Cleavage present but often not visible (continued)

Mineral	Formula	Mohs Hardness	Color	Specific Gravity	Streak	Habit	Other/Special
Scheelite	$CaWO_4$	4.5 - 5	Colorless to white, greenish, reddish, orange-yellow	6.1	White	Octahedral crystals; massive	1 distinct cleavage; adamantine luster; strong bluish-white fluorescence
Smithsonite	$ZnCO_3$	4 - 4.5	White to grayish white, greenish, bluish, yellowish tints	4.43	White, often with brown tints	Massive, botryoidal; rhombohedral or scaleneohedral crystals	3 cleavages, not at 90° (rhombohedral) that are just short of perfect; reacts to hot acid
"Apatite"	$Ca_5(PO_4)_3(OH,Cl,F)$	5	Colorless to white, green, brown, blue, yellow	3.1 - 3.2	White	Prismatic or tabular hexagonal crystals; massive	2 indistinct cleavages; conchoidal fracture; vitreous luster; end members are hydroxylapatite, chlorapatite, fluorapatite

Softer than glass; Cleavage absent or very poorly visible

Mineral	Formula	Mohs Hardness	Color	Specific Gravity	Streak	Habit	Other/Special
Chrysocolla	$(Cu,Al)_2H_2Si_2O_5(OH)_4 \cdot nH_2O$	2 - 4	Pale bluish-white to greenish blue	2.0 - 2.4	White	Massive; botryoidal; coatings	Conchoidal fracture; vitreous to earth luster
"Serpentine"	$Mg_3Si_2O_5(OH)_4$	2 - 5	White to green, yellow, blue, black mottling	2.2 - 2.6	White	Massive; fibrous; botryoidal	Chrysotile is fibrous form with no observed cleavage; antigorite is platy form with 1 perfect and 1 good cleavage; lizardite is massive form with 1 perfect cleavage and waxy, mottled surface; uneven fracture; silky to waxy luster
Vanadinite	$Pb_5(VO_4)_3Cl$	2.5 - 3	Red to orange-red, yellow to brown	6.7 - 7.1	Red to brownish-yellow	Tabular or prismatic hexagonal crystals; massive	Conchoidal to uneven fracture; resinous to adamantine luster; member of apatite group

About as hard as glass; Cleavage present & usually visible

Mineral	Formula	Mohs Hardness	Color	Specific Gravity	Streak	Habit	Other/Special
"Wolframite"	(Fe,Mn)WO$_4$	4 - 5.5	Yellowish-brown to reddish-brown to black	7 - 7.5	Yellowish-brown to reddish-brown to black	Prismatic or tabular crystals with striations on one face	1 perfect cleavage; parting sometimes seen; uneven fracture; contact twinning common; ferberite (Fe) and huebnerite (Mn) make up a solid-solution series
Monazite	(Ce,La,Th,Y)PO$_4$	5 - 5.5	Brown to reddish brown or yellow	4.6 - 5.4	White to light brown	Tabular, prismatic crystals; massive	1 distinct cleavage; uneven to conchoidal fracture; resinous or waxy to vitreous or adamantine luster
Natrolite	Na$_2$Al$_2$Si$_3$O$_{10}$•2H$_2$O	5 - 5.5	Colorless to white, yellowish tints	2.25	White	Acicular or fibrous; massive	2 perfect cleavages; uneven fracture; vitreous luster
Titanite	CaTiSiO$_5$	5 - 5.5	Brown to yellow, green	3.4 - 3.5	White to light brown	Wedge-shaped crystals; massive	2 distinct cleavages; conchoidal fracture; adamantine luster; twinning and parting common
Amphibole Group							
Actinolite	Ca$_2$(Mg,Fe)$_5$Si$_8$O$_{22}$(OH)$_2$	5 - 6	Green to light green, white	3.0 - 3.3	White	Prismatic crystals; massive	*2 perfect cleavages at ≈120° and ≈60°, producing splintery fragments*; conchoidal to uneven fracture; vitreous luster
Tremolite	Ca$_2$Mg$_5$Si$_8$O$_{22}$(OH)$_2$	5 - 6	White to gray	3.0 - 3.3	White	Prismatic crystals; massive	*2 perfect cleavages at ≈120° and ≈60°, producing splintery fragments*; conchoidal to uneven fracture; vitreous luster
Anthophyllite	Mg$_7$Si$_8$O$_{22}$(OH)$_2$	5.5 - 6	Light brown to brown, grayish or greenish tints	2.9 - 3.4	White	Massive; fibrous	*2 perfect cleavages at ≈120° and ≈60°, producing splintery fragments*; conchoidal to uneven fracture; vitreous luster
Glaucophane	Na$_2$(Mg$_3$,Al$_2$)Si$_8$O$_{22}$(OH)$_2$	5.5 - 6	Blue to purplish-blue or black	3.0 - 3.4	White to gray	Massive; fibrous	*2 perfect cleavages at ≈120° and ≈60°, producing splintery fragments*; conchoidal to uneven fracture; vitreous luster

About as hard as glass; Cleavage present & usually visible (continued)

Mineral	Formula	Mohs Hardness	Color	Specific Gravity	Streak	Habit	Other/Special
Pyroxene Group							
Diopside	$CaMgSi_2O_6$	5 - 6	White, light to dark green; brownish	3.3 - 3.5	White	Crystal aggregates; prismatic crystals; massive	**2 perfect cleavages at ≈90°, producing blocky fragments**; conchoidal fracture; vitreous luster
Enstatite "Orthopyroxene"	$Mg_2Si_2O_6$	5.5 - 6	Green to yellowish green, grayish, brownish, bronze, black	3.2 - 3.9	White	Crystal aggregates; massive	**2 perfect cleavages at ≈90°, producing blocky fragments**; uneven fracture; vitreous to silky luster
Nepheline	$(Na,K)AlSiO_4$	5.5 - 6	Colorless to white, gray, reddish	2.5 - 2.6	White	Hexagonal tabular, massive	3 distinct cleavages; conchoidal fracture; *greasy luster*
Rhodonite	$MnSiO_3$	5.5 - 6	Pink to grayish, weathers to black coating	3.4 - 3.7	White to light pink	Massive; prismatic crystals	2 perfect cleavages at ≈90°, rarely seen; splintery fracture; vitreous luster
"Scapolite"	$(Ca,Na)_4Al_6Si_6O_{24}(CO_3,Cl)$	5.5 - 6	Colorless to white, with pink, yellow, purple tints, gray	2.5 - 2.7	White	Prismatic crystals; crystal aggregates	2 distinct, 2 poor cleavages; conchoidal fracture; vitreous luster; meionite (Ca, CO_3) and marialite (Na, Cl) make up a solid-solution series
Plagioclase Feldspar Group							
Albite	$NaAlSi_3O_8$	6	White to light gray, yellow	2.6	White	Tabular crystals; massive	2 good, 2 poor cleavages at ≈90°; conchoidal fracture; vitreous luster; **omnipresent polysynthetic twinning appears as parallel striations on one cleavage face**
Anorthite	$CaAl_2Si_2O_8$	6	Colorless, light to dark green, light to dark gray, black	2.8	White to gray	Massive; tabular crystals	2 good, 2 poor cleavages at ≈90°; conchoidal fracture; vitreous luster; **omnipresent polysynthetic twinning appears as parallel striations on one cleavage face**

About as hard as glass; Cleavage present & usually visible (continued)

Mineral	Formula	Mohs Hardness	Color	Specific Gravity	Streak	Habit	Other/Special
Potassium Feldspar Group							
Microcline	$KAlSi_3O_8$	6	Colorless to white, orange to pink, green	2.6	White	Short prismatic crystals; massive	2 good, 2 poor cleavages at ≈90°; conchoidal fracture; vitreous luster; frequent wavy exsolution lamellae of albite seen in cleavage faces
Orthoclase	$KAlSi_3O_8$	6	Colorless to white, orange to pink, yellow	2.6	White	Short prismatic crystals; massive	2 good, 2 poor cleavages at ≈90°; conchoidal fracture; vitreous luster; frequent wavy exsolution lamellae of albite seen in cleavage faces; contact and penetration twins common
Sanidine	$KAlSi_3O_8$	6	Colorless to yellowish white	2.6	White	Prismatic crystals, usually as phenocrysts	2 good, 2 poor cleavages at ≈90°; conchoidal fracture; vitreous luster

About as hard as glass; Cleavage present but often not visible

Mineral	Formula	Mohs Hardness	Color	Specific Gravity	Streak	Habit	Other/Special
"Wolframite"	$(Fe,Mn)WO_4$	4 - 5.5	Yellowish-brown to reddish-brown to black	7 - 7.5	Yellowish-brown to reddish-brown to black	Prismatic or tabular crystals with striations on one face	1 perfect cleavage; parting sometimes seen; uneven fracture; contact twinning common; ferberite (Fe) and huebnerite (Mn) make up a solid-solution series
Turquoise	$CuAl_6(PO_4)_4(OH)_8 \cdot 4H_2O$	5 - 6	Pale bluish-green to blue or green	2.6 - 2.8	White, with blue or green tints	Massive; as crusts or botryoids	1 perfect, 1 good cleavage; conchoidal fracture; vitreous to waxy luster
Goethite	$FeO(OH)$	5 - 5.5	Yellow to reddish-brown, blackish-brown	3.3 - 4.3	Brownish yellow	Massive; prismatic crystals; botryoidal or stalactiticic masses of radiating crystals	1 perfect cleavage; uneven to splintery fracture; referred to as 'limonite' when in yellowish masses; masses frequently powdery or earthy
Titanite	$CaTiSiO_5$	5 - 5.5	Brown to yellow, green	3.4 - 3.5	White to light brown	Wedge-shaped crystals; massive	2 distinct cleavages; conchoidal fracture; adamantine luster; twinning and parting common

About as hard as glass; Cleavage present but often not visible (continued)

Mineral	Formula	Mohs Hardness	Color	Specific Gravity	Streak	Habit	Other/Special
Glaucophane	$Na_2(Mg_3,Al_2)Si_8O_{22}(OH)_2$	5.5 - 6	Blue to purplish-blue or black	3.0 - 3.4	White to gray	Massive; fibrous	*2 perfect cleavages at ≈120° and ≈60°, producing splintery fragments*; conchoidal to uneven fracture; vitreous luster; part of the amphibole group
Leucite	$KAlSiO_4$	5.5 - 6	Colorless to white or gray	2.4 - 2.5	White	Trapezohedral crystals; massive	6 imperfect cleavages; conchoidal fracture; vitreous to dull luster
Sodalite	$Na_8(Al_6Si_6O_{24})Cl_6$	5.5 - 6	Blue, colorless to white, purple, pinkish	2.2 - 2.3	White	Massive; rarely dodecahedral crystals	6 poor cleavages; conchoidal fracture; vitreous luster

About as hard as glass; Cleavage absent or very poorly visible

Mineral	Formula	Mohs Hardness	Color	Specific Gravity	Streak	Habit	Other/Special
Analcime	$NaAlSi_2O_6 \cdot H_2O$	5 - 5.5	Colorless to white, grayish, greenish	2.3	White	Trapezohedral crystals; massive	Conchoidal fracture; vitreous luster
Hematite	Fe_2O_3	5 - 6	Maroon to reddish-brown, violet-black	5.3	Red to reddish brown	Massive; tabular crystals	Conchoidal to uneven fracture; parting due to twinning common
Opal	$SiO_2 \cdot nH_2O$	5 - 6	Colorless to white, but may appear any color due to impurities	1.9 - 2.2	White	Massive; botryoidal or stalactitic; pseudomorphs common	Conchoidal fracture; vitreous to resinous luster; may show a rainbow play of colors when rotated in light (precious opal)

Harder than glass; Cleavage present & usually visible

Mineral	Formula	Mohs Hardness	Color	Specific Gravity	Streak	Habit	Other/Special
Kyanite	Al$_2$SiO$_5$	5 - 7	Blue, white, grayish, greenish	3.6 - 3.7	White	Tabular crystals; aggregates or radiating	2 perfect cleavages; splintery fracture; vitreous luster; *hardness of 5 parallel to and hardness of 7 perpendicular to, crystal length*
Prehnite	Ca$_2$Al(AlSi$_3$O$_{10}$)(OH)$_2$	6 - 6.5	White, light or yellow-green to green	2.8 - 2.9	White	Masses of tabular, curved crystals; botryoidal; massive	1 distinct cleavage; uneven fracture; vitreous luster
Rutile	TiO$_2$	6 - 6.5	Reddish-brown, commonly with silvery tint	4.2 - 4.4	Pale brown to black	Prismatic to acicular crystals; massive	4 distinct cleavages; conchoidal to uneven fracture; adamantine to metallic luster; contact twinning common
Sillimanite	Al$_2$SiO$_5$	6 - 7	Colorless to white, brown, gray	3.2 - 3.3	White	Fibrous masses; massive; prismatic crystals	2 perfect cleavages; splintery to conchoidal fracture; satiny to vitreous luster
Spodumene	LiAlSi$_2$O$_6$	6.5 - 7	Colorless to white, light tan, purplish, yellowish, greenish	3.1 - 3.2	White	Prismatic crystals; crystal aggregates	*2 perfect cleavages at ≈90°, producing blocky fragments*; uneven fracture; vitreous luster; parting common; member of pyroxene group
Andalusite	Al$_2$SiO$_5$	7.5	Gray to brown, white, pinkish an greenish tints	3.1 - 3.2	White	Prismatic crystals with square cross-section	2 good cleavages; conchoidal fracture; vitreous luster; commonly has oriented inclusions of graphite producing a cruciform cross-section
Topaz	Al$_2$SiO$_4$(OH,F)$_2$	8	Colorless to white, yellow, brown, blue, green	3.5 - 3.6	White	Prismatic crystals, frequently terminated with small faces; massive	1 perfect cleavage, perpendicular to crystal length; conchoidal fracture; vitreous luster

Harder than glass; Cleavage present but often not visible

Mineral	Formula	Mohs Hardness	Color	Specific Gravity	Streak	Habit	Other/Special
Rutile	TiO_2	6 - 6.5	Reddish-brown, commonly with silvery tint	4.2 - 4.4	Pale brown to black	Prismatic to acicular crystals; massive	4 distinct cleavages; conchoidal to uneven fracture; adamantine to metallic luster; contact twinning common
Cassiterite	SnO_2	6 - 7	Yellowish-brown to reddish-brown to brownish black	7.02	White to brown or gray	Massive, botryoidal, concretions; prismatic or pyramidal crystals	1 imperfect cleavage; distinct parting; conchoidal to uneven fracture; adamantine to metallic luster; contact and penetration twins common
Epidote	$Ca_2(Fe,Al)_3(SiO_4)_3(OH)$	6 - 7	Light green to pistachio green to greenish black, brown, yellow	3.4 - 3.5	White with greenish tints	Massive; crusts and masses of fine crystals; prismatic crystals	1 perfect cleavage; uneven fracture; vitreous to pearly luster
Vesuvianite	$Ca_{10}(Fe,Mg)_2Al_4(Si_2O_7)_2(OH,F)_4$	6.5	Colorless to yellow, greenish yellow to brown, blue	3.4 - 3.5	White	Prismatic crystals; massive	2 poor cleavages; conchoidal to uneven fracture; vitreous luster
Zircon	$ZrSiO_4$	7.5	Brown to reddish brown, gray, white	4 - 4.7	White to brownish	Prismatic crystals, often terminated by pyramid of 4 smaller faces	2 poor cleavages; conchoidal fracture; adamantine luster
Beryl	$Be_3Al_2Si_6O_{18}$	8	Light green to yellow, blue, red, white	2.6 - 2.8	White	Hexagonal prismatic crystals; massive	1 poor cleavage; conchoidal fracture; vitreous luster

Harder than glass; Cleavage absent or very poorly visible

Mineral	Formula	Mohs Hardness	Color	Specific Gravity	Streak	Habit	Other/Special
"Garnet"	$A_3B_2(SiO_4)_3$	6 - 7.5	Red to brown, white, orange, green, yellow, black	3.5 - 4.3	White	Dodecahedral or trapezohedral crystals; massive	Conchoidal to uneven fracture; vitreous luster; A = Mg, Fe, Mn and B= Al for pyralspite (pyrope, almandine, spessartine) garnets; A= Ca, B = Al, Fe, Cr for ugrandite (uvarovite, grossular, andradite) garnets
"Olivine"	$(Mg,Fe)_2SiO_4$	6.5 - 7	Green to yellowish green	3.3 - 3.4	White with greenish tint	Granular masses; prismatic crystals	Conchoidal fracture; vitreous luster; forsterite (Mg) and fayalite (Fe) form a solid solution series
Quartz	SiO_2	7	Colorless to white, pink, purple, brown, gray, black, yellow	2.65	White	Hexagonal prismatic crystals; massive	3 indistinct cleavages; conchoidal fracture; vitreous luster; contact twinning common
"Tourmaline"	$AB_3C_6(BO_3)_3(Z)O_{18}(OH,F)_4$	7 - 7.5	Colorless, white, pink, green, brown, black	3.0 - 3.3	White with varying tints	Prismatic crystals with triangular cross-section; radiating needles	Conchoidal fracture; vitreous luster; A=Ca,Na,K; B=Al,Fe,Li,Mg,Mn; C=Al,Cr,Fe,V; Z=Si,Al,B for more than a dozen named varieties
Spinel	$MgAl_2O_4$	7.5 - 8	White, red, blue, brown, purple, black	3.5 - 4.1	White	Octahedral crystals; massive	Conchoidal fracture; vitreous luster; twinning common; parting may occur
Corundum	Al_2O_3	9	White to blue or red, green, violet or yellow less common	4.02	White	Hexagonal prismatic or tabular crystals; massive	Extreme hardness (scratches glass with light touch); uneven fracture; parting common

Dark Colored Minerals With a Non-Metallic Luster

Softer than glass; Cleavage present & usually visible

Mineral	Formula	Mohs Hardness	Color	Specific Gravity	Streak	Habit	Other/Special
Talc	$Mg_3Si_4O_{10}(OH)_2$	1	White to greenish or grayish	2.7 - 2.8	White, with greenish tints	Massive	1 perfect cleavage; pearly to greasy luster; *soapy feel*
Pyrophyllite	$Al_2Si_4O_{10}(OH)_2$	1 - 2	White to tan, greenish, silvery, brown	2.8 - 2.9	White	Massive; thin radiating crystals	1 perfect cleavage; pearly to greasy luster
Gypsum	$CaSO_4 \cdot 2H_2O$	2	Colorless to white, gray, yellow, brown or gray	2.3	White	Massive; tabular or prismatic crystals	1 perfect, 1 distinct cleavage into flexible fragments; conchoidal fracture; vitreous luster; contact twinning common
Halite	$NaCl$	2	Colorless, white, blue, pink, purple, yellow, gray	2.16	White	Cubic crystals; massive	*3 perfect cleavages at 90°*; conchoidal fracture; vitreous luster; salty taste
"Chlorite"	$A_{4-6}Z_4O_{10}(OH,O)_8$	2 - 2.5	Green to pale green, pink, yellow, white	2.6 - 3.0	White	Massive; tabular crystals	*1 perfect cleavage into flexible fragments*; vitreous to pearly luster; A = Al, Fe, Mn, Mg and Z= Al, B, Fe, Si for at least ten named varieties
Cinnabar	HgS	2 - 2.5	Red, with gray or metallic tints	8.1	Red	Massive; tabular, prismatic or rhombohedral crystals	1 perfect cleavage; conchoidal fracture; adamantine luster; may have coatings of liquid mercury

Softer than glass; Cleavage present & usually visible (continued)

Mineral	Formula	Mohs Hardness	Color	Specific Gravity	Streak	Habit	Other/Special
					Mica Group		
"Biotite"	K(Fe,Mg)$_3$AlSi$_3$O$_{10}$(OH)$_2$	2.5 - 3	Dark brown to black	2.8 - 3.4	Pale brown	Pseudo-hexagonal or parallelogram-shaped crystals; massive	*1 perfect cleavage into elastic fragments*; vitreous to pearly luster; name used for intermediate compositions between annite and phlogopite and eastonite and siderophyllite
Phlogopite	KMg$_3$AlSi$_3$O$_{10}$(OH)$_2$	2.5 - 3	Light, coppery-brown to dark brown	2.7	White, with brownish tints	Pseudo-hexagonal or parallelogram-shaped crystals; massive	*1 perfect cleavage into elastic fragments*; vitreous to pearly luster
Calcite	CaCO$_3$	3	Colorless to white, green, yellow, purple, blue, red, gray, black	2.71	White	Scalenohedral, prismatic, rhombohedral, tabular crystals	*3 perfect cleavages, not at 90° (rhombohedral)*; *reacts strongly to acid*; significant double refraction in colorless crystals; commonly twinned
Barite	BaSO$_4$	3 - 3.5	Colorless to white, yellow, brown	4.5	White	Tabular, prismatic or equant crystals; massive	1 perfect cleavage; others less perfect; uneven fracture; vitreous to resinous luster
Dolomite	CaMg(CO$_3$)$_2$	3.5 - 4	Colorless to white, pink, tan, brown, gray, black	2.85	White	Rhombohedral or prismatic crystals, commonly curved into saddle shapes	*3 perfect cleavages at 90° (rhombohedral)*; conchoidal fracture; vitreous to pearly luster; *reacts weakly to acid unless powdered*
Sphalerite	ZnS	3.5 - 4	Metallic black to translucent brown, yellow, green	3.9 - 4.1	Brown to pale yellow-white	Tetrahedral or dodecahedral crystals; massive	6 perfect cleavages; conchoidal fracture; resinous to adamantine luster
Siderite	FeCO$_3$	3.5 - 4.2	Brown to yellowish or reddish brown, gray, white	3.96	Pale brown	Rhombohedral, tabular, prismatic or scalenohedral crystals; massive	*3 perfect cleavages, not at 90° (rhombohedral)*; reacts to hot acid

Softer than glass; Cleavage present & usually visible (continued)

Mineral	Formula	Mohs Hardness	Color	Specific Gravity	Streak	Habit	Other/Special
Wavellite	$Al_3(OH)_3(PO_4)_2 \cdot 5H_2O$	3.5 - 4	Greenish-white to green, yellow, brown, black	2.36	White	Globular or botryoidal masses of radiating fibrous crystals; crusts or coatings	2 perfect, 2 good cleavages; uneven fracture; vitreous to resinous luster
Fluorite	CaF_2	4	Colorless, white, blue, pink, purple, yellow or mixtures	3.18	White	Cubic or octahedral crystals, sometimes combined; massive	*4 perfect cleavages*; conchoidal to uneven fracture; vitreous luster; penetration twinning common

Softer than glass; Cleavage present but often not visible

Mineral	Formula	Mohs Hardness	Color	Specific Gravity	Streak	Habit	Other/Special
Talc	$Mg_3Si_4O_{10}(OH)_2$	1	White to greenish or grayish	2.7 - 2.8	White, with greenish tints	Massive	1 perfect cleavage; pearly to greasy luster; *soapy feel*
Graphite	C	1 - 1.5	Metallic gray to black	2.2	Black	Flat, hexagonal sheets; massive	1 perfect cleavage; *greasy feel, writes on paper*
Pyrophyllite	$Al_2Si_4O_{10}(OH)_2$	1 - 2	White to tan, greenish, silvery, brown	2.8 - 2.9	White	Massive; thin radiating crystals	1 perfect cleavage; pearly to greasy luster
Gypsum	$CaSO_4 \cdot 2H_2O$	2	Colorless to white, gray, yellow, brown or gray	2.3	White	Massive; tabular or prismatic crystals	1 perfect, 1 distinct cleavage into flexible fragments; conchoidal fracture; vitreous luster; contact twinning common
"Chlorite"	$A_{4-6}Z_4O_{10}(OH,O)_8$	2 - 2.5	Green to pale green, pink, yellow, white	2.6 - 3.0	White	Massive; tabular crystals	*1 perfect cleavage into flexible fragments*; vitreous to pearly luster; A = Al, Fe, Mn, Mg and Z= Al, B, Fe, Si for at least ten named varieties

Softer than glass; Cleavage present but often not visible (continued)

Mineral	Formula	Mohs Hardness	Color	Specific Gravity	Streak	Habit	Other/Special
Cinnabar	HgS	2 - 2.5	Red, with gray or metallic tints	8.1	Red	Massive; tabular, prismatic or rhombohedral crystals	1 perfect cleavage; conchoidal fracture; adamantine luster; may have coatings of liquid mercury
"Serpentine"	$Mg_3Si_2O_5(OH)_4$	2 - 5	White to green, yellow, black mottling	2.2 - 2.6	White	Massive; fibrous; botryoidal	Uneven fracture; silky to waxy luster; chrysotile is fibrous form, antigorite is platy; lizardite is massive with waxy, mottled surface
Anhydrite	$CaSO_4$	3.5	Colorless to white, gray, yellow, brown or gray	3	White	Massive; equant or tabular crystals	1 perfect, 1 nearly perfect, 1 good cleavage; uneven to splintery fracture; pearly to vitreous or greasy luster
Azurite	$Cu_3(CO_3)_2(OH)_2$	3.5 - 4	Blue to dark blue	3.77	Blue	Tabular, prismatic or equant crystals; massive	1 perfect, 1 fair cleavage; conchoidal fracture; adamantine luster; reacts strongly to acid
Malachite	$Cu_2CO_3(OH)_2$	3.5 - 4	Green	4.05	Green	Massive, especially botryoidal or fibrous	1 perfect, 1 fair cleavage; uneven to conchoidal fracture; vitreous to adamantine luster; reacts strongly to acid
Pyromorphite	$Pb_5(PO_4)_3Cl$	3.5 - 4	Yellowish-green to dark green, gray, brown	6.5 - 7.1	White, with greenish tints	Prismatic or tabular hexagonal crystals; massive	1 indistinct cleavage; conchoidal to uneven fracture; resinous to adamantine luster; member of apatite group
"Apatite"	$Ca_5(PO_4)_3(OH,Cl,F)$	5	Colorless to white, green, brown, blue, yellow	3.1 - 3.2	White	Prismatic or tabular hexagonal crystals; massive	2 indistinct cleavages; conchoidal fracture; vitreous luster; end members are hydroxylapatite, chlorapatite, fluorapatite

Softer than glass; Cleavage absent or very poorly visible

Mineral	Formula	Mohs Hardness	Color	Specific Gravity	Streak	Habit	Other/Special
"Serpentine"	$Mg_3Si_2O_5(OH)_4$	2 - 5	White to green, yellow, black mottling	2.2 - 2.6	White	Massive; fibrous; botryoidal	Uneven fracture; silky to waxy luster; chrysotile is fibrous form, antigorite is platy; lizardite is massive with waxy, mottled surface
Hematite	Fe_2O_3	2 - 6	Steel grey to black	5.3	Red to reddish brown	Massive; tabular crystals	Conchoidal to uneven fracture; parting due to twinning common
"Psilomelane"	Mixed Mn oxides and hydroxides	2 - 6.5	Black	3.7 - 4.8	Brownish black	Massive; botryoidal or stalactitic; dendritic masses	Often in hard or powdery masses; commonly a mixture of pyrolusite, romanechite, cryptomelane and other difficult to characterize Mn minerals
Vanadinite	$Pb_5(VO_4)_3Cl$	2.5 - 3	Red to orange-red, yellow to brown	6.7 - 7.1	Red to brownish-yellow	Tabular or prismatic hexagonal crystals; massive	Conchoidal to uneven fracture; resinous to adamantine luster; member of apatite group

About as hard as glass; Cleavage present & usually visible

Mineral	Formula	Mohs Hardness	Color	Specific Gravity	Streak	Habit	Other/Special
"Wolframite"	$(Fe,Mn)WO_4$	4 - 5.5	Yellowish-brown to reddish-brown to black	7 - 7.5	Yellowish-brown to reddish-brown to black	Prismatic or tabular crystals with striations on one face	1 perfect cleavage; parting sometimes seen; uneven fracture; contact twinning common; ferberite (Fe) and huebnerite (Mn) make up a solid-solution series
Monazite	$(Ce,La,Th,Y)PO_4$	5 - 5.5	Brown to reddish brown or yellow	4.6 - 5.4	White to light brown	Tabular, prismatic crystals; massive	1 distinct cleavage; uneven to conchoidal fracture; resinous or waxy to vitreous or adamantine luster
Titanite	$CaTiSiO_5$	5 - 5.5	Brown to yellow, green	3.4 - 3.5	White to light brown	Wedge-shaped crystals; massive	2 distinct cleavages; conchoidal fracture; adamantine luster; twinning and parting common

About as hard as glass; Cleavage present & usually visible (continued)

Mineral	Formula	Mohs Hardness	Color	Specific Gravity	Streak	Habit	Other/Special
Amphibole Group							
Actinolite	$Ca_2(Mg,Fe)_5Si_8O_{22}(OH)_2$	5 - 6	Green to light green, white	3.0 - 3.3	White	Prismatic crystals; massive	*2 perfect cleavages at ≈120° and ≈60°, producing splintery fragments*; conchoidal to uneven fracture; vitreous luster
"Hornblende"	$Ca_2[(Mg,Fe)_4(Al,Fe)]Si_7O_{22}(OH)_2$	5 - 6	Dark green to black	3.0 - 3.4	Gray with greenish tints	Prismatic crystals; massive	*2 perfect cleavages at ≈120° and ≈60°, producing splintery fragments*; conchoidal to uneven fracture; vitreous luster; ferrohornblende and magnesiohornblende make up a solid-solution series
Anthophyllite	$Mg_7Si_8O_{22}(OH)_2$	5.5 - 6	Light brown to brown, grayish or greenish tints	2.9 - 3.4	White	Massive; fibrous	*2 perfect cleavages at ≈120° and ≈60°, producing splintery fragments*; conchoidal to uneven fracture; vitreous luster
Glaucophane	$Na_2(Mg_3,Al_2)Si_8O_{22}(OH)_2$	5.5 - 6	Blue to purplish-blue or black	3.0 - 3.4	White to gray	Massive; fibrous	*2 perfect cleavages at ≈120° and ≈60°, producing splintery fragments*; conchoidal to uneven fracture; vitreous luster
Pyroxene Group							
Augite	$(Ca,Na)(Mg,Fe,Al,Ti)(Si,Al)_2O_6$	5 - 6	Dark green to black	3.2 - 3.4	Gray with greenish tints	Prismatic crystals; massive	*2 perfect cleavages at ≈90°, producing blocky fragments*; conchoidal fracture; vitreous luster
Diopside	$CaMgSi_2O_6$	5 - 6	White, light to dark green; brownish	3.3 - 3.5	White	Crystal aggregates; prismatic crystals; massive	*2 perfect cleavages at ≈90°, producing blocky fragments*; conchoidal fracture; vitreous luster
Enstatite "Orthopyroxene"	$Mg_2Si_2O_6$	5.5 - 6	Green to yellowish green, grayish, brownish, black	3.2 - 3.9	White	Crystal aggregates; massive	*2 perfect cleavages at ≈90°, producing blocky fragments*; uneven fracture; vitreous to silky luster
Nepheline	$(Na,K)AlSiO_4$	5.5 - 6	Colorless to white, gray, reddish	2.5 - 2.6	White	Hexagonal tabular; massive	3 distinct cleavages; conchoidal fracture; *greasy luster*

About as hard as glass; Cleavage present & usually visible (continued)

Mineral	Formula	Mohs Hardness	Color	Specific Gravity	Streak	Habit	Other/Special
"Scapolite"	$(Ca,Na)_4Al_6Si_6O_{24}(CO_3,Cl)$	5.5 - 6	Colorless to white, with pink, yellow, purple tints, gray	2.5 - 2.7	White	Prismatic crystals; crystal aggregates	2 distinct, 2 poor cleavages; conchoidal fracture; vitreous luster; meionite (Ca, CO$_3$) and marialite (Na, Cl) make up a solid-solution series
Anorthite	$CaAl_2Si_2O_8$	6	Light to dark gray, black	2.8	White to gray	Massive; tabular crystals	2 good, 2 poor cleavages at ≈90°; conchoidal fracture; vitreous luster; **omnipresent polysynthetic twinning appears as parallel striations on one cleavage face**; member of plagioclase group
Potassium Feldspar group							
Microcline	$KAlSi_3O_8$	6	Colorless to white, pinkish orange, green	2.6	White	Short prismatic crystals; massive	2 good, 2 poor cleavages at ≈90°; conchoidal fracture; vitreous luster; frequent wavy exsolution lamellae of albite seen in cleavage faces
Orthoclase	$KAlSi_3O_8$	6	Colorless to white, pinkish orange, yellow	2.6	White	Short prismatic crystals; massive	2 good, 2 poor cleavages at ≈90°; conchoidal fracture; vitreous luster; frequent wavy exsolution lamellae of albite seen in cleavage faces; contact and penetration twins common

About as hard as glass; Cleavage present but often not visible

Mineral	Formula	Mohs Hardness	Color	Specific Gravity	Streak	Habit	Other/Special
Goethite	FeO(OH)	2 - 5.5	Yellow to reddish-brown, blackish-brown	3.3 - 4.3	Brownish yellow	Massive; prismatic crystals; botryoidal or stalacticic masses of radiating crystals	1 perfect cleavage; uneven to splintery fracture; referred to as 'limonite' when in yellowish masses; masses frequently powdery or earthy
"Wolframite"	(Fe,Mn)WO$_4$	4 - 5.5	Yellowish-brown to reddish-brown to black	7 - 7.5	Yellowish-brown to reddish-brown to black	Prismatic or tabular crystals with striations on one face	1 perfect cleavage; parting sometimes seen; uneven fracture; contact twinning common; ferberite (Fe) and huebnerite (Mn) make up a solid-solution series
Titanite	CaTiSiO$_5$	5 - 5.5	Brown to yellow, green	3.4 - 3.5	White to light brown	Wedge-shaped crystals; massive	2 distinct cleavages; conchoidal fracture; adamantine luster; twinning and parting common
Allanite	(Ca, Ce, Y, La)$_2$(Al,Fe)$_3$(SiO$_4$)$_3$(OH)	5.5 - 6	Dark brown to black	2.7 - 4.2	Gray with brown tints	Prismatic or tabular crystals; massive	Poor cleavages; conchoidal to uneven fracture; vitreous to resinous luster; member of epidote group
Glaucophane	Na$_2$(Mg$_3$,Al$_2$)Si$_8$O$_{22}$(OH)$_2$	5.5 - 6	Blue to purplish-blue or black	3.0 - 3.4	White to gray	Massive; fibrous	*2 perfect cleavages at ≈120° and ≈60°, producing splintery fragments*; conchoidal to uneven fracture; vitreous luster; member of the amphibole group

About as hard as glass; Cleavage absent or very poorly visible

Mineral	Formula	Mohs Hardness	Color	Specific Gravity	Streak	Habit	Other/Special
Hematite	Fe$_2$O$_3$	2 - 6	Steel grey to black	5.3	Red to reddish brown	Massive; tabular crystals	Conchoidal to uneven fracture; parting due to twinning common
"Psilomelane"	Mixed Mn oxides and hydroxides	2 - 6.5	Black	3.7 - 4.8	Brownish black	Massive; botryoidal or stalactitic; dendritic masses	Often in hard or powdery masses; a mixture of pyrolusite, romanechite, cryptomelane and other difficult to characterize Mn minerals
Opal	SiO$_2$•nH$_2$O	5 - 6	Colorless to white, but may appear any color due to impurities	1.9 - 2.2	White	Massive; botryoidal or stalactitic; pseudomorphs common	Conchoidal fracture; vitreous to resinous luster; may show a rainbow play of colors when rotated in light (precious opal)

Harder than glass; Cleavage present & usually visible

Mineral	Formula	Mohs Hardness	Color	Specific Gravity	Streak	Habit	Other/Special
Kyanite	Al_2SiO_5	5 - 7	Blue, white, grayish, greenish	3.6 - 3.7	White	Tabular crystals; aggregates or radiating	2 perfect cleavages; splintery fracture; vitreous luster; *hardness of 5 parallel to and hardness of 7 perpendicular to, crystal length*
Rutile	TiO_2	6 - 6.5	Reddish-brown, commonly with silvery tint	4.2 - 4.4	Pale brown to black	Prismatic to acicular crystals; massive	4 distinct cleavages; conchoidal to uneven fracture; adamantine to metallic luster; contact twinning common
Columbite/ Tantalite	$(Fe,Mn)(Nb,Ta)_2O_6$	6 - 6.5	Black to brownish black with reddish internal reflections	5.2 - 8	Black to dark red	Prismatic, tabular, equant or pyramidal crystals; massive	1 distinct cleavage; conchoidal to uneven fracture; contact twinning common; Forms two series: ferrocolumbite to manganocolumbite and ferrotantalite to manganotantalite
Sillimanite	Al_2SiO_5	6 - 7	Colorless to white, brown, gray	3.2 - 3.3	White	Fibrous masses; massive; prismatic crystals	2 perfect cleavages; splintery to conchoidal fracture; satiny to vitreous luster
Staurolite	$(Fe,Mg,Zn)_2Al_9(Si,Al)_4O_{22}(OH)_2$	7 - 7.5	Brown to dark brown	3.6 - 3.7	Pale brown	Prismatic crystals	2 distinct cleavages; conchoidal fracture; vitreous luster; very commonly penetration twinned, often in a cruciform shape (fairystones)
Andalusite	Al_2SiO_5	7.5	Gray to brown, white, pinkish an greenish tints	3.1 - 3.2	White	Prismatic crystals with square cross-section	2 good cleavages; conchoidal fracture; vitreous luster; commonly has oriented inclusions of graphite producing a cruciform cross-section

Harder than glass; Cleavage present but often not visible

Mineral	Formula	Mohs Hardness	Color	Specific Gravity	Streak	Habit	Other/Special
Rutile	TiO_2	6 - 6.5	Reddish-brown, commonly with silvery tint	4.2 - 4.4	Pale brown to black	Prismatic to acicular crystals; massive	4 distinct cleavages; conchoidal to uneven fracture; adamantine to metallic luster; contact twinning common
Cassiterite	SnO_2	6 - 7	Yellowish-brown to reddish-brown to brownish black	7.02	White to brown or gray	Massive, botryoidal, concretions; prismatic or pyramidal crystals	1 imperfect cleavage; distinct parting; conchoidal to uneven fracture; adamantine to metallic luster; contact and penetration twins common
Epidote	$Ca_2(Fe,Al)_3(SiO_4)_3(OH)$	6 - 7	Light green to pistachio green to greenish black, brown, yellow	3.4 - 3.5	White with greenish tints	Massive; crusts and masses of fine crystals; prismatic crystals	1 perfect cleavage; uneven fracture; vitreous to pearly luster
Staurolite	$(Fe,Mg,Zn)_2Al_9(Si,Al)_4O_{22}(OH)_2$	7 - 7.5	Brown to dark brown	3.6 - 3.7	Pale brown	Prismatic crystals	2 distinct cleavages; conchoidal fracture; vitreous luster; very commonly penetration twinned, often in a cruciform shape (fairystones)
Zircon	$ZrSiO_4$	7.5	Brown to reddish brown, gray, white	4 - 4.7	White to brownish	Prismatic crystals, often terminated by pyramid of 4 smaller faces	2 poor cleavages: conchoidal fracture; adamantine luster

Harder than glass; Cleavage absent or very poorly visible

Mineral	Formula	Mohs Hardness	Color	Specific Gravity	Streak	Habit	Other/Special
"Garnet"	$A_3B_2(SiO_4)_3$	6 - 7.5	Red to brown, white, orange, green, yellow, black	3.5 - 4.3	White	Dodecahedral or trapezohedral crystals; massive	Conchoidal to uneven fracture; vitreous luster; A = Mg, Fe, Mn and B= Al for pyralspite (pyrope, almandine, spessartine) garnets; A= Ca, B = Al, Fe, Cr for ugrandite (uvarovite, grossular, andradite) garnets
"Olivine"	$(Mg,Fe)_2SiO_4$	6.5 - 7	Green to yellowish green	3.3 - 3.4	White with greenish tint	Granular masses; prismatic crystals	Conchoidal fracture; vitreous luster; forsterite (Mg) and fayalite (Fe) form a solid solution series
Quartz	SiO_2	7	Colorless to white, pink, purple, brown, gray, black, yellow	2.65	White	Hexagonal prismatic crystals; massive	3 indistinct cleavages not at 90° (rhombohedral); conchoidal fracture; vitreous luster; contact twinning common
"Tourmaline"	$AB_3C_6(BO_3)_3(Z)O_{18}(OH,F)_4$	7 - 7.5	Colorless, white, pink, green, brown, black	3.0 - 3.3	White with varying with triangular cross-section; radiating needles	Prismatic crystals	Conchoidal fracture; vitreous luster; A=Ca,Na,K; B=Al,Fe,Li,Mg,Mn; C=Al,Cr,Fe,V; Z=Si,Al,B for more than a dozen named varieties
Spinel	$MgAl_2O_4$	7.5 - 8	White, red, blue, brown, purple, black	3.5 - 4.1	White	Octahedral crystals; massive	Conchoidal fracture; vitreous luster; twinning common; parting may occur
Corundum	Al_2O_3	9	White to blue or red, green, violet or yellow less common	4.02	White	Hexagonal prismatic or tabular crystals; massive	Extreme hardness (scratches glass with light touch); uneven fracture; parting common

Appendix B
Identifying Common Rocks in Hand Specimens

Rocks are classified as igneous, sedimentary, or metamorphic types. Igneous rocks crystallize from a melt. Sedimentary rocks form under surface conditions through accumulation of (1) chemical and biochemical precipitates, (2) fragments of rocks, minerals, and/or fossils, or (3) both types of materials. Metamorphic rocks form from:

- igneous or sedimentary rocks, or preexisting metamorphic rocks, when those rocks have been changed mineralogically and/or texturally without undergoing melting,
- heat,
- pressure,
- deviatoric stress, and
- chemically active fluids within Earth.

Because each class of rock has a different general history, each has its own characteristics.

Minerals, textures, or both are used for assigning a rock to the proper class (table B.1). Three major textural types—clastic, crystalline, and aphanitic—are used for an initial assignment of a rock to a broad group. As noted in chapter 9, clastic textures are those dominated by fragments of rock, mineral, glass, or organic remains. In clastic rocks, the grains may be angular or rounded, or may be a mix, and these grains are joined together by (1) slightly recrystallized intergrown grain boundaries, (2) finer grained fragmental material called matrix, or (3) precipitated minerals called cement. All three major classes of rock include clastic textured rocks, but such rocks are predominantly of sedimentary origin.

Recall that interlocking grains characterize crystalline textures. The interlocking nature of crystalline textures results either from recrystallization (i.e., reorganization of the grain boundaries of preexisting mineral grains) or from growth of crystals from a melt, solution, or gas. All classes of rock include types with crystalline textures.

Aphanitic textures are textures in which the grains are too small to recognize or see. These include both clastic and crystalline types. Rocks with aphanitic textures generally have a dull surface appearance or a grainy "sugary" character. All classes of rock include aphanitic types.

Volcanic rocks are glassy, aphanitic, aphanitic-porphyritic, vesicular, or grainy-porphyritic. Volcanic rocks are now customarily classified based on their chemistry (figure 7.19). In the field and in hand specimens, however, visible grains (the phenocrysts) must be used for preliminary classification. Grain percentages may be estimated by comparing the grain numbers to the patterns in figure B.1 (on pp. 307–308). Using visible minerals only, one can make preliminary identifications using charts like those of the IUGS (International Union of Geological Sciences) (figure B.2 on p. 309) and Raymond (figure 7.19, reproduced here as figure B.3 on p. 310). If aphanitic rocks have no distinguishing features, their identification is speculative. Nevertheless, a guess may be made based on structures or the association of the unknown rock with other known rock types. Dark-shaded aphanitic rocks associated with known volcanic rocks are best referred to as *mafite*. Light-shaded aphanitic rocks associated with known volcanic rocks are best referred to as *felsite*.

Igneous rocks with grains large enough to be seen and identified by eye or with a hand lens are plutonic. These rocks typically have crystalline, nonfoliated textures. Plutonic igneous rocks are identified and classified in hand specimens on the basis of mineral percentages and ratios. The most widely used classification for rocks rich in feldspars, quartz, or feldspathoid minerals is that of the IUGS (figure 7.16). Where quantitative analysis and calculation of mineralogical ratios are possible (typically in the laboratory), that classification is used. The difficulty of using such a classification in the field is discussed by Raymond (2007). Field classification of plutonic rocks is more

Table B.1 Chart for Distinguishing the Three Classes of Rock

I. CLASTIC texture (rounded to angular fragments stuck together)
   A. Clasts are dominantly angular, scratched, and polished and of similar composition—the rock is probably *METAMORPHIC*.
   B. Clasts are glass and/or volcanic (e.g., vesicular = full of holes; or porphyritic)—the rock is *IGNEOUS*.
   C. Clasts are rounded or angular and may vary in composition—the rock is *SEDIMENTARY*.

II. CRYSTALLINE texture—grains generally visible to the unaided eye (phaneritic) and have interlocking grain boundaries.
   A. Foliated or lineated rocks are *METAMORPHIC*.
   B. Nonfoliated and nonlineated rocks may be of any class.
   Check hardness.
   1. Rocks composed primarily of minerals harder than 5±.
       a. Most abundant minerals are feldspars; plus quartz, feldspathoids, hornblende, or augite—the rock is probably *IGNEOUS*.
       b. Most abundant minerals include epidote, actinolite, tremolite, kyanite, wollastonite, cordierite, or andalusite—rock is *METAMORPHIC*.
       c. Most abundant minerals are olivine and pyroxenes—rock may be *IGNEOUS* or *METAMORPHIC*. Presence of minor minerals such as talc or anthophyllite suggests metamorphism.
   2. Rocks composed primarily of minerals softer than 5±.
       a. Serpentine minerals abundant—rock is *METAMORPHIC*.
       b. Small (fine) grains of carbonate minerals characterize *SEDIMENTARY* rocks.
       c. Large (coarse) grains of carbonate minerals suggest *METAMORPHIC* or *IGNEOUS* rocks. Minor accessory minerals such as graphite, phlogopite, and epidote suggest a metamorphic origin. Field relations may be necessary to distinguish some igneous and metamorphic carbonates rocks.
       d. Halides, sulfates, and borates predominate in some *SEDIMENTARY* rocks.

III. APHANITIC textures are textures in which minerals are too small to see or recognize. Rocks may be igneous, sedimentary, or metamorphic.
   A. Foliated rocks are *METAMORPHIC*.
   B. Glassy rocks are *IGNEOUS*.
   C. Rocks with vesicles, amygdules, and/or phenocrysts are *IGNEOUS*.
   D. Rocks with primary layering are *SEDIMENTARY*.
   E. Carbonate rocks are probably *SEDIMENTARY* (there are rare exceptions).
   F. Waxy rocks of various colors and conchoidal fracture, harder than 5½, are probably the *SEDIMENTARY* rock chert.
   G. Aphanitic rocks with no distinctive features are difficult to classify. Field evidence is generally necessary to determine the fundamental class to which such rocks belong.

(*Source:* Modified from Raymond, 2009.)

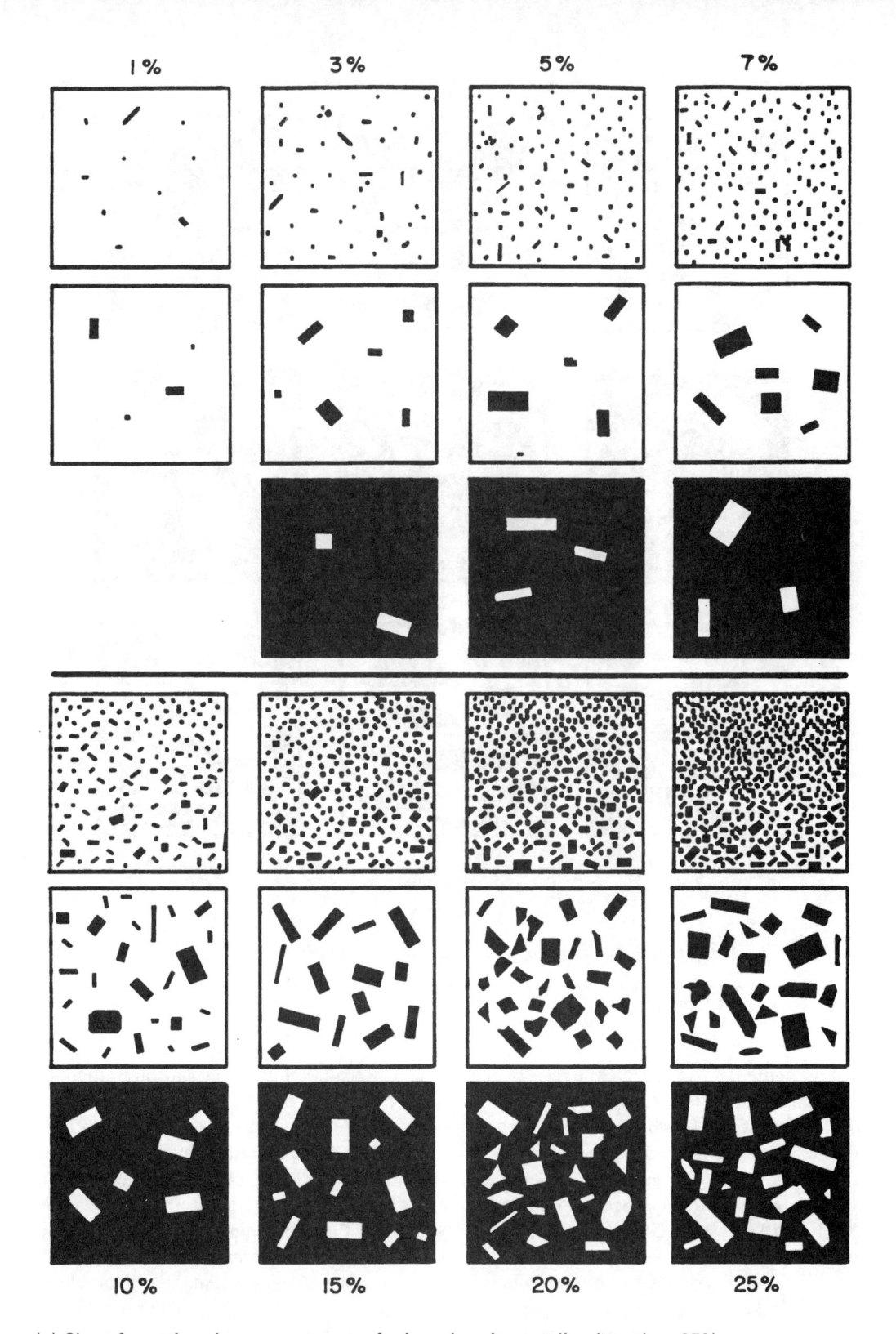

(a) Chart for estimating percentages of mineral grains totaling less than 25%.

Figure B.1
Grain percentages.
(*Sources:* (a) after figure 14a, Raymond, 2009; (b) after figure 14b, Raymond, 2009.)

30%   40%   50%

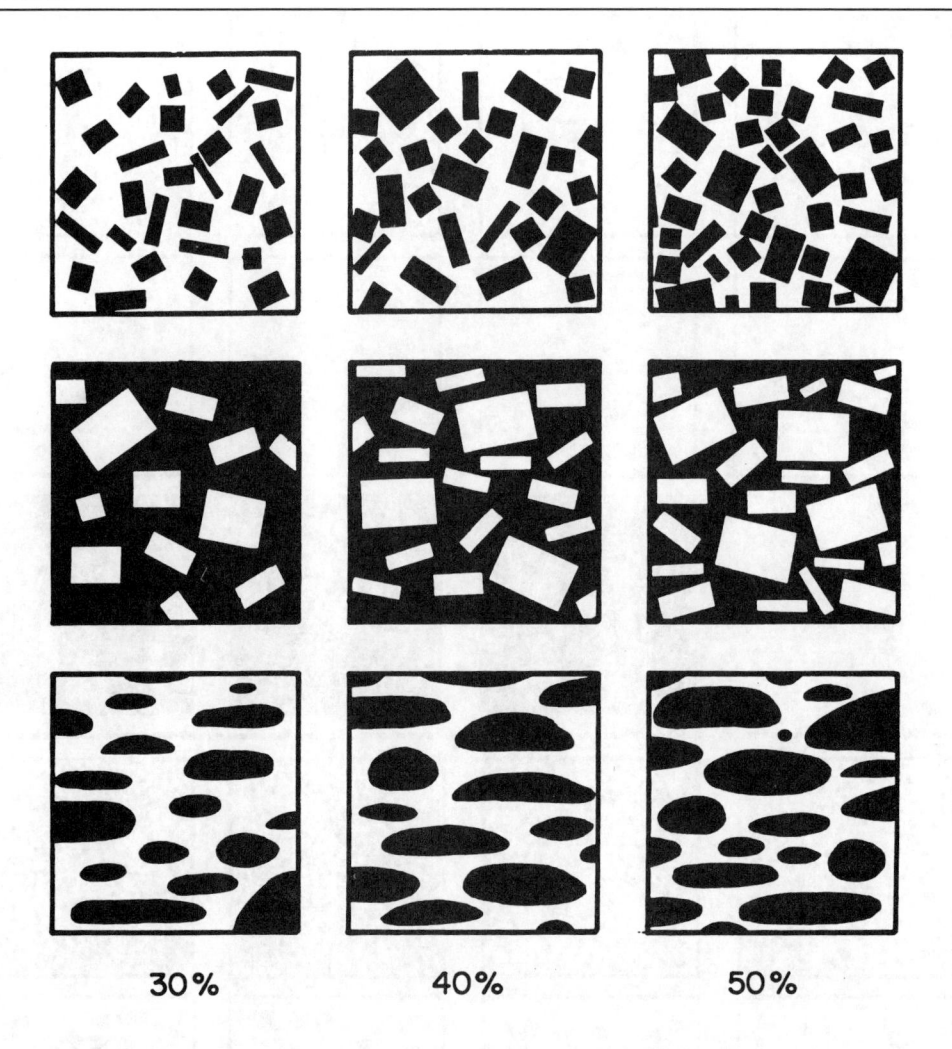

(b) Chart for estimating mineral percentages between 30% and 50%.

Figure B.1 *(cont'd.)*

readily based on figure B.4 (on p. 311), the rectangular IUGS classification, or on the alternative classification for fieldwork of Raymond (1984b; 1993; 2007), based on principles of Shand (1949) and others (figure 7.15, reproduced here as figure B.5 on p. 311). Classifications similar to that in figure 7.15 are recommended for the gabbroic rocks by the IUGS (figure 7.17). In these classifications, rocks containing a combination of plagioclase feldspar + pyroxene + olivine totaling between 90–95% of the rock and plagioclase-pyroxene rocks with less than 5% olivine are assigned the root name gabbronorite. This term is now common in the literature, but does not appear in the IUGS classification of figure B.4 or the RSO classification (figure B.5).

Recall that *plotting points* on triangular diagrams is a multi-step process, as discussed in chapter 7. It involves (1) measuring or calculating the percentages of each of the minerals in the unknown rock that are shown at the three corners of the appropriate classification triangle, (2) summing those mineral percentages, (3) normalizing each percentage to 100% by dividing each percentage by the total, and (4) plotting each normalized percentage on the classification triangle as a line (box 7.1). The point of intersection of the lines shows the appropriate name.

Minerals such as olivine, augite (clinopyroxene), hypersthene (orthopyroxene), hornblende, and other related minerals dominate some plutonic rocks. The

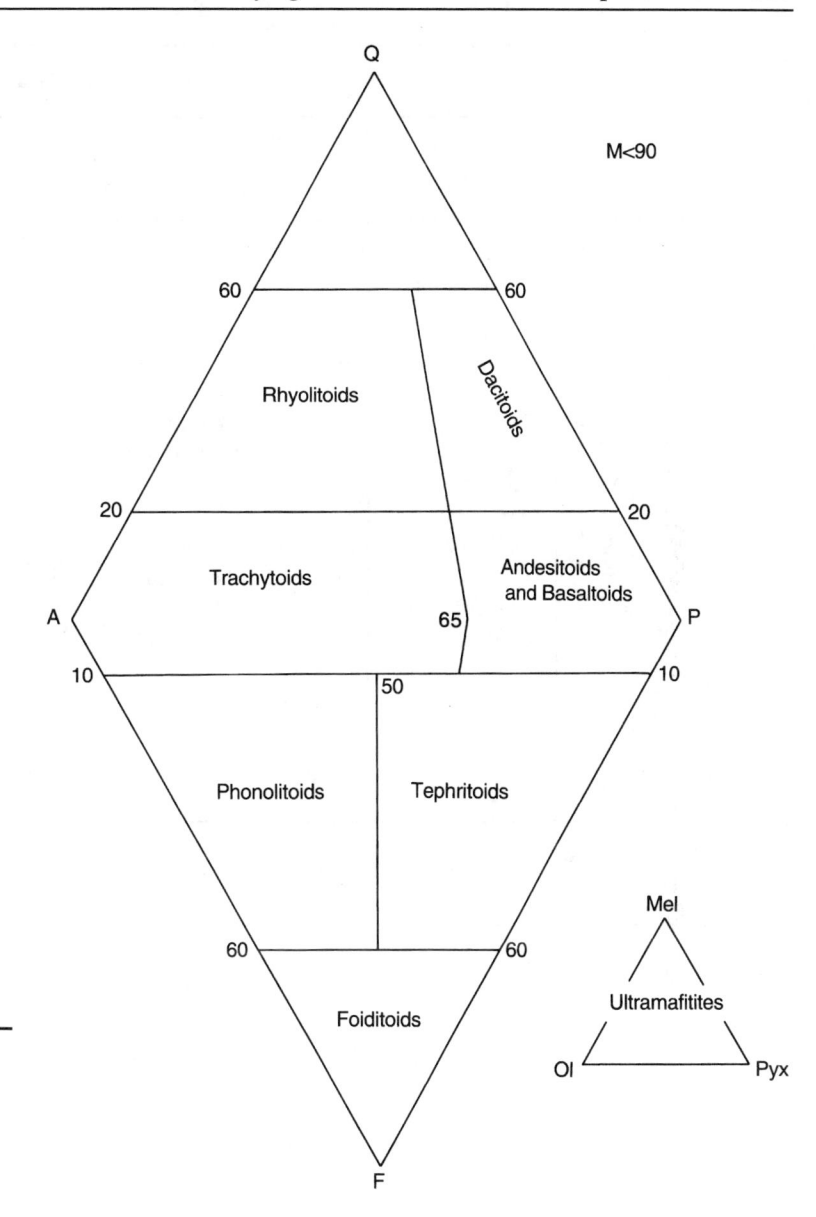

Identifying Common Rocks in Hand Specimens 309

Figure B.2
IUGS classification of volcanic rocks for field use.
(*Source:* After figure 3.17b, Raymond, 2007, as modified from Streckeisen, 1979.)

IUGS recommends use of the classification shown in figure 7.18 for such rocks. An alternative classification for field use is shown in figure B.6 (on p. 312).

The **metamorphic rocks** are commonly, but not in every case, foliated phaneritic rocks. Classification of these rocks is based on textures and minerals, just as are the igneous rocks. Some minerals characteristically or exclusively occur in metamorphic rocks, so recognition of these strongly implies a metamorphic origin. Few structured and comprehensive classifications of metamorphic rocks have been designed, although the IUGS recently adopted a classification scheme (Fettes and Desmonds, 2007). This classification is a flow chart rather than a structured classification. Here, table 10.4, reproduced in this Appendix as table B.2 (on p. 313), is suggested for hand-specimen identification and classification of metamorphic rocks. Rocks are divided into crystalline and clastic types, then subdivided on the basis of texture and mineralogy. Aphanitic rocks are described as individual types.

Sedimentary rocks, described in chapter 9, were divided into (1) siliciclastic rocks, (2) allochemical or biogenic rocks, and (3) precipitates. Recall that allochemical rocks are predominantly limestones and dolostones, but other types exist. Also recall that car-

			Other phenocrysts or grains					
			Alkali-feldspar ±biotite	Plagioclase ±alkali-feldspar ±biotite	Hornblende ±plagioclase	Pyroxene	Olivine	Hornblende ±plagioclase
Porphyritic texture	Essential phenocrysts or grains	Quartz feldspar	f – Rhyolite	f – Dacite			///////	///////
^	^	Feldspar only	f – Trachyte	f – Andesite (Light feldspar)		f – Basalt (Dark feldspar)	
^	^	Felspathoids ±feldspar	f – Phonolite	f – Feldspathoidal andesite (Light feldspar)	^	f – Feldspathoidal basalt (Dark feldspar)	^	f – Basanite
^	^	None	f – Biotite felsite		f – Andesite (Light feldspar)		f – Basalt
Aphanitic texture	No visible grains Glass<50%		Felsite (Light colored)			Mafite (Dark colored)		
^	Glass>50%		Obsidian*					

* Obsidian is actually a textural term and obsidians are dominantly rhyolites that plot at the left end of the obsidian space.

Figure B.3
Raymond's classification of volcanic rocks, for field and laboratory use.
(*Source:* After figure 23, Raymond, 2009.)

bonate rock types of both precipitated and clastic character are classified together. The classification used here—based on the suggestions of Dunham (1962), with modifications by Raymond (2007, p. 286)—is one in which there are five types of limestones, four of which are based on the amounts of lime mud versus grains (figure 9.17, reproduced here as figure B.7a.). See chapter 9 for explanations relevant to this classification. Other precipitated rocks are classified on the basis of the dominant mineral component (see table 9.7).

The siliciclastic rocks, the conglomerates and breccias, sandstones, and mudrocks each have their own classification. Primary division into these types is based on grain size (figure 9.18), with the two divisions between the three groups at 2 mm and 1/16 mm, respectively. For hand-specimen work, in this book we provide the coarse clastic classification of Raymond (2009), the sandstone classification of Gilbert-Dott (Dott, 1964), and the mudrock classification of Raymond (2007) (figure B.7 on pp. 314–315).

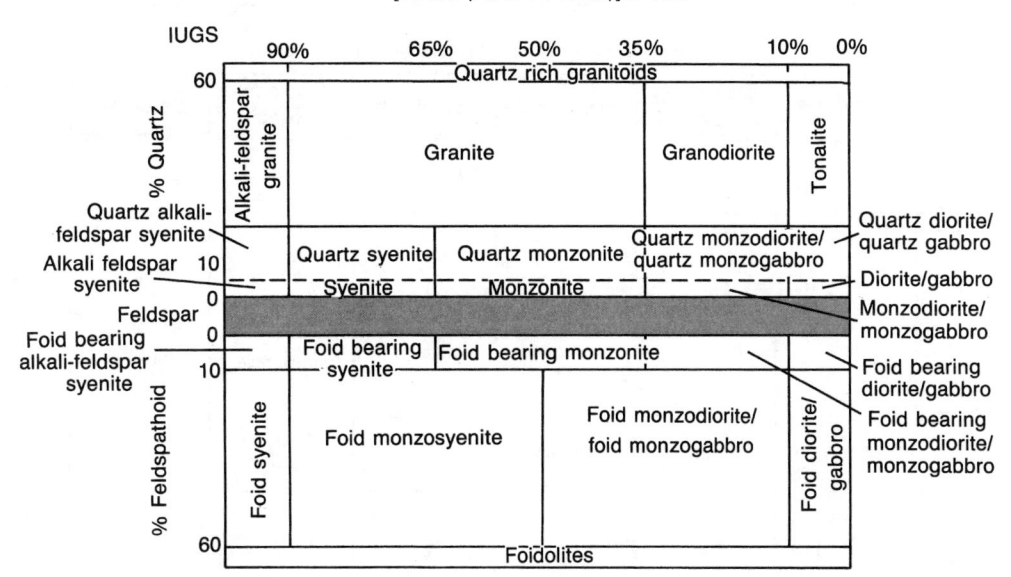

Figure B.4
Classification of feldspar- and quartz-rich plutonic rocks: the rectangular form of IUGS classification.
(*Sources:* After figure 3.6c, Raymond, 2007 as modified from Streckeisen et al., 1973, and Streckeisen, 1974.)

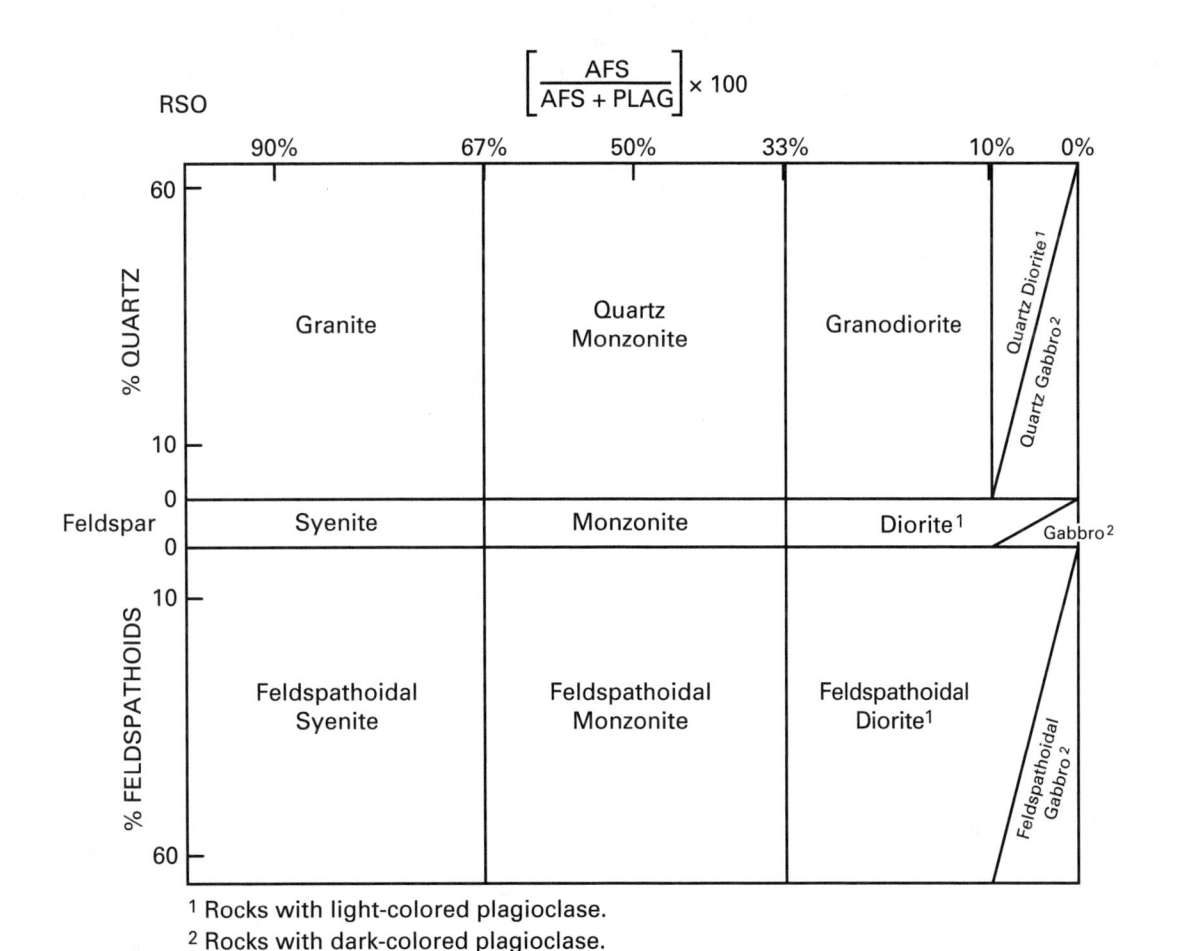

Figure B.5
The RSO classification of phaneritic plutonic rocks.
(*Source:* After figure 15a, Raymond, 2009.)

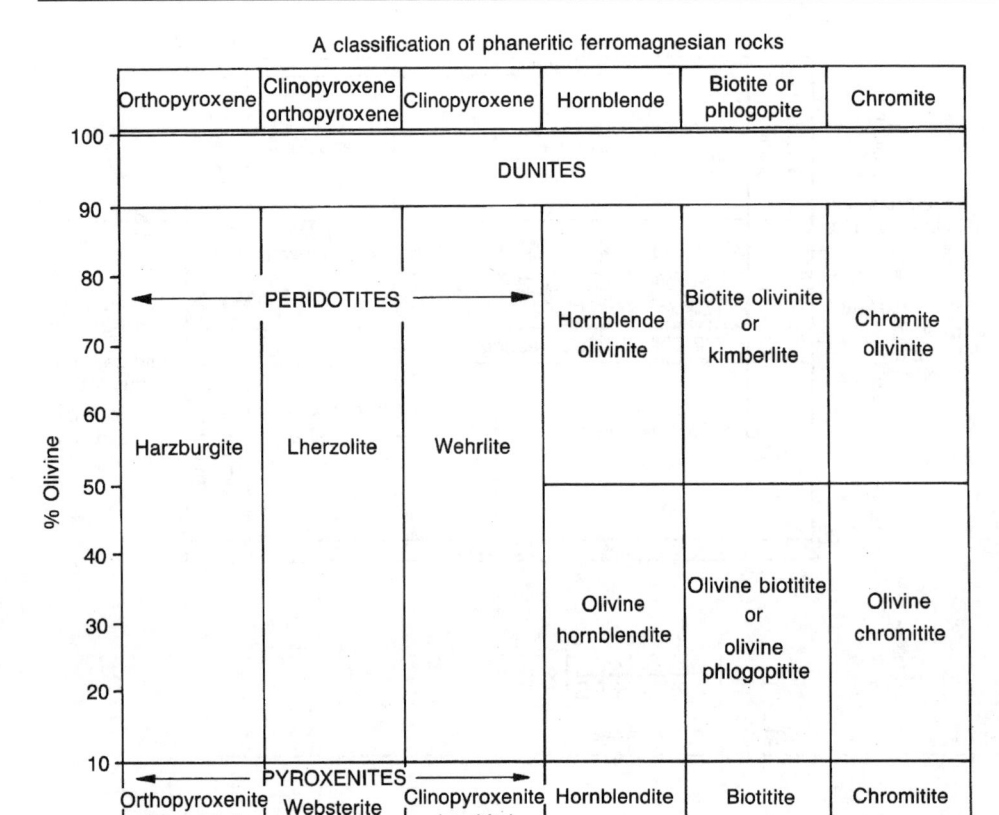

A classification of phaneritic ferromagnesian rocks

% Olivine	Orthopyroxene	Clinopyroxene orthopyroxene	Clinopyroxene	Hornblende	Biotite or phlogopite	Chromite
100–90	DUNITES					
80–50	PERIDOTITES — Harzburgite / Lherzolite / Wehrlite			Hornblende olivinite	Biotite olivinite or kimberlite	Chromite olivinite
40–10				Olivine hornblendite	Olivine biotitite or olivine phlogopitite	Olivine chromitite
10–0	PYROXENITES — Orthopyroxenite (bronzitite) / Websterite / Clinopyroxenite (augitite)			Hornblendite	Biotitite	Chromitite

©Loren A. Raymond

Figure B.6
A classification of phaneritic, ultramafic (olivine-pyroxene-rich and related) rocks. The classification is based on the percentage of olivine versus the percentages of the one or two other most abundant ferromagnesian minerals (listed across the top of the chart). Root names are modified by adding, before the root name, the name of a characterizing accessory mineral (e.g. chromite lherzolite).
(*Source:* Modified from figure 21, Raymond, 1984c.)

The *procedure* for rock identification is simple but involves multiple steps. Use of a hand lens (pocket magnifier) and other simple tools is essential for a good identification. The procedure involves the following steps.

1. Examine the texture and the structures of the unknown rock (and, where necessary, select a general classification based on the texture).
2. List *all* minerals and their percentages, placing each in the appropriate category (essential for the name versus accessory for the description).
3. Select the appropriate rock classification chart.
4. Use the minerals and mineral ratios to select the proper name from the classification selected.

For some complete descriptions, it is also useful to:

1. Name the color (use of a rock color chart may be helpful).
2. Estimate whether the specific gravity of the sample is light, average, or heavy.
3. Give the color index for igneous rocks by adding up the percentages of ferromagnesian minerals.
4. Describe, where possible, what minerals have weathered and altered to produce secondary minerals.

Make a complete name as follows. List in order:

1. the weathered color (where appropriate)
2. the rock color
3. the structures (where appropriate)
4. the textures not implied by the name (e.g., you would not say a rock is a glassy obsidian because by definition obsidian is glassy)
5. the characterizing accessory minerals in order of increasing abundance (the most abundant last)
6. the root name

Thus a typical rock name might be *orange-weathered, pinkish-white, porphyritic, hornblende-biotite granodiorite*. Note that textures like holocrystalline and phaneritic are implied by names such as granite or gabbro, whereas pegmatitic or fine-grained are not. The latter are included in the name, whereas the former are not.

Table B.2 Abbreviated Classification of Metamorphic Rocks

Texture and Composition	Root Name	Examples of Names
CRYSTALLINE ROCKS		
Strongly Foliated		
Non-banded		
Slaty	*Slate*	Black slate
Phyllitic	*Phyllite*	Quartz-chlorite phyllite
Schistose	*Schist*	Biotite-quartz-white mica schist
Mylonitic	*Mylonite*	Quartz-chlorite mylonite
Serpentine-rich	*Serpentinite*	Talc-bearing serpentinite
Banded		
Gneissose	*Gneiss*	Biotite-quartz-plagioclase gneiss
Mylonitic	*Mylonite*	Quartz-muscovite mylonite
Weakly Foliated		
Semi-schistose	*Semi-schist*	Quartz-white mica-jadeite metawacke
		Actinolite semi-schist
Granoblastic		
Aphanitic (near igneous contacts)	*Hornfels*	Tourmaline hornfels
Aphanitic (with relict texture)	*Meta*-plus protolith name	Metabasalt
Compositional types, generally phaneritic		
Serpentine-rich	*Serpentinite*	Magnetite-bearing serpentinite
Carbonate-rich	*Marble*	Diopside dolomite marble
Calc-silicate-rich	*Skarn*, *Granofels*, or *Granoblastite*	Garnet epidote skarn
		Epidote-quartz granoblastite
Garnet-omphacite rock	*Eclogite*	Kyanite eclogite
Quartz-feldspar-rich	*Granofels*	Quartz-feldspar granofels
Diablastic		
Aphanitic or *Phaneritic*		
Serpentine-rich	*Serpentinite*	Magnetite-bearing serpentinite
Any composition	*Diablastite*	Actinolite-chlorite diablastite
CATACLASTIC ROCKS		
Brecciated texture (any composition)	*Cataclastic breccia*	Quartz-feldspar cataclastic breccia
Aphanitic, clay-like	*Gouge*	Gray gouge
Aphanitic, typically black, with strained grains ± glass	*Pseudotachylite*	Black, quartz pseudotachylite

(*Source:* Modified from Raymond, 2009, pp. 116–117.)

314　Appendix B

DUNHAM CLASSIFICATION		EXTENSION OF DUNHAM CLASSIFICATION FOR DOLOMITIC ROCKS			
	Predominantly calcite (Cc >95%)	Dominantly calcite (95% > Cc > 50%)	Dominantly dolomite (Do >50%)	Thoroughly recrystallized rocks with some relict structures	
				Dominantly dolomite	Dominantly calcite
Lime mud, <10% grains	Lime mudstone	Dolomitic lime mudstone	Dolomudstone	Crystalline dolostone	Crystalline limestone
Mud supported, >10% grains	Wackestone	Dolomitic wackestone	Dolowackestone		
Grain supported, contains mud	Packstone	Dolomitic packstone	Dolopackstone		
Grain supported, lacks mud	Grainstone	Dolomitic grainstone	Dolograinstone		
Original components bound together	Boundstone	Dolomitic boundstone	Doloboundstone		

(a) Dunham-style classification of limestones and dolostones, modified by Raymond (2007).

CLASSIFICATION OF CONGLOMERATES, BRECCIAS, AND DIAMICTITES			
MATRIX / SUPPORT	CLAST SHAPE	CLAST COMPOSITION	NAME*
Gravelly or Sandy (Generally clast-supported)	Rounded	• Single composition 　Quartz ± chert 　Calcareous • Varied composition	Name of clast types and "conglomerate" 　Quartzitic conglomerate 　Calcirudite or limestone conglomerate 　Polymict (Petromict) conglomerate
	Angular	• Single composition 　Quartz ± chert 　Calcareous • Varied composition	Name of clast types and "breccia" 　Quartzitic breccia 　Limestone breccia 　Polymict breccia
Muddy (clay ± silt) and mud-supported	Rounded, angular, or both	• Single composition • Varied composition	Oligomict diamictite** Polymict diamictite**

* Prefix the word conglomerate, breccia or diamictite with the clast size designation (e.g., pebble conglomerate; boulder breccia).
** Where the rock is known to be of glacial origin, the term tillite is substituted for diamictite.

(b) Raymond's classification of coarse-grained sedimentary rocks (conglomerates, diamictites, and breccias). Oligomict means clasts are all of the same kind of rock or mineral, whereas polymict means that clasts are of several different types.

Figure B.7
Classifications of common sedimentary rocks.
(*Sources:* (a) Modified from figure 40, Raymond, 2009; (b) Modified from figure 33, Raymond, 2009; (c) Modified from Dott, 1964, and figure 35, Raymond, 2009; based on Dott, 1964; (d) from figure 37, Raymond, 2009.)

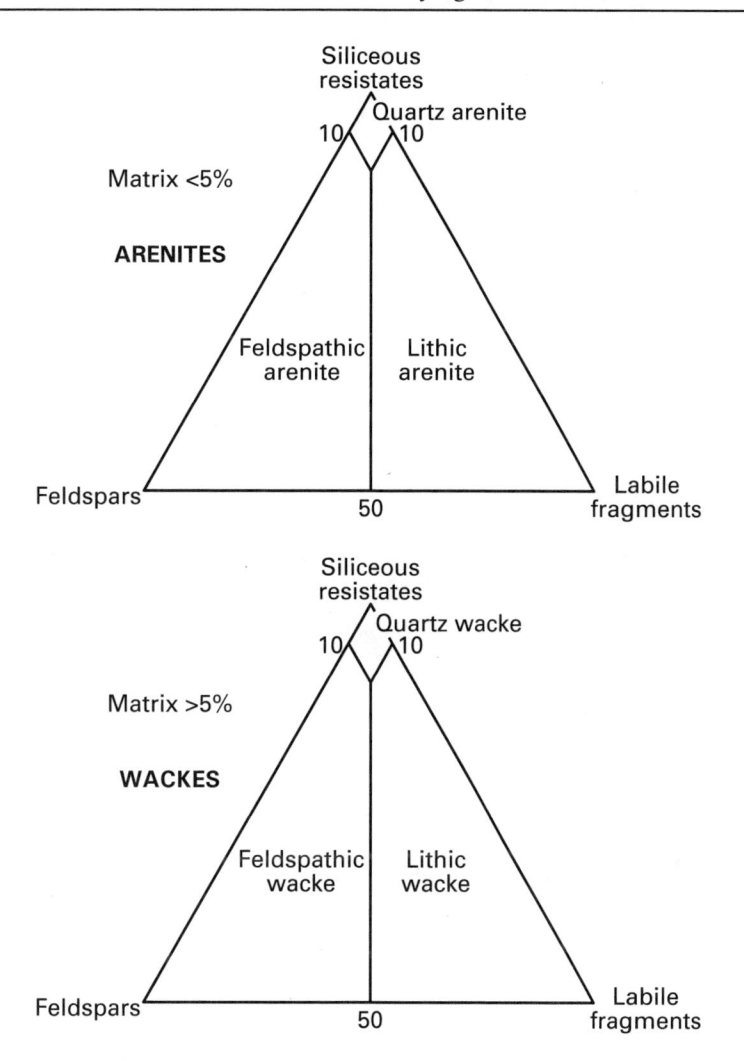

Identifying Common Rocks in Hand Specimens 315

(c) Gilbert-Dott-type classification of sandstones into wackes (with significant matrix) and arenites (without significant matrix).

	MUDROCKS			
	Rocks containing >50%mud			Rocks with <50% mud
	Silt dominant (>67% mud)	Clay and silt	Clay dominant (>67% mud)	Sand-sized or larger grains dominant
Nonlaminated	Siltstone	Mudstone	Claystone	Conglomerates, breccias, diamictites, and sandstones
Laminated	Laminated siltstone	Mudshale	Clayshale	

(d) Raymond's classification of mudrocks.

Figure B.7 *(cont'd.)*

Appendix C
Optical Mineralogy and Petrography in Thin Sections

The petrographic microscope was described in chapter 6 and its unusual features were noted. The microscope sketched in figure C.1 is different than the petrographic microscope shown in chapter 6. In doing routine petrographic work for identification of minerals and rocks and the evaluation of rock textures, thin sections are used as the subject material placed on the rotating stage of the microscope. Preparation of thin sections was described in chapter 6 and these sections ideally have a fixed thickness of 0.030 mm.

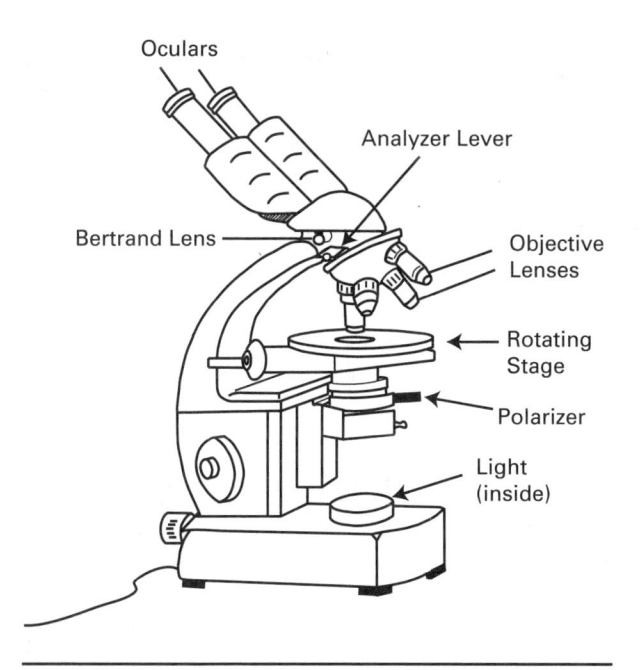

Figure C.1
Sketch of a petrographic microscope with parts identified.
(*Source:* Raymond, 2009).

Recall that polarized light is restricted to vibrating in only one direction. The polarizer below the stage constrains the light coming from the source (a light or a mirror) and passing through the hole in the center of the stage, to vibrate in one plane. Hence, the light passing upward through the polarizer is polarized, typically in an east-west direction, and all other light is blocked out. Where observations of thin sections of minerals and rocks are made in such polarized light, the light is called "plane light," designated as PL in figure C.2. The chart of figure C.2 illustrates the appearances and lists the colors of common minerals as they typically appear in sections of standard thickness.

Recall that if the Nichols are crossed—that is, if the analyzer is inserted into the optic path above the stage and nothing is on the stage of the microscope—the polarized light from the polarizer passes upwards to the analyzer, where it is all blocked. This occurs because the polarizer allows light vibrating in only a single plane (e.g., in an east-west direction) through to the stage, whereas the analyzer only allows passage of light vibrating at 90° to that plane (e.g., in the north-south direction) to pass, but there is no such light coming from below. The field of view will be dark. If the analyzer is "in," it is said that the Nichols are crossed (designated as XN on figure C.2). Obviously, if the analyzer is removed from the path of the light, then the plane polarized light (PL) leaving the polarizer again can be observed.

A thin section placed on the stage of a petrographic microscope can be viewed under plain light (PL) or crossed Nichols (XN). In general practice, both are done. Under PL, two groups of minerals can be distinguished—minerals that transmit light and those that do not. Minerals that do not transmit light and appear black in plane light are referred to collec-

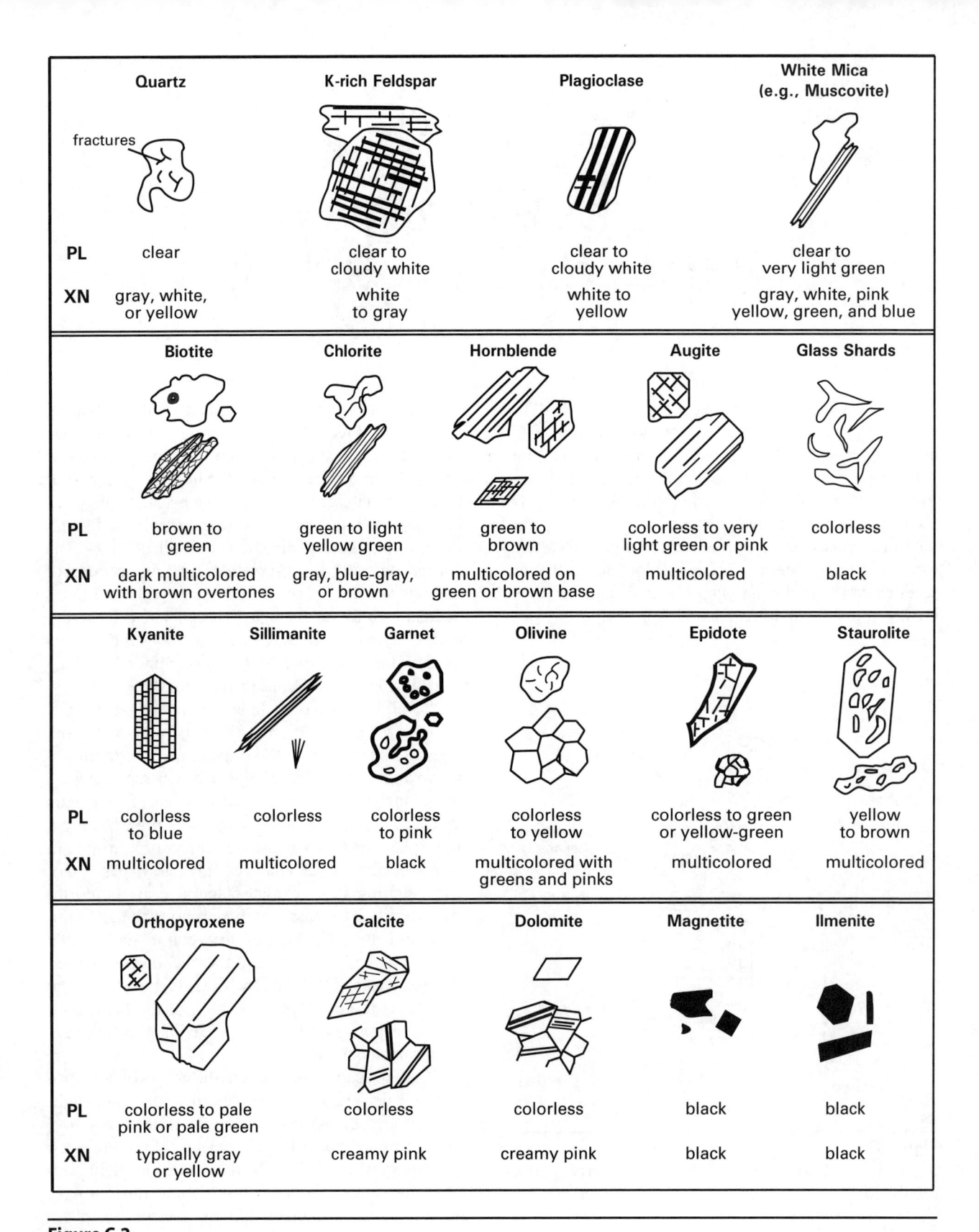

	Quartz	K-rich Feldspar	Plagioclase	White Mica (e.g., Muscovite)
PL	clear	clear to cloudy white	clear to cloudy white	clear to very light green
XN	gray, white, or yellow	white to gray	white to yellow	gray, white, pink yellow, green, and blue

	Biotite	Chlorite	Hornblende	Augite	Glass Shards
PL	brown to green	green to light yellow green	green to brown	colorless to very light green or pink	colorless
XN	dark multicolored with brown overtones	gray, blue-gray, or brown	multicolored on green or brown base	multicolored	black

	Kyanite	Sillimanite	Garnet	Olivine	Epidote	Staurolite
PL	colorless to blue	colorless	colorless to pink	colorless to yellow	colorless to green or yellow-green	yellow to brown
XN	multicolored	multicolored	black	multicolored with greens and pinks	multicolored	multicolored

	Orthopyroxene	Calcite	Dolomite	Magnetite	Ilmenite
PL	colorless to pale pink or pale green	colorless	colorless	black	black
XN	typically gray or yellow	creamy pink	creamy pink	black	black

Figure C.2
Chart showing the typical appearances of common minerals as viewed in thin section. Colors in plane light (PL) and crossed Nichols (XN) are listed under each mineral.

tively as *opaque minerals*. Using another kind of specially designed, reflected light microscope, these opaque minerals can be studied and various minerals identified, but in the routine study of rocks, this is not typically done. In PL, not only can minerals be separated into opaque and non-opaque types, the non-opaque minerals may be observed for variations in color as the stage of the microscope is rotated. In PL, many minerals appear in colors different from the color the mineral displays to the unaided eye. In addition, as the stage of the microscope is rotated, some of the colored minerals change colors. Such minerals are said to be *pleochroic* (many colored). Pleochroic colors are a distinctive property of the mineral and aid in mineral identification.

Inserting the analyzer into the optic path—that is, "crossing the Nichols"—allows one to distinguish among additional groups of minerals. The *isotropic minerals* transmit PL, but do not transmit light under XN (note: glass is also isotropic.). Recall from chapter 6 that the light passing through isotropic minerals travels in all directions with equal velocity and the vibration planes are not changed, so that the analyzer blocks out all light after it passes through the polarizer and the mineral sample. Garnet is an example of an isotropic mineral. Garnet typically appears pink or colorless in PL, but goes dark in XN.

Anisotropic minerals change the vibration direction of transmitted light. As the stage of the microscope is rotated through 360° with an anisotropic mineral in a section on the stage of the microscope, these minerals become extinct (go dark) at four positions 90° from one another. One exception is that if one is viewing directly down the optic axis of an anisotropic mineral, the mineral appears to be isotropic, remaining dark through a 360° rotation of the stage. Anisotropic minerals include *uniaxial* and *biaxial* types. In the uniaxial minerals, the extinction positions are parallel to crystal faces and cleavage directions. In the biaxial minerals, the extinction positions are commonly, but not always, at an angle to cleavage directions and crystal faces.

Using the Bertrand-Amici lens and a converging lens below the stage, one can make somewhat more sophisticated observations. Discrimination of the two groups of non-opaque, non-isotropic minerals—the uniaxial minerals and the biaxial minerals—is possible using this lens. With XN, the Bertand-Amici lens, and the converging lens "in" (inserted in the optic path), uniaxial minerals may show a black cross in the field of view. A cautionary note is that since the mineral may be oriented with its one optic axis at an angle to the optic path, some grains may show some fuzzy curved black bands called isogyres that move within the field of view, rather than the cross, as the stage is rotated. Under the same optical arrangement, biaxial minerals typically show one or two curved isogyres moving within the field of view. Biaxial minerals rarely show a cross and, if they do, the cross will split apart as the stage is rotated.

In summary, the four optical groups of minerals are as follows.

1. Opaque minerals are minerals that do not transmit plane light and remain black in plane light and under XN.
2. Isotropic minerals are minerals that transmit plane polarized light, but do not transmit doubly polarized light, i.e., these minerals remain dark under XN as the stage is rotated.
3. Uniaxial minerals are minerals that transmit PL, generally turn dark at four positions as the stage of the microscope is rotated 360°, and may reveal a black cross in XN, under converging light, with the Bertrand-Amici lens inserted.
4. Biaxial minerals are minerals that transmit PL and typically show one or two curved black bands in XN, under converging light, with the Bertrand-Amici lens inserted.

Each type of mineral corresponds to one or more of the various crystal systems.

The optic sign of uniaxial and biaxial minerals (either + or −) may be determined by inserting the wavelength (λ) plate in the slot on the tube above the stage with the Bertand-Amici lens inserted. Your instructor may explain the theory behind the observations, but typically, one seeks a grain that remains dark upon rotation of the stage but is thought to be uniaxial. (1) Under XN, (2) the high power objective lens is inserted and focused, (3) the converging lens below the stage is also inserted, (4) the Bertrand-Amici lens is inserted, and (5) the full wavelength plate is inserted in the slot in the tube above the stage. If a cross appears before the λ-plate is inserted, then when the plate is inserted the northwest and southeast quadrants change to yellow on positive uniaxial minerals and blue in negative minerals (near the cross center). In biaxial minerals, a curved isogyre is sought. When a curved or single straight isogyre is found that remains centered in the field of view, but rotates as the stage is rotated, the five steps listed above are followed and the isogyre is oriented in a position that trends southwest to northeast across the field of view. If the isogyre is straight, the 2V (figure 6.3b) is large (near or at 90°). If the isogyre has a curve, the 2V is in the range 0° to 80+°. When the λ-plate is inserted into the slot on the tube above the stage, with the isogyre oriented in a position that trends southwest to northeast, positive minerals typically will show yellow on the concave side of the curve, whereas negative minerals will show blue. Some microscopes may have alternative optical designs that your instructor can explain, if that is the case. Your instructor may also explain how to estimate the 2V of minerals using the curve of the isogyre.

The optical petrographer, who studies thin sections, uses this group of properties observed in PL and XN—color in PL, pleochroism, color in XN, optical mineral type, optical sign and other interference properties; plus shape, cleavage, and other characteristics—to identify and characterize the mineral(s). Various books have tables, charts, and detailed descriptions of minerals and their properties that are useful for identification of minerals using the petrographic microscope (e.g., Phillips and Griffin, 1981; Deer et al., 1992; Dyar and Gunter, 2008; Raymond, 2009). Figure C.2 shows the appearance of a few common minerals as they typically appear in thin sections. Textures may be compared to those illustrated in books such as Bard (1986), Scholle (1978), or Scholle (1979).

Glossary

A

Aa lava A type of lava composed of moderate to moderately large blocks of low-silica lava, welded together by congealed volcanic rock of similar composition.

Abrasion The grinding away of rock and mineral surfaces, primarily by wind, running water, and ice drag, and to a minor extent, as a result of animal activities.

AFC (Assimilation and Fractional Crystallization) A magma modification process involving combined processes of assimilation and fractional crystallization (see definitions for each, below).

Aggregational state The tendency for a group of mineral crystals to develop into a particular shape.

Alkali olivine basalt A type of low-silica volcanic rock, typically with 45% to 49% silica and containing olivine and calcium pyroxene (e.g., augite).

Allochem A carbonate sedimentary clast (a fragment of a preexisting rock).

Allochemical rock A sedimentary rock consisting of allochems, i.e., clasts of preexisting rocks and fossils. Some workers call these rocks biogenic rocks.

Amorphous material A material with a true lack of internal atomic order.

Amygdule A small (typically < 1 cm), somewhat spherical to oblong mass of mineral material filling a hole in lava that was created by escaping gas.

Analyzer A removable polarizing filter mounted above the stage of a petrographic microscope.

Andesite A common type of volcanic rock of intermediate composition that is a typical constituent of composite volcanoes.

Anhedral A mineral specimen that lacks crystal faces.

Anion A negatively charged ion, with a charge due to possession of one or more extra electrons.

Anisotropic A designation for a material that has more than one index of refraction, with different indices depending on the direction light passes through the material.

Aphanitic A rock texture consisting of glass or grains too small to see with the naked eye or a low-power hand magnifier.

Aphanitic-porphyritic texture An texture typical of igneous rock in which there are two main sizes of grains (a bimodal population of grains)—a set of fewer larger grains and a larger set of typically aphanitic surrounding grains, referred to as the groundmass.

Arc In tectonics, a linear array of volcanoes located on the edge of a continent or in an ocean basin.

Arenite A sandstone lacking significant matrix.

Aridosol A soil of arid regions typified by a high calcium content.

Ash Fine, fragmented volcanic (pyroclastic) rock material with grains smaller than 2.0 mm in diameter.

Assimilation A magma modification process in which magmas dissolve or absorb surrounding rocks, changing the initial magma composition to a new composition.

Asthenosphere A plastic layer of the mantle lying immediately below the rigid lithosphere.

Atoms A basic component of matter consisting of a nucleus of protons plus, in nearly all cases, neutrons, and a surrounding array of electrons.

Authigenesis A term used in studies of diagenesis of sedimentary rocks to designate the process of new mineral formation. This process is also called neocrystallization.

Axial ratio The ratio of the lengths of the unit cell axes a, b, and c in a crystal.

B

Banded iron formation *See* BIF below.

Basalt A common type of volcanic rock composed of essential calcium-rich plagioclase feldspar and pyroxenes.

Basin The locale in which sediment accumulates. Not all "basins" of sedimentation have the basinal form implied by the name.

Basis vector The vector (the line of a particular size and direction) used as a repeating unit of length in symmetry operations.

Batholith A large body of plutonic igneous rock, traditionally defined as one being exposed over an area of 100 square kilometers or more.

Bed A sedimentary layer 1 cm thick or thicker that is distinguished from adjoining layers by differences in color, composition, or texture.

Bed-load transport The sediment transport process that includes the bouncing, rolling, and sliding of grains or fragments along a surface.

Biaxial A material with two axes perpendicular to which all refractive indices are equal. These are anisotropic materials with orthorhombic, monoclinic, or triclinic symmetry.

BIF (Banded Iron Formation) A kind of Precambrian-aged, iron-rich sedimentary rock, characterized by a series of alternating bands of specular hematite and red hematite-rich chert, some of which are associated with volcanic rocks, some of which are associated with glacial sedimentary rocks, and the bulk of which are associated with passive margin sedimentary rocks.

Biological activity A process of disintegration involving forces exerted by parts of organisms (e.g., tree roots, animal hoofs) and yielding break-up of rocks.

Bioturbation structure Mottled or otherwise disturbed layering produced by burrowing organisms in sedimentary rocks.

Body-centered unit cell A unit cell with a lattice point entirely enclosed in the unit cell.

Boudins Metamorphic structures that consist of a series of sausage-like lenses of material.

Bouma sequence A five-layered bed sequence in sediments or sedimentary rocks that fines upwards, has distinctive structures in each layer, and represents turbidity current deposition and the intervening deposition of mud.

Breccia texture A texture in rocks consisting of larger, very angular fragments surrounded by and included within a finer-grained matrix of materials. Breccia texture occurs in igneous, sedimentary, and metamorphic rocks and the matrix material differs in character among the different types of rock.

Brittle A property of most minerals. The property of breaking easily without obvious plastic deformation.

Burial metamorphism *See* Static metamorphism.

Burrows Irregular to cylindrical filled tubes in sedimentary rocks, produced by burrowing organisms.

C

Carbonate hosted lead-zinc deposit A type of ore deposit formed by connate fluids in carbonate rocks that were originally deposited on marine margins of stable cratons or along failed rift zones.

Carbonatites Igneous rocks that are composed essentially of carbonate minerals, such as calcite and ankerite.

Cataclasis The process of metamorphism involving the crushing and breaking of grains in rocks.

Cataclastic texture A metamorphic texture characterized by broken grains or rock fragments that are simply stuck together. These textures include breccia textures, similar in appearance to sedimentary breccias.

Cation An ion with a positive charge (due to loss of one or more electrons).

Cement Crystalline material precipitated in the voids between the grains, *after* deposition of a sediment.

Cementation The process by which minerals precipitate within the pores of the rock and bind the grains together.

Centered unit cell A unit cell containing more than one lattice point.

Chelation A chemical process in which complex molecules called chelates, with charged surfaces, form in aqueous fluids and during diagenesis, extract metals from minerals with which they come in contact.

Chemically active fluids Fluids that are activated by moving from regions in which they have thermodynamic stability (including chemical stability) into areas where they lack thermodynamic stability.

Chert A sedimentary rock composed of microcrystalline silica minerals (opal, chalcedony, quartz), formed by abiotic chemical or biologically mediated processes of precipitation.

Cinder cone A small volcanic structure, generally steep sided and circular in plan, composed of loose fragments of silica-poor volcanic rock of small (< 1 cm) to moderate (< 16 cm) size.

Cleavage (1) In minerals, the tendency of a mineral to break along repeating parallel planes due to a systematic weakness in the structure. (2) In rocks, the tendency of a rock to break in a particular or "preferred" direction.

Cleavage face A flat surface on a crystal produced by breaking of the crystal.

Closed form A group of crystal faces that are related by a symmetry operation and enclose space.

Coal A rock in the broad sense, composed mostly of altered, compressed fragments of plants.

Colloid formation and base exchange A combined set of reactions representing a type of decomposition process in which finely divided compounds (colloids) form through the breakdown of preexisting minerals, and these colloids have charged surfaces that facilitate exchange of ions between the colloid surfaces and minerals with which the rock or soil fluids come in contact.

Columnar joint A pencil-like structure formed in a lava flow during cooling and oriented generally at a high angle to the flow layer.

Compaction A process in which pressure on sediment reduces the pore space volume between the grains of the rock.

Compatible element An element that substitutes freely for another element in a mineral structure.

Composite cone *See* Composite volcano.

Composite volcano A generally steep-sided volcanic structure that is crudely circular in plan and composed of both lava flows and pyroclastic layers. The rocks of the composite volcanoes are typically of variable composition and range from low- to high-silica types.

Concentration factor A number multiplier by which an elemental abundance must be multiplied to increase its normal crustal abundance to one of ore grade, so that the element can be considered to be commercially extractable.

Concretion A typically rounded, spherical, or irregular to disk-shaped mass of cemented sedimentary rock of small to large size (1 cm to 2 m+) that occurs within a host sedimentary rock.

Conglomerate A siliciclastic sedimentary rock composed of small to large rounded clasts and usually a sandy matrix, with clasts comprising 25% or more of the rock.

Conrad Discontinuity A boundary within some parts of the Earth's crust separating more dense materials below from lighter materials above.

Contact metamorphism Metamorphism under conditions where increased temperature dominates as the driver of metamorphic change.

Convolute bedding Sedimentary layers greater than 1 cm thick that have the form of little basins, domes, and folds.

Convolute laminations Sedimentary layers less than 1 cm thick that have the form of little basins, domes, and folds.

Core The innermost part of the Earth consisting of an outer liquid zone and an inner solid zone, both composed primarily of iron and nickel.

Corona texture A texture in crystalline rocks in which a core mineral has a rim of one or more other minerals, usually a rim of multiple grains.

Country rock The rock surrounding an intrusion of igneous rock. Typically this rock is metamorphosed by the heat and fluids from the intruding magma.

Covalent bond Chemical bonds in which atoms share electrons between them.

Critical point A point on a phase diagram beyond which (at higher pressure and temperature than which) liquids and gases are indistinguishable.

Crossbeds Sets of inclined sedimentary layers, each of which is greater than 1 cm thick, that are locally truncated by other sets of layers inclined in a different direction or at a lower or higher angle.

Cross-laminations Sets of inclined sedimentary layers, with each layer less than 1 cm thick, that are locally truncated by other sets of layers inclined in a different direction or at a lower or higher angle.

Crust The relatively low density, rigid, and brittle outermost layer of Earth.

Cryptocrystalline Material composed of crystals smaller in size than those normally identifiable with optical microscopes.

Crystal (1) A piece of a chemical compound or element that internally has an ordered, three-dimensional arrangement of atoms. (2) A solid mass of mineral material bounded partially or entirely by flat surfaces called "crystal faces."

Crystal face A flat, smooth surface, partially bounding a solid crystalline substance, which has formed during growth of that solid substance.

Crystal form A group of crystal faces that are related by a symmetry operation.

Crystalline texture A texture consisting of masses of interlocking crystals precipitated from solutions or fluids or modified to crystalline form by intergrain reactions.

Crystal system A group of unit cell types.

Cyclic In crystallography, a state of complex twinning, in which the twin planes are not parallel.

Cyclosilicates Silicate minerals with SiO_4 tetrahedra that share two corners with other SiO_4 tetrahedra, forming a closed ring.

D

Darcy's law The amount of water (i.e., the discharge, designated Q) through a layer is proportional to the difference in height between the two ends of the layer multiplied by the cross-sectional area and is inversely proportional to the length of the layer through which flow occurs. Mathematically, $Q = KA(\Delta h)/L$, where K is a constant called the hydraulic conductivity, Q is discharge, Δh is the difference in height between the two ends of the layer, A is the cross-sectional area of the layer, and L is the length of layer.

Decomposition Chemical processes of transformation of rock into soil.

Decompression melting Melting that occurs because the pressure is substantially reduced on already hot rocks.

Density The mass divided by the unit volume of a material.

Derivative magma Magma that has been changed in composition through one or more of several magma-modification processes.

Deviatoric stress A condition in which a principal stress in a system is not equal in size to other principle stresses; that is, in which one or more stresses deviates from the mean (average) stress.

Diabasic texture An igneous texture characterized by randomly oriented, rectangular (lath-shaped) plagioclase crystals with intergranular grains of pyroxene, opaque minerals, or other minerals.

Diablastic texture A metamorphic texture characterized by needle-like, flaky minerals, or both, that are not aligned but rather are arranged in a random fashion.

Diagenesis A collective term for all of the post-depositional processes—physical, chemical, and biological—that result in the transformation of sediment into rock. Diagenetic processes may also affect igneous and metamorphic rocks.

Diamagnetic A material with paired electrons that is repelled by magnetic fields.

Diapir An upside down, teardrop-shaped mass of material that rises through overlying rocks due to its lower density than that of surrounding rocks.

Diatreme A type of volcanic structure consisting of a vertical pipe–like feature developed where CO_2 gas-rich magma forced its way up through overlying strata.

Diffuse The action of elements as they migrate from atomic site to atomic site within solid crystals.

Dike Tabular plutons that cut across surrounding rock structures.

Disintegration Physical processes of transformation of rock into soil.

Distance The difference in position, measured in length, that a body occupies at two different times.

Dissolution A process of decomposition in which those parts of a mineral at the grain-fluid interface are drawn away into solution.

Dome In volcanology, a mound of lava, which is typically silica-rich.

Drainage basin The total land area drained by a stream and its tributaries, measured at a particular discharge point.

Dynamic metamorphism Metamorphism predominantly driven by deviatoric stresses.

Dynamothermal metamorphism Metamorphism in which pressure, temperature, deviatoric stress, and fluids are all important, and which affects entire regions (sometimes called regional metamorphism).

E

Earth materials Earth materials are all of the solid, liquid, and gaseous substances that constitute the Earth.

Economic resource An Earth material that can be extracted and sold for a profit.

Eh The redox potential, which is depicted as a number representing the relative ability of solutions to produce oxidation or reduction reactions.

Elastic A descriptive term for materials, such as certain mineral fragments, capable of bending without breaking and returning to their original shape after the bending force is removed.

Electron A negatively charged, subatomic particle of low mass that is one constituent of atoms.

Element One of the more than 100 basic chemical constituents of matter that cannot be separated into simpler substances with the same properties. Each element has a definitive number of protons in the nucleus of its atoms.

End-centered unit cell Unit cells with lattice points in two opposing faces.

Enthalpy The sum of the energy from all possible sources within a chemical compound, such as a mineral.

Entropy That part of the enthalpy (energy) in a chemical compound that is not available to participate in chemical reactions.

Epiclastic texture A sedimentary texture consisting of rounded to angular grains that are stuck together by cement, matrix, or grain intergrowths.

Epigenetic deposits Ore deposits that form after the host rock formed.

Epitaxial intergrowth A descriptive term for crystal intergrowths of different mineral species, joined along a similar plane of atoms.

Erosion The combined processes of fragmentation and removal of solid Earth materials by wind, water, ice, or gravity from surface exposures.

Euhedral A descriptor for a mineral specimen that is bounded entirely by crystal faces.

Eutectic The lowest possible melting point of certain mixtures of minerals or chemicals.

Evaporite A sedimentary rock formed by precipitation of minerals as a result of evaporation or chemical supersaturation. Minerals typical of evaporites include halite, sylvite, calcite, dolomite, gypsum, and various boron minerals, such as borax and ulexite.

Exfoliation joints A set of fractures along which a rock breaks due to expansion.

Exsolution The breakdown of a homogeneous solid-solution mineral into two separate solid mineral species.

Extinct A condition observed through a petrographic microscope objective in which a mineral appears dark and does not seem to transmit any light to the observer.

Extinction A condition that occurs in a petrographic microscope, when the analyzer absorbs all of the light that has passed through the polarizer and the specimen on the stage.

F

Face-centered unit cell A unit cell with lattice points within all faces.

Facies series A group of metamorphic facies that occur sequentially in a region of Earth and reflect or suggest progressively increasing conditions of metamorphism.

Fault In rocks, the fracture or zone along which movement has occurred parallel to the surface or zone of breaking.

Flame structures Deformed mudrock layers that extend up from one bed into another in a curved way, ending with a point—a form reminiscent of the shape of a flame in the wind.

Flexible A term used to describe mineral fragments that can be bent, but do not return to their original shape when the bending force is removed. (*See* Elastic for contrast).

Flow (1)(noun) A crudely tabular to elliptical body of lava. (2)(verb) To move as a fluid.

Fluids Earth materials that can flow at relatively rapid rates, including liquids and gases, are collectively called fluids.

Fluorescence A kind of luminescence (light emission) that ends when incoming higher energy irradiation ends.

Flute A sole mark in the form of an oblong to irregular protruding cast that filled a furrow carved into underlying sediment by a current.

Fluvial environment The dominant continental environment of sedimentation, in and along streams and rivers.

Flux melting Melting that occurs when fluids mix with hot rocks and lower the melting point of the rocks.

Fold A layer that has been bent into a curviplanar shape.

Foliated textures Rock textures characterized by leaf-like to banded arrangements of minerals that create a tendency in the rock to break into somewhat flat pieces.

Force Mass times acceleration ($m \times a$).

Formation In the study of rock bodies, a mass of rocks that is thick enough to map at a scale of 1:24000, is characterized by a distinctive rock type or group of rock types, and occupies a particular stratigraphic position.

Fractional crystallization A magma-modification process in which part of a magma (a fraction) is crystallized and the remaining melt is separated from the crystallized rock.

Fracture (1) In minerals, a term used to describe breaking in random directions. (2) In structural geology and hydrogeology, a crack in a rock body (also called a joint in some cases). Fractures are openings in rock masses parallel to which no significant movement has occurred, but across which there is a separation and loss of cohesion (i.e., fractures open, but opposing rock masses do not slide past one another).

Frost heaving A process of disintegration involving formation of a lens of ice within soils.

Frost wedging A process of disintegration involving the formation of a wedge of ice within a crack of rock, which results in new fracturing or widening of old cracks.

G

Gabbro A common type of silica-poor plutonic rock composed of pyroxenes and CA-rich plagioclase feldspar.

Gibbs free energy That part of the total energy (the enthalpy) in a chemical compound that is available for chemical reaction (the entropy), multiplied by the temperature, i.e., $H - TS = G$, where H is the enthalpy, S is the entropy, T is the temperature, and G is the Gibbs free energy.

Glassy texture A texture that most commonly occurs in volcanic rocks and is characterized by materials not structured or crystallized into minerals. Glassy textures with phenocrysts are called vitrophyric.

Gneiss Common type of phaneritic metamorphic rock composed of minerals arranged in dark and light bands.

Gneissose texture Phaneritic metamorphic texture characterized by distinct dark and light bands of minerals.

Goniometry The process of measuring the angles between faces of a crystal.

Gouge A clay-like cataclastic metamorphic rock composed of broken fragments smaller than 1/16th mm in diameter or length.

Graded bedding Layering of 1 cm thick or more in thickness in which there is a change in grain size from coarse to fine within the individual layer. In normal grading the grains get smaller upward. In reverse grading, the grains get larger upward.

Granite A common kind of plutonic rock composed of essential quartz and alkali feldspar, commonly with lesser amounts of plagioclase feldspar. The exact proportion of the minerals used to define the rock type depends on the classification used.

Granoblastic texture A metamorphic texture characterized by relatively equant grains.

Graywacke An anachronistic sandstone name—not a part of most commonly used classifications—applied to sandstones with large amounts of matrix (>15%), some of which was likely produced by diagenetic processes after sedimentation.

Groundwater Water that exists in soils and in pores and fractures in the rocks below the surface.

Growth In crystallography, the process of addition of atoms to crystal surfaces resulting in an increase in the size of the crystal.

H

Habit The particular shape that mineral crystals tend to develop into, when they crystallize.

Histosol A soil type common to swamps and alpine tundra, characterized by having a thick A-Horizon with a high carbon concentration.

Hornfels A metamorphic rock that is generally fine-grained to aphanitic, lacks foliation, and occurs and forms at or near the contact between an igneous intrusion and the surrounding country rock.

Hornito A small volcanic structure that has an inverted cone shape, is located atop a lava flow, and is composed of lava flow rock.

H,P facies series A series of metamorphic facies on a petrogenetic grid representing a geothermal gradient of low temperature, but low to high pressure. The Franciscan (H,P) Facies Series consists of the zeolite, prehnite-pumpellyite, blueschist, and eclogite facies. The Sanbagawa (H,P) Facies Series contains the zeolite, prehnite-pumpellyite, blueschist, greenschist, amphibolite, and locally, the eclogite facies.

Hydration In diagenetic studies, a process of decomposition in which water combines with a preexisting compound in a mineral to yield a new hydrated mineral.

Hydrogen bond A chemical bond formed when a molecule containing a hydrogen ion creates a polar molecule (a molecule with a positive charge on one end and a negative charge on the other) and the positive and negative ends of individual molecules are attracted and bond together as a result of their opposing charges.

Hydrolysis A process of decomposition in which an excess of H^+ or $(OH)^-$ is produced in solutions in rock by a reaction that yields these ions.

Hydrothermal fluids Hot fluids, heated by magmas or hot rocks.

Hypidiomorphic-granular texture An igneous texture in which, on average, grains are equant (granular) and subhedral.

Hypogene Ore deposits that form below the surface as a result of the activity of fluids ascending from deeper levels of the crust.

Hypothetical resource A resource that possibly exists, shares similarities with known deposits, and is thought to exist in the same districts or regions containing known deposits.

I

Identified resource A resource for which the location and grade, as well as the quality and quantity of the valuable material, are known or are carefully estimated based on well-specified geologic evidence.

Incompatible element An element that does not substitute freely in a mineral structure for other elements.

Index of refraction The ratio of the speed of light in a vacuum to the speed of light in a given medium; it is a measure of how effective the medium is in slowing light down.

Industrial minerals Minerals mined for their commercial value, excluding the fuel minerals and metallic minerals.

Industrial rocks Rocks mined for their commercial value, excluding (1) rocks that contain fuel minerals or those, such as coal, that are themselves fuel sources, and (2) rocks rich in and mined for their included metallic minerals.

Inferred resource A resource known at a level of low confidence due to the fact that continuity between or beyond measured or inferred ore-bearing rock may not be fully supported by existing measurements.

Inner core The inner part of the innermost major component of the Earth, composed of solid Nickel-iron materials, with minor additional amounts of a few other elements.

Intergranular texture A volcanic texture in which rectangular grains of plagioclase feldspar are arranged somewhat randomly and the spaces between the plagioclase grains are filled with pyroxene and other mineral grains.

Inosilicates Silicate minerals with SiO_4 tetrahedra that share two to three corners with other SiO_4 tetrahedra, forming chains that may be cross-linked.

Inversion point A point about which symmetry operations take place in three-dimensional space.

Ion An atom that has a charge. Negatively charged ions are called anions and positively charged ions are called cations.

Ionic bond A chemical bond formed by the attraction between positively and negatively charged ions (each formed by loss and gain of electrons, respectively).

Isogonal Space groups that have corresponding (the same) symmetry elements.

Isograd Inograds are lines drawn on metamorphic maps or petrogenetic grids that mark a change in mineral assemblage. Initially, such changes were thought to be lines of equal (iso) grade (grad), but in some cases are now known to reflect changes in rock composition.

Isostructural A term used to designate two crystals with the same arrangement of atoms, but containing different kinds of atoms.

Isotope A variant of an element with a different number of neutrons than other atoms of the same element.

Isotropic A material that has one index of refraction.

J, K

Joints Fractures or breaks in rocks, along which the only significant movement is perpendicular to the fracture surface (i.e., the joint has "opened").

Kilogram The weight of a standard platinum-iridium cylinder.

Kimberlite A type of plutonic igneous rock formed through intrusion of unusual ultramafic magma, typically at depth, in diatremes. Kimberlites are a source of diamonds.

L

Laccolith A pluton that has intruded generally parallel to the enclosing layers of rock, i.e., a sill, but a sill with a convex-up top.

Laminar flow Flow characterized by parallel to subparallel flow lines.

Lamination A sedimentary layer less than 1 cm thick that is distinguished from adjoining layers by differences in color, composition, or texture.

Lapilli tuff A volcanic pyroclastic rock composed of ash matrix enclosing larger fragments of between 2 mm and 64 mm in diameter.

Latent heat of crystallization Heat given up by a liquid as it is transformed into a solid by crystallization.

Laterite A clay-, iron oxide-, and aluminum oxide-rich material formed as a B-horizon soil by weathering of rocks in moist, warm climates.

Lattice An ordered arrangement of points in space.

Lava Magma erupted at the surface, typically one that has lost considerable amounts of gas. The term is used for both fluid and rock forms of erupted material.

Lava plain A large, tabular mass of volcanic rock that typically is hundreds to thousands of meters thick and covers hundreds to thousands of square kilometers of area and is composed dominantly of lava flows erupted from point sources such as shield volcanoes.

Lava plateau A large, tabular mass of volcanic rock that typically is hundreds to thousands of meters thick and covers hundreds to thousands of square kilometers of area and is composed dominantly of lava flows erupted from fractures (e.g., continental rift zones).

Law of Bravais A scientific law that states that the most common crystal faces are those that contain the greatest number of lattice points.

Lever rule A rule (used in analyzing phase diagrams) that specifies that the mass and its distance from a fulcrum must equal the opposing mass and its distance from the fulcrum for balance to exist. In phase diagrams, a large mass of small volume must be counterbalanced by a larger volume of a smaller mass to make a total of 100%.

Liesegang rings A series of colored rings, produced by oxidation or reduction, within weathered concretions or other sedimentary rocks.

Line group A closed assemblage of point and one-dimensional translation operations.

LIP A **L**arge **I**gneous **P**rovince consisting of thousands of square kilometers of volcanic rocks, primarily basalt.

Liquid limit In engineering practice, the percentage of water in a soil, above which the soil flows under its own weight and below which the soil behaves as a solid.

Liquidus The curve on a phase diagram representing the boundary separating conditions in which only liquid (fluid) exists from those conditions under which liquids and solids exist together.

Lime mud A carbonate sediment consisting of grains smaller than 0.004 mm.

Lithofacies A body of sedimentary rocks characterized by particular rock types, textures, and structures.

Lithosphere The rigid outer layer of Earth composed of deeper mantle materials and shallower, generally lighter crustal materials.

Littoral-beach environment The near-shore to beach environment of sedimentation, the most common transitional environment represented in the sedimentary record.

Load cast Rounded, bulbous sole marks formed during compaction of sediment.

Lopolith A large, dish-shaped and layered pluton.

Luminescence The production of visible light by a mineral when irradiated by light of higher energy.

Luster A subjective and comparative descriptive quality of the light reflected from the surface of a mineral.

M

Magma The melted rock materials that ultimately crystallize to form igneous rocks. Typically, magma contains gasses that are often lost during eruption at the surface.

Magma mixing A process of formation of a derivative magma, in which two magmas of different composition come into contact and mix to form a magma of a new composition.

Magmatic fluid Fluids of relatively low density and low viscosity that arrive in the crust as part of a magma.

Magnesium number The ratio of $Mg^{++}/(Mg^{++} + Fe^{++})$ in rocks or melts.

Mantle The middle layer of the Earth, between the crust and the core, composed of magnesium- and iron-rich minerals.

Mantle plume A crudely cylindrical, rising mass of plastic mantle material, the top of which typically melts to yield magma.

Marble A metamorphic rock derived from a carbonate protolith, such as limestone or dolostone. Typically the grains in marble are phaneritic, but not in every case.

Massive A descriptor for a mineral specimen that has no particular shape or form.

Mass wasting Gravity-induced erosion, transportation, and deposition yielding landslide masses (*sensu lato*).

Matrix Finer-grained material between larger grains or clasts in a rock. In sedimentary rocks, the fine-grained siliciclastic material that is deposited along with the clasts.

Measured resource A resource of calculated quantity, in which dimensions are determined by direct observations in outcrops, drill holes, trenches, or existing mine workings.

Mesosphere The relatively rigid part of the mantle beneath the asthenosphere.

Metamorphic facies A group (a set) of mineral assemblages reflecting a particular and limited range of pressure and temperature conditions under which the mineral assemblage formed. The term is also sometimes applied to the pressure-temperature conditions of metamorphism.

Metamorphic facies series The sequence of metamorphic facies crossed by a curviplanar line or trajectory through the petrogenetic grid, representing a geothermal gradient within the Earth.

Metamorphic fluids Fluids derived from sources such as magmatic fluids, hydrothermal fluids, groundwater, and the breakdown of minerals during metamorphic reactions.

Metamorphism The process of changing textures, chemical composition, minerals, or some combination of these characters in a rock, in response to changes in the conditions, such as the pressure and temperature, under which the rock exists.

Metasomatism A process of metamorphic change in which the composition of a rock is altered by fluids that bring in or remove particular elements from the rock.

Meteoric water Water derived from precipitation as rain, fog, ice, or snow.

Meter A distance equal to 1,650,763.73 wavelengths of orange-red light of krypton-86.

Micrite A carbonate sediment or sedimentary rock (e.g., a very fine-grained limestone) consisting of grains smaller than 0.004 mm.

Mineral A substance formed by geological processes that is solid and naturally occurring (Nickel, 1995).

Mingle Term for a process in which two magmas come together, but do not mix due to conditions of temperature and chemistry, and form blobs or masses of one magma within another.

Mirror line A line of reflection in two-dimensional space.

Mirror plane A surface of reflection in three-dimensional space.

Mode In petrology, the list of observed minerals and their percentages for a rock.

Moho *See* Mohorovicic Discontinuity.

Mohorovicic Discontinuity A zone of rapid and relatively abrupt increase in seismic wave velocity in the Earth separating the crust above from the mantle below.

MORB **M**id-**O**cean **R**idge **B**asalt.

Morphology The particular shape of a crystal.

Mudcracks Polygonal fractures in muddy layers, formed by shrinkage where drying mud masses contract, and preserved where sand or other sediment fills the crack before lithification of the rock body as a whole.

Mudrock A sedimentary rock composed predominantly of grains less than 1/16 mm.

MVT **M**ississippi **V**alley **T**ype deposit. *See* Carbonate hosted lead-zinc deposit.

Mylonite A weakly to strongly foliated metamorphic rock consisting of highly deformed, but in some cases also undeformed to slightly deformed, porphyroclasts, in a deformed matrix of finer-grained, crystal-plastically deformed material.

Mylonitic texture A metamorphic texture of aphanitic to phaneritic nature characterized by deformed, elliptical to elongated grains reflecting crystal-plastic deformation.

N

Neocrystallization The process of new mineral formation via geochemical reactions.

Nesosilicates Silicate minerals with SiO_4 tetrahedra that are isolated and do not share corners with other SiO_4 tetrahedra.

Neutron An electrically neutral subatomic particle that resides in the nucleus.

Nucleation The process of bonding of a few atoms to from a tiny mass of atoms arranged in the order characteristic of a particular mineral.

Nucleus (1) In Chemical reactions, the positively-charged central part of an atom, where protons and neutrons reside and where most of the atomic mass is located. (2) In crystallization processes, a tiny mass of atoms arranged in the internal order characteristic of a specific mineral.

O

Olistostrome A submarine landslide deposit formed by slope failures and resulting downslope debris flow.

Oncolite A spherical to irregularly rounded and concentrically laminated carbonate mass, typical a few centimeters in diameter, produced by microbial precipitation.

Ooid Small (0.25 to 2.0 mm), very round grains of carbonate sediment formed in very shallow waters.

Oolitic A descriptive term for ooid-bearing carbonate rocks.

Opaque Material that does not transmit light.

Open form A crystal form that does not enclose space.

Ophiolite A type of igneous rock complex that occurs in mountain belts and includes metamorphosed ultramafic rocks at the base overlain by successive layers of ultramafic rock (dunite, peridotite), layered gabbros, various nonlayered plutonic rocks, microgabbro (diabase) dikes, and basaltic (low-silica) volcanic rocks. Typically, oceanic sedimentary rocks such as chert overlie the volcanic rocks.

Ophitic texture A special type of plutonic rock texture consisting of a few larger poikilitic grains of pyroxene containing enclosed grains of plagioclase feldspar.

Ore A material that is naturally enriched—that is, is more abundant in particular rocks than it is in average rocks—and has economic value.

Origin In crystallography, the starting point for generating a lattice using symmetry operations.

Outer core The exterior part of the innermost major part of the Earth, consisting of liquid materials composed predominantly of nickel and iron.

Oxidation The process of ion formation (ionization) in which an electron is lost from an atom.

Oxidation state The charge on an ion (is its oxidation state).

Oxisol A leached soil of the humid tropics, with a strongly leached A-Horizon, significant accumulation of iron and aluminum hydroxides along with some evidence of breakdown of clays in Horizon-B.

P

Pahoehoe lava A type of lava, typically with a smooth to ropy surface and low-silica composition.

Parallel intergrowths In crystallography, interconnected crystals of the same mineral aligned so that mutual symmetry elements are parallel.

Paramagnetic A designation for a material, with unpaired electrons, that is attracted to magnetic fields.

Parental magma Magma that gives rise to derivative magmas via modification.

Parting The weak tendency of a mineral to break along planes that are not due to systematic structural weakness, but a weakness imposed by external factors. The name is also given to the planes.

Ped In soil science, an aggregate of distinctive shape consisting of soil particles.

Pegmatite (1) In economic geology, a pegmatite or pegmatite deposit is a rock body dominated by crystals larger than 3 centimeters in length containing economic minerals. These rocks are typically granitic in composition. (2) In petrology, a term often used loosely to refer to granitoid rocks of pegmatitic texture, i.e., granitoid rocks composed of crystals dominantly larger than 3 cm in diameter.

Pegmatitic texture A texture of plutonic igneous rocks consisting predominantly of grains of 3 cm or larger.

Pellet In sedimentology, a small spherical grain (< 0.2 mm) that represents the feces of mud-eating organisms.

Peridotite A common, general category of very low silica (ultrabasic) rock composed of combinations of olivine and pyroxenes.

Permeability A measure of the ability of a fluid to flow through a rock.

Perthite Intergrowths of small masses or lenses of albite in a larger potassium feldspar crystal.

Petrotectonic assemblage A group or collection of rock types formed at a particular location in or on the crust.

PGE Platinum group element.

PGM Platinum group metal.

pH The negative logarithm of the hydrogen ion concentration in a solution.

Phaneritic texture A texture with grains large enough to see with the unaided eye or a low-power hand magnifier.

Phaneritic-porphyritic texture A texture of plutonic rocks in which all the grains are visible to the unaided eye or the eye aided by a low-power hand magnifier and in which there are two distinct size populations of grains present in the rock.

Phase A unique substance that can, in principal, be physically separated from other substances.

Phase diagram A diagram that depicts the conditions of stability of phases, relative to such parameters as the pressure, temperature, and composition of a system.

Phenocryst A large grain, substantially larger than surrounding grains, in an igneous rock.

Phosphorescence Luminescence that continues for a brief period of time after the higher energy irradiation ends.

Phosphorites Marine sedimentary and other rocks containing more than 15 to 20 weight percent P_2O_5.

Phyllite A common type of foliated metamorphic rock consisting of aligned aphanitic grains.

Phyllitic texture A fine-grained to aphanitic metamorphic texture characterized by layers that exhibit a crenulation cleavage, so that they appear corrugated on the surface.

Phyllosilicates Silicate minerals with SiO_4 tetrahedra that share three corners with other SiO_4 tetrahedra to form sheets.

Pillow lava Volcanic rock that contains structures that have the crude form of pillows or irregular tubes.

Placer Ore deposits created in moving waters by gravity-driven separation (settling) of heavy, chemically and physically resistant minerals, such as gold, platinum, diamond, zircon, rutile, cassiterite, or monazite.

Plastic limit The percentage of water in a soil marking the boundary between brittle, friable behavior and plastic behavior.

Plate In tectonic studies, a curved slab of lithosphere.

Plate tectonics A theory that explains the origin of the major structural features of Earth—the ocean basins, continents, mountain ranges, and great faults—in terms of the interaction of large slabs of lithosphere.

Pleochroism Changes in the color of a mineral grain as the microscope stage is rotated under plane polarized light.

Plutonic rocks Rocks crystallized as significant depths below the surface and characterized by visible and generally interlocking mineral grains.

Pluton A rock body consisting of rock with mostly visible grains; i.e., consisting of plutonic rock.

Poikilitic texture A texture consisting of a larger, individual mineral grain containing inclusions (oikocrysts) of smaller grains of various minerals.

Point group A closed assemblage of point symmetry operations.

Point symmetry Symmetry that results from reflection of a line of objects across a point.

Polarization A restriction of the directions in which light waves can vibrate.

Polarizer A polarizing filter mounted below the stage of a petrographic microscope.

Polymorph A mineral with a different arrangement of atoms than a second mineral, but containing the same kinds of atoms.

Polysynthetic A term applied in mineralogy to complex twins where the twin planes are parallel and multiple.

Pore A space between grains in a rock.

Porosity The percentage of the volume of a rock consisting of pores.

Porphyritic texture A texture in igneous rocks consisting of two distinct size populations of grains. The larger grains are called phenocrysts and the smaller grains are collectively called "the groundmass."

Porphyroblastic texture A metamorphic texture characterized by two groups (populations) of grains that crystallized in the same rock, with one population consisting of a few larger grains surrounded by a generally larger population of substantially smaller grains.

Porphyroclastic texture A metamorphic texture superficially similar to porphyroblastic texture, with larger broken or deformed grains in a finer-grained matrix.

Porphyry ore deposit A type of ore deposit typically rich in copper, molybdenum, tin, and tungsten, that exists as a stockwork of ore-bearing veins within an (igneous) stock or batholith.

Precipitates In sedimentary petrology, sedimentary rocks formed either by precipitation of sediment by organisms or by precipitation by abiotic processes. Limestones and evaporites are major types of precipitate. Some workers call these rocks chemical sedimentary rocks.

Pressure A condition in which principal stresses in a system are all equal in size.

Primary magma Magma formed as a melt derived from any rock type, but a melt that is largely unmodified after it formed.

Primitive magma Magma formed by melting of unaltered mantle rock.

Primitive unit cell A unit cell containing only one lattice point, with each corner of the lattice contributing $1/n$th lattice points to the total.

Prograde metamorphism Metamorphism that generally involves an increase in pressure (P) and temperature (T), i.e., a change in conditions away from 0 °C and 0 GPa.

Protolith A parental or preexisting rock that existed prior to metamorphism.

Proton A positively charged, subatomic component of atoms that occurs in the nucleus.

Provenance The source area of sediment and its particular rocks.

Pseudotachylites Fault rocks (cataclasites) with glass in the matrix.

Pumiceous texture A texture in volcanic igneous rocks created by an array of tiny, closely spaced, tubular holes separated by thin sheets of volcanic glass.

Pyroclastic sheet A tabular mass of fragmented volcanic rock, usually the fine-grained material called ash, covering areas of hundreds or thousands of square kilometers.

Pyroclastic texture A volcanic texture consisting of accumulated fragments of the ash, which in some cases are cemented together.

Q, R

Radioactivity Spontaneously emitted energy resulting from reactions within nuclei of atoms.

Rapikivi texture A special corona texture in granitic rocks, consisting of grains with a core of alkali feldspar and a rim of albite.

Recrystallization A process that occurs where particular pressure, temperature, and fluid phase conditions at locations within a rock or sediment cause reorganization of the crystal lattices of minerals present there, without changing the kinds of minerals that exist.

Redox The combined process of oxidation and reduction.

Reduction The process of ion formation (ionization) in which an electron is gained by an atom.

Replacement A process by which a mineral is simply changed, in place, to another mineral through an exchange of ions.

Reservoir A body of water, either on the surface or within the pores of a rock mass.

Residual Ore deposits that develop in place from the chemical weathering of original rock present close to the surface of the Earth.

Retrograde metamorphism Metamorphism that involves a change in conditions towards zero P,T conditions.

Rhyolite A common type of volcanic rock that is high in silica.

Ripple marks Regularly undulating surfaces produced in sediments either by oscillating waters (waves) or currents; commonly preserved in sedimentary rocks.

Rock A naturally occurring solid, composed of crystalline mineral grains, glass, fragmented or altered organic matter, or combinations of these materials.

Rock cleavage A planar structure in rocks resulting from aligned fractures or the alignment of mineral grains, which gives the rock the tendency to split or break into tabular or sheet-like pieces.

Roll-front A type of arcuate-fronted ore deposit, typically of uranium and vanadium, formed in a medium- to coarse-clastic aquifer between aquicludes as groundwater flows through the aquifer.

Rotation point A point symmetry element that is the point of intersection of two mirror lines.

Rotation axis A line about which objects are rotated in three-dimensional space.

Runoff Water flowing across the surface of the Earth.

S

Sandstone A siliciclastic sedimentary rock with grains that are dominantly in the 2 mm to 1/16 mm range.

Saturated A fluid condition in which the fluid will no longer accept dissolved ions.

Schist A very common type of metamorphic rock composed of flaky or needle-like minerals that are strongly aligned, giving the rock a tendency to break into leaf-like pieces.

Schistose texture A phaneritic metamorphic texture in which flaky, bladed, or acicular grains are arrayed abundantly in a sub-parallel arrangement.

SEDEX (SEDimentary EXhalative) deposits A type of stratabound and stratiform ore deposit developed in continental rift or passive margin basins.

Sedimentary environment A specific site at which sediments are formed, deposited, or both in surface regions of the lithosphere below or above sea level, where a particular set of chemical, physical, and biological characteristics identify the setting.

Seriate texture A texture in plutonic rocks in which there is a range of grain sizes from large to small, but in which the sizes are dispersed more or less evenly over that range.

Serpentinization The process of metamorphism by which mafic and ultramafic rocks, particularly those that are olivine-bearing, are converted to serpentinites.

Shelf environment The marine environment of sedimentation that extends seaward from the littoral/beach environments out to a point, called the shelf break, where the submarine slope steepens and slopes down towards the deep seafloor.

Shield cone *See* Shield volcano.

Shield volcano Large volcanic structure, crudely circular in plan, with gentle slopes, and a composition dominated by lava flows of low-silica (basaltic) character.

Sial Crustal materials relatively rich in the two elements **si**licon and **al**uminum.

Siliciclastic rock A sedimentary rock composed predominantly of silicate mineral and rock fragments.

Sill A pluton that has intruded generally parallel to the enclosing layers of rock.

Sima Crustal materials rich in a combination of **si**licon, iron, and **ma**gnesium.

Skeletal A descriptive term for a sedimentary rock containing fossil fragments or whole fossils.

Slate A common type of metamorphic rock characterized by aphanitic texture and micaceous minerals.

Slaty texture An aphanitic to very fined-grained metamorphic texture with micaceous minerals strongly aligned, causing the rock to break into flat pieces.

Soil Weathered and disaggregated rock and associated organic materials, unmoved from and overlying the bedrock.

Sole marks Bulbous to linear features that are usually observed on the bottom (sole) of a bed, after erosion has removed the weaker, underlying bed.

Solid solution A phenomenon in which a (solid) mineral can have complete substitution of one element for another in its structure.

Solidus A curve on a phase diagram separating conditions under which only solids exist from those under which liquids plus solids exist together.

Solution (1) (noun) A mixture of a solvent material, such as water, and ions. (2) (noun) The process of decomposition involving removal of ions from the surface-fluid interface, also called dissolution. (3) (verb) The action of dissolution.

Sorosilicates Silicate minerals with paired SiO_4 tetrahedra that share one corner.

Sorting A measure of the distribution of grain sizes in a sediment or soil. In geology, well-sorted means the grains are all nearly the same size.

Space lattice A regular, repeated array of points in three dimensions.

Spar In sedimentology, carbonate sediment grains larger than 0.004 mm.

Specific gravity The ratio of the mass of a mineral to an equal volume of water; a measure of the density of a mineral expressed in unit-less form.

Spreading center A zone separating plates of lithosphere, along which plates separate and new crust is formed.

Static metamorphism Metamorphism in which uniform pressures dominate as the cause of metamorphism.

Stock Plutons of somewhat equant shape in surface exposure and exposed over an area of less than 100 square kilometers.

Stockwork vein complex A type of hydrothermal ore deposit or vein network associated with porphyry deposits of copper, molybdenum, tin, and tungsten.

Stoichiometry A balance of charges within an electrically neutral crystal.

Stope The action of a magma in breaking off and engulfing blocks of the country rock along its route toward the surface.

Stratabound A designation for ore deposits that are confined to a very limited number of sedimentary layers.

Stratiform deposits Layered deposits that have a series of layers superficially resembling the layering of sedimentary strata.

Stratigraphy The detailed description and study of the rock layering and associated characteristics.

Streak The color of a powdered mineral, generally as seen on an unglazed porcelain tile.

Stress Force per unit area (F/A).

Stromatolites Laminated carbonate rocks, commonly with a domal form, that are considered to be produced by algal precipitation in very shallow marine environments.

Stylolites Irregular surfaces coated or marked by dark organic or oxide mineral accumulations and commonly appearing as jagged lines in rocks. These are formed after lithification by dissolution of carbonate or other minerals, leaving the dark residue.

Structures Physical features of a rock that are of hand-specimen or larger scale, but do not pervade hand-specimen size samples (i.e., they are not repeated over and over in each small part of a hand specimen).

Subduction zone A zone separating one plate from another, where plates converge and collide, and where one plate descends beneath the other.

Sub-economic resource A resource that cannot be extracted and sold for a profit.

Subhedral A mineral specimen that is partially bounded by crystal faces.

Supergene An ore deposit resulting from remobilized, descending fluids, principally rainwater, that interacts with rocks to produce mineralization.

Supersaturation A somewhat forced, temporary, and nonequilibrium condition in which the fluid contains more than the number of ions necessary for saturation.

Suspension In sedimentology, the process of lifting particles from a surface and carrying them to another location.

Symmetry The property of a material in which exactly identical parts of that material are related in some way to one another (for example by reflection or rotation), but occupy a different position or orientation.

Symmetry operation The process of adding symmetry to a collection of objects.

Syngenetic deposits Ore deposits that form along with host rocks.

System Any part of the natural world that we wish to isolate for the purposes of analysis or discussion, either physically in real time or conceptually in our minds.

T

Tabular A type of ore deposit that is tabular in shape and forms in coarse clastic sediments or sedimentary rocks geographically associated with former playa lake (desert) sediments.

Tectosilicates Silicate minerals with individual SiO_4 tetrahedra that share four corners with other SiO_4 tetrahedra to form a three-dimensional framework.

Temperature A property of a system that renders it in thermal equilibrium with another system, that is in a state in which there is no tendency for the system to change.

Tessellation The mathematical name for filling space with objects.

Texture The combination of the arrangement, sizes, and shapes of mineral grains (and fragments), and the relationships between them.

Thermally induced melting Melting that occurs primarily at the base of the crust and along subduction zones, where increases in temperature due to heat released from magma and created by friction contribute the overall temperature increase, and hence, to melting.

Thermohaline currents Temperature-salinity currents in large bodies of water such as oceans or lakes.

Tholeiite A type of basaltic volcanic rock, typically with 48% to 52% silica and containing low-calcium pyroxene, such as orthopyroxenes or pigeonite.

Tie lines Lines in a phase diagram connecting co-existing phases.

Trachytic texture A volcanic texture in which rectangular grains such as plagioclase grains are crudely aligned (a volcanic equivalent of trachytoidal texture in plutonic rocks).

Trachytoidal texture An igneous texture of plutonic rocks, typically feldspar-rich, that consists of crudely aligned feldspar crystals surrounded by other mineral grains.

Transform fault A large, typically vertical fault in the lithosphere, along which plates slide past one another.

Translating the object In crystallography, the process in which an object is placed on a line, duplicated, and the copy is moved some distance away.

Translational symmetry Symmetry resulting from the process by which an object is placed on a line, duplicated, and the copy is moved some distance away.

Trench environment A major, typically deep marine site of sediment accumulation and the surface expression of a subducting plate.

Triple junction A site at plate corners, where three plates join.

Triple point A point in a phase diagram where three phases are stable together.

Tuff A volcanic pyroclastic rock composed of grains smaller than 2 mm in diameter.

Turbidity current A sediment laden, heavy (dense) current that flows down a slope or canyon and, commonly, onto a basin floor, where it may contribute to the formation of a submarine fan.

Turbulent flow Fluid flow in which the flow lines cross, swirl, and circle.

Twins Intergrowths of crystals of the same mineral, related by a symmetry element not present in either crystal.

U

Unconformity-related uranium deposits A type of uranium ore deposit formed where fluids containing uranium ions, rising through crystalline rocks below an unconformity, interact with fluids of different character in the rocks above the unconformity, resulting in deposition of uranium ore.

Undiscovered resource A resource thought to exist that has yet to be discovered.

Uniaxial A designation for a material with one axis perpendicular to which all refractive indices are equal. These are anisotropic materials with tetragonal, hexagonal, or trigonal symmetry.

Unit cell One of the individual plane figures that fills space.

Unit cell axes The basis vectors for unit cells.

Unloading A disintegration process in which erosion removes surficial materials and the rock load above buried rock, the buried rock expands upward and outward, and the expansion creates tensional forces that break the rock along a set of fractures called exfoliation joints.

Unsaturated The condition of a fluid that will accept ions into solution.

V

Valence electrons Electrons that occur in the outermost orbitals of atoms and are typically involved in chemical bonding.

Van der Waals bond A chemical bond formed when a molecule is polar (with a positive charge on one end and a negative charge on the other) and positive and negative ends of different molecules are attracted and bond together as a result of their opposing charges. Hydrogen bonds are special cases of Van der Waals bonding involving molecules with hydrogen.

Varves Repeated dark and light layers (generally in mudrocks) formed in response to alternating seasonal changes in sedimentation.

Vein A generally tabular, but locally irregular, blob-like, to curviplanar body of mineralogically simple rock, that fills a crack in preexisting rock.

Velocity The distance between two different points divided by the time required for a body to move between those two points.

Vesicle A small (typically < 1 cm), somewhat spherical to oblong hole in lava created by escaping gas.

Viscosity The internal resistance to flow of a fluid.

Vitreous The mineral luster that resembles the appearance of glass.

Vitrophyric texture A volcanic texture in which glass surrounds pheoncrysts.

VMS (Volcanogenic Massive Sulfide) deposit A type of stratabound and stratiform ore deposit developed at mid-ocean ridge spreading centers.

Volcanic breccia A volcanic, pyroclastic or flow rock characterized by an abundance of large fragments (> 64 mm in diameter) in a matrix of finer-grained materials. The U.S. Geological Survey notes that agglomerate is an equivalent rock type, but other workers consider agglomerates to include only rocks composed of accumulated volcanic bombs.

Volcanic rocks Rocks crystallized at or near the surface, in some cases fragmented by explosive eruptions, and characterized by materials such as microscopic crystals and glass (aphanitic materials). They have textures that are glassy, aphanitic, aphanitic-porphyritic, vesicular, pumiceous, or granular-porphyritic.

W, X, Y, Z

Wacke A sandstone with significant amounts of matrix.

Weathering A general term for the group of processes that change rock into soil.

Xenolith A fragment of rock of a type foreign to the enclosing rock. For example, a piece of the surrounding metamorphic "country rock" enclosed within an intruded granite.

XRD (X-ray diffraction) A technique of crystal structure analysis involving generation of X-rays from a source and measuring the reflections of those X-rays when they are directed at a crystalline material.

XRF (X-ray fluorescence) A technique of chemical analysis involving generation of X-rays from a material and measuring the emitted rays with electronic detectors.

References

A

Abily, B., and Ceuleneer, G. 2013. The dunitic mantle-crust transition zone in the Oman ophiolite: Residue of melt-rock interaction, cumulates from high-MgO melts, or both? *Geology*, v. 41, pp. 67–70. doi:10.1130/G33351.1

Adamo, P., Colombo, C., and Violante, P. 1997. Iron oxides and hydroxides in the weathering interface between *Stereocaulon vesuvianum* and volcanic rock. *Clay Minerals*, v. 32, pp. 275–283.

Alexandri, R., Jr., Force, E. R., Cannon, W. F., Spiker, E. C. 1985. Sedimentary manganese carbonate deposits of the Molango District, Mexico. *Geological Society America, Abstracts with Programs*, v. 17.

Amundson, R., Richter, D. D., Humphreys, G. S., Jobbagy, E. G., and Gaillardet, J. 2007. Coupling biota and earth materials in the critical zone. *Elements*, v. 3, pp. 327–332.

Anderson, G. M. 1996. *Thermodynamics of Natural Systems*. Wiley, New York. 382 p.

Anderson, H. S., Yoshinobu, A. S., Nordgulen, Ø., and Chamberlain, K. 2013. Batholith tectonics: Formation and deformation of ghost stratigraphy during assembly of the mid-crustal Andalshatten batholith, central Norway. *Geosphere*, v. 9, pp. 667–690. doi:10.1130/GES00824.1

Anderson, J. L., Osborne, R. H., and Palmer, D. F. 1983. Cataclastic rocks of the San Gabriel Fault: An expression of deformation at deeper crustal levels in the San Andreas Fault Zone. *Tectonophysics*, v. 98, pp. 209–251.

Anderson, R. N., Uyeda, S., and Miyashiro, A. 1976. Geophysical and geochemical constraints at converging plate boundaries—I: Dehydration in the downgoing slab. *Geophysical Journal of the Royal Astronomical Society*, v. 44, pp. 333–357.

Anderson, S. P., von Blanckenbur, F., and White, A. F. 2007. Physical and chemical controls on the critical zone. *Elements*, v. 3, pp. 315–319.

Ausich, W. I., and Meyer, D. L. 1990. Origin and composition of carbonate buildups and associated facies in the Fort Payne Formation (Lower Mississippian, south-central Kentucky): An integrated sedimentologic and paleoecologic analysis. *Geological Society America Bulletin*, v. 102, pp. 129–146.

B

Back, M. E., and Mandarino, J. A. 2008. *Fleischer's Glossary of Mineral Species* (10th ed.). Mineralogical Record, Tucson. 346 p.

Bailey, E. H., Irwin, W. P., and Jones, D. L. 1964. Franciscan and related rocks, and their significance in the geology of Western California. *California Division of Mines and Geology Bulletin*, 183. 177 p.

Baker, D. R. 1996. Granitic melt viscosities: Empirical and configurational entropy models for their calculation. *American Mineralogist*, v. 81, pp. 126–134.

Baker, D. R. 1998. Granitic melt viscosity and dike formation. *Journal of Structural Geology*, v. 20, pp. 1395–1404.

Baldock, J. A., and Nelson, P. N. 2000. Soil organic matter. In Sumner, M. E. (ed.), *Handbook of Soil Science*. CRC Press, Boca Raton, pp. B25–B84.

Balk, R. 1937. *Structural Behavior of Igneous Rocks*. Geological Society of America Memoir 5. 177 p.

Ballard, R. D., Van Andel, T. H., and Holcomb, R. T. 1982. The Galapagos Rift at 86°W: Part 5: Variations in volcanism, structure, and hydrothermal activity along a 30-kilometer segment of the rift valley. *Journal of Geophysical Research*, v. 87, pp. 1149–1161.

Bard, J. P. 1986. *Microtextures of Igneous and Metamorphic Rocks*. D. Reidel, Dordrecht. 264 p.

Barrett, J. 2002. *Structure and Bonding*. Wiley, New York. 181 p.

Barrow, G. 1893. On an intrusion of muscovite-biotite gneiss in the South-eastern Highlands of Scotland, and its accompanying metamorphism. *Quarterly Journal of the Geological Society*, London, v. 49, pp. 330–388.

Bartley, J. M., Coleman, D. S., and Glazner, A. F. 2008. Incremental pluton emplacement by magmatic crack-seal. *Transactions of the Royal Society of Edinburgh–Earth Sciences*, v. 97, pp. 383–396.

Bartley, J. M., Glazner, A. F., and Mahan, K. H. 2012. Formation of pluton roofs, floors, and walls by crack opening at Split Mountain, Sierra Nevada, California. *Geosphere*, v. 8, pp. 1086–1103. doi:10.1130/GES00722.1.

Barton, P. B. 1970. Sulfide Petrology. *Mineralogical Society of America Special Paper 3*, pp. 187–198.

Basaltic Volcanism Study Project (BVSP). 1981. *Basaltic Volcanism on the Terrestrial Planets*. Pergamon Press. 1286 p.

Bass, J. D., and Parise, J. B. 2008. Deep earth and recent developments in mineral physics: *Elements*, v. 4, pp. 157–163.

Basu, A. R. 1975. Hot-spots, mantle plumes and a model for the origin of ultramafic xenoliths in alkali basalts. *Earth and Planetary Science Letters*, v. 28, pp. 261–274.

Bateman, R. 1984. Comment: On the mechanics of igneous diapirism, stoping, and zone melting. *American Journal of Science*, v. 284, pp. 979–980.

Bayliss, P. 2000. *Glossary of Obsolete Mineral Names*. Mineralogical Record, Tucson. 235 p.

Bea, F. 2012. The sources of energy for crustal melting and the geochemistry of heat-producing elements. *Lithos*, v. 153, pp. 278–291. doi:10.1016/j.lithos.2012.01.017.

Beane, R. E., and Titley, S. R. 1981. Porphyry Copper Deposits. Part II. Hydrothermal Alteration and Mineralization. In Skinner, B. J. (ed.), *Economic Geology 75th Anniversary Volume*, pp. 214–269.

Berkland, J. O., Raymond, L. A., Kramer, J. C., Moores, E. M., and O'Day, M. 1972. What is Franciscan? *American Association of Petroleum Geologists Bulletin*, v. 56, pp. 2295–2302.

Best, M. G. 1982. *Igneous and Metamorphic Petrology*. Freeman, San Francisco. 630 p.

Beukes, N. J., and Klein, C. 1990. Geochemistry and sedimentology of a facies transition—from microbanded to granular iron-formation—in the Early Proterozoic Transvaal Supergroup, South Africa. *Precambrian Research*, v. 47, pp. 99–139.

Blake, M. C., Jr., Jayko, A. S., McLaughlin, R. J., and Underwood, M. B. 1988. Metamorphic and tectonic evolution of the Franciscan Complex, northern California. In Ernst, W. G. (ed.), *Metamorphism and crustal evolution of the western United States* (Rubey Volume 7). Prentice-Hall. Englewood Cliffs, NJ, pp. 1035–1060.

Blatt, H., Tracy, R. J., and Owens, B. E. 2006. *Petrology: Igneous, Sedimentary, and Metamorphic* (3rd ed.). Freeman, New York. 530 p.

Blatter, D. L., and Carmichael, I. S. E. 1998. Hornblende peridotite xenoliths from central Mexico reveal the highly oxidized nature of subarc upper mantle. *Geology*, v. 26, pp. 1035–1038.

Bloomer, S. H., Taylor, B., MacLeod, C. J., Stern, R. J., Freyer, P., Hawkins, J. W., and Johnson, L. 1995. Early arc volcanism and the ophiolite problem: A perspective from drilling in the western Pacific. In Taylor, B., and Natlaund, J. (eds.), *Active Margins and Marginal Basins of the western Pacific*. American Geophysical Union Geophysical Monograph 88, pp.1–30.

Bloss, F. D. 1971. *Crystallography and Crystal Chemistry*. Holt, Rinehart and Winston, New York. 545 p.

Bloss, F. D. 1999. *Optical Crystallography*. Mineralogical Society of America, Washington, DC. 239 p.

Boggs, S., Jr. 1992. *Petrology of Sedimentary Rocks*. Macmillan, New York. 707 p.

Boggs, S., Jr. 2009. *Petrology of Sedimentary Rocks* (2nd ed.). Cambridge University Press, New York. 609 p.

Boisen, M. B., and Gibbs, G. V. 1990. Mathematical crystallography (revised). *Reviews in Mineralogy*, v. 15. Mineralogical Society of America, Washington, DC. 460 p.

Bolt, B. A. 1982. *Inside the Earth*. Freeman, San Francisco. 191 p.

Borg, G., Piestrzynski, A., Bachman, G. H., Püttmann, W., Walther, S., and Fiedler, M. 2012. An overview of the European Kupferschiefer deposits. In Hedenquist, J. W., Harris, M., and Camus, F. (eds.), *Geology and Genesis of Major Copper Deposits and Districts of the World: A Tribute to Richard H. Sillitoe*. Society of Economic Geologists Special Publication 16, pp. 455–486.

Boudier, F., and Nicolas, A. 1985. Harzburgite and lherzolite subtypes in ophiolitic and oceanic environments. *Earth and Planetary Science Letters*, v. 76, pp. 84–92.

Bouma, A. H. 1962. *Sedimentology of Some Flysch Deposits*. Elsevier, Amsterdam. 168 p.

Bouma, A. H., and Stone, C. G. (eds.). 2000. *Fine-grained Turbidite Systems*. AAPG Memoir 72 and SEPM Special Publications No. 68. 342 p.

Bouma, A. H., Berryhill, H. L., Brenner, R. L., and Knebel, H. J. 1982. Continental shelf and epicontinental seaways. In Middleton, G. V., and Bouma, A. H. (eds.), *Turbidites and Deep-Water Sedimentation*. Pacific Section SEPM Short Course Notes, pp. 79–118.

Bowen, N. L. 1913. The melting phenomena of the plagioclase feldspars. *American Journal of Science*, Fourth Series, v. 35, pp. 577–599.

Bowen, N. L. 1940. Progressive metamorphism of siliceous limestone and dolomite. *Journal of Geology*, v. 48, pp. 225–274.

Bowen, N. L., and Anderson, O. 1914. The binary system $MgO-SiO_2$. *American Journal of Science*, Fourth Series, v. 37, pp. 487–500.

Bozlaker, A., Prospero, J. M., Fraser, M. P., and Chellam, S. 2013. Quantifying the contribution of long-range Saharan dust transport on particulate matter concentrations in Houston, Texas, using detailed elemental analysis. *Environmental Science and Technology*, v. 47, pp. 10179–10187. doi:10.1021/es4015663.

Brady, N. C, and Weil, R. R. 1999. *The Nature and Properties of Soils* (12th ed.). Prentice Hall, Upper Saddle River, NJ. 881 p.

Brandeis, G., and Jaupart, C. 1987. The kinetics of nucleation and crystal growth and scaling law for magmatic crystallization. *Contributions to Mineralogy and Petrology*, v. 96, pp. 24–34.

Brenan, J. M. 2008. The platinum-group elements: "Admirably adapted" for science and industry. *Elements*, v. 4, pp. 227–232.

Brewer, R. C., Bolton, J. C., and Driese, S. G. 1990. A new classification of sandstone. *Journal of Geological Education*, v. 38, pp. 343–347.

Brodie, K., Fettes, D., and Harte, B. 2007. Structural terms including fault rock terms. In Fettes, D., and Desmons, J. (eds.), *Metamorphic Rocks: A Classification and Glossary of Terms*. Cambridge University Press, New York, pp. 24–31.

Brown, M. 2001. Orogeny, migmatites and leucogranites: A review. *Proceedings of the Indiana Academy of Sciences* (renamed *Journal of Earth System Science*), v. 110, pp. 313–336.

Brown, M. 2013. Granite: Genesis to emplacement. *Geological Society of America Bulletin*, v. 125, pp. 1079–1113.

Brown, M., Averkin, Y. A., McLellan, E. L., and Sawyer, E. W. 1995. Melt segregation in migmatites. *Journal of Geophysical Research*, v. 100, pp. 15,655–15,679. doi:10.1029/95JB00517.

Brown, M., and Rushmer, T. 1997. The role of deformation in the movement of granite melt: Views from the laboratory and the field. In Holness, M. B. (eds.), *Deformation-Enhanced Fluid Transport in the Earth's Crust and Mantle*. The Mineralogical Society Series 8, Chapman and Hall, London, pp. 111–144.

Broz, M. E., Cook, R. F., and Whitney, D. L. 2006. Microhardness, toughness and modulus of Mohs scale minerals. *American Mineralogist*, v. 91, pp. 135–142.

Brush, G. J., and Penfield, S. L. 1911. *Manual of Determinative Mineralogy with an Introduction on Blowpipe Analysis* (16th ed.). Wiley, New York. 312 p.

Bucher, K., and Frey, M. 1994. *Petrogenesis of Metamorphic Rocks* (6th ed.). Springer-Verlag, Berlin. 318 p.

Burchardt, S., Tanner, D., and Krumbholz, M. 2011. The Slaufrudar pluton, southeast Iceland: An example of shallow magma emplacement by coupled cauldron subsidence and magmatic stoping. *Geological Society of America Bulletin*, v. 124, pp. 213–227.

Buerger, M. J. 1945. The genesis of twin crystals. *American Mineralogist*, v. 30, pp. 469–482.

Buerger, M. J. 1978. *Elementary Crystallography: An Introduction to the Fundamental Geometrical Features of Crystals* (Revised). MIT Press, Cambridge, MA. 528 p.

Burov, E., Jaupart, C., and Guillou-Frottier, L. 2003. Ascent and emplacement of buoyant magma bodies in brittle-ductile upper crust. *Journal of Geophysical Research*, v. 108, pp. 2177. doi:10.1029/2002JB001904.

Burov, E., Francois, T., Yamato, P., and Wolf, S. 2014. Mechanisms of continental subduction and exhumation of HP and UHP rocks. *Gondwana Research*, v. 25, pp. 464–493.

Button, A., Brock, T. D., Cook, P. J., Eugster, H. P., Goodwin, A. M., James, H. L., Margulis, L., Nealson, K. H., Nriagu, J. O., Trendall, A. F., and Walter, M. R. 1982. Sedimentary iron deposits, evaporites and phosphorites state of the art report. In Holland, H. D., and Schidlowski, M. (eds.), *Mineral Deposits and the Evolution of the Biosphere*. Springer-Verlag, New York, pp. 259–273.

C

Cai, J., Powell, R. D., Cowan, E. A., and Carlson, P. R. 1997. Lithofacies and seismic-reflection interpretation of temperate glacimarine sedimentation in Tarr Inlet, Glacier Bay, Alaska. *Marine Geology*, v. 143, pp. 5–37.

Callahan, W. H. 1967. Some spatial and temporal aspects of the localization of Mississippi Valley-Appalachian type ore deposits. In Brown, J. S. (ed.), *Genesis of Stratiform Lead-Zinc-Barite-Fluorite Deposits (Mississippi Valley Type Deposits)*. Economic Geology Monograph 3, pp. 14–20.

Cameron E. N., Jahns, R. H., McNair, A. H., and Page, L. R. 1949. *Internal Structure of Granitic Pegmatites*. Economic Geology Publishing Co., Urbana, IL. 115 p.

Cann, J. R., Langseth, M. G., Honnorez, J., Von Herzen, R. P., White, S. M., et al. 1983. Site 505: Sediments and ocean crust in an area of low heat flow south of the Costa Rica Rift. *Initial Reports of the Deep Sea Drilling Project*, v. 69, pp. 75–214.

Carmichael, I. S. E. 1967. The petrology of Thingmuli, a tertiary volcano in eastern Iceland. *American Mineralogist*, v. 125, pp. 1815–1841.

Carswell, D. A., and Zhang, R. Y. 2000. Petrographic characteristics and metamorphic evolution of ultrahigh-pressure eclogites in plate-collision belts. In Ernst, W. G., and Liou, J. G. (eds.), *Ultra-high Pressure Metamorphism and Geodynamics in Collision-type Orogenic Belts*. Bellweather, Columbia, MD, pp. 39–56.

Cashman, K. V., and Sparks, R. S. J. 2013. How volcanoes work: A 25-year perspective. *Geological Society of America Bulletin*, v. 125, pp. 664–690.

Cerny, P. 1982. Anatomy and classification of granitic pegmatites. In Cerny, P. (ed.), *Granitic Pegmatites in Science and Industry*. Mineralogical Association of Canada, Winnipeg, pp. 1–39.

Chadwick, O. A., and Chorover, J. 2001. The chemistry of pedogenic thresholds. *Geoderma*, v. 100, pp. 321–353.

Chinner, G. A. 1960. Pelitic gneisses with varying ferrous/ferric ratios from Glen Clova, Angus, Scotland. *Journal of Petrology*, v. 1, pp. 178–217.

Chinner, G. A. 1978. Metamorphic zones and fault displacement in the Scottish Highlands. *Geological Magazine*, v. 115, pp. 37–45.

Chipping, D. H. 1971. Paleoenvironmental significance of chert in the Franciscan formation of western California. *Geological Society of America Bulletin*, v. 82, pp. 1707–1712.

Choquette, P. W., and James, N. P. 1987. Diagenesis #12. Diagenesis in limestones - 3. The deep burial environment. *Geoscience Canada*, v. 14, pp. 3–35.

Choquette, P. W., and Pray, L. C. 1970. Geologic nomenclature and classification of porosity in sedimentary carbonates. *American Association of Petroleum Geologists Bulletin*, v. 54, pp. 207–250.

Chopin, C. 1986. Phase relationships of ellensbergite, a new high-pressure Mg-Al-Ti silicate in pyrope-coesite-quartzite from the western Alps. In Evans, B. W., and Brown, E. H. (eds.), *Blueschists and Eclogites*. Geological Society of America Memoir 164, pp. 31–42.

Chopin, C., and Schertl, H.-P. 1999. The UHP unit in the Dora-Maira massif, Western Alps. *International Geology Review*, v. 41, pp. 765–780.

Chorover, J., Kretzschmar, R., Garcia-Pichel, F., and Sparks, D. L. 2007. Soil biogeochemical processes within the critical zone. *Elements*, v. 3, pp. 321–326.

Christiansen, R. L., and Blank, H. R., Jr. 1972. *Volcanic Stratigraphy of the Quaternary Rhyolite Plateau in Yellowstone National Park*. U.S. Geological Survey Professional Paper no. 729-B. 18 p.

Christiansen, R. L., and Lipman, P. W. 1972. Cenozoic volcanism and plate-tectonic evolution of the western United States: Part 2, Late Cenozoic. *Philosophical Transactions of the Royal Society of London A.*, v. 271, pp. 249–284.

Clark, C., Fitzsimons, I. C. W., Healy, D., and Harley, S. L. 2011. How does the continental crust get really hot? *Elements*, v. 7, pp. 235–240. doi:10.2113/gselements.7.4.235.

Clarke, D. B. 1992. *Granitoid Rocks*. Chapman and Hall, London. 283 p.

Cloos, M. 1982. Flow mélanges: Numerical modeling and geologic constraints on their origin in the Franciscan subduction complex, California. *Geological Society of America Bulletin*, v. 93, pp. 330–345.

Coleman, R. G. 1977. *Ophiolites: Ancient Ocean Lithosphere?* Springer-Verlag, Berlin. 229 p.

Colombo, C., and Violante, A. 1997. Effect of aging on the nature and interlayering of mixed hydroxy Al-Fe-montmorillonite complexes. *Clay Mineralogy*, v. 32, pp. 55–64.

Compton, R. R. 1962. *Manual of Field Geology*. Wiley, New York. 378 p.

Condie, K. C. 2001. *Mantle Plumes and Their Record in Earth History*. Cambridge University Press, Cambridge, UK. 326 p.

Cook, H. E. 1968. Ignimbrite flows, plugs, and dikes in the southern part of the Hot Creek Range, Nye County, Nevada. In Coates, R. R., Hay, R. L., and Anderson, C. A. (eds.), *Studies in Volcanology*. Geological Society of America Memoir 116, pp. 107–152.

Cook, P. J., and McElhinny, M. W. 1979. A reevaluation of the spatial and temporal distribution of sedimentary phosphate deposits in the light of plate tectonics. *Economic Geology*, v. 74, pp. 315–330.

Cowan, E. A., and Powell, R. D. 1990. Suspended sediment transport and deposition of cyclically interlaminated sediment in a temperate glacial fjord, Alaska, USA. In Dowdeswell, J. A., and Scourse, J. D. (eds.), *Glacimarine Environments: Processes and Sediments*. Geological Society of London, Special Publications 53, pp. 75–89.

Cox, D. P., and Singer, D. A. (eds.). 1986. *Mineral Deposit Models*. U.S. Geological Survey Bulletin 1693. 379 p.

Craig, J. R., Vaughan, D. J., and Skinner, B. J. 2011. *Resources of the Earth* (4th ed.). Prentice-Hall, Upper Saddle River, NJ. 508 p.

Cuadros, J., Afsin, B., Jadubansa, P., Ardakani, M., Ascaso, C., and Wierzchos, J. 2013. Microbial and inorganic control on the composition of clay from volcanic glass alteration experiments. *American Mineralogist*, v. 98, pp. 319–334.

Cullity, B. D. 1967. *Elements of X-Ray Diffraction*. Addison-Wesley, Reading, MA. 514 p.

Cuney, M. 2009. The extreme diversity of uranium deposits. *Mineralium Deposita*, v. 44, No. 1, pp. 3–9.

Curran, K. J., Hill, P. S., Milligan, T. G., Cowan, E. A., Syvitski, J. P. M., Konings, S. M. 2004. Fine-grained sediment flocculation below the Hubbard Glacier meltwater plume, Disenchantment Bay, Alaska. *Marine Geology*, v. 203, pp. 83–94.

Czamanske, G. K., and Zientek, M. L. (eds.). *The Stillwater Complex, Montana: Geology and Guide*. Montana Bureau of Mines and Geology Special Publication 92, pp. 97–117.

D

Daines, M. J., and Kohlstedt, D. L. 1993. A laboratory study of melt migration. In Cox, K. G., McKenzie, D. P., and White, R. S. (eds.), *Melting and Melt Movement in the Earth*. Oxford University Press, Oxford, UK, pp. 43–52.

Dashtgard, S. E., Venditti, J. G., Hill, P. R., Sisulak, C. F., Johnson, S. M., and La Croix, A. D. 2012. Sedimentation across the tidal-fluvial transition in the lower Fraser River, Canada. *The Sedimentary Record*, v. 10, no. 4, pp. 4–9. doi:10.2110/sedred.2012.4.4.

Davis, A. S., and Clague, D. A. 1987. Geochemistry, mineralogy, and petrogenesis of basalt from the Gorda Ridge. *Journal of Geophysical Research*, v. 92, pp. 10,467–10,483.

Davis, B. T. C., and England, J. L. 1964. The melting of forsterite up to 50 kilobars. *Journal of Geophysical Research*, v. 69, pp. 1113–1116.

Dawson, J. B. 1981. The nature of the upper mantle. *Mineralogical Magazine*, v. 44, pp. 1–18.

Deer, W. A., Howie, R. A., and Zussman, J. 1992. *An Introduction to the Rock-forming Minerals*. Longman, Essex, England and Wiley, New York. 528 p.

Degens, E. T. 1965. *Geochemistry of Sediments*. Prentice-Hall, Englewood Cliffs, NJ. 342 p.

De Graaff-Surpless, K., Graham, S. A., Wooden, J. L., and McWilliams, M. O. 2002. Detrital zircon provenance analysis of the Great Valley Group, California: Evolution of an arc-forearc system. *Geological Society of America Bulletin*, v. 114, pp. 1564–1580. doi:10.1130/0016-7606(2002)114<1564:DZPAOT>2.0.CO;2.

De Vivo, B., Lima, A., and Webster, J. D. 2005. Volatiles in magmatic-volcanic systems. *Elements*, v. 1, pp. 19–24.

Dickinson, W. R. 1970. Interpreting detrital modes of graywacke and arkose. *Journal of Sedimentary Petrology*, v. 40, pp. 695–707.

Dickinson, W. R. 1971. Plate tectonics in geologic history. *Science*, v. 174, pp. 107–113.

Dickinson, W. R., and Gehrels, G. E. 2008. U-Pb ages of detrital zircons in relation to paleogeography: Triassic paleodrainage networks and sediment dispersal across southwest Laurentia. *Journal of Sedimentary Research*, v. 78, pp. 745–764. doi:10.2110/jsr.2008.088.

Dijkstra, A. H., Drury, M. R., and Vissers, R. L. M. 2001. Structural petrology of plagioclase-peridotites in the West Othris Mountains (Greece): Melt impregnation in mantle lithosphere. *Journal of Petrology*, v. 42, pp. 5–24.

Dilek, Y., and Furnes, H. 2011. Ophiolite genesis and global tectonics: Geochemical and tectonic fingerprinting of ancient oceanic lithosphere. *Geological Society of America Bulletin*, v. 123, pp. 387–411. doi:10.1130/B30446.1.

Dilek, Y., and Furnes, H. 2014. Ophiolites and their origins. *Elements*, v. 10, pp. 93–100.

Dilek, Y., Moores, E. M., and Furnes, H. 1998. Structure of modern oceanic crust and ophiolites and implications for faulting and magmatism at oceanic spreading centers. In Buck, W. R., Delaney, P. T., Karson, J. A., and Lagabrielle, Y. (eds.), *Faulting and Magmatism at Mid-ocean Ridges*. American Geophysical Union Geophysical Monograph 106, pp. 219–266.

Dobretsov, N. L., and Sobolev, V. S. 1972. The study of metamorphic facies. In Sobolev, V. S. (ed.), *The Facies of Metamorphism*. Australian National University, Canberra, pp. 168–205.

Donnay, J. D. H., and Harker, D. 1937. A new law of crystal morphology extending the Law of Bravais. *American Mineralogist*, v. 22, pp. 446–467.

Dostal, J., Zentilli, M., Caelles, J. C., and Clark, A. H. 1977. Geochemistry and origin of volcanic rocks of the Andes (26°–28°S). *Contributions to Mineralogy and Petrology*, v. 63, pp. 113–128.

Dott, R. H., Jr. 1964. Wacke, graywacke and matrix: What approach to immature sandstone classification? *Journal of Sedimentary Petrology*, v. 34, pp. 625–632.

Dove, M. 2003. *Structure and Dynamics: An Atomic View of Materials*. Oxford University Press, Oxford, UK. 334 p.

Du, Q., Yi, H., Hui, B., Li, S., Xia, G., Yang, W., and Wu, X. 2013. Recognition, genesis and evolution of manganese ore deposits in southeastern China. *Economic Geology*, v. 55, pp. 99–109. doi:10.1016/j.oregeorev.2013.05.001.

Dumitru, T. A., Elder, W. P., Hourigan, J. K., Chapman, A. D., Graham, S. A., and Wakabayashi, J., 2016. Four Cordilleran paleorivers that connected Sevier thrust zones in Idaho to depocenters in California, Washington, Wyoming, and, indirectly, Alaska. *Geology* v. 44, pp. 75–78. doi:10.1130/G37286.1.

Dunham, R. J. 1962. Classification of carbonate rocks according to depositional texture. In Ham, W. E. (ed.), *Classification of carbonate rocks: A symposium*. American Association of Petroleum Geologists Memoir 1, pp. 108–121.

Dyar, M. D., Gunter, M. E., and Tasa, D. 2008. *Mineralogy and Optical Mineralogy*. Mineralogical Society of America, Chantilly, VA. 708 p.

E

Eaton, J. P., and Murata, K. J. 1960. How volcanoes grow. *Science*, v. 132, pp. 925–938.

Eby, G. N. 2004. *Principles of Environmental Geochemistry*. Brooks/Cole, Pacific Grove, CA. 514 p.

Ehlers, E. G. 1972. *The Interpretation of Geological Phase Diagrams*. Freeman, San Francisco. 280 p.

Ehlers, E. G., and Blatt, H. 1982. *Petrology: Igneous, Sedimentary, and Metamorphic*. Freeman, San Francisco. 732 p.

Eichelberger, J. C. 1995. Silicic volcanism: Ascent of viscous magma from crustal reservoirs. *Annual Review of Earth and Planetary Sciences*, v. 23, pp. 41–64.

Eichelberger, J. C., Chertkoff, D. G., Dreher, S. T., and Nye, C. J. 2000. Magmas in collision: Rethinking chemical zonation in silicic magmas. *Geology*, v. 28, pp. 603–606.

Elsasser, W. M. 1971. Sea-floor spreading as thermal convection. *Journal of Geophysical Research*, v. 76, pp. 1101–1112.

Emerman, S. H., Turcotte, D. L., and Spence, D. A. 1986. Transport of magma and hydrothermal solutions by laminar and turbulent fluid fracture. *Physics of the Earth and Planetary Interiors*, v. 41, pp. 249–259.

Engel, C. G., and Fisher, R. L. 1975. Granitic to ultramafic rock complexes of the Indian Ocean Ridge system, western Indian Ocean. *Geological Society of America Bulletin*, v. 86, pp. 1553–1578.

England, G. L., Rasmussen, B., Krapez, B., and Groves, D. I. 2001. The origin of uraninite, bitumen nodules, and carbon seams, in Witwatersrand gold-uranium-pyrite ore deposits, based on a permo-triassic analogue. *Economic Geology*, v. 96, pp. 1907–1930.

Enos, P., and Moore, C. H. 1983. Fore-reef slope environment. In Scholle, P. A. et al. (eds.), *Carbonate Depositional Environments*. AAPG Memoir 33, pp. 507–537.

Erickson, R. L., and Blade, L. V. 1963. *Geochemistry and Petrology of the Alkalic Igneous Complex of Magnet Core, Arkansas*. U.S. Geological Survey Professional Paper 425. 95 p.

Ernst, R. E., Buchan, K., and Campbell, I. H. 2005. Frontiers in large igneous province research. *Lithos*, v. 79, pp. 271–297.

Ernst, W. G. 1971. Metamorphic zonations on presumably subducted lithospheric plates from Japan, California and the Alps. *Contributions to Mineralogy and Petrology*, v. 34, pp. 43–59.

Ernst, W. G. (ed.). 2004. *Serpentine and Serpentinite: Mineralogy, Petrology, Geochemistry, and Tectonics*. Bellweather, Columbia, MD (for the Geological Society of America). 606 p.

Ernst, W. G., and McLaughlin, R. J. 2012. Mineral parageneses, regional architecture, and tectonic evolution of Franciscan metagraywackes, Cape Mendocino-Garberville-

Covelo 30′ × 60′ quadrangles, northwest California. *Tectonics*, v. 31, TC1001. doi:10.1029/2011TC002987.

Eskola, P. 1915. On the relations between the chemical and mineralogical composition in the metamorphic rocks of the Orijarvi region. *Commission Geologique Finlande Bulletin* (Bulletin of the Geological Society of Finland), no. 44, pp. 109–145.

Eskola, P. 1920. The mineral facies of rocks. *Norsk Geol. Tidsskrift*, v. 6, pp. 143–194.

Evans, A. M. 1997. *An Introduction to Economic Geology and Its Environmental Impact*. Blackwell, Malden, MA. 364 p.

Evans, B. W. 1977. Metamorphism of Alpine peridotite and serpentinite. *Annual Reviews of Earth Planetary Sciences*, v. 5, pp. 397–447.

Evans, B. W., Hattori, K., and Baronnet, A. 2013. Serpentinites: What, why, where? *Elements*, v. 9, pp. 99–106.

Evans, K. A., McCuaig, T. C., Leach, D., Angerer, T., and Hagemann, S. G. 2013. Banded iron formation to iron ore: A record of the evolution of Earth environments? *Geology*, v. 41; pp. 99–102. doi:10.1130/G33244.1.

Ewart, A. 1976. Mineralogy and chemistry of modern orogenic lavas: Some statistics and implications. *Earth and Planetary Science Letters*, v. 31, pp. 417–432.

Ewart, A. 1982. The mineralogy and petrology of tertiary-recent orogenic volcanic rocks, with special reference to the andesite-basaltic compositional range. In Thorpe, R. S. (ed.), *Andesites: Orogenic Andesites and Related Rocks*. Wiley, Chichester, UK, pp. 25–87.

Ewing, J. 1969. Seismic model of the Atlantic Ocean. In Hart, P. J. (ed.), *The Earth's Crust and Upper Mantle Structure, Dynamic Processes, and Their Relation to Deep-Seated Geological Phenomena*. American Geophysical Union Geophysical Monograph 13, pp. 220–225. doi:10.1029/GM013p0220.

F

Fan, D., Liu, T., and Ye, J. 2006. The process of formation of manganese carbonate deposits hosted in black shale series. *Economic Geology*, v. 87, pp. 1419–1429.

Farmelo, G. (ed.). 2002. *It Must Be Beautiful: Great Equations of Modern Science*. Granta, London. 283 p.

Faure, G. 1986. *Principles of Isotope Geology* (2nd ed.). Wiley, New York. 608 p.

Faure, G. 1998. *Principles and Applications of Geochemistry* (2nd ed.). Prentice-Hall, Upper Saddle River, NJ. 600 p.

Fenn, P. M. 1986. On the origin of graphic granite. *American Mineralogist*, v. 71, pp. 325–330.

Ferry, J. M., Sorenson, S. S., and Rumble, D. 1998. Structurally controlled fluid flow during contact metamorphism in the Ritter Range Pendant, California, USA. *Contributions to Mineralogy and Petrology*, v. 130, pp. 358–378.

Festa, A., Dilek, Y., Pini, G. A., Codegone, G., and Ogata, K. 2012. Mechanisms and processes of stratal disruption and mixing in the development of mélanges and broken formations: Redefining and classifying mélanges. *Tectonophysics*, v. 568–569, pp. 7–24.

Fetter, C. W. 1994. *Applied Hydrogeology* (3rd ed.). Prentice Hall, Englewood Cliffs, NJ. 691 p.

Fettes, D., and Desmons, J. (eds.). 2007. *Metamorphic Rocks: A Classification and Glossary of Terms*. Cambridge University Press, New York. 244 p.

Fodor, R. V., and Vandermeyden, H. J. 1988. Petrology of gabbroic xenoliths from Mauna Kea Volcano, Hawaii. *Journal of Geophysical Research*, v. 93, pp. 4435–4452.

Folk, R. L. 1974. *Petrology of Sedimentary Rocks*. Hemphill, Austin. 182 p. (Other editions published in 1965 and 1980).

Forsman, J. P., and Hunt, J. M. 1958. Insoluble organic matter (kerogen) in sedimentary rocks. *Geochimica et Cosmochimica Acta*, v. 15, pp. 170–182.

Fralick, P., and Barrett, T. J. 1995. Depositional controls on iron formation associations in Canada. In Plint, A. G. (ed.), *Sedimentary Facies Analysis*. International Association of Sedimentologists, Special Publication No. 22. Blackwell, Oxford, UK, pp. 137–156.

Franklin, J. M., Lydon, J. W., and Sangster, D. F. 1981. Volcanic-associated massive sulfide deposits. In Skinner, B. J. (ed.), *Economic Geology 75th Anniversary Volume*, pp. 485–627.

Fujii, T., and Kushiro, I. 1977. Density, viscosity, and compressibility of basaltic liquid at high pressures. *Carnegie Institution of Washington Yearbook* 76, pp. 419–424.

G

Gaetani, G. A., and Grove, T. L. 2003. Experimental constraints on melt generation in the mantle wedge. In Eiler, J. (ed.), *Inside the Subduction Factory*. American Geophysical Union Geophysical Monograph 138, pp. 107–134. doi:10.1029/138GM07.

Gaines, R. V., Skinner, H. C., Foord, E., Mason, B., Rosenwieg, A., King, V. T., and Dowty, E. 1997. *Dana's New Mineralogy*. Wiley, New York. 1819 p.

Galloway, W. F., and Hobday, D. K. 1983. *Terrigenous Clastic Depositional Systems: Applications to Petroleum, Coal, and Uranium Exploration*. Springer-Verlag, New York. 423 p.

Gansecki, C. A., Mahood, G. A., and McWilliams, M. O. 1998. New ages for the climactic eruptions of Yellowstone: Single-crystal 40AR/39AR dating identifies contamination. *Geology*, v. 26, pp. 343–346.

Garrels, R. M., and Christ, C. L. 1965. *Solutions, Minerals, and Equilibria*. Harper and Row, New York. 450 p.

Garrett, A. B., Lippincott, W. T., and Verhoek, F. H. 1972. *Chemistry: A Study of Matter* (2nd ed.). Xerox, Lexington, MA. 674 p.

Gibson, H. L. Allen, R. L., Riverin, G., and Lane, T. E. 2007. The VMS Model: Advances and application to exploration targeting. In Milkereit, B. (ed.), *Proceeding of Exploration 07: Fifth Decennial International Conference on Mineral Exploration*, Paper 49, pp. 713–730.

Gilbert, G. K., 1880. *Report on the Geology of the Henry Mountains* (2nd ed.). U.S. Government Printing Office, Washington, DC. 170 p.

Gill, R. 1996. *Chemical Fundamentals of Geology.* Chapman and Hall, London. 290 p.

Girarddeau, J., Marcoux, J., Forucade, E., Bassoullet, J. P., and Tang, Y. K. 1985. Xainxa ultramafic rocks, central Tibet, China: Tectonic environment and geodynamic significance. *Geology*, v. 13, pp. 330–333.

Girardeau, J., and Mercier, J.-C. C. 1988. Petrology and texture of the ultramafic rocks of the Xigaze ophiolite (Tibet): Constraints for mantle structure beneath slow-spreading ridges. *Tectonophysics*, v. 147, pp. 33–58.

Glazner, A. F., and Bartley, J. M. 2006. Is stoping a volumetrically significant pluton emplacement process? *Geological Society of America Bulletin*, v. 118, pp. 1185–1195. doi:10.1130/B25738.1.

Glazner, A. F., Bartley, J. M., Coleman, D. S., Gray, W., and Taylor, R. Z. 2004. Are plutons assembled over millions of years by amalgamation from small magma chambers? *GSA Today*, v. 14, pp. 4–11. doi:10.1130/1052-5173(2004)014<0004:APAOMO>2.0.CO;21996.

Goldich, S. S. 1938. A study in rock-weathering. *Journal of Geology*, v. 46, pp. 17–58.

Goldschmidt, V. M. 1923. Geochemical laws of the distribution of the elements. *Norske videnskaps-akademi i Oslo. Matematisk-naturvidensapelig klasse*, v. 2, 117 p.

Goodenough, K, M., Thomas, R. J., Styles, M. T., Schofield, D. I., and MacLeod, C. J. 2014. Records of ocean growth and destruction in the Oman-UAE Ophiolite. *Elements*, v. 10, pp. 109–114.

Gorczyk, W., Gerya, T. V., Connolly, J. A. D., and Yuen, D. A. 2007. Growth and mixing dynamics of mantle wedge plumes. *Geology*, v. 35, pp. 587–590. doi:10.1130/G23485A.1.

Greeley, R. 1982. The Snake River Plain, Idaho: Representative of a new category of volcanism. *Journal of Geophysical Research*, v. 87, pp. 2705–2712.

Green, H. W., II, and Jung, H. 2005. Fluids, faulting, and flow. *Elements*, v. 1, pp. 31–37.

Guilbert, J. M., and Park, C F. 1986. *The Geology of Ore Deposits.* Waveland, Long Grove, IL. 985 p.

Gurney, J. J., Helmstaedt, H. H., LeRoex, A. P., Nowicki, T. E., Richardson, S. H., and Westerlund, K. J. 2005. Diamonds: Crustal distribution and formation processes in time and space and an integrated deposit model. In Hedenquist, J. W., Thompson, J. F. H, Goldfarb, R. J., and Richards, J. P. (eds.), *Economic Geology 100th Anniversary Volume*, pp. 143–147.

Gutzmer, J., and Beukes, N. J. 1995. Fault controlled metasomatic alteration of Early Paleozoic sedimentary manganese ores in the Kalahari manganese field, South Africa. *Economic Geology*, v. 90, pp. 823–844.

H

Habib, P. 1982. *An Outline of Soil and Rock Mechanics.* Cambridge University Press, Cambridge, UK. 149 p.

Hagemann, S., Rosiére, C. A., Gutzmer, J., and Beukes, N. J. (eds.). 2008. Banded iron formation-related high-grade iron ore. *Reviews in Economic Geology*, v. 15, Society of Economic Geologists, pp. 73–106.

Hamilton, W. B., and Myers, W. B. 1967. *The Nature of Batholiths.* U.S. Geological Survey Professional Paper 554-C. 30 p.

Hamilton, W. B., and Myers, W. B. 1974. The nature of the Boulder Batholith of Montana. *Geological Society of America Bulletin*, v. 85, 365378.

Hammond, C. 1997. *The Basics of Crystallography and Diffraction.* International Union of Crystallography, Oxford University Press, Oxford, UK. 249 p.

Hardie, L. A. 1990. The roles of rifting and hydrothermal $CaCl_2$ brines in the origin of potash evaporates: An hypothesis. *American Journal of Science*, v. 290, pp. 43–106.

Hargraves, R. B. (ed.). 1980. *Physics of Magmatic Processes.* Princeton University Press. 585 p.

Harker, A. 1932. *Metamorphism. A Study of the Transformations of Rock-masses.* Meuthen, London. 360 p.

Harlow, G. E., and Davies, R. M. 2005. Diamonds. *Elements*, v. 1, no. 2, pp. 67–70.

Harper, B. E., Miller, C. F., Koteas, G. C., Cates, N. L., Wiebe, R. A., Lazzareschi, D. S., and Cribb, J. W. 2004. Granites, dynamic magma chamber processes and pluton construction: The Aztec Wash pluton, Eldorado Mountains, Nevada, USA. *Transactions of the Royal Society of Edinburgh: Earth Sciences*, v. 95, pp. 277–295.

Harper, G. D. 1984. The Josephine ophiolite, northwestern California. *Geological Society of America Bulletin*, v. 95, pp. 1009–1026.

Harte, B., and Hudson, N. F. C. 1979. Pelite facies series and the temperatures and pressures of Dalradian metamorphism in E. Scotland. In Harris, A. C., Holland, C. H., and Leake, B. E. (eds.), *The Caledonides of the British Isles—Reviewed.* Geological Society (London) Special Publication No. 8, Scottish Academic Press Ltd., Edinburgh, pp. 323–337.

Hawthorne, F. C. (ed.). 1988. *Spectroscopic Methods in Mineralogy and Geology.* In *Reviews in Mineralogy 18*, Mineralogical Society of America, Washington, DC. 698 p.

Hawthorne, J. B. 1975. Model of a kimberlite pipe. First International Kimberlite Conference. *Physics and Chemistry of the Earth*, v. 9, pp. 1–16.

Heath, R. C., and Trainer, F. W. 1981. *Introduction to Ground Water Hydrology.* Water Well Journal Publishing Company, Worthington, OH. 283 p.

Heckel, P. H. 1977. Origin of phosphatic black shale facies in Pennsylvanian cyclothems of mid-continent North America. *American Association of Petroleum Geologists Bulletin*, v. 61, pp. 1045–1068.

Heiße, F., Köhler-Langes, F., Rau, S., Hou, J., Junck, S., Kracke, A., Mooser, A., et al. 2017. High-precision measurement of the proton's atomic mass. *Physical Review Letters*, v. 119, p. 033001. doi:10.1103/PhysRevLett.119.033001.

Hekinian, R., Moore, J. G., and Bryan, W. B. 1976. Volcanic rocks and processes of the Mid-Atlantic Ridge Rift Valley near 36°49′ N. *Contributions to Mineralogy and Petrology*, v. 58, pp. 83–110.

Helmke, P. A. 2000. The chemical composition of soils. In Sumner, M. E. (ed.), *Handbook of Soil Science*. CRC Press, Boca Raton, pp. B3–B24.

Henderson, A. L., Najman, Y., Parrish, R., BouDagher-Fadel, M., Barford, D., Garzanti, E., and Andò, S. 2010. Geology of the Cenozoic Indus Basin sedimentary rock: Paleoenvironmental interpretation of sedimentation from the western Himalaya during the early phases of India–Eurasia collision. *Tectonics*, v. 29, TC6015, doi:10.1029/2009TC002651.

Heron, S. D., Moslow, T. F., Berelson, W. M., Herber, J. R., Steele, G. A., III, and Susman, K. R. 1984. Holocene sedimentation of a wave-dominated barrier-island shoreline: Cape Lookout, North Carolina. *Marine Geology*, v. 60, pp. 413–434.

Higgins, M. W. 1971. *Cataclastic Rocks*. U.S. Geological Survey Professional Paper 687. 97 p.

Hildebrand, R. S., Hoffman, P. F., Housh, T., and Bowring, S. A. 2010. The nature of volcano-plutonic relations and the shapes of epizonal plutons of continental arcs as revealed in the Great Bear magmatic zone, northwestern Canada. *Geosphere*, v. 6, pp. 812–839.

Hildreth, W. 1981. Gradients in silicic magma chambers: Implications for lithospheric magmatism. *Journal of Geophysical Research*, v. 86, pp. 10153–10192.

Hildreth, W., Christiansen, R. L., and O'Neil, J. R. 1984. Catastrophic isotopic modification of rhyolitic magma at times of caldera subsidence, Yellowstone Plateau Volcanic Field. *Journal of Geophysical Research*, v. 89, pp. 8339–8369.

Hildreth, W., Halliday, A. N., and Christiansen, R. L. 1991. Isotopic and chemical evidence concerning the genesis and contamination of basaltic and rhyolitic magma beneath the Yellowstone Plateau Volcanic Field. *Journal of Petrology*, v. 32, pp. 63–138.

Hofmann, H. J. 1973. Stromatolites: Characteristics and utility. *Earth-Science Reviews*, v. 9, pp. 339–373.

Holdaway, M. J., and Mukhopadhyay, B. 1993. A reevaluation of the stability relations of andalusite: Thermochemical data and phase diagram for the aluminum silicates. *American Mineralogist*, v. 78, pp. 298–315.

Holtz, F., Johannes, W., Tamic, N., and Behrens, H. 2001. Maximum and minimum water contents of granitic melts generated in the crust: A reevaluation and implications. *Lithos*, v. 56, pp. 1–14.

Hsu, K. J. 1968. The principles of mélanges and their bearing on the Franciscan-Knoxville paradox. *Geological Society of America Bulletin*, v. 79, pp. 1063–1074.

Huang, W. H., and Kiang, W. C. 1972. Laboratory dissolution of plagioclase feldspars in water and organic acids at room temperature. *American Mineralogist*, v. 57, pp. 1849–1859.

Huebner, J. S., and Flohr, M. J. 1990. *Microbanded Manganese Formations: Protoliths in the Franciscan Complex, California*. U.S. Geological Survey Professional Paper 1502. 72 p.

Hunt, C. B., Averitt, P., and Miller, R. 1953. *Geology and Geography of the Henry Mountains Region, Utah*. U.S. Geological Survey Professional Paper 228. 234 p.

Hurlbut, C., Jr. 1961. *Mineralogy* (17th ed.). John Wiley, New York. 609 p.

Hurlbut, C., Jr., and Klein, C. 1977. *Manual of Mineralogy* (19th ed.). Wiley, New York. 532 p.

Hutko, A. R., Lay, T., Garnero, D. J., and Revenaugh, J. 2006. Seismic detection of folded, subducted lithosphere at the core-mantle boundary. *Nature*, v. 441, pp. 333–336. doi:10.1038/nature04757.

Hyman M. E., Johnson C. E., Bailey S. W., April R. H., and Hornbeck, J. W. 1998. Chemical weathering and cation loss in a base-poor watershed. *Geological Society of America Bulletin*, v. 110, pp. 85–95.

Hyndman, D. W. 1985. *Petrology of Igneous and Metamorphic Rocks* (2nd ed.). McGraw-Hill, New York. 786 p.

I

Isacks, B., Oliver, J., and Sykes, L. R. 1968. Seismology and the new global tectonics. *Journal of Geophysical Research*, v. 73, pp. 5855–5899.

J

Jackson, J. A. 1997. *Glossary of Geology* (4th ed.). American Geological Institute, Alexandria, VA. 769 p.

Jahns, R. H. 1955. The study of pegmatites. In Bateman, A. M. (ed.), *Economic Geology 50th Anniversary Volume*, pp. 1025–1130.

Jain, A. K., Lal, N., Sulemani, B., Awasthi, A. K., Singh, S., Kumar, R., and Kumar, D. 2009. Detrital-zircon fission-track ages from the Lower Cenozoic sediments, NW Himalayan foreland basin: Clues for exhumation and denudation of the Himalaya during the India-Asia collision. *Geological Society of America Bulletin*, v. 121, pp. 519–535. doi:10.1130/B26304.1.

James, H. L. 1983. Reef environment. In Scholle, P. A. et al. (eds.), *Carbonate Depositional Environments*, AAPG Memoir 33, pp. 345–440.

James, H. L. 1984. Reefs. In Walker, R. G. (ed.), *Facies Models* (2nd ed.). Geological Association Canada–Geoscience Canada Reprint Series 1, pp. 229–244.

James, N. P. 1997. The cool-water carbonate depositional realm. In James, N. P., and Clarke, J. L. (eds.), *Cool-water Carbonates*. Society of Economic Mineralogists and Paleontologists Special Publication 56, pp. 1–20.

Jensen, M. L., and Bateman, A. M. 1979. *Economic Mineral Deposits*. Wiley, New York. 593 p.

Johnson, H. D., and Baldwin, C. T. 1986. Shallow siliciclastic seas. In Reading, H. G. (ed.), *Sedimentary Environments and Facies* (2nd ed.). Blackwell, Oxford, UK, pp. 229–282.

Johnson, H. D., and Baldwin, C. T. 1996. Shallow clastic seas. In H. G. Reading (ed.), *Sedimentary Environments: Processes, Facies and Stratigraphy*. Blackwell, Oxford, UK, pp. 236–286.

Johnson, N. E. 2001. X-Ray diffraction simulation using laser pointers and printers. *Journal of Geoscience Education*, v. 49, pp. 346–350.

Jones, P. R. 2001. Review of *Gold aus dem Meer*. *Bulletin of the History of Chemistry*, v. 26, No. 2, pp. 136–137.

K

Karason, H., and van der Hilst, R. D. 2000. Constraints on mantle convection from seismic tomography. In Richards, M. R., Gordon, R., and van der Hilst, R. D. (eds.), *The History and Dynamics of Global Plate Motion*. American Geophysical Union Geophysical Monograph 212, pp. 277–288.

Karig, D. E. 1980. Material transport within accretionary prisms and the "knocker" problem. *Journal of Geology*, v. 88, pp. 27–39.

Kay, B. D., and Angers, D. A. 2000. Soil structure. In Sumner, M. E. (ed.), *Handbook of Soil Science*. CRC Press, Boca Raton, pp. A229–A276.

Kelemen, P. B., Shimizu, N., and Salters, V. J. M. 1995. Extraction of mid-ocean ridge basalt from the upwelling mantle by focused flow of melt in dunite channels. *Nature*, v. 375, pp. 747–753.

Keller, E. A. 2000. *Environmental Geology*. Prentice Hall, Upper Saddle River, NJ. 562 p.

Kelley, S. P., and Wartho, J. A. 2000. Rapid kimberlite ascent and the significance of Ar-Ar ages in xenolith phlogopites. *Science*, v. 289, pp. 609–611.

Kesler, S. E. 2005. Ore-forming fluids. *Elements*, v. 1, pp. 13–18.

Kimura, J-I., Hacker, B. R., van Keken, P. E., Kawabata, H., Yoshida, T., and Stern, R. J. 2009. Arc Basalt Simulator version 2, a simulation for slab dehydration and fluid-fluxed mantle melting for arc basalts: Modeling scheme and application. *Geochemistry, Geophysics, Geosystems*, v. 10, pp. 1525–2027. doi:10.1029/2008GC002217.

King, J. 1961. A new evaluation of hardness standards. *Advanced Materials Technology*, Fall, pp. 7–11.

Kirkpatrick, J. D., and Rowe, C. D. 2013. Disappearing ink: How pseudotachylytes are lost from the rock record. *Journal of Structural Geology*, v. 52, pp. 183–198.

Kirkpatrick, R. J. 1975. Crystal growth from the melt: A review. *American Mineralogist*, v. 60, pp. 798–814.

Kirkpatrick, R. J. 1981. Kinetics of crystallization of igneous rocks. In Lasaga, A. C., and Kirkpatrick, R. J. (eds.), *Kinetics of Geochemical Processes*. In *Reviews in Mineralogy* 8, Mineralogical Society of America, pp. 321–398.

Kjarsgaard, B. A. 2007. Kimberlite pipe models: Significance for exploration. In Milkereit, B. (ed.), *Proceedings of Exploration 07*, Paper 46, Fifth Decennial International Conference on Mineral Exploration, pp. 667–677.

Klein, C. 2005. Some Precambrian banded iron-formations (BIFs) from around the world: Their age, geologic setting, mineralogy, metamorphism, geochemistry, and origins. *American Mineralogist*, v. 90, pp. 1473–1499.

Klein, C. K. 2002. *Manual of Mineral Science* (22nd ed.). Wiley, New York. 641 p.

Klein, C. K., and Dutrow, B. 2008. *Manual of Mineral Science* (23rd ed.). Wiley, New York. 675 p.

Knoop, F., Peters, C. G., and Emerson, W. B. 1939. A sensitive pyramidal diamond tool for indentation measurements. *Journal of Research of the National Bureau of Standards*, v. 23, pp. 39–61.

Koren, I., Kaufman, Y. J., Washington, R., Todd, M. C., Rudich, Y., Martins, J. V., and Rosenfeld, D. 2006. The Bodélé Depression: A single spot in the Sahara that provides most of the mineral dust to the Amazon forest. *Environmental Research Letters*, v. 1, pp. 014–015. doi:10.1088/1748-9326/1/1/014005.

Koster van Groos, A. F. 1988. Weathering, the carbon cycle, and the differentiation of the continental crust and mantle. *Journal of Geophysical Research*, v. 93, pp. 8952–8958.

Koyanagi, R. Y., and Endo, E. T. 1971. *Hawaiian Seismic Events during 1969*. U.S. Geological Survey Professional Paper 750-C, pp. C158–C164.

Kraus, M. J. 1999. Paleosols in clastic sedimentary rocks: Their geologic applications. *Earth-Science Reviews*, v. 47, pp. 41–70.

Krauskopf, K. B. 1967. *Introduction to Geochemistry*. McGraw-Hill, New York. 721 p.

Krauskopf, K. B. 1979. *Introduction to Geochemistry* (2nd ed.). McGraw-Hill, New York. 617 p.

Krauskopf, K. B., and Bird, D. K. 1995. *Introduction to Geochemistry* (3rd ed.). McGraw-Hill, New York. 646 p.

Krueger, S. W., and Jones, D. L. 1989. Extensional fault uplift of regional Franciscan blueschists due to subduction shallowing during the Laramide orogeny. *Geology*, v. 17, pp. 1157–1159.

Krumbein, W. C. 1934. Size frequency distributions of sediments. *Journal of Sedimentary Petrology*, v. 4, pp. 65–77.

Krumbein, W. C., and Garrels, R. M. 1952. Origin and classification of chemical sediments in terms of pH and oxidation-reduction potentials. *Journal of Geology*, v. 60, pp. 1–33.

Krushchov, M. M. 1949. On the introduction of a new hardness scale. *Zavodskaia Laboratoriia*, v. 15, pp. 213–217.

Krynine, P. D. 1948. The megascopic study and field classification of sedimentary rocks. *Journal of Geology*, v. 56, pp. 130–165.

Kuleshov, V. N. 2012. A superlarge deposit—Kalahari manganese ore field (Northern Cape, South Africa): Geochemistry of isotopes ($\delta^{13}C$ and $\delta^{18}O$) and genesis. *Lithology and Mineral Resources*, v. 47, pp. 217–233.

Kump, L. R., Kasting, J. F., and Crane, R. G. 1999. *The Earth System*. Prentice Hall, Upper Saddle River, NJ. 351 p.

Kushiro, I. 1973. Origin of some magmas in oceanic and circum-oceanic regions. *Tectonophysics*, v. 17, pp. 211–222.

Kushiro, I. 1980. Viscosity, density, and structure of silicate melts at high pressures, and their petrological applications. In Hargraves, R. B. (ed.), *Physics of Magmatic Processes*. Princeton University Press, Princeton, NJ, pp. 93–120.

Kushiro, I. 1983. On the lateral variations in chemical composition and volume of Quaternary volcanic rocks across Japanese arcs. *Journal of Volcanology and Geothermal Research*, v. 18, pp. 435–447.

Kushiro, I. 2007. Origin of magmas in subduction zones: A review of experimental studies. *Proceedings of the Japanese Academy, Series B, Physical and Biological Sciences*, v. 83, pp. 1–15. PMC3756732.

L

Lang, H. M., and Rice, J. M. 1985. Metamorphism of pelitic rocks in the Snow Peak area, northern Idaho: Sequence of events and regional implications. *Geological Society of America Bulletin*, v. 96, pp. 731–736.

Langmuir, D. 1978. Uranium solution-mineral equilibria at low temperatures with applications to sedimentary ore deposits. In Kimberly, M. M. (ed.), *Uranium Deposits, Their Mineralogy and Origin*. Mineralogical Association of Canada Short Course Handbook, v. 3, pp. 17–55.

Laporte, D., and Watson, E. B. 1995. Experimental and theoretical constraints on melt distribution in crustal sources: The effect of crystalline anisotropy on melt interconnectivity. *Chemical Geology*, v. 124, pp. 161–184.

Lasaga, A. C. 1990. Atomic treatment of mineral-water surface reactions. In Hochella, M. F. Jr., and White, A. F. (eds.), *Mineral-water Interface Geochemistry*. In *Reviews in Mineralogy* v. 23, pp. 17-85.

Law, J. D. M., and Phillips, G. N. 2005. Hydrothermal replacement model for Witwatersrand gold. In Hedenquist, J. W., Thompson, J. F. H, Goldfarb, R. J., and Richards, J. P. (eds.), *Economic Geology 100th Anniversary Volume*, Society of Economic Geologists, pp. 799–811.

Leach, D. L., Sangster, D. F., Kelley, K. D., Large, R. R., Garven, G., Allen, C. R., Gutzmer, J., and Walters, S. 2005. Sediment-hosted lead-zinc deposits: A global perspective. In Hedenquist, J. W., Thompson, J. F. H, Goldfarb, R. J., and Richards, J. P. (eds.), *Economic Geology 100th Anniversary Volume*, pp. 561–607.

LeBas, M. J., LeMaitre, R. W., Streckeisen, A., and Zanettin, B. 1986. A chemical classification of volcanic rocks based on the total alkali-silica diagram. *Journal of Petrology*, v. 27, pp. 745–750.

Leeder, M. 1999. *Sedimentology and Sedimentary Environments*. Blackwell, London. 602 p.

Leet, L. D., and Judson, S. 1958. *Physical Geology* (2nd ed.). Prentice-Hall, Englewood Cliffs, NJ. 502 p.

Leopold, L. B. 1997. *Water, Rivers and Creeks*. University Science Books, Sausalito, CA. 185 p.

Le Pichon, X., 1968. Sea-floor spreading and continental drift. *Journal of Geophysical Research*, v. 73, pp. 3661–3697.

Le Pichon, X., Francheteau, J., and Bonnin, J. 1973. *Plate Tectonics*. Elsevier, New York. 300 p.

Lewis, D. W. 1984. *Practical Sedimentology*. Hutchinson Ross, Stroudsburg, PA. 227 p.

Li, J., Lowenstein, T. K., and Blackburn, I. R. 1997. Responses of evaporite mineralogy to inflow water sources and climate during the past 100 k.y. in Death Valley, California. *Geological Society of America Bulletin*, v. 109, pp. 1361–1371.

Lide, D. R. (ed.) 2008. *CRC Handbook of Chemistry and Physics* (88th ed.). Taylor and Francis, Boca Raton, FL. 2640 p.

Lillie, R. J. 1999. *Whole Earth Geophysics*. Prentice Hall, Upper Saddle River, NJ. 361 p.

Lindgren, W. 1913. *Mineral Deposits*. McGraw-Hill, New York. 883 p.

Liou, J. G. 2000. Petrotectonic summary of less intensively studied UHP regions. In Ernst, W. G., and Liou, J. G. (eds.), *Ultra-high Pressure Metamorphism and Geodynamics in Collision-type Orogenic Belts*. Bellweather, Columbia, MD, pp. 20–35.

Liou, J. G., Maruyama, S., and Cho, M. 1987. Very low-grade metamorphism of volcanic and volcaniclastic rocks: Mineral assemblages and mineral facies. In Frey, M. (ed.), *Low Temperature Metamorphism*. Blackie, Glasgow, pp. 59–113.

Liou, J. G., Tsugimori, T., Zhang, R. Y., Katayama, I., and Maruyama, S. 2006. Global UHP metamorphism and continental subduction/collision: The Himalayan model. In J. G. Liou and M. Cloos (eds.), *Phase Relations, High-pressure Terranes, P-T-ometry, and Plate Pushing: A Tribute to W. G. Ernst*. The Geological Society of America, International Book Series, v. 9, pp. 53–79.

Lister, J. R., and Kerr, R. C. 1991. Fluid-mechanical models of crack propagation and their application to magma transport in dykes. *Journal of Geophysical Research*, v. 96, pp. 10,049–10,077.

Lofgren, G. 1980. Experimental studies on the dynamic crystallization of silicate melts. In Hargraves, R. B. (ed.), *Physics of Magmatic Processes*. Princeton University Press, Princeton, NJ, pp. 487–551.

Lomas, S. A., and Joseph, P. (eds.). 2004. *Confined Turbidite Systems*. The Geological Society of London Special Publication 222. 328 p.

London, D., and Kontak, D. J. 2012. Granitic pegmatites: Scientific wonders and economic bonanzas. *Episodes*, v. 8, pp. 257–261.

London, D., and Morgan, G. B. 2012. The pegmatite puzzle. *Elements*, v. 8, pp. 263–268.

Longman, M. W. 1980. Carbonate diagenetic textures from near surface diagenetic environments. *AAPG Bulletin*, v. 64, pp. 461–487.

Longman, M. W. 1981. A process approach to recognizing facies of reef complexes. In Toomey, D. F. (ed.), *European Fossil Reef Models*. SEPM Special Publication No. 30, pp. 9–40.

Longman, M. W. 1982. Carbonate diagenesis as a control of stratigraphic traps with examples from the Williston Basin. *American Association of Petroleum Geologists Continuing Education Course Notes Series*, v. 21. 159 p.

Lowe, D. R. 1976. Subaqueous liquefied and fluidized sediment flows and their deposits. *Sedimentology*, v. 23, pp. 285–308.

Lowell, J. D., and Guilbert, J. M. 1970. Lateral and vertical alteration-mineralization zoning in porphyry ore deposits. *Economic Geology*, v. 65, pp. 373–408.

Luttge, A., Arvidson, R. S., and Fischer, C. 2013. A stochastic treatment of crystal dissolution kinetics. *Elements*, v. 9, pp. 183–188.

Lydon, J. W. 1988. Volcanogenic massive sulphide deposits. Part 2: Genetic models. In Roberts, R. G., and Sheahan, P. A. (eds.), *Ore Deposit Models*. Geoscience Canada Reprint Series 3, pp. 155–181.

M

Maaloe, S. 1981. Magma accumulation in the ascending mantle. *Journal of the Geological Society* (London), v. 138, pp. 223–236.

MacGregor, I. D. 1968. Mafic and ultramafic inclusions as indicators of the depth of origin of basaltic magmas. *Journal of Geophysical Research*, v. 73, pp. 3737–3745.

MacGregor, I. D. 1974. The system $MgO-Al_2O_3-SiO_2$: Solubility of Al_2O_3 in enstatite for spinel and garnet peridotite compositions. *American Mineralogist*, v. 59, pp. 110–119.

MacKenzie, W. S., Donaldson, C. H., and Guilford, C. 1982. *Atlas of Igneous Rocks and Their Textures*. Wiley, New York. 148 p.

Manheim, F., Rowe, G. T., and Jipa, D. 1975. Marine phosphorite formation off Peru. *Journal of Sedimentary Petrology*, v. 45, pp. 243–251. doi:10.1306/212F6D20-2B24-11D7-8648000102C1865D.

Manheim, F. T., and Gulbrandsen, R. A. 1979. Marine phosphorites. In Burns, R. G. (ed.), *Marine Minerals*. Mineralogical Society of America, *Reviews in Mineralogy*, v. 6, pp. 151–173.

Mantle, G. W., and Collins, W. J. 2008. Quantifying crustal thickness variations in evolving orogens: Correlation between arc basalt composition and moho depth. *Geology*, v. 36, pp. 87–90. doi:10.1130/G24095A.1.

Marsh, B. D. 1979. Island-arc volcanism. *American Scientist*, v. 67, pp. 161–172.

Marsh, B. D. 1984. Reply (to Comment: On the mechanics of igneous diapirism, stoping, and zone melting). *American Journal of Science*, v. 284, pp. 981–984.

Marsh, B. D., and Kantha, L. H. 1978. On the heat and mass transfer from an ascending magma. *Earth and Planetary Science Letters*, v. 39, pp. 435–443.

Marshall, T. J., Holmes, J. W., and Rose, C. W. 1996. *Soil Physics* (3rd ed.). Cambridge University Press, New York. 453 p.

McBirney, A. R., and White, C. M. 1982. The Cascade Province. In Thorpe, R. S. (ed.), *Andesites: Orogenic Andesites and Related Rocks*. Wiley, Chichester, UK, pp. 115–135.

McKelvey, V. E. 1972. Mineral resource estimates and public policy. *American Scientist*, v. 60, pp. 32–40.

McKelvey, V. E., Wright, N. A., and Bowen, R. W. 1983. *Analysis of the World Distribution of Metal-Rich Subsea Manganese Nodules*. U.S. Geological Survey Circular 886. 55 p.

McKinney, F. K. 2007. *The Northern Adriatic Ecosystem: Deep Time in a Shallow Sea*. Columbia University Press, New York. 299 p.

McKinney, F. K., and Gault, H. W. 1980. Paleoenvironment of Late Mississippian fenestrate bryozoans, eastern United States. *Lethaia*, v. 13, pp. 127–146.

McLeish, A. 1992. *Geological Science*. Nelson Thornes, Cheltenham, UK. 312 p.

McLellan, E. L. 1985a. Metamorphic reactions in the kyanite and sillimanite zones of the Barrovian Type area. *Journal of Petrology*, v. 26, pp. 789–818.

McLellan, E. L. 1985b. Staurolite breakdown and the formation of kyanite and sillimanite in the Barrovian type area (Abs.). *Journal of the Geological Society* (London), v. 142, p. 5.

McManus, D. A. 1963. A criticism of certain usage of the Phi-notation. *Journal of Sedimentary Petrology*, v. 33, pp. 670–674.

McMillan, W. J., and Panteleyev, A. 1988. Porphyry copper deposits. In Roberts, R. G., and Sheahan, P. A. (eds.), *Ore Deposit Models*. Geoscience Canada Reprint Series 3, pp. 45–58.

McNulty, B. A., Tobisch, O. T., Cruden, A. R., and Gilder, S. 2000. Multistage emplacement of the Mount Givens pluton, central Sierra Nevada batholith, California. *Geological Society of America Bulletin*, v. 112, pp. 119–135.

Mehnert, K. R. 1968. *Migmatites and the Origin of Granitic Rocks*. Elsevier, Amsterdam. 393 p.

Merritt, F. S. (ed.). 1986. *Civil Engineering Reference Guide*. McGraw-Hill, New York. 608 p.

Merritt, F. S., and Gardner, W. S. 1986. Geotechnical engineering. In Merritt, F. S. (ed.), *Civil Engineering Reference Guide*. McGraw-Hill, New York, pp. 6-1–6-61.

Merschat, A. J., Hatcher, R. D., Jr., and Davis, T. L. 2005. The northern Inner Piedmont, southern Appalachians,

USA: Kinematics of transpression and SW-directed mid-crustal flow. *Journal of Structural Geology*, v. 27, pp. 1252–1281.

Merschat, A. J., Hatcher, R. D., Jr., Bream, B. R., Miller, C. F., Byars, H. E., Gatewood, M. P., and Wooden, J. L. 2010. Detrital zircon geochronology and provenance of southern Appalachian Blue Ridge and Inner Piedmont crystalline terranes. In Tollo, R. P., Bartholomew, M. J., Hibbard, J. P., and Karabinos, P. M. (eds.), *From Rodinia to Pangea: The Lithotectonic Record of the Appalachian Region*. Geological Society of America Memoir 206, pp. 661–699. doi:10.1130/2010.1206(26).

Metcalf, R., Rochelle, C. A., Savage, D., and Higgo, J. 1994. Fluid-rock interactions during continental red bed diagenesis: Implications for theoretical models of mineralization in sedimentary basins. In Parnell, J. (ed.), *Geofluids: Origin, Migration and Evolution of Fluids in Sedimentary Basins*. Geological Society Special Publications v. 78, pp. 301–324.

Metcalf, R. V., and Shervais, J. W. 2008. Suprasubduction zone ophiolites: Is there really an ophiolite conundrum? In Wright, J. E., and Shervais, J. W. (eds.), *Ophiolites, Arcs, and Batholiths: A Tribute to Cliff Hopson*. Geological Society of America Special Paper 438, pp. 191–222.

Metz, P., and Puhan, D. 1971. Korrektur zur Arbeit, Experimentelle untersuchung der metamorphose von kieselig dolomitischen sedimenten: Part 1, Die gleichgewichtsdaten der reaktion 3 dolomit + 4 quarz + 1 H_2O 1 talk + 3 calcit + 3 CO_2 fur die Gesamtgasdrucke von 1000, 3000 und 5000 bar. *Contributions to Mineralogy and Petrology*, v. 31, pp. 169–170.

Miller, C. F., and Wark, D. A. 2008. Supervolcanoes and their explosive supereruptions. *Elements*, v. 4, pp. 11–16.

Miller, W., III, Webb, F., Jr., and Raymond, L. A. 2009. Clustering and morphologic variation in *Arthrophycus alleghaniensis* (Lower Silurian of Virginia, USA) as evidence of behavioral paleoecology. *Neues Jarbuch Fur Geologie und Palaontologic-Abhandlungen*, v. 251, no. 1, pp. 109–117.

Miyashiro, A. 1961. Evolution of metamorphic belts. *Journal of Petrology*, v. 2, pp. 277–311.

Miyashiro, A. 1973. The troodos ophiolitic complex was probably formed in an island arc. *Earth and Planetary Science Letters*, v. 19, pp. 218–224.

Miyashiro, A. 1994. *Metamorphic Petrology*. Oxford University Press, New York. 404 p.

Monnier, C., Girardeau, J., Pubellier, M., and Permana, H. 2000. L'ophiolite de la Chaîne Central d'Irian Jaya (Indonésie): Évidences pétrologiques et géochimiques pour une origine dans un bassin arrière arc. C. R. Académie Sciences, Paris, *Sciences de la Terre et des planètes*, v. 331, pp. 691–699. doi:10.1016/S1251-8050(00)01479-8.

Moore, D. M., and Reynolds, R. C., Jr. 1989. *X-Ray Diffraction and the Identification and Analysis of Clay Minerals*. Oxford University Press, Oxford, UK. 332 p.

Moore, J. M., Jr. 1960. Phase relations in the contact aureole of the Aureole Pluton, Maine. PhD Dissertation, MIT, Cambridge. 217 p.

Moores, E. M., 1969. *Petrology and structure of the Vourinos ophiolite complex, northern Greece*. Geological Society of America Special Paper 118. 74 p.

Moores, E. M. 1973. Geotectonic significance of ultramafic rocks. *Earth-Science Reviews*, v. 9, pp. 241–258.

Moores, E. M. 1982. Origin and emplacement of ophiolites. *Reviews in Geophysics and Space Physics*, v. 20, pp. 735–760.

Moores, E. M., and Jackson, E. D. 1974. Ophiolites and oceanic crust. *Nature*, v. 250, pp. 136–138.

Moores, E. M., Kellogg, L. H., and Dilek, Y. 2000. Tethyan ophiolites, mantle convection, and tectonic "historical contingency": A resolution of the "ophiolite conundrum". In Dilek, Y., Moores, E. M., Elthon, D., and Nicolas, A. (eds.), *Ophiolites and Oceanic Crust: New Insights from Field Studies and the Ocean Drilling Program*. Geological Society of America Special Paper 349, pp. 3–12.

Moores, E. M., and Twiss, R. J. 1995. *Tectonics*. Freeman, New York. 415 p.

Morgan, W. J. 1968. Rises, trenches, great faults, and crustal blocks. *Journal of Geophysical Research*, v. 73, pp. 1959–1982.

Morgan, W. J. 1972. Plate motions and deep mantle convection. In Shagam, R., Hargraves, R. B., Morgan, W. J., Van Houten, F. B., Burk, C. A., Holland, H. D., and Hollister, L. C. (eds.), *Studies in Earth and Space Sciences: A Memoir in Honor of Harry Hammond Hess*. Geological Society of America Memoir 132, pp. 7–22.

Morisawa, M. 1968. *Streams: Their Dynamics and Morphology*. McGraw-Hill, New York. 175 p.

Morris, R. C. 1987. Iron ores derived by enrichment of banded iron formation. In Hein, J. R. (ed.), *Siliceous Sedimentary Rock-Hosted Ores and Petroleum*. Van Nostrand Reinhold, New York, pp. 231–267.

Morris, R. C., and Horwitz, R. C. 1993. The origin of the iron-formation-rich Hamersley Group of western Australia—Deposition on a platform. *Precambrian Research*, v. 21, pp. 273–297.

Morse, S. A. 1980. *Basalts and Phase Diagrams*. Springer-Verlag, New York. 493 p.

Mottana, A., Crespi, R., and Liborio, G., Prinz, M., Harlow, G., and Peters, J., eds. 1978. *Simon and Schuster's Guide to Rocks and Minerals*. Simon and Schuster, New York. 607 p.

Moulton, K. L., and Berner, R. A. 1998. Quantification of the effect of plants on weathering: Studies in Iceland. *Geology*, v. 26, pp. 895–898.

Mungall, J. E., and Naldrett, A. J. 2008. Ore deposits of the platinum group elements. *Elements*, v. 4, pp. 253–258. doi:10.2113/GSELEMENTS.4.4.253.

Munteanu, M., Marincea, S., Kasper, H. U., Zak, K., Alexe, V., Trandafir, V., Saptefrati, G., and Mihalache, A.

2004. Black chert-hosted manganese deposits from the Bistritei Mountains, Eastern Carpathians (Romania): Petrography, genesis and metamorphic evolution. *Ore Geology Reviews*, v. 24, pp. 45–65.

Muntener, O., and Piccardo, G. B. 2003. Melt migration in ophiolitic peridotites: The message from Alpine-Apennine peridotites and implications for embryonic ocean basins. In Dilek, Y., and Robinson, P. T. (eds.), *Ophiolites in Earth History*. Geological Society of London Special Publication No. 218, pp. 69–90. doi:10.1144/GSL.SP.2003.218.01.05.

Murphy, D. V., Sparling, G. P., and Fillery, I. R. P. 1998. Stratification of microbial biomass C and N and gross N mineralization with soil depth in two contrasting Western Australian Agricultural soils. *Australian Journal of Soil Research*, v. 36, pp. 45–55.

Murphy, M. D., Sparks, R. S. J., Barclay, J., Carroll, M. R., and Brewer, T. S. 2000. Remobilization of andesite magma by intrusion of mafic magma at the Soufrière Hills Volcano, Montserrat, West Indies. *Journal of Petrology*, v. 41, pp. 21–42.

Mutti, E., and Ricci Lucchi, F. 1972 (1978). *Turbidites of the Northern Apennines: Introduction to Facies Analysis*. AGI Reprint Series 3, AGI, Falls Church, VA, pp. 125–166.

Myers, J. D., and Johnston, A. D. 1996. Phase equilibria constraints on models of subduction zone magmatism (Overview). In Bebout, G. E., Scholl, D. W., Kirby, S. H., and Platt, J. P. (eds.), *Subduction Top to Bottom*. American Geophysical Union Geophysical Monograph 96, pp. 229–249.

N

Naldrett, A. J. 1973. Nickel sulfide deposits—Their classification and genesis, with special emphasis on deposits of volcanic association. *Canadian Institute of Mining and Metallurgy Transactions*, v. 76, pp. 183–201.

Nardin, T. R., Hein, F. J., Gorsline, D. S., and Edwards, B. D. 1979. A review of mass movement processes, sediment and acoustic characteristics, and contrasts in slope and base-of-slope systems versus canyon-fan-basin floor systems. In Doyle, L. J., and Pilkey, O. H. (eds.), *Geology of Continental Slopes*. SEPM Special Publication No. 27, pp. 61–73.

Nash, W. P., and Wilkinson, J. F. G. 1970. Shonkin Sag Laccolith, Montana: I. Mafic minerals and estimates of temperature, pressure, oxygen fugacity and silica activity. *Contributions to Mineralogy and Petrology*, v. 25, pp. 241–269.

Nelson, S. B. 1986. Water engineering. In Merritt, F. S. (ed.), *Civil Engineering Reference Guide*: McGraw-Hill, New York, pp. 7-1–7-79.

Nesbitt, H. W., Fedo, C. M., and Young, G. M. 1997. Quartz and feldspar stability, steady and non-steady state weathering, and petrogenesis of siliciclastic sands and muds. *Journal of Geology*, v. 105, pp. 173–191.

Nesse, W. D. 2000. *Introduction to Mineralogy*. Oxford University Press, Oxford, UK. 442 p.

Nickel, E. H. 1995. The definition of a mineral. *Canadian Mineralogist*, v. 33, pp. 689–690.

Nicolas, A. 1986. A melt extraction model based on structural studies in mantle peridotites. *Journal of Petrology*, v. 27, pp. 999–1022.

Nicolas, A., and Boudier, F. 2003. Where ophiolites come from and what they tell us. In Dilek, Y., and Newcomb, S. (eds.), *Ophiolite concept and the evolution of geological thought*. Geological Society of America Special Paper 373, pp. 137–152.

Nordstrom, D. K., and Munoz, J. L. 1986. *Geochemical Thermodynamics*. Blackwell, Oxford, UK. 477 p.

O

Ogasawara, Y. 2005. Microdiamonds in ultrahigh-pressure metamorphic rocks. *Elements*, v. 1, pp. 91–96.

O'Hanley, D. S. 1996. *Serpentinites: Records of Tectonic and Petrological History*. Oxford University Press, New York. 277 p.

Ohtani, E. 2005. Water in the mantle. *Elements*, v. 1, pp. 25–30.

Ohtani, E., and Kumazawa, J. 1981. Melting of forsterite Mg_2SiO_4 up to 15 GPa. *Physics of the Earth and Planetary Interiors*, v. 27, pp. 32–38.

Okita, P. M., and Shanks, W. C., III 1992. Origin of stratiform sediment-hosted manganese carbonate ore deposits: Examples from Molango, Mexico, and TaoJiang, China. *Chemical Geology*, v. 99, pp. 139–163.

Open University Course Team. 1989. *Ocean Circulation*. The Open University, England. http://www3.open.ac.uk/study/undergraduate/qualification/science/geology/index.htm.

P, Q

Padovani, E. R., and Carter, J. L. 1977. Aspects of the deep crustal evolution beneath south central New Mexico. In Heacock, J. G. et al. (eds.), *The Earth's Crust: Its Nature and Physical Properties*. American Geophysical Union Geophysical Monograph 20, pp. 19–55.

Page, B. M. 1966. Geology of the coast ranges of California. In Bailey, E. H. (ed.), *Geology of Northern California*. California Division of Mines and Geology Bulletin 190, pp. 255–322.

Palache, C., Berman, H., and Frondel, C. 1944. *The System of Mineralogy of James Dwight Dana and Edward Salisbury Dana* (7th ed.). Wiley, New York. 834 p.

Panteleyev, A. 1989. A Canadian cordilleran model for epithermal gold-silver deposits. In Roberts, R. G., and Sheahan, P. A. (eds.), *Ore Deposit Models*. Geoscience Canada Reprint Series 3, pp. 31–44.

Parnell, J. 1983. Ancient duricrusts and related rocks in perspective: A contribution from the old red sandstone. In Wilson, R. C. L. (ed.), *Residual Deposits: Surface Related Weathering Processes and Materials*. Blackwell, Oxford, UK, pp. 197–209.

Parsons, T., Christensen, N. I., and Wilshire, H. G. 1995. Velocities of southern basin and range xenoliths: Insights on the nature of lower crustal reflectivity and composition. *Geology*, v. 23, pp. 129–132.

Paterson, S. R., and Fowler, T. K., Jr. 1993. Re-examining pluton emplacement processes. *Journal of Structural Geology*, v. 15, pp. 191–206.

Paterson, S. R., and Vernon, R. H. 1995. Bursting the bubble of ballooning plutons: A return to nested diapirs emplaced by multiple processes. *Geological Society of America Bulletin*, v. 107, pp. 1356–1380.

Paton, T. R. 1978. *The Formation of Soil Material*. Allen & Unwin, London. 143 p.

Peacock, S. M. 1987a. Creation and preservation of subduction-related inverted metamorphic gradients. *Journal of Geophysical Research*, v. 92, pp. 12,763–12,781.

Peacock, S. M. 1987b. Serpentinization and infiltration metasomatism in the Trinity peridotite, Klamath province, northern California: Implications for subduction zones. *Contributions to Mineralogy and Petrology*, v. 95, pp. 55–70.

Peacock, S. M. 1989. Numerical constraints on rates of metamorphism, fluid production, and fluid flux during regional metamorphism. *Geological Society of America Bulletin*, v. 101, pp. 476–485.

Peacock, S. M. 1990a. Fluid processes in subduction zones. *Science*, v. 248, pp. 329–337.

Peacock, S. M. 1990b. Numerical simulation of metamorphic pressure-temperature-time paths and fluid production in subducting slabs. *Tectonics*, v. 9, pp. 1197–1212.

Peacock, S. M. 1993. The importance of the blueschist ? eclogite dehydration reactions in subducting oceanic crust. *Geological Society of America Bulletin*, v. 105, pp. 684–694.

Peacock, S. M. 1996. Thermal and petrologic structure of subduction zones. In Bebout, G. E., Scholl, D. W., Kirby, S. H., and Platt, J. P. (eds), *Subduction Top to Bottom*. American Geophysical Union Geophysical Monograph 96, pp. 119–133.

Peacock, S. M. 2003. Thermal structure and metamorphic evolution of subducting slabs. In Eiler, J. (ed.), *Inside the Subduction Factory*. American Geophysical Union Geophysical Monograph 138, pp. 7–22. doi:10.1029/138GM02.

Peacock, S. M., Rushmer, T., and Thompson, A. B. 1994. Partial melting of subducting oceanic crust. *Earth and Planetary Science Letters*, v. 121, pp. 227–244.

Pearce, J. A., and Peate, D. W. 1995. Tectonic implications of the composition of volcanic arc magmas. *Annual Reviews of Earth and Planetary Sciences*, v. 23, pp. 251–285.

Perfit, M. R., Fornari, D. J., Smith, M. C., Bender, J. F., Langmuir, C. H., and Haymon, R. M. 1994. Small-scale spatial and temporal variations in mid-ocean ridge crest magmatic processes. *Geology*, v. 22, pp. 375–379.

Perkins, D. 2002. *Mineralogy* (2nd ed.). Prentice-Hall, Upper Saddle River, NJ. 483 p.

Perkins, D., and Henke, K. R. 2004. *Minerals in Thin Section* (2nd ed.). Pearson Education, Upper Saddle River, NJ. 163 p.

Peterson, J. A., and Osmond, J. C. (eds.) 1961. *Geometry of Sandstone Bodies*. American Association of Petroleum Geologists, Tulsa, OK. 240 p.

Petford, N. 1996. Dykes or diapirs? *Transactions of the Royal Society of Edinburgh: Earth Sciences*, v. 87, pp. 105–114.

Pettijohn, F. J., Potter, P. E., and Siever, R. 1987. *Sand and Sandstone* (2nd ed.). Springer-Verlag, New York. 553 p.

Philbrick, S. S. 1936. The contact metamorphism of the Onawa Pluton, Piscataguis County, Maine. *American Journal of Science*, v. 231, pp. 1–40.

Phillips, F. C. 1963. *An Introduction to Crystallography*. Wiley, New York. 340 p.

Phillips, W. R., and Griffen, D. T. 1981. *Optical Mineralogy: The Nonopaque Minerals*. Freeman, San Francisco. 677 p.

Philpotts, A. R. 1989. *Petrography of Igneous and Metamorphic Rocks*. Prentice-Hall, Englewood Cliffs, NJ. 178 p.

Phipps Morgan, J., Blackman, D. K., and Sinton, J. M. (eds.). 1992. *Mantle Flow and Melt Generation at Mid-ocean Ridges*. American Geophysical Union Geophysical Monograph 71. 361 p.

Phipps Morgan, J., Harding, A., Orcutt, J., Kent, G., and Chen, Y. J. 1994. An observational and theoretical synthesis of magma chamber geometry and crustal genesis along a mid-ocean ridge spreading center. In Ryan, M. P. (ed.), *Magmatic Systems*. Academic Press, San Diego, pp. 139–178.

Pichler, H., and Zeil, W. 1972. The Cenozoic rhyolite-andesite association of the Chilean Andes. *Bulletin Volcanologique*, v. 35, no. 2, pp. 424–452.

Pilla, S., Ali, M. Y., Watts, A. B., and Searle, M. P. 2016. UAE-Oman Mountains give clues to oceanic crust and mantle rocks. *EOS*, v. 97, pp. 8–12.

Pitcher, W. S. 1997. *The Nature and Origin of Granite*. Chapman and Hall, London. 387 p.

Plendle, J. N., and Gielisse, P. J. 1963. Atomistic expression of hardness. *Zeitschrift für Kristallographie*, v. 118, pp. 404–421.

Plummer, C. C., Carlson, D. H., and McGeary, D. 2007. *Physical Geology* (11th ed.). McGraw-Hill, Boston. 648 p.

Pohl, W. L. 2011. *Economic Geology Principles and Practice*. Wiley-Blackwell, Hoboken, NJ. 663 p.

Porter, J. P., Schroeder, K., and Austin, G. 2012. Geology of the Bingham Canyon porphyry Cu-Mo-Au deposit, Utah: Geology and genesis of major copper deposits and districts of the world. In Hedenquist, J. W., Harris, M., and Camus, F. (eds.), *A Tribute to Richard H. Sillitoe*. Society of Economic Geologists Special Publication 16, pp. 127–146.

Postma, G. 1986. Classification for sediment gravity-flow deposits based on flow conditions during sedimentation. *Geology*, v. 14, pp. 291–294.

Potter, P. E. 1963. Late Paleozoic sandstones of the Illinois Basin. *Illinois Geological Survey, Report of Investigations*, 217. 92 p.

Poulton, S. W., and Canfield, D. E. 2011. Ferruginous conditions: A dominant feature of the ocean through Earth's history. *Elements*, v. 7, pp. 107–112.

Powell, C. M. 1979. A morphological classification of rock cleavage. *Tectonophysics*, v. 58, pp. 21–34.

Powell, R. D., and Molnia, B. F. 1989. Glacimarine sedimentary processes, facies and morphology of the south-southeast Alaska shelf and fjords. *Marine Geology*, v. 85, pp. 359–390.

Powers, M. C. 1953. A new roundness scale for sedimentary particles. *Journal of Sedimentary Petrology*, v. 23, pp. 117–119.

Press, F., and Siever, R. 1986. *Earth* (4th ed.). Freeman, New York. 656 p.

Pretorius, D. A. 1981. Gold and uranium in quartz pebble conglomerates. In Skinner, B. J. (ed.), *Economic Geology 75th Anniversary Volume*, pp. 117–138.

Prothero, D. R., and Schwab, F. 2004. *Sedimentary Geology: An Introduction to Sedimentary Rocks and Stratigraphy* (2nd ed.). Freeman, New York. 557 p.

Pryor, W. A. 1973. Permeability-porosity patterns and variations in some Holocene sand bodies. *American Association of Petroleum Geologists Bulletin*, v. 57, pp. 162–189.

Putnis, A. 1992. *Introduction to Mineral Sciences*. Cambridge Press, Cambridge, UK. 457 p.

Putnis, A. 2002. Mineral replacement reactions: From macroscopic observations to microscopic mechanisms. *Mineralogical Magazine*, v. 66, pp. 689–708.

Putnis, C. V., and Ruiz-Agudo, E. 2013. The mineral-water interface: Where minerals react with the environment. *Elements*, v. 9, pp. 177–182.

R

Rabinowicz, M., and Vigneresse, J. L. 2004. Melt segregation under compaction and shear channeling: Application to granitic magma segregation in a continental crust. *Journal of Geophysical Research*, v. 109, B04407. doi:10.1029/2002JB002372.doi:10.1029/95JB00517.

Raia, F., and Spera, F. J. 1997. Simulations of crustal anatexis: Implications for the growth and differentiation of continental crust. *Journal of Geophysical Research*, v. 102, pp. 22,629–22,648.

Raymond, L. A. 1984a. Classification of mélanges. In Raymond, L. A. (ed.), *Mélanges: Their Nature, Origin, and Significance*. Geological Society of America Special Paper 198, pp. 7–20.

Raymond, L. A. 1984b. *Petrography Laboratory Manual: Volume 1, Handspecimen Petrography*. GEOSI, Boone, NC. 170 p.

Raymond, L. A. 1993. *Petrography Laboratory Manual: Part 1, Handspecimen Petrography* (2nd ed.). GEOSI, Boone, NC. 154 p.

Raymond, L. A. 1995. *Petrology: The Study of Igneous, Sedimentary, and Metamorphic Rocks*. Brown, Dubuque, IA. 742 p.

Raymond, L. A. 2007. *Petrology: The Study of Igneous, Sedimentary, and Metamorphic Rocks* (2nd ed.). Waveland Press, Long Grove, IL. 720 p.

Raymond, L. A. 2009. *Petrography Laboratory Manual: Handspecimen and Thin Section Petrography* (3rd ed.). Waveland Press, Long Grove, IL. 170 p.

Raymond, L. A., and Love, A. B., 2006. Pseudobedding, primary structures and thrust faults in the Grandfather Mountain Formation, NW North Carolina, USA. *Southeastern Geology*, v. 44, no. 2, pp. 53–71.

Raymond, L. A., Merschat, A., and Vance, R. K. 2016. Metaultramafic schists and dismembered ophiolites of the Ashe Metamorphic Suite of northwestern North Carolina, USA. *International Geology Review*, v. 58. doi.org/10.1080/00206814.2015.1129515.

Raymond, L. A., Swanson, S. E., Love, A. B., and Allan, J. F. 2003. Cr-spinel compositions, metadunite petrology, and the petrotectonic history of Blue Ridge ophiolites, southern Appalachian Orogen, USA. In Dilek, Y., and Robinson, P. T. (eds.), *Ophiolites in Earth History*. Geological Society of London Special Publication 218, pp. 253–278.

Raymond, L. A., Webb, F., Jr., and Love, A. B. 2014. The Rose Hill formation and associated Tuscarora and Keefer sandstones of Clinch Mountain Wildlife Management Area, southwestern Virginia, USA: Issues of stratigraphic variation and diagenesis. *Southeastern Geology*, v. 50, pp. 61–82.

Reading, H. G. (ed.) 1996. *Sedimentary Environments: Processes, Facies and Stratigraphy* (3rd ed.). Blackwell, Oxford, UK. 688 p.

Retallack, G. J. 1990. *Soils of the Past*. Unwin Hyman, Boston. 520 p.

Richards, T. A., and McTaggart, K. C. 1976. Granitic rocks of the southern coast Plutonic Complex and northern Cascades of British Columbia. *Geological Society of America Bulletin*, v. 87, pp. 935–953.

Ridley, J. 2013. *Ore Deposit Geology*. Cambridge University Press, Cambridge, UK, 398 p.

Rivers, J. M., James, N. P., Kyser, T. K., and Bone, Y. 2007. Genesis of palimpsest cool-water carbonate sediment on the continental margin of southern Australia. *Journal of Sedimentary Research*, v. 77, pp. 480–494. doi:10.2110/jsr2007.046.

Ringwood, A. E. 1974. The petrological evolution of island arc systems. *Journal of the Geological Society* (London), v. 130, pt. 3, pp. 183–204.

Robb, L. 2005. *Introduction to Ore-Forming Processes*. Blackwell, Oxford, UK. 373 p.

Robinson, P. T., Dick, H. J. B., Natland, J. H., and ODP Leg 176 Shipboard Party. 2000. Lower oceanic crust formed at an ultra-slow-spreading ridge: Ocean Drilling Program Hole 735B, Southwest Indian Ridge. In Dilek, Y., Moores, E. M., Elthon, D., and Nicolas, A. (eds.), *Ophiolites and Oceanic Crust: New Insights from Field Studies and the Ocean Drilling Program.* Geological Society of America Special Paper 349, pp. 75–86.

Roedder, E. 1971. Fluid inclusion studies on the porphyry-type ore deposits at Bingham, Utah, Butte, Montana, and Climax, Colorado. *Economic Geology*, v. 66, pp. 98–120.

Rona, P. A., Bostrom, K., Laubier, L., and Smith, K. L., Jr. (eds.). 1983. *Hydrothermal Processes at Seafloor Spreading Centers.* Plenum, New York. 796 p.

Ronov, A. B., and Yaroshevsky, A. A. 1969. Chemical composition of the earth's crust. In Hart, P. J. (ed.), *The Earth's Crust and Upper Mantle.* American Geophysical Union Geophysics Monograph no. 13, pp. 37–57.

Rubin, A. M. 1995. Propagation of magma-filled cracks. *Annual Reviews of Earth and Planetary Sciences*, v. 23, pp. 287–336.

Rudnick, R. L., and Gao, S. 2005. Composition of the continental crust. In Rudnick, R. L., Holland, H. D., and Turekian, K. K. (eds.), *The Crust.* Elsevier, Amsterdam, pp. 1–64.

Ruppel, S. C., and Loucks, R. G. 2008. Black mudrocks: Lessons and questions from the Mississippian Barnett Shale in the southern Midcontinent. *The Sedimentary Record*, v. 6, no. 2, pp. 4–8.

Rychert, C. A., and Shearer, P. M. 2009. A global view of the lithosphere-asthenosphere boundary. *Science*, v. 324, pp. 495–498. doi:10.1126/science.1169754.

S

Sawyer, E. W. 2001. Melt segregation in the continental crust: Distribution and movement of melt in anatectic rocks. *Journal of Metamorphic Geology*, v. 19, pp. 291–309.

Scaillet, B., Holtz, F., Pichevant, M., and Schmidt, M. 1996. Viscosity of Himalayan leucogranites: Implications for mechanisms of granitic magma ascent. *Journal of Geophysical Research*, v. 101, pp. 27,691–27,699.

Scarfe, C. M., Mysen, B. O., and Virgo, D. 1987. Pressure dependence of the viscosity of silicate magmas. In Mysen, B. O. (ed.), *Magmatic Processes: Physicochemical Principles.* Geochemical Society Special Publication No. 1, University Park, PA, pp. 59–67.

Schmid, R., Fettes, D., Harte, B., Davis, E., and Desmons, J. 2007. Classification and nomenclature scheme. In Fettes, D., and Desmons, J. (eds.), *Metamorphic Rocks: A Classification and Glossary of Terms.* Cambridge University Press, New York, pp. 3–15.

Scholle, P. A. 1978. *A Color Illustrated Guide to Carbonate Rock Constituents, Textures, Cements, and Porosities.* American Association of Petroleum Geologists Memoir 27. 241 p.

Scholle, P. A. 1979. *A Color Illustrated Guide to Constituents, Textures, Cements, and Porosities of Sandstones and Associated Rocks.* American Association of Petroleum Geologists Memoir 28. 201 p.

Scholle, P. A., Bebout, D. G., and Moore, C. H. (eds.). 1983. *Carbonate Depositional Environments.* American Association of Petroleum Geologists Memoir 33. 708 p.

Schreiber, B. C., and El Tabakh, M. 2000. Deposition and early alteration of evaporates. *Sedimentology*, v. 47, pp. 215–238.

Schwab, A. P. 2000. The soil solution. In Sumner, M. E. (ed.), *Handbook of Soil Science.* CRC Press, Boca Raton, pp. B85–B122.

Scoffin, T. P. 1987. *An Introduction to Carbonate Sediments and Rocks.* Blackie, Glasgow. 274 p.

Sears, F. W., and Zemansky, M. W. 1964. *University Physics* (3rd ed.). Addison-Wesley, Boston. 1028 p.

Selles-Martinez, J. 1996. Concretion morphology, classification and genesis. *Earth Science Reviews*, v. 41, pp. 177–210.

Selley, R. C. 1976. *An Introduction to Sedimentology.* Academic Press, London. 408 p.

Shacklette, H. T., and Boerngen, J. G. 1984. *Element Concentrations in Soils and Other Surficial Materials of the Conterminous United States.* U.S. Geological Survey Professional Paper 1270. 105 p.

Shand, S. J. 1949. *Eruptive Rocks* (3rd ed). Wiley, New York. 488 p.

Sheldon, R. P. 1987. Association of phosphatic and siliceous marine sedimentary deposits. In Hein, J. R. (ed.), *Siliceous Sedimentary Rock-Hosted Ores and Petroleum.* Van Nostrand Reinhold, New York, pp. 58–80.

Shervais, J. W., Evans, J. P., Schmitt, D. R., Christiansen, E. H., and Prokopenko, A. 2014. Drilling in the track of the Yellowstone Hot Spot. *EOS*, v. 95, pp. 85–86.

Sheth, H. C. 2007. Large Igneous Provinces (LIPs): Definition, recommended terminology, and a hierarchical classification. *Earth-Science Reviews*, v. 85, pp. 117–124. doi:10.1016/j.earscirev.2007.07.005.

Sibson, R. H. 1977. Fault rocks and fault mechanisms. *Journal of the Geological Society* (London), v. 133, pp. 191–213.

Sienko, M. J., and Plane, R. A. 1961. *Chemistry* (2nd ed.). McGraw-Hill, New York. 623 p.

Simmons, S. F., White, N. C., and John, D. A. 2005. Geological characteristics of epithermal precious and base-metal deposits. In Hedenquist, J. W., Thompson, J. F. H, Goldfarb, R. J., and Richards, J. P. (eds.), *Economic Geology 100th Anniversary Volume*, pp. 485–522.

Simmons, W. B. 2007. Gem-bearing pegmatites. In Groat, L. A. (ed.), *Geology of Gem Deposits.* Mineralogical Association of Canada Short Course, v. 37, pp. 169–206.

Simmons, W. B., Pezzotta, F., Shigley, J. E., and Beurlen, H. 2012. Granitic pegmatites as sources of colored gemstones. *Elements*, v. 8, pp. 281–287.

Simonson, B. M. 2011. Iron ore deposits associated with Precambrian iron formations. *Elements*, v. 7, pp. 119–120.

Singer, M. J. 2003. Soil science. *Geotimes*, v. 48, no. 7, pp. 20–21.

Sisson, T. W., Grove, T. L., and Coleman, D. S. 1996. Hornblende gabbro sill complex at Onion Valley, California, and a mixing origin for the Sierra Nevada batholith. *Contributions to Mineralogy and Petrology*, v. 126, pp. 81–108.

Smulikowski, W. Desmons, J., Fettes, D., Harte, B., Sassi, F., and Schmid, R. 2007. Types, grade and facies of metamorphism. In Fettes, D., and Desmons, J. (eds.), *Metamorphic Rocks: A Classification and Glossary of Terms*. Cambridge University Press, New York, pp. 16–23.

Sobolev, A. V., and Chaussidon, M. 1996. H_2O concentrations in primary melts from supra-subduction zones and mid-ocean ridges: Implications for H_2O storage and recycling in the mantle. *Earth and Planetary Science Letters*, v. 137, pp. 45–55.

Soil Survey Staff. 1975. *Soil Taxonomy*. U.S. Department of Agriculture, Agriculture Handbook No. 436. 754 p.

Solar, G. S., and Brown, M. 2001. Petrogenesis of migmatites in Maine, USA: Possible source of peraluminous leucogranite in plutons. *Journal of Petrology*, v. 42, pp. 789–823. doi:10.1093/petrology/42.4.789.

Sorensen, H. (ed.). 1974. *The Alkaline Rocks*. Wiley, London. 622 p.

Sours-Page, R., Johnson, K. T. M., Neilsen, R. L., and Karsten, J. L. 1999. Local and regional variation of MORB parent magmas: Evidence from melt inclusions from the Endeavour segment of the Juan de Fuca Ridge. *Contributions to Mineralogy and Petrology*, v. 134, pp. 342–363.

Sparks, D. W., and Parmentier, E. M. 1994. The generation and migration of partial melt beneath oceanic spreading centers. In Ryan, M. P. (ed.), *Magmatic Systems*. Academic Press, San Diego, pp. 55–76.

Spear, F. S. 1993. *Metamorphic Phase Equilibria and Pressure-Temperature-Time Paths*. Mineralogical Society of America Monograph. 799 p.

Spera, F. J. 1980. Aspects of magma transport. In Hargraves, R. B. (ed.), *Physics of Magmatic Processes*. Princeton University Press, Princeton, NJ, pp. 265–323.

Spiegelman, M. 1993. Physics of melt extraction: Theory, implications and applications. In Cox, K. G., McKenzie, D. P., and White, R. S. (eds.), *Melting and Melt Movement in the Earth*. Oxford University Press, Oxford, UK, pp. 23–41.

Spock, L. E. 1962. *Guide to the Study of Rocks*. Harper, New York. 298 p.

Spry, A. 1969. *Metamorphic Textures*. Pergamon, Oxford, UK. 350 p.

Stern, C. R., Huang, W. L., and Wyllie, P. J. 1975. Basalt-andesite-rhyolite-H_2O: Crystallization intervals with excess H_2O and H_2O-undersaturated liquidus surfaces to 35 kilobars, with implications for magma genesis. *Earth and Planetary Science Letters*, v. 28, pp. 189–196.

Streckeisen, A. 1974. Classification and nomenclature of plutonic rocks: Recommendations of the IUGS subcommission on the systematics of igneous rocks. *Geologische Rundschau*, v. 63, pp. 773–786.

Streckeisen, A. 1976. To each plutonic rock its proper name. *Earth-Science Reviews*, v. 12, pp. 1–33.

Streckeisen, A. 1979. Classification and nomenclature of volcanic rocks, lamprophyres, carbonatites, and melilitic rocks: Recommendations and suggestions of the IUGS subcommission on the systematics of igneous rocks. *Geology*, v. 7, pp. 331–335.

Streckeisen, A. 1980. Classification and nomenclature of volcanic rocks, lamprophyres, carbonatites and melilitic rocks. *Geologische Rundschau*, v. 69, pp. 194–207.

Streckeisen, A. et al. 1973. Plutonic rocks: Classification and nomenclature recommended by the IUGS subcommission on the systematics of igneous rocks. *Geotimes*, v. 18, pp. 26–30.

Sugioka, I., and Bursik, M. 1995. Explosive fragmentation of erupting magma. *Nature*, v. 373, pp. 689–692. doi:10.1038/373689a0.

Sumner, M. E. (ed.). 2000. *Handbook of Soil Science*. CRC Press, Boca Raton, FL. 2112 p.

Sverjensky, D. A. 1989. Chemical evolution of basinal brines that formed sediment-hosted Cu-Pb-Zn deposits. In Boyle, R. W., Brown, A. C., Jefferson, C. W., Jowett, E. C., and Kirkham, R. V. (eds.), *Sediment-Hosted Stratiform Copper Deposits*. Geological Association of Canada Special Paper 36, pp. 127–134.

Swanson, D. A., Dzurisin, D., Holcomb, R. T., Iwatsubo, E. Y., Chadwick, W. W., Jr., Casadevell, T. J., Ewert, J. W., and Heliker, C. C. 1987. Growth of the lava dome at Mount St. Helens, Washington (USA), 1981–1983. In Fink, J. H. (ed.), *The Emplacement of Silicic Domes and Lava Flows*. Geological Society of America Special Paper 212, pp. 1–16.

Swanson, S. E. 1977. Relation of nucleation and crystal-growth rate to the development of granitic textures. *American Mineralogist*, v. 62, pp. 966–978.

Sylvester, A. G., Oertel, G., Nelson, C. A., and Christie, J. M. 1978. Papoose Flat Pluton: A granitic blister in the Inyo Mountains, California. *Geological Society of America Bulletin*, v. 89, pp. 1205–1219.

Symmes, G. H., and Ferry, J. M. 1995. Metamorphism, fluid flow and partial melting in pelitic rocks from the Onawa contact aureole, central Maine, USA. *Journal of Petrology*, v. 36, pp. 587–612.

T

Tan, K. H. 1994. *Environmental Soil Science*. Marcel Dekker, New York. 304 p.

Tan, K. H. 1998. *Principles of Soil Chemistry* (3rd ed.). Marcel Decker, New York. 521 p.

Tatsumi, Y. 1989. Migration of fluid phases and genesis of basalt magmas in subduction zones. *Journal of Geophysical Research*, v. 94, B4, pp. 4697–4707. doi:10.1029/88JB03904.

Tatsumi, Y., and Ishizaka, K. 1981. Existence of andesitic primary magma: An example from southwest Japan. *Earth and Planetary Science Letters*, v. 53, pp. 124–130.

Tatsumi, Y., and Ishizaka, K. 1982. Magnesian andesite and basalt from Shodo-Shima Island, southwest Japan, and their bearing on the genesis of calc-alkaline andesites. *Lithos*, v. 15, pp. 161–172.

Taylor, E. W. 1949. Correlation of the Mohs scale of hardness with the Vickers' hardness numbers. *Mineralogical Magazine*, v. 28, pp. 718–721.

Taylor, K. G., and Macquaker, H. S. 2011. Iron minerals in marine sediments record chemical environments. *Elements*, v. 7, pp. 113–118.

Teng, H. H. 2013. How ions and molecules organize to form crystals. *Elements*, v. 9, pp. 189–194.

Textor, A. 2010. Top 16 uses of petroleum: *GasOilEnergy Magazine*. http://gasoilenergy.com/2010/04/top-16-uses-of-petroleum/.

Thompson, A. B. 1976. Mineral reactions in pelitic rocks: Part 1, Prediction of P-T-X (Fe-Mg) phase relations. *American Journal of Science*, v. 276, pp. 401–424.

Thurman, H. V. 1997. *Introductory Oceanography* (8th ed.). Prentice Hall, Upper Saddle River, NJ. 544 p.

Thurman, H. V., and Trujillo, A. P. 2004. *Introductory Oceanography* (10th ed.). Prentice Hall, Upper Saddle River, NJ. 624 p.

Thy, P., and Dilek, Y. 2003. Development of ophiolitic perspectives on spreading center magma chambers. In Dilek, Y., and Newcomb, S. (eds). *Ophiolitie Concept and the Evolution of Geological Thought*. Geological Society of America Special Paper 373, pp. 188–226.

Titley, S. R. 1978. Geologic history, hypogene features, and processes of secondary sulfide enrichment at the Plesyumi copper prospect, New Britain, Papua New Guinea. *Economic Geology*, v. 73, pp. 768–784.

Touret, J., and Dietvorst, P. 1983. Fluid inclusions in high-grade anatectic metamorphites. *Journal of the Geological Society* (London), v. 140, pp. 635–649.

Trask, P. D. et al. 1950. *Geologic Description of the Manganese Deposits of California*. California Division of Mines Bulletin no. 152. 378 p.

Turcotte, D. L., and Phipps Morgan, J. 1992. The physics of magma migration and mantle flow beneath a mid-ocean ridge. In Phipps Morgan, J., Blackman, D. K., and Sinton, J. M. (eds.), *Flow and Melt Generation at Mid-ocean Ridges*. American Geophysical Union Geophysical Monograph 71, pp. 155–182.

Turcotte, D. L., and Schubert, G. 1973. Frictional heating of the descending lithosphere. *Journal of Geophysical Research*, v. 78, pp. 5876–5886.

Turner, F. J. 1958. Concept of metamorphic facies. In Fyfe, W. S., Turner, F. J., and Verhoogen, J. (eds.), *Metamorphic Reactions and Metamorphic Facies*. Geological Society of America Memoir no. 73, pp. 3–20.

Turner, F. J. 1968. *Metamorphic Petrology: Mineralogical and Field Aspects*. McGraw-Hill, New York, 403 p.

Turner, F. J. 1981. *Metamorphic Petrology: Mineralogical, Field, and Tectonic Aspects* (2nd ed.). McGraw-Hill, New York. 524 p.

U

Underwood, M. B., and Bachman, S. B. 1982. Sedimentary facies associations within subduction complexes. In Leggett, J. K. (ed.), *Trench-Forearc Geology: Sedimentation and Tectonics on Modern and Ancient Active Plate Margins*. Geological Society (London) Special Publication No. 10. Blackwell, Oxford, UK, pp. 537–550.

Underwood, M. B., Bachman, S. B., and Schweller, W. L. 1980. Sedimentary processes and facies associations within trench and trench-slope settings. In Field, M. E. et al. (eds.), *Quaternary Depositional Environments of the Pacific Coast*. Pacific Section Society of Economic Paleontologists and Mineralogists, pp. 211–229.

U.S. Geological Survey. 1980. *Principles of a Resource/Reserve Classification for Minerals*. U.S. Geological Survey Circular 831. 5 p.

U.S. Geological Survey Water-Data Report 2013. http://wdr.water.usgs.gov/wy2013/pdfs/11467000.2013.pdf

V

Velbel, M. A. 1999. Bond strength and the relative weathering rates of simple orthosilicates. *American Journal of Science*, v. 299, pp. 679–696.

Velbel, M. A., Basso, C. L. Jr., and Zieg, M. J. 1996. The natural weathering of staurolite: Crystal-surface textures, relative stability, and the rate-determining step. *American Journal of Science*, v. 296, pp. 453–472.

Vergniolle, S. 1996. Bubble size distribution in magma chambers and dynamics of basaltic eruptions. *Earth and Planetary Science Letters*, v. 140, pp. 269–279.

Verhoogen, J., Turner, F. J., Weiss, L. E., Wahrhaftig, C., and Fyfe, W. S. 1970. *The Earth*. Holt, Rinehart and Winston, New York. 748 p.

Viccaro, M., Giuffrida, M., Nicotra, E., and Cristofolini, R. 2016. Timescales of magma storage and migration recorded by olivine crystals in basalts of the March–April 2010 eruption at Eyjafjallajökull Volcano, Iceland. *American Mineralogist*, v. 101, pp. 222–230. doi:10.2138/am-2016-5365.

W

Wadge, G. 1980. Output rate of magma from active central volcanoes. *Nature*, v. 288, pp. 253–255.

Wager, L. R., and Brown, G. M. 1968. *Layered Igneous Rocks*. Oliver and Boyd, Edinburgh and London, UK. 588 p.

Wahlstrom, E. E. 1979. *Optical Crystallography* (5th ed.). Wiley, New York. 488 p.

Wakabayashi, J., and Unruh, J. R. 1995. Tectonic wedging, blueschist metamorphism, and exposure of blueschists: Are they compatible? *Geology*, v. 23, pp. 85–88.

Wakita, K. 2012. Mappable features of mélanges derived from ocean plate stratigraphy in the Jurassic accretionary complexes of Mino and Chichibu terranes in Southwest Japan. *Tectonophysics*, v. 568–569, pp. 74–85.

Walker, R. G. 1984. Shelf and shallow marine sands. In Walker, R. G. (ed.), *Facies Models* (2nd ed.). Geological Association of Canada, Geoscience Canada Reprint Series 1, pp. 141–170.

Walker, R. G., and Cant, D. J. 1984. Sand fluvial systems. In Walker, R. G. (ed.), *Facies Models* (2nd ed.). Geological Association of Canada, Geoscience Canada Reprint Series 1, pp. 71–89.

Walker, R. G., and James, N. P. (eds.). 1992. *Facies Models: Response to Sea Level Change*. Geological Association of Canada, St. John's, Newfoundland. 409 p.

Wallace, P., and Anderson, A. T. 2000. Volatiles in magmas. In Sigurdsson, H., Houghton, B. F., McNutt, S. R., Rymer, H., and Stix, J. (eds.), *Encyclopedia of Volcanoes*. Academic Press, San Diego, pp. 149–170.

Wang, S., Anderson, W., and Raymond, L. A. 2014. Hydrogeology and water quality at the Tater Hill Groundwater Monitoring and Research Station, Watauga County, North Carolina. *North Carolina Division of Water Resources Groundwater Bulletin* 2014-01. 53 p. http://portal.ncdenr.org/web/wq/aps/gwp/groundwater-study-publications.

Wark, D. A., and Watson, E. B. 2000. Effect of grain size on the distribution and transport of deep-seated fluids and melts. *Geophysical Research Letters*, v. 27, pp. 1–4.

Wasklewicz, T. A. 1994. Importance of environment on the order of mineral weathering in olivine basalts, Hawaii. *Earth Surface Processes and Landforms*, v. 19, pp. 715–734.

Wentworth, C. K. 1922. A scale of grade and class terms for clastic sediments. *Journal of Geology*, v. 30, pp. 377–392.

Whetten, J. T. 1966. Sediments from the lower Columbia River and origin of greywacke. *Science*, v. 152, pp. 1057–1058.

White, N. C., and Hedenquist, J. W. 1995. Epithermal gold deposits: Styles, characteristics and exploration. *Society of Economic Geologists Newsletter* 23, pp. 9–13.

White, S. M., Crisp, J. A., and Spera, F. J. 2006. Long-term volumetric eruption rates and magma budgets. *Geochemistry, Geophysics, Geosystems*, v. 7, no. 3, Q03010. doi:10.1029/2005GC001002.

Wieser, M. E., Holden, N., Coplen, T. B., Böhlke, J. K., Berglund, M., Brand, W. A., De Bièvre, P., and Gröning, M. et al. 2013. Atomic weights of the elements 2011. IUPAC Technical Report. *Pure and Applied Chemistry*, v. 85, pp. 1047–1078.

Wilkinson, J. F. G. 1982. The genesis of Mid-Ocean Ridge Basalt. *Earth-Science Reviews*, v. 18, pp. 1–57.

Williams, H., and McBirney, A. R. 1979. *Volcanology*. Freeman, Cooper & Co., San Francisco. 397 p.

Wilson, J. L. 1974. Characteristics of carbonate-platform margins. *American Association of Petroleum Geologists Bulletin*, v. 58, pp. 810–824.

Wilson, J. T. 1965. A new class of faults and their bearing on continental drift. *Nature*, v. 207, pp. 343–347.

Winkler, H. G. F. 1979. *Petrogenesis of Metamorphic Rocks* (5th ed.). Springer-Verlag, New York. 348 p.

Winter, M. J. 1994. *Chemical Bonding*. Oxford University Press, New York. 92 p.

Wolfe, C. W. 1953. *Manual for Geometrical Crystallography*. Edwards Brothers, Ann Arbor, MI. 263 p.

Wood, E. A. 1977. *Crystals and Light: An Introduction to Optical Crystallography* (2nd ed.). Dover, New York. 156 p.

Wright, V. P., and Burchette, T. P. 1996. Shallow-water carbonate environments. In Reading, H. G. (ed.), *Sedimentary Environments: Processes, Facies and Stratigraphy*. Blackwell, Cambridge, UK, pp. 325–394.

Wylie, J. J., Voight, B., and Whitehead, J. A. 1999. Instability of magma flow from volatile-dependent viscosity. *Science*, v. 285, pp. 1883–1885.

Wyllie, P. J. 1971. *The Dynamic Earth: Textbook in Geosciences*. Wiley, New York. 416 p.

Wyllie, P. J. 1983. Experimental studies on biotite- and muscovite-granites and some crustal magmatic sources. In Atherton, M. P., and Gribble, C. D. (eds.), *Migmatites, Melting and Metamorphism*. Shiva, Cheshire, UK, pp. 12–26.

X, Y, Z

Yardley, B. W. D., MacKenzie, W. S., and Guilford, C. 1990. *Atlas of Metamorphic Rocks and Their Textures*. Longman/Wiley, New York. 120 p.

Yoder, H. S., Jr. 1976. *Generation of Basaltic Magma*. National Academy of Sciences, Washington, DC. 265 p.

Zhang, Y. 1999. H_2O in rhyolitic glasses and melts: Measurement, speciation, solubility, and diffusion. *Reviews of Geophysics*, v. 37, pp. 493–516.

Figure Credits

2.11, 2.14
from E. G. Ehlers, 1972, *The Interpretation of Geological Phase Diagrams,* reprinted 1987 by Dover. Used with permission of Dover Publications.

5.12, 5.17
from F. Donald Bloss, 1971, *Crystallography and Crystal Chemistry.* Used with permission of F. Donald Bloss.

7.3
from Stearns A. Morse, 1980, *Basalts and Phase Diagrams.* Used with permission of Stearns A. Morse.

7.7
from A. G. Sylvester et al., 1978, "Papoose Flat Pluton: A Granitic Blister in the Inyo Mountains, California," *Geological Society of America Bulletin* 89(8). Used with permission of the Geological Society of America.

7.16, 7.17, 7.18
from A. Streckeisen, 1974, "Classification and Nomenclature of Plutonic Rocks: Recommendations of the IUGS Subcommission on the Systematics of Igneous Rocks," *Geologische Rundschau* 63(2). Used with permission of Springer.

7.20
from M. J. LeBas et al., 1986, "A Chemical Classification of Volcanic Rocks Based on the Total Alkali-Silica Diagram," *Journal of Petrology* 27. Used with permission of Oxford University Press.

8.7, 8.8
from F. S. Merritt (ed.), 1986, *Civil Engineering Reference Guide.* Used with permission of McGraw-Hill Education.

9.3
from J. Verhoogen et al., 1970, *The Earth: An Introduction to Physical Geology.* Used with permission of Cengage Learning.

9.5
from R. M. Garrels and W. C. Krumbein, 1952, "Origin and Classification of Chemical Sediments in Terms of pH and Oxidation-Reduction Potentials," *Journal of Geology* 60(1). Used with permission of University of Chicago Press.

9.10
from M. C. Powers, 1953, "A New Roundness Scale for Sedimentary Particles," *Journal of Sedimentary Petrology* 23. Used with permission of SEPM, the Society for Sedimentary Geology.

9.21
from R. H. Dott, 1964, "Wacke, Graywacke, and Matrix: What Approach to Immature Sandstone Classification," *Journal of Sedimentary Petrology* 34(3). Used with permission of SEPM, the Society for Sedimentary Geology.

10.3(a)
from C. M. Powell, 1979, "A Morphological Classification of Rock Cleavage," *Tectonophysics* 58. Used with permission of Elsevier.

10.10
from Norman L. Bowen, 1940, "Progressive Metamorphism of Siliceous Limestone and Dolomite," *Journal of Geology* 48(3). Used with permission of University of Chicago Press.

11.3, 11.19, 11.20, 11.21
from John Ridley, 2013, *Ore Deposit Geology.* Used with permission of Cambridge University Press.

11.4(b)
from W. L. Pohl, 2011, *Economic Geology Principles and Practice.* Used with permission of Wiley Books.

11.17, 11.25, 11.26
from James R. Craig, David J. Vaughan, and Brian J. Skinner, 2011, *Earth Resources and the Environment,* 4th ed. Used with permission of Pearson Education, Inc.

Index

Aa lava, 118
Abrasion, 146
Abyssal plain, 201
Acceleration, 14, 15
Accessory minerals, 129–130
Acicular habit, 79
Acid reaction, 91
Åckermanite, 77
Acrospirifer brachiopods, 183
Active margin basins, 11
Adamantine luster, 86
AFM phase diagrams, 225, 227
Aggregational state, 78
A-horizon, 150
Alcohol lamp, 95
Algoman deposits, 256
Alkali olivine basalt, 135
Alkaline igneous magmas, 271
Allochem, 173
Allochemical rocks, 183
Alluvial fans, 167
Aphanitic-porphyritic feldspathic volcanic rocks, 136
Alpine quartz diorite, 125
Alpine ultramafic rocks, 228–230
Aluminum-rich laterites, 255
Amethyst, 3
Amorphous, 50
Amphibolites, 228
Amygdules, 118
Analyzer, 98
Andesitic magma, 139
Anion, 18
Aniostropic materials, 97–98
Anorogenic igneous rocks, 140–141
Antiferromagnetic material, 91
Aphanitic textures, 121
Aphanitic-porphyritic texture, 124, 126, 127
Aragonite, 74
Arc, 137, 228
Arc crust, 8

Arc-root plutons, 270
Area, 14
Arenites, 188
Aridosols, 152
Ash, 118
Assimilation, 114
Assimilation and fractional crystallization (AFC), 116, 139
Asthenosphere, 4
Atmophile, 30
Atmosphere, 2
Atomic number, 17
Atoms, 16–18
Authigenesis, 146, 202, 205
Axial ratios, 95

Banded aggregation, 80
Banded iron formations (BIF), 190, 256, 257
Banding/layering, 212
Barrovian Facies Series, 226, 228
Basalt(s), 135–137
Basaltic magma, 110, 139, 267, 268
Base exchange, 148
Basin, 164
Basis vector, 55
Batholiths, 117
Bauxites, 255
Bed(s), 174, 175, 179, 181
Bedding surface, 176
Bed-load transport, 162, 164
Benitoite, 77
Beryl crystal, 69
Besshi-type deposits, 248
B-horizon, 150
Binary eutectic diagram, 24, 26
Binary solid solution diagrams, 28–29
Biofacies, 167
Biological activity, 146
Biological agents, 11
Biosphere, 2
Biotite-feldspar-quartz gneiss, 215

Biotite-hornblende granodiorite, 122
Bioturbation, 180, 202
Bitterns, 259
Black smoker, 248
Bladed habit, 79
Blowpipe, 95
Body-centered unit cells, 62
Bond, 18, 20
    ionic vs. covalent, 20–22
    metallic and dipolar interactions, 22
Botryoidal aggregation, 80
Boudins, 213
Bouma sequence, 176, 199, 201
Boundstone, 188
Bowen's Reaction Series, 116, 149
Boyle's law, 3, 102, 103
Bravais lattices, 63
Breccia, 188, 191, 231
Breccia texture, 126, 218
Brittle, 87
Buchan Facies Series, 228
Burial metamorphism, 206
Burrows, 180

Calcite, 74
Cambrian Nolichucky formation, 178
Capillary habit, 79
Ca-poor pyroxene, 135
Carbonate minerals, 91, 271
Carbonate-hosted lead-zinc deposit, 250–252
Carbonatites, 141, 242–243
Cataclasis, 205
Cataclastic textures, 218
Cation, 18
Celsius (C), 14
Cementation, 188, 202
Center of symmetry, 62
Centered unit cells, 59
Chalcophile, 30
Chelation, 148

Index

Chemical end products, 149
Chemical potential gradient, 113
Chemical reaction, 22–23
Chemical sediments, 267
Chemically active fluids, 206
Chemistry
    atoms and structure, 16–18
    bonding, 18, 20
        ionic vs. covalent, 20–22
        metallic and dipolar interactions, 22
    chemical reactions, stability and direction, 22–23
    ions and ionization, 18
    phase diagrams, 23–28
        binary solid solution diagrams, 28–29
        eutectic systems, earth history and rock formation, 29–30
        solid solutions and elemental compatibility, 30
Chert, 190
Chert breccia, 191
Chlorine, 18
Chlorite, 225
C-horizon, 150
Chrysoberyl, 83
Cinder cones, 118
Circulation, 11
Classic vein deposit, 247
Clastic rocks, 218
Clastic textures, 170
Clay mineral, 91
Clayey, 154
Clayshale, 190
Cleavage, 69, 87–88, 213, 214
    faces, 65
    surfaces, 65
Climate, 146, 166
Clinopyroxene, 133
Closed forms, 69
Coal, 190, 263
Colloid, 148, 152
Collophane, 258
Color, 83–85
Columnar joints, 118, 120
Compaction, 201
Compatibility, 30
Complex lower crust, 4
Complex twinning, 82
Composite cone, 118
Concentration factor, 235
Concentric aggregation, 80
Concretions, 180
Conduits, 39–41
Cone sheet, 117

Conglomerate, 188, 191
Connate fluids, 249–252
Conrad discontinuity, 4
Consistence, soils, 155
Constructive interference, 102
Contact Facies Series, 221, 228
Contact metamorphism, 206, 225–226, 270
Contact twins, 82
Continental crust, 8, 112
Continental environments, 193–196
Continuous cleavage, 212
Convolute bedding, 174
Copper, 251
Copper-rich shale, 252
Carbon, 31
Core, 4
Coriolis parameter, 41
Corona texture, 123
Country rocks, 206, 225, 226
Covalent bonding, 20–22
Crenulation cleavage, 213
Cretaceous Panoche Formation sandstone, 177, 178
Critical angle, 85
Critical mineral assemblages, 222, 227
Critical point, 24
Critical zone, 145
Cross-bed(s), 174
Cross-bedded sand, 199
Crust, 4, 7
Crustal Earth materials
    Earth
        structure and chemistry, 4–8
        systems at and near surface, 11–12
    fluids, 3
    minerals, 2–3
    plate tectonics, 8–11
    rocks, 3
    soils, 3–4
Cryptocrystalline, 258
Crystal
    faces, 51
    form, 69
    nucleus, 50
    systems, 59
    tuffs, 128
Crystalline, 2, 49
    dolostone, 188
    limestone, 188
    rocks, 218
    texture, 38, 170, 173
Crystallization, 49
    crystal features, 65
        faces, morphologies, and forms, 67–70

        lines and planes within crystals, 65–67
    latent heat, 27
    of a melt, 240–243
    order, disorder, crystallinity, nucleation, crystal growth and morphology, 49–54
    solid solution, 54
    solidification, 121–124
    symmetry, 54
        one dimension, 54–56
        three dimension, 60–65
        two dimension, 56–60
Crystallography, 69, 71
Cyclic twins, 82
Cyclosilicates, 76
Cyprus-type deposits, 248

Darcy's Law, 38
Debris avalanche, 163
Decomposition, 145
Decompression melting, 110
Deltas, 267
Density (ρ), 33
Derivative magmas, 109
Desert environments, 194
Destructive interference, 102
Detrital deposits, 260–262
Deviatoric stress, 16
Diabase, 138
Diabasic texture, 123, 125
Diablastic rocks, 220
Diablastic texture, 218
Diagenesis, 40
Diamagnetic materials, 91
Diamictites, 188
Diamond, 49
Diamond-bearing kimberlites, 243, 244
Diapir(s), 110
    ballooning, 118
    melt, 114
Diatremes, 137
Diffraction, 101–102
Diffuse, 54
Dike(s), 117
Dike propagation, 114
Dipolar interactions, 22
Disintegration, 145
Dissolution, 148
Distance, 14
Dolostones, 183, 189
Domes, 118
Drainage basin, 34
Ductile shear zones, 213
Ductility, 87

Index

Dunham-based classification, 189
Dynamic metamorphism, 206
Dynamic processes, 271
Dynamic viscosity, 36
Dynamoblastic rocks, 230, 232
Dynamothermal metamorphism, 206

Earth
   chemical differences, 7
   elemental abundances, 7
   fluid-mediated processes, 42–45
   history, 29–30
   interior, 5
   layers, 6
   materials, 1
   plate boundary, three types, 6
   rock-forming minerals, crustal
      abundances, 7–8
   structure and chemistry, 4–8
   systems at and near surface,
      11–12
Earth sciences, 1
Earth-material resource, 263
Eclogite, 112
Economic deposits, 236, 242
Eddy viscosity, 41
Effervescence, 91
Eh-pH fence diagram, 164, 166, 253
Elastic behavior, 87
Electromagnetic radiation, 83
Electron, 17
Electron-beam microprobe (EM),
   104–105
Electron density, 96
Element, 16
Elemental compatibility, 30
End-centered unit cells, 62
Energy dispersive spectroscopy
   (EDS), 105
Enthalpy, 22
Entropy, 23
Eogenesis, 202
Epiclastic texture, 170, 171
Epitaxial intergrowth, 78, 81
Equant habit, 79
Ergs, 168
Erosion, 11, 161, 167
Estuarine mudrocks, 197
Estuarine settings, 196
Euhedral crystal, 76
Eutectic systems, 25, 29–30, 113
Evaporites, 190, 258
Exfoliation joints, 146, 147
Exsolution, 54, 84
Extraordinary ray, 98
Extrinsic variables, 201

Fabric, 150
Face-centered unit cells, 62
Facies series, 220, 221
Failed intracontinental rift zones, 251
Fat clays, 155
Fault(s), 213
Fault zone rocks, 230, 231
Ferrimagnetic material, 91
Ferromagnetic material, 91
Ferrous iron, 257
Fibrous aggregate, 80
Fibrous habit, 79
Fizzing, 91
Flame structure, 177, 180
Flexible sheets, 87
Flow(s), 34, 118
   fractures and conduits, 39–41
   through porous rocks, 38–39
   streams, stream channels and
      water, 34–38
   velocity, 35
Fluid, 3, 9, 206, 270
   flow, 34
      fractures and conduits, 39–41
      through porous rocks, 38–39
      streams, stream channels and
         water, 34–38
   fluid-mediated processes, 42–45
   kinds and sources, 42–45
   nature and distribution of, 33–34
   oceans, seas, and lakes, 41–42
   separation, 239–243
Fluid-rich magmas, 271
Fluorescence, 87
Flute, 180
Fluvial environments, 167
Fluvial stratigraphy, 197
Flux melting, 109, 270
Folds, 213
Foliated textures, 216
Force, 14
Forearc basins, 11
Fore-reef, 199
Formation, rock, 174
Fractional crystallization, 114, 116,
   270
Fracture, 39–41, 87–88
Franciscan Facies Series, 221, 227
Frost heaving, 146
Frost wedging, 146
Funnel, 117

Gabbroic rocks, 122, 123, 133
Galena (PbS) crystals, 52, 74, 75, 90
Garnet, 53, 112
Geology, 1

Geothermal gradient, 4
German Rancho Formation, 191
Gibbs free energy, 23, 24
Gilbert-Dott classification, 192
Glacial stratigraphy, 194
Glassy textures, 121, 124
Gneiss, 214, 220
Gneissose texture, 216, 217
Goniometry, 95
Gouge, 218, 231
Graded bedding, 174
Grain, 172
   roundness, 173
   shapes, 173
   sorting, 172
Grainstones, 188
Granite, 111
Granoblastic texture, 217, 218, 220
Granular aggregation, 80
Graphic texture, 124, 125
Gravity, 161
Graywacke, 189
Groundmass, 109
Groundwater, 42–44, 158, 252–256
Growth, 122
Gypsum, 82, 89, 167

Habit of mineral, 67, 78
Halite, 74, 75, 90
Hardness, 89–90
Hastingsite, 140
Heterogeneous nucleation, 122
Hexagons, 58
High field-strength elements (HFSE),
   30
High refractive index, 86
High-Mg calcite, 202
High-pressure belt (HP belt), 226
High-sulfidation epithermal veins,
   247
Histosols, 152
Hjulstrom's diagram, 165
Homogeneous nucleation, 122
Horizon, 150, 156
Hornblende, 134
Hornfels, 208, 225
Hornitos, 118
Hot-spot magmas, 136
Hydration, 148
Hydraulic conductivity, 38
Hydraulic radius, 37
Hydrocarbon fuels, 263
Hydrogen bond, 22
Hydrologic cycle, 11
Hydrolysis, 148
Hydrosphere, 2, 4, 33

Index

Hydrothermal fluid, 40–44, 243–249, 268
Hydrothermal metamorphism, 225–226
Hydrous fluid, 3, 34, 139
   deposits
      connate fluids and stratabound deposits, 249–252
      groundwater/meteoric fluids and associated deposits, 252–256
      hydrothermal fluids, 243–249
Hypidiomorphic-granular texture, 122–125
Hypogene deposit, 254
Hypothetical resources, 237

Iceland Spar, 98
Identified resource, 236
Igneous rock, 3, 109, 267, 271
   anorogenic and LIPS, 140–141
   basalts and significance, 135–137
   chemical composition and classification, 124–135
   formation, 117–120
   magma, 109
      crystallization and solidification, 121–124
      emplacement and eruption, 117–120
      generation, 110
      modification, 114–116
      movement, 113–114
      origin, 109–113
   minerals, 129
   MORB, 138–139
   subduction zones and orogenic belts, 139–140
Ijolite, 140–141
Inclusions, 84
Incompatibility, 30
Index of refraction, 96
Indicated resource, 237
Industrial rocks, 262
Inertinite, 190
Inferred resources, 237
Inner core, 4
Inner hornfels zone, 226
Inosilicates, 76
Instantaneous equivalent, 14
Intergranular texture, 124, 127
International Union of Geological Sciences (IUGS), 126
Intrinsic variables, 201
Inversion, 62, 64
Ion, 18

Ionic bonding, 20–22
Iron deposits, 256
Ironstones, 190
Isogonal, 65
Isograd, 222
Isostructural crystal, 73, 75
Isotopes, 30, 31
Isotropic material, 97

Jacupirangite, 140
Joints, 39–41, 213
Jurassic Navajo Sandstone, 199
Jurassic pillow lavas, 120

Kelvin (K), 14
Kerogen, 190
Kilogram, 13
Kimberlites, 137, 141
Konnarock Formation, 197
Kuroko-type deposits, 248
Kyanite, 228

Laccoliths, 117
Lacustrine stratigraphies, 194
Lagoonal settings, 196
Lakes, 41–42
Laminar flow, 34, 35
Laminated siltstone, 190
Laminations, 174
Lapilli tuffs, 128
Large igneous provinces (LIPs), 140–141
Large-ion lithophile elements (LILE), 30
Latent heat of crystallization, 27
Laterite, 156, 255
Lattice, 55
Lava
   flow, 120
   plain, 118
   plateaus, 118
Law of Bravais, 52
Lever rule, 27
Liesegang rings, 180
Light wave interaction, 102
Lime mud, 171, 199
Lime mudstone, 188
Limestones, 183, 189
Line group, 56
Linear symmetry, 55
Liptinite, 190
Liquid, 90
Liquid crystals, 49–50
Liquid limit (LL), 155
Liquidus, 25, 121
Lithified rock, 201

Lithium, 17
Lithofacies, 167, 168, 193
Lithophile, 30
Lithosphere, 4, 5, 9
Littoral-beach environment, 167
Load casts, 180
Loamy, 154
Long-range order, 49
Lopoliths, 117
Low field-strength elements (LFSE), 30
Low-density fluids, 38
Low-sulfidation epithermal veins, 247
Luminescence, 86–87
Luster, 85–86

Macerals, 190
Magma, 33, 42, 43, 109, 267
   crystallization and solidification, 121–124
   emplacement and eruption, 117–120
   generation, 110
   mixing, 114
   modification, 114–116
   movement, 113–114
   origin, 109–113
Magmatic fluid, 42, 43
Magmatic process deposits
   crystal and fluid separation, 239–243
   volcanic process deposits, 243
Magmatic segregation, 239, 242
Magnesium number, 113
Magnetism, 91
Magnetite, 91
Malleability, 87
Manganese, 258, 259
Mantle, 4
Mantle plumes, 9
Marbles, 208
Marginal, 236
Marine environments, 167, 168, 197–201
Mass wasting, 162
Massive, 76
Massive aggregation, 79
Meandering river environment, 168
Measured resources, 236–237
Melanges, 227
Melt diapirs, 114
Melted rock, 3
Melting, 25, 205
Mesogenesis, 202
Mesosphere, 5
Metabasites, 207
Metaharzburgite, 230

Metallic bonding, 22
Metamorphic facies, 220
Metamorphic fluid, 41, 43, 44, 268
Metamorphic phase diagrams, 222
Metamorphic processes, 205–207
Metamorphic rocks, 3, 271
   chemical compositions, 207–208
   classification, 218–220
   facies, facies series, and depictions, 220–225
   metamorphism
      arcs and moderate-to-high-temperature metamorphic belts, 228
      contact and hydrothermal, 225–226
      faults, 230–232
      and metamorphic processes, 205–207
      static and, 226–228
      ultrabasic rocks, 228–230
   minerals, 209–211
   structures and textures, 208–218
Metapelites, 207
Meta-quartz arenite boudins, 215
Meta-quartzo-feldspathic rocks, 208
Metasiltstone, 215
Metasomatism, 205
Meteoric fluid, 42–44, 252–256
Meter, 13
Methane, 3
Micrite, 171
Microbes, 267
Microlithons, 231
Middle phase diagram, 225–226
Mid-ocean ridge basalts (MORBs), 30, 111, 113, 137–139, 268
Migmatites, 228
Miller indices, 65–67, 81
Mineral, 2–3, 73, 262, 271
   aggregations, 80
   analytical methods, 103
      electron-beam microprobe, 104–105
      EM, 104–105
      XRF, 106
   classification, 73–76
   cleavage forms, 70
   habits
      aggregation and, 79
      forms vs., 68
   hand specimens, identifying
      acid reaction, 91
      cleavage and fracture, 87–88
      color and coloring phenomena, 83–85
      crystal forms, habits, and aggregational states, 76–78
      crystallographic intergrowths, 78–83
      fragments, 151
      hardness, 89–90
      luminescence, 86–87
      luster, 85–86
      magnetism, 91
      odor, water solubility and feel, 91–92
      parting, 88–89
      specific gravity, 90–91
      streak, 85
      tenacity, 87
   historical methods, 95
   igneous rocks, 129
   metamorphic rocks, 209–211
   optical mineralogy
      light interactions with matter, 96–98
      petrographic microscope, 98–101
      polarization of waves, 96
   sedimentary rocks, 185–186
   XRD
      procedures and applications, 102–104
      x-rays and diffraction, 101–102
Mirror line, 56, 57, 64
Mirror plane, 62
Miscellaneous metamorphic rocks, 208
Mississippi Valley Type (MVT) deposits, 250
Moderate-to-high-temperature metamorphic belts, 228
Modes, 181, 187–188
Mohorovicic Discontinuity, 1, 4
Mohs hardness, 89
Mudcrack, 180, 181, 189, 197, 192
Mudshale, 190
Mudstone, 190
Mullions, 213
Multiplicity factor, 69
Mylonite, 10, 213, 216, 220, 231
Mylonitic texture, 218

Natural waters, 258
Neocrystallization. *See* Authigenesis
Neoproterozoic Rapitan-type deposits, 257
Nepheline, 140
Nepheline-pyroxene rocks, 140–141
Nesosilicates, 76
Neutrons, 17

Newton (N), 14
Non-silicate minerals, 130, 209–211
Nonterminal reactions, 223, 224
Normal stress, 15
Nuclear reactions
   isotopic systems, geologic uses, 31
   reaction process, 30–31
Nucleation, 51, 122
Nucleus, 17, 51, 121, 267

Objective lenses, 98
Obsidian, 124
Oceanic crust, 4
Oceanic layers, 4
Oceanic plate, 8
Oceans, 41–42
O-horizon, 150
Oil-field brines, 249
Oligoclase, 83
Olistostromal mélanges, 270
Olistostrome, 199, 200
Olivine, 3, 77, 127, 133, 134, 255
Oncolites, 180
Ooids, 171, 174
Oolitic, rocks, 173
Oolitic sandstone iron ore, 257
Opaque material, 84, 97
Open forms, 69
Ophiolites, 139
Ophitic texture, 123
Optical indicatrices, 97
Optical mineralogy
   light interactions with matter, 96–98
   petrographic microscope
      design of, 98–99
      use of, 99–101
   polarization of waves, 96
Orbitals, 17
Orders, 156
Ordinary ray, 98
Ordovician Bowen Formation, 181
Ordovician Knobs Formation sandstone, 182
Ore, 235
Ore-grade deposits, 235
Organic component, 151–152
Organic rocks, 208
Orogenic belts, 139–140
Orthoclase, 82
Orthopyroxene, 133
Outer core, 4
Oxidation, 18, 146–147
Oxidation state, 18
Oxidized uranium (U^{6+}), 251
Oxisols, 152

Index

Packstones, 188
Pahoehoe lava, 118
Paleoplacer deposit, 261
Paleosols, 158–159
Pans, 154
Parallel intergrowths, 78
Parallelograms, 58
Paramagnetic material, 91
Parental magmas, 109
Parting, 88–89
Pascal (Pa), 15
Passive margin basins, 11
Pedalfers, 156
Pedocals, 156
Peds, 154
Pegmatites, 241–242
Pegmatitic granite, 43
Pegmatitic textures, 123
Pelagic environments, 168
Pellets, 171
Penetration twins, 82
Pentagons, 58
Peridotites, 112
Periodic table, 18, 19
Permeability, 38
Perthite, 54
Petrogenetic grid, 220
Petrographic microscope, 98–99
    design of, 98–99
    use of, 99–101
Petroleum, 263
Petrotectonic assemblages, 9, 268–271
pH, 34, 164
Phaneritic rocks, 126
Phaneritic textures, 121
Phaneritic plutonic rocks
    IUGS classification, 132
    RSO classification, 131
Phaneritic-porphyritic, 123
Phase diagrams, 23–28
    binary solid solution diagrams, 28–29
    eutectic systems, earth history and rock formation, 29–30
    solid solutions and elemental compatibility, 30
Phenocrysts, 123
Philippine Sea plate, 9
Phosphorescence, 87
Phosphorites, 258, 260
Phosphorus, 258
Phyllitic texture, 216
Phyllosilicates, 76
Phyric rocks, 124
Physical end products, 149

Physics, 13
    derivative measurement, 14–16
    measurement, 13–14
Pillows, 118
Pipes, 243
Pisolitic aggregate, 80
Placer deposits, 260–262
Placer ore, 270
Plagioclase, 112, 133
Plane groups, 61
Plane lattice, 60
Plane polarized light (PL), 100
Plastic limit (PL), 155
Plasticity Index (PI), 155
Plate settings, 268–271
Plate tectonic, 6, 8–11
Platinum group elements (PGE), 239, 240
Platinum group metal (PGM), 239, 240
Playa lakes, 167
Pleochroism, 100
Plutonic rock, 109, 267
Plutons, 117
Poikilitic, 123
Point group, 56
Point symmetry, 56
Polarization, 96
Polarizer, 98
Polymict cobble-boulder conglomerate, 191
Polymorph crystal, 73, 75
Polysynthetic, 82
Pores, 38
Porosity, 38
Porous rocks, 40
Porphyroblastic texture, 217, 218
Porphyry deposit, 246
Powder diffraction method, 103
Precambrian tillite, 197
Precipitation, 11, 183
Press, 155
Pressure (P), 16, 206
Pressure solution, 201
Primary magmas, 109
Primitive magmas, 109
Primitive unit cell, 58
Prismatic habit, 79
Prograde metamorphism, 206
Proterozoic unconformity-related uranium deposit, 253
Protolith, 205
Protons, 17
Provenance, 164, 166
Pseudocleavage planes, 88
Pseudotachylite, 218, 231

Pumiceous texture, 121, 124
Pure quartz sand, 263
Pyrite, 69
Pyritohedra, 69
Pyroclastic rocks, 128
Pyroclastic sheets, 118
Pyroclastic texture, 124, 127
Pyroxene, 133, 134

Quantum numbers, 17, 18
Quartz, 49, 82
Quasimylonites, 231

Radiating aggregate, 80
Radioactivity, 31
Rapikivi texture, 123
Rapitan deposits, 257
Rare-earth elements (REE), 30, 243
Reaction front, 254
Recrystallization, 201, 205
Rectangles, 58
Red beds, 251
Redox, 18
Redox potential, 164
Reduced uranium (U^{4+}), 251
Reduction, 18, 148
Reef environments, 167–168, 199
Reflected light, 84
Regional metamorphism, 206
Relief, refractive indices, 100
Replacement, 202
Reservoir, 34
Residual deposit, 254, 261
Resource geology, 235
    resources and deposits, 235–236
        knowledge and accessibility, 236–238
        value, 236
Resource deposits, 238–239
    hydrous fluid deposits
        connate fluids and stratabound deposits, 249–252
        groundwater/meteoric fluids and associated deposits, 252–256
        hydrothermal fluids, 243–249
    magmatic process deposits
        crystal and fluid separation, 239–243
        volcanic process deposits, 243
    minerals, rocks and energy resources, 262–263
    sedimentary deposits
        detrital deposits, 260–262
        surface water precipitation, 256–260

Index

Retrograde metamorphism, 206
Rhyolite magma, 139
Rhyolite obsidian breccia, 126
Ribbon quartz grains, 231
Riebeckite, 140
Ripple marks, 180
Rise environments, 168
Rocks, 3
    analytical methods, 103
        electron-beam microprobe, 104–105
        EM, 104–105
        XRF, 106
    cleavage, 212
    deformation, 14
    formation, 29–30
    flour, 167
    historical methods, 95
    optical mineralogy
        light interactions with matter, 96–98
        petrographic microscope, 98–101
        polarization of waves, 96
    triangular plot, composition, 135
    types, 267
    XRD
        procedures and applications, 102–104
        x-rays and diffraction, 101–102
Roll-front deposit, 252–253, 255
Root traces, 159
Ropy pahoehoe lava, 118, 120
Rotation axis, 62, 64
Rotation point, 56
Rotoinversion, 62
Roughness, 37
RSO classification, 131
Runoff, 34

Sabkhas, 259
Sanbagawa Facies Series, 227
Sand dunes, 165
Sandstone, 175, 192
Saturated fluid, 51
Scanning electron microscope (SEM), 104–105
Scarn, 208
Scattered light, 84
Schistose porphyroblastic texture, 216, 217
Schists, 220
Scratchability, 89
Seas, 41–42
Secondary porosity, 38, 201
Sectility, 87

Sedimentary environments, 167–170
Sedimentary exhalative (SEDEX) deposits, 247, 249
Sedimentary deposits
    detrital deposits, 260–262
    surface water precipitation and associated ore formation, 256–260
Sedimentary rocks, 3, 140, 267
    chemical compositions, 185
    classifications, 183–193
    compositions, 181–183
    diagenesis and lithification, 201–202
    environments, 167–170
        continental environments, 193–196
        marine environments, 197–201
        transitional environments, 196–197
    erosion, transportation and deposition, 161–164
    minerals, 185–186
    modes, 187–188
    rock types and characteristics, 169
    structures, 174–181
    textures, 170–174
Sedimentary textures, 170–174
Seriate texture, 123, 126
Serpentinization, 229, 230
Shear stress, 15–16
Shelf break, 197
Shelf environment, 197
Shelf limestone-shale sequence, 200
Shelf-shallow sea environments, 167
Shield volcanoes/cones, 118
Sial crust, 4
Sialic crust, 4
Siderophile, 30
Silica-poor igneous rocks, 10
Silicate minerals, 77, 209–211
Siliceous magma, 271
Siliceous silicate magmas, 240
Silicic acid molecule, 149
Siliciclastic rocks, 183
Sills, 117
Siltstone, 190
Sima crust, 4
Simple twins, 82
Sinuous stream channel, 36
Skeletal, 154, 173
Slaty texture, 216
Slope environment, 168, 199
Sodium metal, 18
Soil, 3–4, 145, 150–151
    abundant elements, 152
    chemistry, 151–153

    classification, 155–156
        USC, 156, 158
        USDA, 156, 157
    origins, 156–158
    paleosols, 158–159
    physical properties, 153–155
    structure, 150, 154
    texture, 153
Solar system, 13
Sole marks, 180
Solid potassium chloride, 54
Solid solution, 28–30, 54
Solidus, 25
Sorosilicates, 76
Sorting, 171, 172
Space lattice, 60
Spaced cleavage, 212, 214
Spar, 171
Specific gravity, 90–91
Speculative resources, 237
Sphalerite, 53
Spreading centers, 8
Squares, 58
Stability, 22–23
Static metamorphism, 206, 226–228
Staurolite, 82
Stocks, 117
Stockwork vein complex, 242, 246
Stoichiometry, 51
Stope, 118
Stratabound, 247
Stratabound deposits, 249–252
Stratiform deposits, 240, 247
Stratigraphy, 193
Streak, 85
Stream and stream channels, 34–38
Stress, 15
Stromatolites, 178, 180
Stylolites, 180
Subduction zones, 8, 139–140, 268, 270
Sub-economic, 236
Subhedral crystal, 76
Submetallic mineral, 86
Sulfide liquids, 241
Sulfide mineral, 91
Supergene, 254, 256
Superior deposits, 256
Super-saturation, 51
Surface water precipitation, 256–260
Suspension, 162
Symmetrical-compositional units, 267
Symmetry, 54
    center of, 62
    one dimension, 54–56
    operation, 55

Index

three dimension, 60–65
two dimension, 56–60
Synthetic gems, 2
System, 13

Tabular deposits, 252
Tabular habit, 79
Tectonites, 138
Tectosilicates, 76
Telogenesis, 202
Temperature (T), 14, 205–206
Tenacity, 87
Terminal reactions, 223
Terrace deposit, 261
Tessellation, 56
Texture, 109, 122. *See also* Rock
    aphanitic, 121
    aphanitic-porphyritic, 124, 126, 127
    breccia, 126, 218
    cataclastic, 218
    clastic, 170
    corona, 123
    diabasic, 123, 125
    diablastic, 218
    epiclastic, 170, 171
    foliated, 216
    glassy, 121, 124
    gneissose, 216, 217
    granoblastic, 217, 218, 220
    graphic, 124, 125
    hypidiomorphic-granular, 122–125
    intergranular, 124, 127
    mylonitic, 218
    ophitic, 123
    pegmatitic, 123
    phaneritic, 121
    phyllitic, 216
    porphyroblastic, 217, 218
    pumiceous, 121, 124
    pyroclastic, 124, 127
    rapikivi, 123
    schistose porphyroblastic, 216, 217
    sedimentary, 170–174
    seriate, 123, 126
    slaty, 216
    trachytic, 124, 127
    trachytoidal, 122, 123
Thermal equilibrium, 14

Thermally induced melting, 111
Thermohaline, 41
Thin-section analysis, 101
Tholeiite, 135
Tholeiitic basalt, 135
Three point symmetry, 64
Tidal flats, 197
Tidal-fluvial setting, 196
Tie lines, 27
Tool mark, 180
Tourmaline, 77
Trachytic texture, 124, 127
Trachytoidal texture, 122, 123
Transform faults, 8, 268
Transitional environments, 196–197
Translational symmetry, 55
Transmitted light, 83
Transmutation, 31
Trans-tensional fault zone, 10
Transparent material, 84
Transportation, 161–164
Trench environments, 168
Triangles, 58
Triceratops scapula, 184
Triple junctions, 9
Triple point, 14, 24
Tubular burrows, 177
Tuff, 118
Turbidity currents, 167
Turbulence, 37
Turbulent flow, 35
Twins, 81

Ultrabasic rocks, 228–230
Ultra-high pressure (UHP) rocks, 226–228, 271
Ultramafic igneous rocks, 134
Unconformity-related uranium deposits, 251
Undiscovered resources, 237
Uniaxial indicatrix, 97
Unified Soil Classification (USC), 153, 155, 156, 158
Unit cell axes, 58
United States Department of Agriculture (USDA) classification, 156, 157
Unloading, 146
Unsaturated fluid, 51

Uraniferous humate deposits, 252
Uranium-vanadium minerals, 254

Valence electrons, 17–18
Van der Waals bond, 22
Varves, 194
Veins, 44, 212, 243–246
Velocity, 4, 14
Vertical tensional joints, 39, 40
Vesicles, 118, 121
Viscosity, 33
Vitric tuffs, 128
Vitrinite, 190
Vitrophyric, 124
Volcanic arc, 8
Volcanic breccias, 128
Volcanic process deposits, 243
Volcanic rocks, 109, 128, 137, 139
Volcanic-hosted massive sulfide (VHMS) deposits, 247
Volcano, 120
Volcanogenic massive sulfide (VMS) deposits, 247, 248

Wackes, 188
Wackestone, 188
Water, 33–38, 109–110
Weathering processes, 145–149
Wentworth Scale, 171, 189
White quartz vein, 215

Xenocrysts, 243
Xenoliths, 112, 114, 115
X-ray(s), 101–102
X-ray diffraction (XRD)
    procedures and applications, 102–104
    x-rays and diffraction, 101–102
X-ray fluorescence (XRF), 106

Zone
    accumulation, 150
    critical, 145
    ductile shear, 213
    failed intracontinental rift, 251
    inner hornfels, 226
    leaching, 150
    subduction, 8, 139–140, 268, 270
    trans-tensional fault, 10